ary+

be returned on or before

Introduction to Biological Physics
for the Health and Life Sciences

Introduction to Biological Physics for the Health and Life Sciences

Kirsten Franklin
Paul Muir
Terry Scott
Lara Wilcocks
Paul Yates

Staff at the University of Otago,
New Zealand

WILEY

A John Wiley and Sons, Ltd., Publication

This edition first published 2010
© 2010 John Wiley and Sons Ltd

Registered office

John Wiley & Sons Ltd, The Atrium, Southern Gate, Chichester, West Sussex, PO19 8SQ, United Kingdom

For details of our global editorial offices, for customer services and for information about how to apply for permission to reuse the copyright material in this book please see our website at www.wiley.com.

Library of Congress Cataloging-in-Publication Data

Introduction to biological physics for the health and life sciences / Kirsten Franklin ... [etal.]. p. cm.
Summary: "This book aims to demystify fundamental biophysics for students in the health and biosciences required to study physics and to understand the mechanistic behaviour of biosystems. The text is well supplemented by worked conceptual examples that will constitute the main source for the students, whilst combining conceptual examples and practice problems with more quantitative examples and recent technological advances"–Provided by publisher.
Includes bibliographical references and index.
ISBN 978-0-470-66592-3 (hardback)–ISBN 978-0-470-66593-0 (paper) 1. Biophysics–Textbooks. 2. Physics–Textbooks. 3. Medical physics–Textbooks.
I. Franklin, Kirsten.
QH505.I68 2010
571.4–dc22   2010023304

A catalogue record for this book is available from the British Library.
ISBN 978-0-470-66592-3 (Cloth) 978-0-470-66593-0 (Paper)
Typeset by the authors.
Printed in Singapore by Markono Print Media Pte Ltd

CONTENTS

VII Appendices   431

PREFACE

Physics is central to an understanding of biomedical science. We are aware that many students studying for a career in biomedicine are not primarily motivated by physics; they are interested in other areas of science. We are also aware that no currently available first-year textbook takes the physics needs of health-science students seriously.

In this textbook we have several goals. Firstly, we are trying to present the necessary base concepts of physics as clearly as possible. Secondly, the textbook is designed to remove any unnecessary conceptual load from students by removing all physics that is not absolutely necessary for health-science students. The decision as to which parts of physics are necessary has been determined in close collaboration with the physicists and teachers of the Department of Physics and the professional clinicians and academics in the Faculty of Health Science at the University of Otago. Thirdly, we are keenly aware that student motivation is always an issue in the study of physics for the health sciences. We have tried to add as many applications to the biomedical sciences as possible to the text in an attempt to aid this motivation. The companion website for this book is available at www.wiley.com/go/biological_physics.

The production of a textbook is an enormous task and this textbook is no exception. In writing this book, we have relied on the expertise and goodwill of a large group of academic colleagues. We would like to express our gratitude to Mr Gordon Sanderson of the Ophthalmology Department and Professor Terence Doyle of the Radiology Department in the University of Otago Medical School. We would like to thank Dai Redshaw for the many hours he has spent reading through the text and working through the problem sets and Dr Phil Sheard from the Department of Physiology for his inspiring review lectures on bioelectricity. We would particularly like to thank Dr Don Warrington for his diligent and careful reading of the entire manuscript, and for his many corrections and suggestions. Finally we would like to thank the staff of the Department of Physics at Otago for the time and support that they have rendered over the past years. While the staff of the Department of Physics are listed as authors of this textbook we would particularly like to thank Gerry Carrington, Pat Langhorne, Craig Rodger, Rob Ballagh, Neil Thomson and Bob Lloyd.

Finally, the goal of this textbook is to provide for the needs of our students. In order to achieve this goal, we have depended on the feedback provided by our students. There will of course still be errors which have escaped our editing process, and for these we apologise in advance, and we welcome feedback from our readers.

I

Mechanics

Mechanics is the study of motion. It may be divided into two related areas: kinematics and dynamics. Kinematics is the study of the fundamental properties of motion: displacement, acceleration, velocity, distance, and speed. These concepts allow us to quantify motion and this allows for its scientific study. Dynamics is the study of force as described by Newton's three laws. Forces produce accelerations and thus cause changes in the motion of objects.

Mechanics is the most fundamental subject in physics: it shows how the forces of nature produce the changes which are observed in nature. The concepts introduced in mechanics will be used throughout the rest of this book.

Mechanics is of central importance in the health sciences. The applications of mechanics in biological systems appear whenever the concepts of force, energy or momentum appear. There are, however, many more direct applications of the ideas of mechanics. The working of the musculoskeletal system in humans and other vertebrates cannot be understood without an understanding of mechanical concepts such as torque, force, levers and tension. The energy and forces required for everyday activities in nature – jumping, flying, accelerating to elude capture – can only be evaluated using the techniques introduced in this section.

KINEMATICS

1.1 Introduction

Kinematics is that part of mechanics which is concerned with the description of motion. This is a vital first step in coming to an understanding of motion, since we will not be able to describe its causes, or how it changes, without a clear understanding of the properties of motion. Kinematics is about the definition and clarification of those concepts necessary for the complete description of motion. Only six concepts are needed: time, distance, displacement, speed, velocity and acceleration.

We will begin by focussing on linear motion in one dimension. Later we will expand this to include motion in two and three dimensions, and we will then look at three particularly important special cases of motion in one and two dimensions: circular motion, simple harmonic motion, and wave motion.

Key Objectives

- To develop an understanding of the concepts used to describe motion: time, distance, displacement, speed, velocity and acceleration.

- To understand the relationships between time, displacement, velocity and acceleration.

- To understand the distinction between average and instantaneous velocity and acceleration.

- To understand that the horizontal and vertical components of vector quantities, such as acceleration and velocity, may be treated independently.

1.2 Distance and Displacement

Motion is characterised by the direction of movement, as well as the amount of movement involved. It is not surprising that we must use vector quantities in kinematics. The **distance** an object travels is defined as the length of the path that the object took in travelling from one place to another. Distance is a scalar quantity. **Displacement**, on the other hand, is the distance travelled, but with a direction associated. Thus a road trip of 100 km to the north covers the same distance as a road trip of 100 km to the south, but these two trips have quite different displacements. The use of displacement rather than distance to give directions is commonplace.

1.3 Speed and Velocity

We are accustomed to talking about the speed at which an object is moving. We also talk about the velocity with which an object is moving. In normal usage these two words mean the same thing. I can talk about the speed with which a car is travelling, or I can talk about its velocity. In physics, we redefine these two words, **speed** and **velocity**, so that they have similar, but distinct meanings.

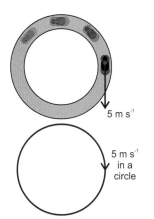

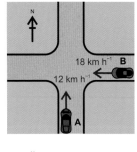

Figure 1.1 A toy car on a race track. How do we characterise its motion?

Vector equations vs. scalar equations

When demonstrating numerical calculations the vector character that many quantities possess will not be explicitly addressed in the equation itself. Most numerical examples will be treated as scalar problems without any attempt to represent the various quantities used as vectors. This is to keep problems simple and readable. Please note that this does NOT mean that vector properties are ignored, but rather that they are addressed in the process of constructing the problem.

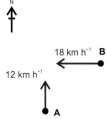

Figure 1.2 Linear motion in two directions. The cars are travelling at different speeds and in different directions.

Key concept:
The **velocity** of an object is the change in its position, divided by the time it took for this change to occur. Velocity is a vector and has both a magnitude and a direction.

Mathematically, the velocity of an object is

$$\boldsymbol{v} = \frac{\Delta \boldsymbol{x}}{\Delta t} \tag{1.1}$$

where $\boldsymbol{v}$ is the velocity vector, $\Delta \boldsymbol{x}$ is the displacement vector and Δt is the time interval over which the displacement occurs. Note that we will use bold symbols, such as $\boldsymbol{v}$, for vectors and normal-weight symbols, such as v, for scalar quantities. Note also that the Greek letter Δ (capital delta) represents the change in a quantity. In the above expression, Eq. (1.1), for example, the change in the position of an object is its final position minus its initial position:

$$\Delta x = x_\mathrm{f} - x_\mathrm{i} \tag{1.2}$$

Key concept:
Speed is the magnitude of the velocity. Speed is a scalar, and it does not have a direction.

The speed of an object is the distance travelled, divided by the time it took to travel that distance:

$$v = \frac{\Delta x}{\Delta t} \tag{1.3}$$

Note the differences between Eq. (1.1) and Eq. (1.3). In Eq. (1.1), we use bold symbols for both the $\boldsymbol{v}$ and the $\boldsymbol{x}$, indicating that we are referring to the velocity and the displacement in this equation. In Eq. (1.3) we use normal weight symbols, v and x, indicating that we are referring to the speed and distance in this equation.

Many textbooks use d to represent distances and $\boldsymbol{d}$ to represent displacements rather than Δx and $\Delta \boldsymbol{x}$. We will often follow this practice when specific reference to the initial and final positions is not called for.

Consider Figure 1.1. A toy car is travelling in a circle around a toy race track and we wish to characterise its motion. If we are interested only in how fast the car is going, we could say it is travelling at 5 m s^{-1} ($= 18 \text{ km h}^{-1}$). Two cars travelling on the same circle will be perfectly well distinguished by noting the different lengths of the circle they traverse in the same time.

Now consider the situation illustrated in Figure 1.2. In this case, two cars approach the same intersection from different directions. In this situation, we might point out that one of the cars is travelling at 18 km h^{-1}, while the other is travelling at 12 km h^{-1}. However, this will not cover all of the differences between the two cars. Another important fact about them is that they are travelling in different directions. If we wanted to predict where these two cars would be in an hour (for example) it would not be enough to just use the magnitude of their velocity; we would also need to take into account their directions.

1.4 Acceleration

In kinematics, the **acceleration**, $\boldsymbol{a}$, is a vector which quantifies changes in velocity. In everyday conversation we use the word acceleration to mean that the speed of an object is increasing. If an object was slowing down we would say that the object was decelerating. The concept of acceleration in physics is more general and applies to a larger set of situations. In physics, acceleration is defined to be the rate of change (in time) of the velocity:

$$\boldsymbol{a} = \frac{\Delta \boldsymbol{v}}{\Delta t} \tag{1.4}$$

This definition implies several characteristics of the acceleration:

1. Acceleration is a vector: it has a direction as well as a magnitude. The acceleration is the rate of change of the velocity, and velocity is a vector, therefore acceleration must also be a vector.

2. The acceleration vector of an object may point in the opposite direction to that object's velocity vector. When this happens, the object's velocity will decrease and may even reverse direction. This means that deceleration (slowing down) is just another acceleration, but in a particular direction.

3. An object may have an acceleration without its speed changing at all. Should the acceleration vector point in a direction perpendicular to the velocity vector, the direction of the velocity vector will change, but its length will not. A good example of this is when an object moves in a circle. In this case, the acceleration is always perpendicular to the velocity, so the speed of the object is constant, but its velocity is constantly changing.

To illustrate these ideas, consider a car which starts from rest ($v_i = 0$) and accelerates along a straight road so that its velocity increases by 2 m s^{-1} every second. The velocity of this car is illustrated at a series of later times in Figure 1.3.

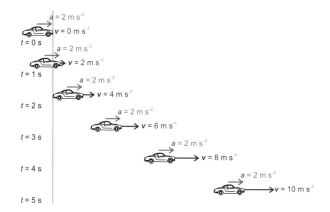

Figure 1.3 A car accelerating at 2 m s^{-2} for 5 s.

Since the velocity changes by the same amount every second (2 m s^{-1}), the acceleration of the car is constant. The velocity is changing at a rate of 2 m s^{-1} per second, or 2 metres per second per second. This acceleration would normally be written as $a = 2$ m s^{-2} (or 2 m/s^2) to the right.

We can calculate the velocity at any time. Since we know how much the velocity increases every second and we also know that the car was initially stationary, we just multiply this rate by the time elapsed since the acceleration began, i.e., we use the equation

$$v = at \tag{1.5}$$

Note that this is a vector equation, so that the velocity is in the same direction as the acceleration. For the car in this example, which is accelerating in a straight line at a constant rate of 2 m s^{-2} from rest, after 4 s the speed is $v = at = 2$ m s$^{-2} \times 4$ s $= 8$ m s^{-1}, and so on.

What if the car had not been at rest initially? Suppose that the car in the previous example had been travelling at a constant velocity of 5 m s^{-1} for some unspecified length of time, and then began to accelerate at 2 m s^{-2}. Figure 1.4 shows this car at a sequence of later times. Compare this figure with Figure 1.3.

In one most important respect, the situation has not changed. The velocity of the car still increases at the same rate, so that the *change* in velocity after the acceleration begins is given by the equation

$$\Delta v = at \tag{1.6}$$

The difference between Eq. (1.5) and Eq. (1.6) is that we now explicitly recognise that it is the *change* in velocity that we are calculating. In the previous example we calculated the change in velocity, but since the car started at rest, the velocity of the car was the same as how much the velocity had increased. Since we now have a nonzero initial velocity, we must recognise that the change in velocity is the final velocity minus the initial velocity, so

$$v_f - v_i = at$$

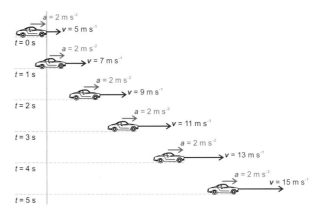

Figure 1.4 A car accelerates from an initial velocity of 5 m s^{-1} with a constant acceleration of 2 m s^{-2}.

Thus after 5 s we would find that 15 m s^{-1} − 5 m s^{-1} = 2 m s^{-2} × 5 s. We are normally interested in calculating the final velocity, so we write the above equation in the form

$$\boldsymbol{v}_\text{f} = \boldsymbol{v}_\text{i} + \boldsymbol{a}\,t \tag{1.7}$$

We can use this equation to find the final velocity at any later time, so long as the acceleration has not changed.

Note that the velocity calculated using this formula is the *instantaneous* velocity of the car at that time. This is the velocity that you would read off the car's speedometer. We will discuss instantaneous and average velocities in more detail next.

1.5 Average Velocity or Speed

The discussion above allows us to calculate the instantaneous velocity of an object moving with constant acceleration. This is the velocity of the object at a particular instant of time. It is also often useful, when we are dealing with motion in a straight line, to use the average velocity of an object to solve problems. The average velocity is

$$\boldsymbol{v}_\text{av} = \frac{\text{total displacement}}{\text{total time}} = \frac{\boldsymbol{d}}{t} \tag{1.8}$$

If you drive a car from Dunedin to Christchurch (370 km away) in 5 h, your average speed is given by

$$\boldsymbol{v}_\text{av} = \frac{d}{t} = \frac{370\,\text{km}}{5\,\text{h}} = 74\,\text{km h}^{-1}$$

It is important to realise that this calculation does not require any knowledge of the details of your trip. You may have traveled at a constant 74 km h^{-1} the whole way, or (more likely) you may have varied your speed significantly. You may even have stopped to look at the view and eat lunch for half an hour. These details are not needed for the calculation of the average velocity.

We will now derive a general relationship between the distance travelled by an object, its initial velocity, and its constant acceleration. If an object, for example a car, a plane or a soccer ball, has constant acceleration, then the displacement, $\boldsymbol{d}$, occurring in some given time, t, is

$$
\begin{aligned}
\boldsymbol{d} = \boldsymbol{v}_\text{av}\,t \;\; &= \;\; \frac{1}{2}\,(\boldsymbol{v}_\text{i} + \boldsymbol{v}_\text{f})\,t \\
&= \;\; \frac{1}{2}\,(\boldsymbol{v}_\text{i} + \boldsymbol{v}_\text{i} + \boldsymbol{a}\,t)\,t \\
&= \;\; \left(\boldsymbol{v}_\text{i} + \frac{1}{2}\,\boldsymbol{a}\,t\right) t
\end{aligned}
$$

$$\boxed{\boldsymbol{d} = \boldsymbol{v}_\text{i}\,t + \frac{1}{2}\,\boldsymbol{a}\,t^2} \tag{1.9}$$

Often $v_i = 0$, i.e., the object is starting from rest, so then

$$d = \frac{1}{2}at^2 \qquad (1.10)$$

We will now investigate an acceleration which is particularly important for the motion of objects near the surface of the earth: the *acceleration due to gravity*.

Example 1.1 *Falling ball (1D kinematics)*

Problem: If you drop a cricket ball from a 125 m high tower, how far will it fall in 5 s?

Solution: We can solve this problem in two different ways. We can find the average velocity of the ball over the first 5 s and use this average velocity to calculate a displacement, or we can calculate a displacement directly.

(a) The acceleration due to gravity is $10 \, \text{m s}^{-2}$ downwards and so the velocity increases by $10 \, \text{m s}^{-1}$ in the downwards direction every second. The initial velocity is $0 \, \text{m s}^{-1}$ so the final velocity must be $5 \, \text{s} \times 10 \, \text{m s}^{-2} = 50 \, \text{m s}^{-1}$ in the downwards direction. The average velocity of the cricket ball is therefore

$$v_{av} = \frac{0 \, \text{m s}^{-1} + 50 \, \text{m s}^{-1}}{2} = 25 \, \text{m s}^{-1} \text{ downwards}$$

Using this average velocity, the distance that the cricket ball will fall in 5 s is $25 \, \text{m s}^{-1} \times 5 \, \text{s} = 125 \, \text{m}$.

(b) The second technique uses Eq. (1.10) (since the initial velocity is zero) using g for the acceleration. The change in displacement of the ball is

$$d = \frac{1}{2}gt^2 = \frac{1}{2} \times 10 \, \text{m s}^{-2} \times (5 \, \text{s})^2 = 125 \, \text{m}$$

which is the same answer as we found with the previous method.

1.6 The Acceleration Due to Gravity

Galileo found (and countless experiments since have also shown) that all objects falling freely towards the Earth have the same acceleration. (In order to see this effect, we must take into account the effect of air resistance when this is significant.) Thus every object in free fall has its downward speed increased by $10 \, \text{m s}^{-1}$ in every second regardless of its mass. Galileo claimed that this was an experimental fact and is reported to have shown it by dropping two balls of unequal mass from the top of the Leaning Tower of Pisa. Later we will discuss the theoretical explanation for this experimental fact when we investigate the relationship between acceleration and the concept of *force*. The value of this constant acceleration is given by

$$g = 9.81 \, \text{m s}^{-2} \approx 10 \, \text{m s}^{-2} \qquad (1.11)$$

This quantity, g, is called the **acceleration due to gravity**. (The value of $9.8 \, \text{m s}^{-1}$ is the value at sea level on the surface of the Earth; the value will change with altitude.)

Consider an object released from rest and accelerating in free fall. Assuming that the air resistance is negligible, we are able to calculate its velocity after 5 s:

$$v = gt = 10 \, \text{m s}^{-2} \times 5 \, \text{s} = 50 \, \text{m s}^{-1}$$

Note that the mass of this object is not even mentioned in the original question. Also note that we have not considered the vectorial character of either the acceleration due to gravity or of the velocity achieved by this object after 5 seconds. These quantities are of course vectors, but their directions may be assumed to be towards the centre of the Earth and need not be considered in this problem. This is not always the case.

Air resistance

Actually, the acceleration is only the same in a vacuum. Objects falling in air are affected by air resistance which reduces the acceleration. Often this is small enough to be ignored.

Precision and *g*

In mathematical examples where we wish to find an accurate result, we will often use a value for g of 9.8 or $9.81 \, \text{m s}^{-2}$. However, in cases where we are more interested in the concepts and methods than the exact answer we will frequently approximate the value as $10 \, \text{m s}^{-2}$ to keep things simpler.

Example 1.2 *A ball thrown straight up (I) (1D kinematics)*

Problem: If you throw a cricket ball straight up at 12 m s^{-1}, how high will it go?

Solution: This problem is very similar to the previous one, except now the cricket ball has an initial velocity. This initial velocity is in the opposite direction to the acceleration due to gravity. This means that, at first, gravity will *reduce* the upward velocity of the cricket ball by 10 m s^{-1} every second. At some point the upward velocity of the cricket ball will have been reduced to zero – the cricket ball has stopped travelling up so it has reached its maximum height.

To calculate the maximum height that the ball reaches we need to find the time taken for the upward velocity to decrease to zero as this is also the time for the ball to reach its maximum height. We then calculate the average velocity and use this combined with the elapsed time to calculate the change in displacement of the ball.

For this problem we will define the upwards direction as positive, and define $d = 0$ m to be the height at which the ball was released.

The change in velocity of the ball is $v_f - v_i = -12$ m s^{-1} (i.e., the velocity goes from $+12$ m s^{-1} initially to 0 m s^{-1} at its highest point). The time it takes the acceleration due to gravity (-10 m s^{-2}, as it points in the downwards direction) to cause this change in velocity is,

$$\Delta v = gt$$
$$t = \frac{\Delta v}{g} = \frac{-12\,\mathrm{m\,s^{-1}}}{-10\,\mathrm{m\,s^{-1}}} = 1.2\,\mathrm{s}$$

The average velocity is 6 m s^{-1} ($v_{\mathrm{av}} = \frac{1}{2}\left(v_f + v_i\right) = \frac{1}{2}\left(12\,\mathrm{m\,s^{-1}} + 0\,\mathrm{m\,s^{-1}}\right) = 6\,\mathrm{m\,s^{-1}}$) and using this we calculate the change in displacement of the cricket ball during the 1.2 seconds it takes to reach the maximum height,

$$d = v_{\mathrm{av}}t = 6\,\mathrm{m\,s^{-1}} \times 1.2\,\mathrm{s} = 7.2\,\mathrm{m}$$

The highest point the ball reaches is 7.2 m above the point at which it was released.

Example 1.3 *A ball thrown straight up (II) (1D kinematics)*

Problem: How long does it take the ball in Example 1.2 to fall to its original position from its maximum height?

Solution: The ball reaches a maximum height of 7.2 m above the point at which it is thrown. At its maximum height it has a velocity of 0 m s^{-1}. The time taken to fall a distance of 7.2 m back to its original position can be found using Eq. (1.9):

$$d = v_i t + \frac{1}{2}at^2$$

Since the initial velocity, v_i, is the velocity of the ball at its maximum height, i.e., $v_i = 0$ m s^{-1} this can be reduced to

$$d = \frac{1}{2}at^2$$

The time in this equation, t, is the time the ball takes to fall back to its original position. We rearrange this equation to solve for t where (using the same sign convention as in Example 1.2) $d = -7.2$ m and $a = g = -10$ m s^{-2}

$$t = \sqrt{\frac{2d}{g}} = \sqrt{\frac{2 \times -7.2\,\mathrm{m}}{-10\,\mathrm{m\,s^{-2}}}}$$
$$= 1.2\,\mathrm{s}$$

Note that this is the same time it took the ball to reach its maximum height from the point at which it was initially thrown in Example 1.2. This is a useful general result. Projectile motion is symmetrical about the point of maximum height. It takes the same amount of time to reach the maximum height from a starting height as it does to get back to that height from the maximum height.

Example 1.4 *A ball thrown straight up (III)*

Problem: What is the velocity of the ball in Examples 1.2 and 1.3 when it falls back to the height at which it was released?

Solution: As in the previous problem we can think of the ball starting at its maximum height and falling 7.2 m under the influence of gravity to its original height, which we found would take 1.2 s, the same as the time taken to reach its maximum height from the point at which it was released. The change in velocity over the three seconds in which it is falling from its maximum height is

$$\Delta v = at = -10 \, \text{m s}^{-2} \times 1.2 \, \text{s}$$
$$= -12 \, \text{m s}^{-1}$$

The ball is travelling at the same speed at it was when released, but in the opposite direction! Again this is a useful general result we can apply to many kinematics problems without going through an extensive derivation.

Example 1.5 *A ball thrown straight up (IV) (1D kinematics)*

Problem: If the ball in the Example 1.2 was released (travelling upwards) at a height of 1.2 m above the ground, what is the velocity of the ball just before it hits the ground?

Solution: We will first need to find the time it takes for the ball to hit the ground and then use $\Delta v = at$ to find the change in velocity, and hence the final velocity. Initially it is tempting to try to use Eq. 1.9 to solve this problem directly.

$$d = v_i t + \frac{1}{2} a t^2$$

We know the change in position of the ball (as, in this case, the ball ends up 1.2 m below it's starting point), the initial velocity of the ball ($v_i = 12 \, \text{m s}^{-1}$), and acceleration ($g = -9.8 \, \text{m s}^{-2}$). In order to use this equation, however, we would need to solve for t, which would require solving a quadratic equation. Even using the shortcuts highlighted in the previous two examples does not allow us to avoid this quadratic as we still end up with two terms featuring t. If you're confident doing this, that is great. If you are not very confident at solving quadratic equations however, all is not lost.

We can simplify the problem by using the fact that we know a bit more about the situation than is apparent from the question. From Example 1.2, we know that the ball reaches a maximum height of 7.2 m above the point at which it was released. This means that the ball will reach a height of $7.2 + 1.2 = 8.4$ m above the ground. At this maximum height, the velocity of the ball is $0 \, \text{m s}^{-1}$, and by ignoring the first part of the ball's motion, we can simplify Eq. (1.9) to $d = \frac{1}{2} a t^2$, where d is the change in displacement of the ball as it moves from its maximum height to the ground, $d = -8.4$ m, and $a = g = -9.8 \, \text{m s}^{-2}$:

$$d = \frac{1}{2} a t^2$$
$$t = \sqrt{\frac{2d}{g}} = \sqrt{\frac{2 \times -8.4 \, \text{m}}{-9.8 \, \text{m s}^{-2}}}$$
$$= 1.3 \, \text{s}$$

This gives a change in velocity of $\Delta v = gt = -9.8 \, \text{m s}^{-2} \times 1.3 \, \text{s} = -13 \, \text{m s}^{-1}$, so the ball will be travelling at 13 m s^{-1} in the downwards direction as it hits the ground.

1.7 Independence of Motion in 2D

In the previous section we considered objects which move vertically up and down. This means that in these cases the velocity vector is always parallel to the acceleration vector. These objects would go straight upward and fall straight downward. They will not move horizontally since there is no initial velocity in the horizontal direction, and no acceleration in the horizontal direction to cause a non-zero horizontal velocity to develop. What would happen if the initial velocity was not straight upward? What would have happened if the initial velocity was at some angle to the vertical? This is the situation shown in Figure 1.5, using the example of a cricket ball launched upward at an angle. In Figure 1.5, the cricket ball has a vertical velocity and a horizontal velocity and

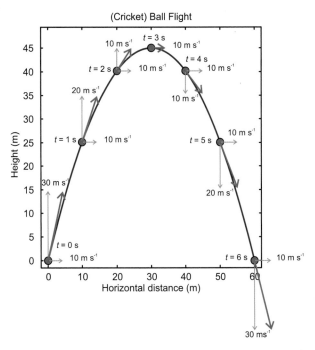

Figure 1.5 The trajectory of a cricket ball initially launched upward at an angle to the vertical. The vertical and horizontal components of the ball's velocity are shown at a number of times in its trajectory.

this results in a net velocity at an angle to the vertical. The acceleration due to gravity, however, acts only in the vertical direction, and changes only the vertical component of the velocity. The horizontal component of the velocity is initially 10 m s^{-1} and is still 10 m s^{-1} when the ball reaches its maximum height at 45 m after 3 s, and is 10 m s^{-1} when the ball reaches the ground again after a total of 6 s. There is no acceleration in the horizontal direction, so the velocity component in this direction *cannot change*.

The vertical component of the velocity is changed by the acceleration due to gravity. This can be seen in Figure 1.5 as well. In point of fact, the vertical component of the velocity behaves in exactly the same way as it did in the examples above. The vertical velocity is initially 30 m s^{-1} upward. After 3 s this has dropped to 0 m s^{-1} when the ball reaches its maximum height. The vertical velocity is again 30 m s^{-1} just before the ball hits the ground after 6 s, but now the velocity is in the downward direction.

This is what we mean when we say that the horizontal and vertical components of the velocity vector are independent. These components are acted on separately by the accelerations in those directions. An acceleration in the horizontal direction would not change the velocity component in the vertical direction. This means that when we are attempting to solve a kinematics problem in three dimensions, we may look at the components of the velocity and acceleration in a given direction independently of their components in the other two directions.

This effect may be seen in the following experiment. Suppose that we have two identical balls which are held on a platform in a darkened room. (The balls do not need to have the same mass for this experiment to work, but we will simplify the discussion by assuming that they do.) Now suppose that we drop one ball directly downward from a platform and at the same instant fire the other ball horizontally out from the same platform. As the balls fall, a strobe light flashes at regular intervals and the trajectory of the two balls is recorded on a camera with a very long exposure time. Figure 1.6 is an example of the sort of image that would be obtained from this experiment.

In this figure, we observe that the balls are at the same height at each interval, i.e., at each flash of the strobe light. This means that their vertical velocity components are the same at each time. Their horizontal velocity components are quite different, however. The ball which is simply dropped has no horizontal velocity component, whereas

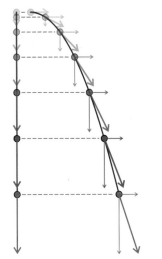

Figure 1.6 Independence of the vertical and horizontal components of velocity.

the horizontal velocity component of the other ball is constant.

Example 1.6 *Projectile motion (2D kinematics)*

Problem: A modern artist throws a bottle of paint towards the wall of a nearby building. The bottle leaves the artist's hand at a height of 2.00 m, a speed of 16.0 m s^{-1}, and at an angle of 30.0° above the horizontal. If the building is 15.0 m away

(a) **At what height H does the bottle hit the wall?**

(b) **At what velocity v is the bottle travelling as it hits the wall?**

Solution:

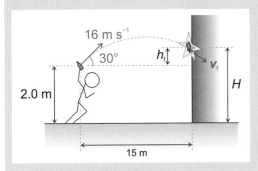

Figure 1.7 An artist throws a bottle against a wall.

A good first step for this kind of problem is to draw a diagram like Figure 1.7. Remember that when dealing with 2D kinematics you can always separate out the horizontal and vertical motions. Note also that since we are given numerical quantities to two significant figures, we will use this level of numerical precision throughout the problem, i.e., we will use $g = -9.8$ m s^{-2}.

In order to answer both of the questions, we will need to know how long after the bottle leaves the artist's hand it hits the wall. We can calculate this time by looking at the horizontal motion of the bottle. As the acceleration due to gravity is in the vertical direction only, we know that the horizontal velocity is constant and has a magnitude of $v_x = 16.0$ m s$^{-1} \times \cos 30° = 13.9$ m s^{-1}. So the time it takes the bottle to travel the 15 m horizontally to the wall is

$$d_x = v_x t$$

$$t = \frac{d_x}{v_x} = \frac{15 \text{ m}}{13.9 \text{ m s}^{-1}} = 1.08 \text{ s}$$

(a) The difference in height between the bottle's initial height of 2.0 m and its final height of H is h_f. We can calculate this height by using the initial vertical velocity of the ball ($v_{yi} = 16$ m s$^{-1} \times \sin 30° = 8.0$ m s^{-1}) and the fact that the ball is accelerating in the vertical direction at a rate of $a = g = -9.8$ m s^{-2}.

$$h_f = v_i t + \frac{1}{2} g t^2$$

$$= 8.0 \text{ m s}^{-1} \times 1.08 \text{ s} + \frac{1}{2} \times \left(-9.8 \text{ m s}^{-2}\right) \times (1.08)^2$$

$$= 2.93 \text{ m}$$

So the bottle must hit the wall a total of 4.9 m above the ground (giving solution to two significant figures).

(b) In order to find the final velocity of the bottle we will have to add together the vertical and horizontal components of the bottle's velocity as it hits the wall. We already know that the horizontal velocity of the bottle is constant as there is no acceleration in the horizontal direction, therefore the horizontal component of the final velocity is $v_x = 14$ m s^{-1}. To find the final vertical velocity we can use $v_f = v_i + at$ where $v_i = v_{yi} = 8.0$ m s^{-1} and $a = g = -9.8$ m s^{-2}.

$$v_{yf} = v_{yi} + g t$$

$$= 8.0 \text{ m s}^{-1} + \left(-9.8 \text{ m s}^{-2}\right) \times 1.08 \text{ s} = -2.58 \text{ m s}^{-1}$$

By the time the bottle has hit the wall, it has reached its maximum height (at which $v_y = $ m s^{-1}) and has started moving back down, hence the negative vertical velocity.

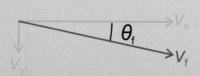

Figure 1.8 The velocity vector components of the bottle as it hits the wall.

We can get the magnitude of the final velocity by vector addition of the two components v_x, and v_{yf}:

$$|v_f| = \sqrt{v_x^2 + v_{yf}^2}$$
$$= \sqrt{(13.9 \text{ m s}^{-1})^2 + (-2.58 \text{ m s}^{-1})^2} = 14.14 \text{ m s}^{-1}$$

The direction in which the bottle is traveling can be found by using trigonometry.

$$\tan\theta_f = \frac{v_{yf}}{v_x}$$

$$\theta_f = \tan^{-1}\left(\frac{v_{yf}}{v_x}\right)$$

$$= \tan^{-1}\left(\frac{2.58 \text{ m s}^{-1}}{13.9 \text{ m s}^{-1}}\right) = 10.52°$$

So as the bottle hits the wall it is travelling at a speed of 14 m s^{-1}, 11° below the horizontal (to 2 s.f.).

1.8 Summary

Key Concepts

elapsed time (Δt) The time interval between two events

distance (d or Δx) The length of a path between two spatial positions.

displacement ($\boldsymbol{d}$ or $\Delta\boldsymbol{x}$) The vector equivalent of distance, which specifies the distance and direction of one point in space relative to another. It depends only on the initial and final spatial positions, and is independent of the path taken from one position to the other.

speed (v) A scalar measure of the rate of motion. The SI unit of speed is metres per second (m/s or m s^{-1}).

velocity ($\boldsymbol{v}$) A vector measure of the rate of motion, which specifies both the magnitude and direction of the rate of motion.

acceleration ($\boldsymbol{a}$) A measure of the rate of change of the velocity. Acceleration is a vector quantity. The SI units of acceleration are m/s^2 or m s^{-2}.

Equations

$$\boldsymbol{d} = \boldsymbol{v}_{av}t$$
$$\Delta\boldsymbol{v} = \boldsymbol{a}t$$
$$\boldsymbol{v}_{av} = \frac{1}{2}(\boldsymbol{v}_i + \boldsymbol{v}_f)$$
$$\boldsymbol{d} = \boldsymbol{v}_i t + \frac{1}{2}\boldsymbol{a}t^2$$

www.wiley.com/go/biological_physics

1.9 Problems

1.1 A dog chasing a ball starts at rest and accelerates uniformly over a distance of 5 meters. It takes the dog 1 s to cover that first 5 m. What is the dog's acceleration, and what speed is the dog travelling when it reaches the 5 m point?

1.2 During a particular car crash, it takes just 0.18 s for the car to come to a complete stop from $50 \, \text{km h}^{-1}$.

  (a) At what rate is the car accelerating during the crash?

  (b) How many times larger than the acceleration due to gravity is this?

1.3 A jogger starting their morning run accelerates from a standstill to their steady jogging pace of $8.0 \, \text{km h}^{-1}$. They reach a speed of $8.0 \, \text{km h}^{-1}$, 5 s after starting. How long does it take the jogger to reach the end of their 20 m driveway?

1.4 A driver in a blue car travelling at $50 \, \text{km h}^{-1}$ sees a red car approaching in his rear-view mirror. The red car is travelling at $60 \, \text{km h}^{-1}$ and is 30 m behind the blue car when first spotted.

  (a) How many seconds from the time the driver of the blue car first noticed it until the red car passes the blue car?

  (b) How much farther down the road will the blue car travel in this time?

1.5 You are abducted by aliens who transport you to their home world in a galaxy far far away. Oddly, the only thing you can think of doing is measuring the acceleration due to gravity on this strange new world. You drop an alien paperweight from a height of 12 m and use an alien stopwatch to measure the interval of 1.36 s it takes the paperweight to hit the ground below. What is the acceleration due to gravity on the alien home world?

1.6 In a bid to escape from your alien captors you hurl your paperweight straight up towards the door switch on a space ship above you. If the switch is 25 m above you how fast does the paperweight need to leave your hand?

1.7 An initially stationary hovercraft sits on a large lake. When a whistle blows the hovercraft accelerates due north at a rate of $1.2 \, \text{m s}^{-2}$ for 10 s, does not accelerate at all for the next 10 s, and

then accelerates at a rate of $0.6 \, \text{m s}^{-2}$ due east for another 10 s. The hovercraft then coasts for another 10 s without any acceleration.

  (a) What is the velocity of the hovercraft 40 s after the whistle blows?

  (b) What is the displacement of the hovercraft 40 s after the whistle blows?

1.8 A jogger takes the following route to the entrance of their local park: north 120 m, west 100 m, south 35 m, and finally west 50 m. It takes them 2 minutes 18 seconds to reach the park entrance.

  (a) What distance did the jogger travel?

  (b) What is the displacement of the jogger as she enters the park?

  (c) What is the average speed of the jogger?

  (d) What is the average velocity of the jogger?

1.9 A tennis ball is hit down at an angle of 30° below the horizontal from a height of 2 m. It is initially travelling at $5.0 \, \text{m s}^{-1}$. What is the velocity of the ball when it hits the ground if we can neglect air resistance?

1.10 A stunt rider is propelled upward from his motorbike by a spring loaded ejector seat. The rider was travelling horizontally at $60 \, \text{km h}^{-1}$ when the ejector seat was triggered, and as they leave the seat they are travelling with a vertical velocity of $15 \, \text{m s}^{-1}$. The seat is 1.0 m off the ground.

  (a) What is the initial velocity of the stunt rider (in km h^{-1})?

  (b) How high does the stunt rider reach?

  (c) How far along the track does the stunt rider land on the ground?

  (d) What is the velocity of the stunt rider when they hit the ground (in km h^{-1})?

1.11 A bullet is fired horizontally from a gun that is 1.5 m from the ground. The bullet travels at $1000 \, \text{m s}^{-1}$ and strikes a tree 150 m away. How far up the tree from the ground does the bullet hit? [Neglect air resistance.]

FORCE AND NEWTON'S LAWS OF MOTION

2.1   Introduction

We now have a clear and complete description of motion. This description relates the primary properties of motion – displacement, velocity, acceleration and elapsed time – to each other. The question now arises: what causes a change in the motion of an object? This amounts to asking: 'Where does acceleration come from?' The answer to this question is deceptively simple: 'Accelerations are caused by the application of forces.' The purpose of this chapter is to explain clearly what a 'force' is in physics and how it can be used to solve problems relating to the motion of objects. When forces are included in the discussion of motion, it is called dynamics.

Key Objectives

- To understand the concept of force.

- To understand the relationship between force and motion.

- To be able to identify action–reaction pairs of forces.

- To understand normal, friction and tension forces.

- To be able to solve straightforward problems in dynamics.

2.2   The Concept of Force

In everyday conversation we use the word 'force' quite liberally. The Oxford English Dictionary gives a number of definitions of the noun 'force':

1. physical strength or energy accompanying action or movement.

2. (Physics) a measurable influence that causes something to move.

3. pressure to do something backed by the use of threat of violence.

4. influence or power.

5. a person or thing having influence: *a force for peace*.

6. an organised group of soldiers, police or workers.

The word is also used as a verb, as in sentences like: 'She forced the committee to consider her application seriously.' In physics, the word 'force' has a very precise meaning, and this is given by Newton's laws. These laws define force by listing the essential properties of a force. If some phenomenon does not have all of these properties, then it is not a force. In this section we will go through Newton's Laws and explain each of them in turn.

Force vectors in figures

In this text *force vectors* are represented as acting on the *centre of mass* of an object. It should be noted that other texts use differing conventions for contact force vectors in which a contact force is shown acting at the surface between two objects. The distinction between the two ways of representing contact force vectors is partially style, although when dealing with contact forces that may act to rotate an object the exact position of action of a force is important. Such cases will not be dealt with in this text and so the simpler standard in which all force vectors are shown to act at the centre of mass of an object has been used in the interests of clarity.

Newton's First Law

In the list of Oxford English Dictionary definitions of force given above, the second item is closest to the definition used by physicists. A force is essentially anything that is measurable and causes a *change* in the motion of an object. For example, if an object is at rest, then we must apply a force to it to cause it to accelerate and to develop a non-zero velocity.

Newton's first law states that any object continues at rest, or at constant velocity (constant speed in a straight line), unless an external force acts on it. Figure 2.1 illustrates the effect of a force on the motion of an object. This object is travelling in a straight line until an external force acts on it in a direction perpendicular to its motion. This causes the object to be deflected from its straight-line motion.

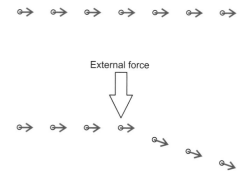

Figure 2.1 A pair of objects initially travel in the same direction. The lower object is subject to a force and its motion then deviates from a straight line while the force acts on it, but continues on in a new straight line after the force ceases to act.

Newton's Second Law

> **Key concept:**
> An external force gives an object an acceleration. The acceleration produced is proportional to the force applied, and the constant of proportionality is the **mass**.

Newton's second law can be summarised with the following equation:

$$F = ma \tag{2.1}$$

In this equation, a is the acceleration (in m s^{-2}) as usual, m is the mass (in kg), and F is the force (in N) (N = newton). The SI unit of force is the newton; one newton (1 N) is the force which would accelerate a 1 kg mass at 1 m s^{-2} (i.e., would cause its velocity to increase by 1 m s^{-1} in every second).

As is illustrated in Figure 2.2, if the mass of an object on which the force is applied does not change, but the force is doubled, then the acceleration of this object will also be doubled (see diagram to the right in Figure 2.2). If the applied force is not changed, but the mass of the object is doubled, then the acceleration will be halved (see centre diagram of Figure 2.2). For numerical examples of the relationship between mass, force and acceleration, see Table 2.1.

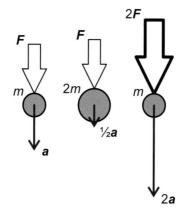

Figure 2.2 If the force applied to an object does not change, but the mass of the object is doubled, then the acceleration is halved. If the mass does not change and the force is doubled, then the acceleration is doubled also.

Weight and Mass

Since any object of mass m near the surface of the Earth falls with acceleration g downwards, it must be acted on by a force:

$$
\begin{aligned}
F &= ma \\
&= mg \text{ downwards}
\end{aligned}
$$

This downward force is due to the gravitational attraction between the Earth and the object and is often called the object's **weight**, i.e., the magnitude of the force, W, is

$$W = mg \tag{2.2}$$

> **Key concept:**
> The weight of an object is a force, not a mass.

When we step on our bathroom scales and are told our mass in kilograms, we are effectively being given our weight in newtons with a zero removed. The scales do not directly measure our mass or weight – they measure the magnitude of the contact force (also referred to as the normal force or support force) between the scales and our feet. If we were to step onto our scales while resident on the International Space Station (ISS), we would read a weight of nearly zero kilograms. Our mass has not changed; what has happened is that both the scales and ourselves are in free fall, and there is no significant contact force between the scales and our feet.

Interestingly, because the ISS orbits at an altitude of only 350 km (6.34×10^6 m + 0.35×10^6 m = 6.69×10^6 m from the centre of the Earth), the difference in the gravitational force, and hence acceleration due to gravity, between an object on the surface of the Earth and on the ISS is quite small ($F_{\text{grav, ISS}} \approx 0.90 \times F_{\text{grav, sea level}}$). (The strength of the gravitational force is discussed in Section 2.4.

m (kg)	a (m s^{-2})	F (N)
1	1	1
1	2	2
2	1	2
2	2	4
4	5	20

Table 2.1 Newton's second law and the relationship between force, mass and acceleration.

Forces are Vectors

When I push against an object, say, an apple, I push hard or not so hard, and I push in a particular direction. For example, I could be pushing the apple across the table towards you, or I could be pulling the apple toward myself. It seems that forces have both magnitude and direction; this means that we must represent forces as *vectors*. This fits with what we have said so far about forces. Thus to find the total or net force on an object, we must find the *vector sum* of all of the individual forces on that object. It is the net force acting on an object that produces an acceleration. Since acceleration is a vector, and the mass is not, the force that acts on the object must also be a vector (see Eq. (2.1).) This means that if several forces are acting on the same object, we find the total force using vector addition to add up all of the applied forces.

As in the case of the velocity and acceleration vectors, we may treat the components of a force vector as separate independent vectors. Thus, in calculations we are able to look at the behaviour of the horizontal components of the force, acceleration, velocity and displacement, and then the vertical components of these vectors. Once we have completed these separate calculations we can combine the components of the relevant vector quantities to find the total force, velocity or acceleration.

Newton's Third Law

> **Key concept:**
> For every action there is an equal and opposite reaction.

Newton's third law states that forces come in pairs – for every force that is applied to a body, there is a force applied *by* that body. The 'action' referred to in the box above is the force applied by one object on the other. Suppose that the 'action' is the force exerted by object 1 on object 2. Newton's third law then states that object 2 will exert a force of equal size, but in the opposite direction, on object 1. This second force is the 'reaction' to the force exerted by object 1.

Forces act in pairs and each force acts between a pair of objects. These force pairs are called **action–reaction pairs** or **third-law force pairs**. It is important to correctly identify the action–reaction pairs in a problem. For example, the weight force of an object is due to the gravitational interaction between that object and the Earth. Thus the Earth exerts a force on the object and the object exerts an equal, but opposite, force on the Earth. The object will accelerate toward the Earth under the influence of the

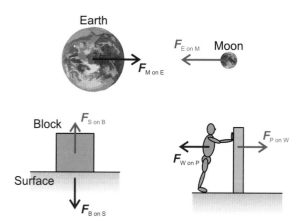

Figure 2.3 Three examples of action–reaction pairs. The gravitational force between the Earth and the Moon, the downward contact force a block exerts on a surface and the upwards support force the surface exerts on the block, and a person pushing against a wall.

gravitational force exerted by the Earth – this is source of the acceleration due to gravity. These equal and opposite forces exist whether the object under discussion is in free fall toward the Earth or is sitting on a set of bathroom scales being weighed; in both of these cases the action–reaction pair is the force of the Earth on the object and the force of the object on the Earth. Later we will discuss the electric force, the force described by Coulomb's Law. This is a force which again acts between two objects (which are charged), with the force on one being equal and opposite to the force on the other.

In Figure 2.3 we illustrate Newton's third law with three examples. In the first example, the Earth (E) exerts an attractive gravitational force on the Moon (M), $F_{E \text{ on } M}$. In turn, the Moon exerts and equal, but opposite (and hence also attractive), force on the Earth, $F_{M \text{ on } E}$. In the second example, the surface (S) on which a block (B) is sitting exerts an upward (normal or support) force $F_{S \text{ on } B}$, on the block, which exerts an equal and opposite force, $F_{B \text{ on } S}$, on the surface. Note that the downward force exerted by the block is not necessarily the weight force of the block. If I push down on the block, the downward force is the sum of the forces acting downward, and the support force provided by the surface will increase to equal it. Finally, a person (P) pushing against a wall (W) applies a force $F_{P \text{ on } W}$ to the wall, and the wall applies an equal and opposite force $F_{W \text{ on } P}$ to the person. It is important to note that each force in an action–reaction pair acts on a different object. It is *always* the case that action–reaction pairs act on different objects.

Example 2.1 *A box on a box*

Problem:

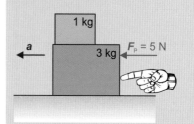

Figure 2.4 Two boxes being pushed along a frictionless surface

As shown in Figure 2.4, a large 3 kg box is being pushed with a horizontal force of $F_P = 5$ N and as a result is accelerating along the horizontal frictionless surface upon which it rests. The large box has a smaller 1 kg box resting on top of it. This box does NOT slide from the top of the big box as it accelerates.

(a) At what rate are the boxes accelerating?

(b) What are the magnitudes and directions of the friction forces acting on each box?

(c) What are the magnitudes and directions of the normal forces acting on each box?

Solution: Because the small box does not slide around on top of the large one, the small box accelerates at the same rate as the large one $a_S = a_L$. Drawing a diagram of all the forces acting on each box will be helpful at this point. Because each box is accelerating horizontally, there must be a net force on each box which is horizontal. The sum of all vertical forces on each box must be zero.

N_S and W_S, and f_S and f_L are each Newton's third-law force pairs. The large box pushes up on the small box with a force of N_S and so the small box pushes down on the large box with a force W_S which is of equal magnitude. Similarly the frictional force between the two boxes pushes the small box forward with a magnitude of f_L and the large box backwards with a force f_L of the same magnitude.

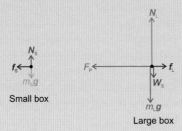

Figure 2.5 Forces acting on each box

(a) The 5 N pushing force accelerates a total mass of 4 kg at a rate of $a = \frac{F_P}{(m_L + m_S)} = \frac{5\,\text{N}}{4\,\text{kg}} = 1.25\,\text{m s}^{-2}$.

(b) The net force on the small box is equal to the frictional force (there are no other horizontal forces). $f_S = m_S a = 1\,\text{kg} \times 1.25\,\text{m s}^{-2}$ to the left $= 1.25\,\text{N}$ to the left. As f_S and f_L are a Newton's third-law pair, $f_L = 1.25\,\text{N}$ to the right.

(c) As the small box is not accelerating in the vertical direction, the normal force must have the same magnitude as the weight force of the box. $N_S = m_S g = 1\,\text{kg} \times 10\,\text{m s}^{-2} = 10\,\text{N}$, and it is directed in the opposite direction to the weight force, which is upwards. The normal force acting on the large box must be equal in magnitude to the sum of the two downwards forces acting on the box, which are the weight force of the large box itself and W_S. $N_L = m_l g + W_S = 3\,\text{kg} \times 10\,\text{m s}^{-2} + 10\,\text{N} = 40\,\text{N}$ upwards. This is equal in magnitude to the combined weight forces of both boxes.

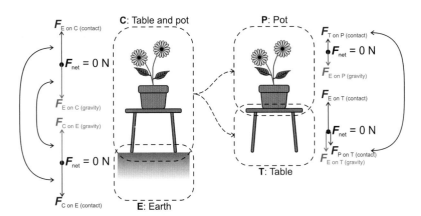

Figure 2.6 A flowerpot sits on a table which itself sits on the ground. Several systems of interest are shown. On the left the Earth (system E) and the combination of the flowerpot and table (system C) constitute two systems. System C is then split into two systems: the pot (system P) and the table (system T). The net force on each system (and subsystem) is zero, but each object is exerting forces on other objects and having forces exerted on them.

In order to clarify the analysis of collections of objects which are exerting forces on each other, it is often useful to begin by defining the 'system of interest'. The system of interest contains the objects whose behaviour we are interested in analysing, and the net forces acting on them. Sometimes both objects in an action–reaction pair will be in the system of interest, but in many cases only half of an action–reaction pair is relevant to a given problem.

We are at liberty to define the system of interest in the way which is most convenient for the problem at hand, and we will illustrate this point with an example. Figures 2.6 and 2.7 show a flowerpot at rest upon a table, which is itself at rest. In other words, the

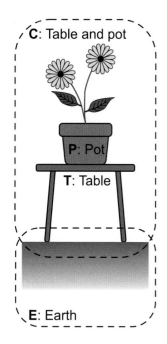

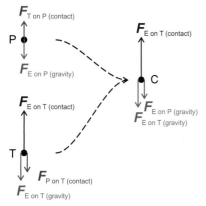

Figure 2.7 Detail of the forces acting in system C of Figure 2.6. $F_{\text{T on P}}$ and $F_{\text{P on T}}$ are third-law pair forces and as such *always* equal and opposite. Thus when considering the motion of system C as a whole by adding together all of the forces acting upon it these two forces, $F_{\text{T on P}}$ and $F_{\text{P on T}}$ will always cancel each other out and only the three remaining forces need be considered.

table and flowerpot are not accelerating with reference to the Earth.

Figure 2.7 shows the forces acting on each object in system C from Figure 2.6. The forces $F_{\text{T on P}}$ and $F_{\text{P on T}}$ are an action–reaction pair and thus are *always* equal and opposite. As they both act on objects which are inside system C, they will cancel each other out and not contribute to the motion of system C as a whole.

The remaining forces, $F_{\text{E on P (gravity)}}$, $F_{\text{E on T (gravity)}}$ and $F_{\text{E on T (contact)}}$, are independent of each other. $F_{\text{E on P (gravity)}}$ and $F_{\text{E on T (gravity)}}$ are the weight forces of the pot and table respectively These two forces obviously have no bearing on each other, while $F_{\text{E on T (contact)}}$ is the force with which the ground is pushing up on the table. In this example, this force is equal and opposite to the sum of the $F_{\text{E on P (gravity)}}$ and $F_{\text{E on T (gravity)}}$ and so system C as a whole does not accelerate. This need not be the case, however, as $F_{\text{E on T (contact)}}$ could be higher or lower than the combined weight force of the pot and table (perhaps during an earthquake or if the table were placed in an elevator), resulting in acceleration of the system C.

Newton's laws define the meaning of the word 'force' in physics, and allow us to analyse relatively simple situations in which we are able to clearly define the forces acting on an object. We will now look at the kinds of forces that arise between pairs of objects.

2.3 Kinds of Force

The Fundamental Forces

There are four fundamental kinds of force: the **electromagnetic force**, the **gravitational force**, and the **strong** and **weak nuclear forces**. All forces which will be important in understanding the mechanics of the human body and other biological systems may ultimately be reduced to these fundamental forces. Two of these fundamental forces are most important in everyday biological applications:

- **The electromagnetic force.** At the most basic level, this force holds atoms and molecules together, and is thus responsible for the rigidity of solids and the fluid properties of liquids. It is the force which results in all chemical behaviour. In a later topic we will consider this force in more detail and discuss the vital role of electricity in the functioning of every biological system, from nerves to muscle tissue to the functioning of vital cellular systems like the cell membrane. As we shall see, the human body is an electrical machine. Furthermore, the greatest and most easily available source of energy in the earth biosphere is electromagnetic: solar radiation (sunlight).

- **The gravitational force.** Every organism on Earth is subject to a gravitational force, and all of these organisms have evolved body structures to take advantage of, or minimise the effects of, gravity. We will discuss the gravitational force in more detail in a later section.

The importance of the strong and weak nuclear forces in biological systems is not as obvious. However, these are the forces which maintain the structure of the atomic nucleus on the one hand and produce nuclear radiation on the other hand. Nuclear radiation used as a medical tool is becoming more important as medical imaging techniques become more sophisticated. Thus an understanding of the risks and advantages of the use of nuclear radiation in medicine is becoming more important to health professionals.

These four forces are in an important respect simple, in that they act between pairs of objects and are not the result of the vector addition of a number of more fundamental forces. There are a number of important forces which are not simple in this sense: they are forces which are the result of combinations of more fundamental forces.

Derived Forces

We will now consider three specific examples of derived forces: the normal and friction forces, and tension. These forces are the result of a large number of more fundamental interactions, but may be treated as single forces in their own right. When treated in this way, they have characteristic behaviours which at first sight may seem counterintuitive, but which are the result of the combination of more fundamental forces on which they depend.

Tension

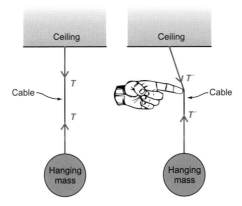

Figure 2.8 A cable exerts the same magnitude force at either end, but the flexibility of the cable ensures that these forces need not lie along a line.

Tension is the name given to the forces which exist in the body of a flexible cable or line. The cable is assumed to be infinitely flexible, meaning that no lateral forces are supported. Any forces which are not directed along the cable result in bending of the cable (see Figure 2.8). Thus the only forces which exist inside the cable are forces directed along the cable. The cable is also assumed to be inelastic: it does not stretch, and it is unbreakable. These are simplifications; real cables will always have some elasticity and will be breakable. The elasticity and breakability of real cables is of considerable importance in biological systems and will be discussed in a later chapter. However, for our present purposes, we will discuss the simpler system of the perfectly flexible, inelastic and unbreakable cable.

Tension exists at each point in the cable when forces are applied in opposite directions at each end of the cable. If the cable is stationary these forces are transmitted along its length. At each point along the cable there will exist equal and opposite forces acting on an imagined cross section at that point. These forces are the sum of the microscopic forces acting between the component molecules of the cable.

The Normal Force and Friction

Consider a situation in which a solid block sits on an inclined surface, as shown in Figure 2.9. The block and the surface are stationary, so Newton's second law tells us that the weight of the block is balanced by an upward force produced by the surface. Note that for the sake of simplicity these forces are shown as acting at a single point in the block. In actual fact, the contact forces act at the points of contact between the block and the surface, and we then sum these forces to obtain the resultant upward contact force. This simplification does not alter the fundamental physics of the situation.

The upward force which balances the weight force of the block is ultimately electrostatic in that it is the result of the electrical forces between the atoms of the surface and the atoms of the block, as well as the electrostatic forces which give the block and surface their rigidity. This force is known as a contact force, as it occurs at the contact points between the block and the surface. We know that there must be a resultant upward force from this contact, since this is the force which keeps the block from accelerating toward the centre of the Earth in response to the weight force.

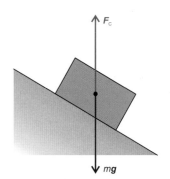

Figure 2.9 Weight force and contact force acting on a block on an inclined plane.

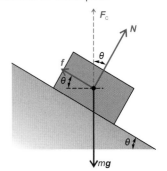

Figure 2.10 Surface contact force resolved into components perpendicular and parallel to the surface.

To analyse the situation shown in Figure 2.9, it is useful to think of the reaction force produced by the surface as having two components, one perpendicular to the surface, and one parallel to it. These components are shown in Figure 2.10 and are labelled N and f. The component of force which is perpendicular to the surface is called the **normal force** (the word 'normal' means perpendicular to a line or surface in mathematics), and the component of force which is parallel to the surface is called the **friction force**, or often just **friction**. In Figures 2.10 and 2.9 the only force acting downward is the gravitational weight force on the block. This is not the only possibility. We could, for example, push down on the block. The normal and friction forces would then be the perpendicular and parallel components of a contact force equal and opposite to the *total* downward force.

If the block were to begin to slide down the slope, this motion would be due to the component of the weight force which is directed parallel to the surface. If the block does not slide down the slope this is because there is a force produced by the surface which is great enough to cancel the parallel component of the block's weight force. This opposing force is the component of the surface reaction force which is directed along the surface, i.e., the friction force. Thus, the friction force is the force which prevents the block from sliding down the slope.

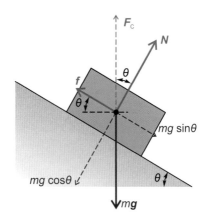

If the block is stationary on the inclined surface, then the forces perpendicular and parallel to the surface will balance: the perpendicular and parallel components of the block's weight force will be equal and opposite to the perpendicular and parallel components of the surface reaction forces. This is illustrated in Figure 2.11. From this figure we can see that the magnitudes of the normal and friction forces are given by

$$N = mg\cos\theta \qquad (2.3)$$
$$f = mg\sin\theta \qquad (2.4)$$

where θ is the angle of incline of the surface. These two forces are linked by the fact that they are components of a single force. We are thus able to write the friction force in terms of the normal force and the angle of incline of the surface,

$$f = N\tan\theta \qquad (2.5)$$

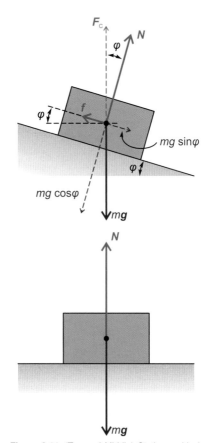

As the angle of incline is increased, there will generally be a point at which the block will begin to slide. Let us call this angle the critical angle, θ_c. We can then write the relationship between the normal and friction forces as

$$f_{\text{max}} = \mu_{\text{max}}N \qquad (2.6)$$

where the coefficient μ_{max} is given by

$$\mu_{\text{max}} = \tan\theta_c \qquad (2.7)$$

This coefficient is defined for the point of maximum tilt of the surface, so it is the **maximum coefficient of friction** for this block and this surface. We are able to extend the idea of the coefficient of friction to those cases where the block is inclined at an angle less than θ_c and also to cases in which the block is sliding. In these cases the friction force must be determined empirically, i.e., by measuring the parallel force which must be applied to start the block moving, and then the force required to keep the block moving with constant velocity. The coefficients calculated from $f = \mu N$ are known as the coefficients of **static** and **kinetic friction** respectively. The coefficient of kinetic friction is always smaller than the coefficient of static friction.

Now consider what happens as the angle of incline θ decreases to zero, i.e., when the surface is horizontal, as in Figure 2.11 (bottom). In this case, the friction force goes to zero (since $\sin 0 = 0$) and the normal force goes to $N = mg$ (since $\cos 0 = 1$). This situation is particularly easy to interpret; the surface will deform downwards until the net upward force is equal to the downward force of the block. Often the downward force due to the block will be due to its weight force alone. However, this is not always the case. When there are forces other than the weight force acting downward on the block, the upward normal reaction force of the surface will increase until the downward force is again balanced by the upward normal reaction force (or the surface breaks).

Figure 2.11 (Top and Middle) Stationary block with normal and friction forces balancing weight of the block. (Bottom) As the incline of the surface on which the block is resting becomes more horizontal, the components of the weight and contact forces directed along the slope approach zero.

Note that the action–reaction pair consisting of the solid block and the Earth is the source of the weight force of the block, but the Earth is not part of the system of interest shown in Figures 2.10 and 2.9. The weight force of the block is *not* part of an action–reaction pair involving any force produced by the inclined surface. The normal and friction forces are reaction forces, however, in that they are the result of action–reaction pairs between atoms on the surface and atoms on the block – these forces are electrostatic in character, not gravitational. Thus the normal reaction force of the surface is not necessarily equal to the weight force of the block.

Drag Forces

In our previous discussion of contact forces, we were concerned with the forces produced by the contact between two solid, rigid objects. We will now consider the forces which occur when solid objects move through gases and liquids. An object moving through a fluid must move fluid out of the way; this will be easier or harder depending on the viscosity of the fluid. (We will discuss fluid viscosity in Chapter 15.) The fluid in which an object is moving therefore exerts a force on that object in the direction opposite to its motion. This force is known as **drag**. In the case of an object moving through air this drag is often called **air resistance** and is the reason for the streamlining of modern cars. The reduction of drag forces is also a dominant factor in the evolutionary pressure selecting for more streamlined forms in aquatic animals.

The magnitude of the drag force on an object moving through air can be modelled as being proportional to the square of its speed, i.e.,

$$f = k v^2 \tag{2.8}$$

The constant of proportionality, k, is determined by the shape of the object, and by the density of the air through which it is moving.

2.4 Newtonian Gravity

In this section we will briefly introduce the force of gravity as described by Newton. The study of gravity advanced considerably in the twentieth century with the discovery of general relativity by Albert Einstein. Newton's theory of gravity is a very good approximation to the behaviour of massive objects in everyday situations, and general relativity is needed only in situations where extreme accuracy is required, such as in the calculations performed by GPS units, or when the force of gravity is extremely strong, such as the effects of black holes and other massive astronomical bodies.

Newton's theory of gravity is based on the recognition that the force of gravity exists between two bodies by virtue of their mass and decreases with the separation between these bodies. The exact form of the gravitational force law is

$$F = G\frac{m_1 m_2}{r^2} \tag{2.9}$$

The constant $G = 6.67 \times 10^{-11}$ N m^2 kg^{-2} and is called the universal gravitational constant. This constant is extremely small and this means that the force of gravity is extremely weak unless one or both of the masses involved is extremely large. The symbol r in the above equation is the distance between the centres of mass of the two bodies and m_1 and m_2 are the masses of the two bodies in kilograms. The force of gravity is attractive and directed from one body to the other, and the gravitational force on m_1 is equal in magnitude to the gravitational force on m_2. The gravitational force on m_1 is directed from m_1 toward m_2 and the gravitational force on m_2 is directed from m_2 toward m_1.

An object of mass m on or near the surface of the Earth will experience an attractive gravitational force exerted by the Earth, mass M_E. The magnitude of this force is given by

$$F = G\frac{M_{\mathrm{E}}m}{r^2} = ma$$
$$\Rightarrow a = \frac{GM_{\mathrm{E}}}{r^2} = g$$

where g is the acceleration due to gravity discussed in an earlier section. This shows that the acceleration due to gravity will vary with distance from the centre of the Earth. This variation is very slight, however, as the typical changes in height which we consider are of the order of hundreds of metres, whereas the radius of the earth is $R_{\mathrm{E}} = 6.37 \times 10^6$ m. Using this radius and the mass of the Earth, $M_{\mathrm{E}} = 5.97 \times 10^{24}$ kg gives

$$
\begin{aligned}
g &= G\frac{M_{\mathrm{E}}}{R_{\mathrm{E}^2}} \\
&= 6.67 \times 10^{-11} \text{ N m}^2 \text{ kg}^{-2} \times \left(\frac{5.97 \times 10^{24} \text{ kg}}{\left(6.37 \times 10^6 \text{ m}\right)^2} \right) \\
&= 9.81 \text{ m s}^{-2}
\end{aligned}
$$

2.5 Summary

Key Concepts

mass (m) A measure of a body's resistance to acceleration. The SI unit of mass is the kilogram (kg), which gives the ratio of the mass of an object to a standard mass, a cylinder of platinum-iridium alloy kept at the International Bureau of Weights and Measures near Paris, France.

force (F) A vector quantity that produces an acceleration of a body in the direction of its application. The SI unit of force is the newton (N). $1 \text{ N} \equiv 1 \text{ kg m s}^{-2}$.

net force The vector sum of all the forces acting on a body or system.

Newton's first law Any object continues at rest, or at constant velocity, unless an external force acts on it.

Newton's second law An external force gives the object an acceleration which is proportional to the force. $F = ma$.

Newton's third law For every action there is an equal and opposite reaction.

friction force (f) A force that resists the relative motion between two surfaces in contact.

normal force (N) The perpendicular component of the contact force between two objects in physical contact with each other.

action–reaction pair Newton's third law states that for every reaction, there is an equal and opposite reaction. These two forces form an 'action–reaction' pair that are equal in magnitude and opposite in direction. An alternative name is Newton's third-law pair.

tension force (T) A force that tends to stretch intermolecular bonds.

drag The resistance an object encounters moving through a fluid.

Equations

$$F = ma$$
$$W = mg$$
$$f_{\max} = \mu N$$
$$f = kv^2$$
$$F = G\frac{m_1 m_2}{r^2}$$

2.6 Problems

2.1 A courier is delivering a 5 kg package to an office high in a tall building.

(a) What upwards force does the courier apply to the package when carrying it horizontally at a constant velocity of $2\,\mathrm{m\,s^{-1}}$ into the building?

(b) The courier uses the elevator to reach the office. While the elevator (containing the courier who is holding the package) is accelerating upwards at $0.11\,\mathrm{m\,s^{-2}}$ what upwards force is the courier applying to the package?

(c) When the elevator is traveling upwards at a constant speed of $6\,\mathrm{m\,s^{-1}}$ what upwards force does the courier apply to the package?

(d) In order to stop at the correct floor the elevator accelerates downwards (decelerates) at a rate of $0.20\,\mathrm{m\,s^{-2}}$. What is the upwards force the courier applies to the package during the deceleration?

2.2 You live at the top of a steep (a slope of 15° above the horizontal) hill and must park your 2200 kg car on the street at night.

(a) You unwisely leave your car out of gear one night and your handbrake fails. Assuming no significant frictional forces are acting on the car, how quickly will it accelerate down the hill?

(b) The increase in insurance premiums due to the results of your mistake mean that you cannot afford to fix your handbrake properly. You resolve to always leave your car in gear when parked on a slope. If the rolling frictional force caused by leaving the drive-train connected to the wheels is 5000 N, at what rate will your car accelerate down the hill if the handbrake fails again?

2.3 You are pulling your younger sister along in a small wheeled cart. You weigh 65.0 kg and the combined mass of your sister and the cart is 35.0 kg. You are pulling the cart via a short rope which you pull horizontally. You hold one end of the rope and your sister holds the other end. If you are accelerating at a rate of $0.10\,\mathrm{m\,s^{-2}}$, the rope is inelastic, and the frictional force acting upon the cart is 30 N:

(a) What is the tension in the rope?

(b) What force are you applying to the ground in order to produce this acceleration?

2.4 Two flexible balls rolling along a frictionless horizontal surface collide with each other. The larger of the balls weighs 50 g and the smaller weighs 30 g. Immediately after the balls first touch each other (the beginning of the collision), the center of mass of the larger ball is accelerating at a rate of $5\,\mathrm{m\,s^{-2}}$ to the right. What is the acceleration of the center of mass of the smaller ball?

2.5 The Earth has a mass of 5.97×10^{24} kg and the Moon has a mass of 7.36×10^{22} kg. The average distance between the center of the Earth and the center of the Moon is 3.84×10^8 m.

(a) What is the gravitational force acting on the Moon due to the Earth?

(b) What is the gravitational force acting on the Earth due to the Moon?

(c) How far away would the Moon need to be for the magnitude of the gravitational force acting on it due to the Earth be the same as the magnitude of the gravitational force of a 72 kg student sitting at their desk on the surface of the earth?

(c) A supermassive black hole passes through the edge of the solar system 1.20×10^{13} m away. The gravitational force between an observant 79 kg astronomer and the black hole is 1 N. What is the mass of the black hole?

2.6 A 4 kg vase of flowers is placed directly in the middle of a glass table. The glass tabletop itself weighs 8 kg. With what force do each of the four legs of the table push on the glass after the vase has been placed on top?

2.7 In order to drink from your glass you first need to lift it to your mouth. As you begin to lift your 0.20 kg glass of water it is accelerating upwards at $0.090\,\mathrm{m\,s^{-2}}$.

(a) What is the net force acting on the glass?

(b) What force are you applying to the glass?

(c) What force is the glass applying to you?

2.8 During a car crash a 65 kg person's head goes from travelling at $50\,\mathrm{km\,h^{-1}}$ to stationary in 0.15 s.

(a) What is the magnitude of the average net force acting on the head of a person with a 4.5 kg head?

(b) How does this compare with the weight force acting on the person?

2.9 A 10 kg box is being pushed up a slippery ramp as shown in Figure 2.12. The coefficient of friction between the box and the ramp is just $\mu = 0.1$.

(a) What force does the man need to apply to the box to keep it traveling up the ramp at a steady speed?

(b) What fraction of the weight force of the box is this?

If the angle of the ramp is raised to 45° then:

(c) What force does the man need to apply to the box to keep it traveling up the ramp at a steady speed now?

(d) What fraction of the weight force of the box is this?

Figure 2.12 A box is pushed up a ramp

2.10 A 1.6 kg chicken is blown into a wall by a strong gust of wind, and held there as shown in Figure 2.13. If the maximum coefficient of friction between the chicken and the wall is $\mu_{max} = 0.25$, what minimum force must the gust of wind be applying to the chicken?

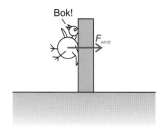

Figure 2.13 A chicken is blown into a wall.

MOTION IN A CIRCLE

3.1 Introduction

The problem of describing circular motion itself is not of great importance to the biological sciences. However, the tools for describing and understanding circular motion are important to developing an understanding of oscillatory systems and waves.

Key Objectives

- To understand the concepts of angular displacement and velocity, and the use of radian measure.

- To understand circular velocity and acceleration and the relationship between these quantities and linear velocity.

- To understand the concept of centripetal force.

- To be able to use centripetal force to calculate properties of the circular motion produced by that force.

3.2 Description of Circular Motion

Angular Displacement and Radians

In Figure 3.1, the angle θ is subtended by the arc s. The size of the angle θ in **radians** is defined by

$$\theta = \frac{s}{r} \tag{3.1}$$

θ is the **angular displacement** of an object that moves from point A to point B on the circle shown in this figure. The angle subtended by the complete circle is

$$\theta = \frac{2\pi r}{r} = 2\pi$$

Thus $360° = 2\pi$ radians and 1 radian $= 360/2\pi \approx 57.3°$.

Angular Velocity

If it takes time t to move a distance s round the circle, and this distance subtends an angle θ, then using Eq. (3.1),

$$\frac{\theta}{t} = \frac{s}{rt}$$

The ratio $\frac{\theta}{t}$ is given the symbol ω and is the rate at which the angular displacement changes, and ω is therefore the **angular velocity** (or angular frequency). The ratio $\frac{s}{t}$ is the speed, v, around the circumference of the circle. Therefore

$$\omega = \frac{v}{r} \tag{3.2}$$

The angular velocity is measured in radians s^{-1}. The rotational equivalent of $d = vt$ is

$$\theta = \omega t \tag{3.3}$$

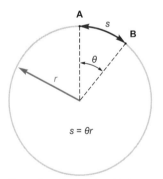

Figure 3.1 The relationship between the subtended angle and arc length.

$s = \theta r$

which will give the angle in radians moved in time t.

If an object rotates around a circle at f cycles (i.e., complete rotations) per second, then it moves $2\pi f$ radians per second. The relationship between the frequency in hertz (that is, cycles per second) and the angular velocity in radians per second is therefore

$$\omega = 2\pi f \qquad (3.4)$$

For example, 1 Hz $\equiv$ 6.28 radians s^{-1}.

Example 3.1 *Angular displacement and angular velocity*

Problem:
A juggler throws a bowling pin as shown in Figure 3.2.

(a) What is the angular velocity of the pin?

(b) What will the angular displacement of the pin be at a time of 0.9 s?

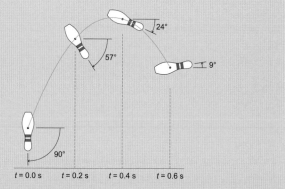

Figure 3.2 A juggling pin is thrown in an arc through the air.

Solution:

(a) The angular displacement of the pin changes 33° anticlockwise every 0.2 s. $33° = \frac{33°}{360°} \times 2\pi = 0.576$ radians. The angular velocity of the pin is $\omega = \frac{\Delta\theta}{\Delta t} = \frac{0.576\,\text{radians}}{0.2\,\text{s}} = 2.88$ radians s^{-1} anticlockwise.

(b) At $t = 0$ s, $\theta = 90\deg = \frac{\pi}{2}$ radians clockwise from right-horizontal. At $t = 0.9$ s the angular displacement will be $\theta = \theta_i + \omega t = \frac{\pi}{2} + \left(-2.88\,\text{radians s}^{-1}\right) \times 0.9\,\text{s} = -1.02$ radians or 1.02 radians (58.5°) clockwise from right-horizontal.

3.3   Circular Velocity and Acceleration

If an object travels around a circle once, the distance travelled is the circumference of that circle. The time taken is the known as the **period**, T, of the motion. The speed of the object is therefore

$$v = \frac{2\pi r}{T} \qquad (3.5)$$

Since this is the speed of the object, the direction of motion is not important. If the object is travelling at a constant speed, the magnitude of the velocity is constant. The direction that the object is moving in, however, is always changing.

Key concept:
The velocity of an object in circular motion changes continuously since the direction in which the object is moving is changing continuously.

The velocity of the object will always be tangential to the circle. Since the velocity is changing, the object must be continuously accelerating. The acceleration is the change in the velocity from one point to the next, divided by the time taken to get there. Figure 3.3 shows that this change in velocity points toward the centre of the circle.

Key concept:
If an object is travelling in a circle at constant speed, its instantaneous acceleration is always pointed exactly toward the centre of the circle.

This acceleration is called the **centripetal** acceleration. It can also be shown that the magnitude of the centripetal acceleration is given by

$$a = \frac{v^2}{r} \qquad (3.6)$$

We will not cover the details of the derivation of the direction and magnitude of the centripetal acceleration.

3.4 Centripetal Force

Since there is a centripetal acceleration associated with the circular motion of an object, there must be a **centripetal force** to produce that acceleration. Whenever an object is in circular motion, there must be a force directed toward the centre of the circular path of the object. We may find an expression for the magnitude of this force by substituting the expression for the centripetal acceleration (Eq. (3.6)) into Newton's second law; for an object of mass m this gives

$$F = m\frac{v^2}{r} \qquad (3.7)$$

3.5 Sources of Centripetal Force

Given that an object is moving in a circular path, and that there is therefore a centripetal force, where does this force come from? A few of the more common everyday sources of the centripetal force are listed below.

- When a car drives in a circle on a flat stretch of road, or in a car park, the centripetal force is supplied by friction between the rubber tyres and the road surface.

- When a car travels along an arc (i.e., turns a corner) on the motorway, the road is typically banked – the road is angled. This allows the normal reaction component of the contact force between the car tyres and the road to contribute to the centripetal force.

- When a ball or some other object is swung in a circle from a string, the tension in string provides the necessary centripetal force.

- A satellite in a circular orbit about the Earth is maintained in this circular path by the Earth's gravitational attraction, i.e., gravity provides the centripetal force.

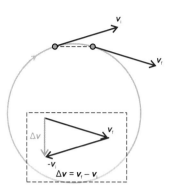

Figure 3.3 The velocity of an object travelling in a circle at constant speed is always tangential to the circle. As the object travels around the circle the change in velocity will point towards the centre (provided this is measured over a small time). This implies that the object is accelerating towards the centre of the circle.

Example 3.2 *Angular velocity and centripetal force*

Problem: A 45 kg child is riding the rollercoaster at an amusement park. A portion of the track is made up of two semicircular sections as shown in Figure 3.4, one with a radius of 10 m and the other with a radius of 16 m. The rollercoaster travels along this section of track at a *constant speed* of 7 m s⁻¹.

(a) **In what direction is the child accelerating at point A?**

(b) **In what direction is the child accelerating at point B?**

(c) **If the child was on bathroom scales, what would they read (in kg) at point A?**

(d) **If the child was on bathroom scales, what would they read (in kg) at point B?**

Solution:

(a) As the child is travelling along a circular path at a constant speed, we know that the velocity at each point along their path must be tangential to the path. The direction of the acceleration of the child at each point can be found by thinking about the change in direction of the acceleration vector as it moves around the path (refer back to Figure 3.3). At point A the rollercoaster must be accelerating straight down.

(b) Using the same reasoning as that described in part (a), the acceleration of the rollercoaster at point B can be identified as in the vertical upwards direction.

(c) The apparent weight of the child can be found by considering the forces acting upon them. The centripetal force (on the child) is the same as the net force acting on the child. We can find the centripetal force by using

$$F_{\text{net}} = F_c = m \frac{v^2}{r} = 45 \,\text{kg} \times \frac{\left(7 \,\text{m s}^{-1}\right)^2}{10 \,\text{m}} = 220.5 \,\text{N}$$

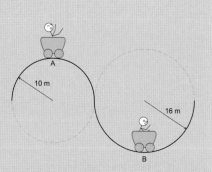

Figure 3.4 A section of a roller coaster is shown. The cart travels along this section at a *constant speed*. The apparent weight of the child in the cart will change depending upon where they are along the track.

This indicates that at point A the net force on the child is 220 N downwards. As $F_{\text{net}} = mg - N$, the normal force on the child must be $N = 450 \,\text{N} - 220 \,\text{N} = 230 \,\text{N}$ and the scales will read 23 kg.

(d) As in part (c) the net force acting on the child is

$$F_{\text{net}} = 45 \,\text{kg} \times \frac{\left(7 \,\text{m s}^{-1}\right)^2}{16 \,\text{m}} = 138 \,\text{N}$$

This force is directed in the upwards direction so $F_{\text{net}} = N - mg$, the normal force on the child is $450 \,\text{N} + 138 \,\text{N} = 588 \,\text{N}$ and the reading on the scales will be 59 kg.

3.6 Summary

Key Concepts

angular displacement (θ) The angle subtended by an object moving around the circumference of a circle.

angular velocity (ω) The rate of change of the angular displacement. The unit of angular velocity (or angular frequency) is the radian per second.

centripetal acceleration The acceleration of an object due to the change in its velocity required for it to travel around a circle.

centripetal force The force required to produce the centripetal acceleration.

Equations

$$\theta = \frac{s}{r}$$

$$\omega = \frac{v}{r}$$

$$\theta = \omega t$$

$$\omega = 2\pi f$$

$$v = \frac{2\pi r}{T}$$

$$a = \frac{v^2}{r}$$

$$F = m \frac{v^2}{r}$$

3.7 Problems

3.1 (a) Convert the following values from radians to degrees: (i) $\frac{\pi}{6}$ (ii) $\frac{\pi}{4}$ (iii) $\frac{\pi}{2}$ (iv) 0.1 (v) $\frac{3\pi}{4}$
(b) Convert the following values from degrees to radians: (i) 1° (ii) 45° (iii) 60° (iv) 180° (v) 360°

3.2 In a particular rear-end car collision the driver's head rotates 45° backward before being stopped by the headrest. What is the average angular velocity of the driver's head if the duration of the collision was 0.1 s?

3.3 When an athlete throws a javelin her forearm snaps through an angle of approximately π radians in 0.20 s. The athlete's hand moves with approximately constant speed, the length of her forearm is 45 cm, and the combined mass of her forearm and javelin is 2.0 kg. Assuming that the system is well approximated by a mass of 2.0 kg located 45 cm from the pivot, what force do the ligaments holding the forearm to the elbow need to exert?

3.4 (a) A 3800 kg car travels round an unbanked corner (i.e. a horizontal road) at the recommended speed of 65 km h^{-1}. The radius of curvature is 80 m. What is the force that the road exerts on the car to keep it in motion around the corner?
(b) What force would the road need to exert if the car was travelling at 100 km h^{-1}?

3.5 A car is traveling around a circular race track at 180 km h^{-1}. If a single lap of the track is 2.4 km long, what is the angular velocity of the car (in rad s^{-1})?

3.6 If the car in problem 3.5 weighs 2500 kg, what is the centripetal force acting on the car as it travels around the track?

3.7 An adventurous ant finds herself at the end of a fan blade when it is switched on. It is a high speed fan with blades measuring 0.20 m long. If she has a mass of 0.20 g and can hold on to the fan blade with a maximum force of 0.0124 N. What is the maximum number of *revolutions per minute* the fan can run at before she will be flung off?

3.8 An 8.0 m radius merry-go-round completes one revolution every 7.0 s.

(a) What is the angular velocity of the merry-go-round?

(b) With what speed are children moving when they ride on the merry-go-round?

(c) What is the centripetal *acceleration* these children feel when riding on the merry-go-round?

(d) What is the average acceleration of each child over the course of half a revolution of the merry-go-round?

(e) What is the average acceleration of each child over the course of a full revolution turn of the merry-go-round?

STATICS

4

4.1 Introduction

Statics is the study of stability in systems. Generally, we are primarily interested in the stability of an object or objects under the influence of gravity. Statics is of central importance in biomechanics. The essential ideas on which statics is based allow us to understand the anatomical mechanisms by which humans and other terrestrial animals are able to stand upright and move. Statics also provides the basis for understanding the body shapes necessary to allow fish and other aquatic animals to remain upright while swimming.

Key Objectives

- To understand the concepts of static, dynamic, stable and unstable equilibrium.

- To understand the concept of torque.

- To understand the principle of moments.

- To be able to solve lever problems using the concept of torque and the principle of moments.

4.2 Equilibrium

A system is said to be in **equilibrium** when the net force on that system is zero and the net torque on the system is also zero. A system is in **static equilibrium** when it is in equilibrium and stationary. A system is in **dynamic equilibrium** when it is in equilibrium and also in motion, which implies that the system is travelling at constant velocity and/or rotating at a constant rate.

Static equilibrium occurs when a seesaw is perfectly balanced, or in the ankle when an individual stands on tip toe, or in the elbow joint when an individual holds aloft an apple or some other object. Dynamic equilibrium occurs when a car travels along the motorway at constant velocity, and the combined friction forces are balanced by the driving force of the car's engine.

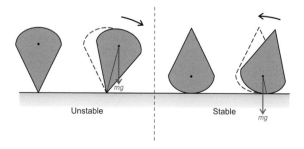

Figure 4.1 An object with a wide base and a pointed top will be stable if placed on its base, but unstable if placed on its point. This is because the torque produced by the weight force will act to rotate the object either back to the middle (as in the stable case) or further from upright (as in the unstable case).

A system is in **stable equilibrium** if it will return to equilibrium after it has been subject to a small displacement. A system is in an **unstable equilibrium** if it will not return to this equilibrium having been subject to a small displacement. As an example, consider Figure 4.1. The object displayed is in a stable equilibrium position when placed on its broad base, but in an unstable equilibrium when placed on its narrow top. When placed on its base and displaced slightly, as on the right-hand side of Figure 4.1, it will tend to fall back to its original equilibrium position. When placed on its narrow top, a slight displacement will cause it to topple onto its side, as on the left-hand side of Figure 4.1. These two positions are relatively stable and unstable. Clearly, an object with a much narrower top and broader base will be more unstable when placed on its top and more stable when placed on its base.

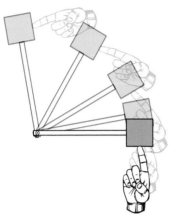

Figure 4.2 A force applied to an object fixed to a pivot point via a rigid rod will cause an *angular* acceleration. That is, the angular velocity of the object increases over time.

Definition of Moments

In physics, the word *moment* means 'the product of a quantity and its perpendicular distance (or some power of the distance) from a given point.' In the case of the moment of a force about an axis, it is the product of the force, and the perpendicular distance between the axis and the line of action of the force.

4.3 Torque

In the previous chapter, we discussed the acceleration and force required to keep an object moving in a circle. The physical quantity which causes an object to begin to rotate or move in a circle, or (more generally) to change its rate of rotation, is a **torque**. A torque is *not* a force in the Newtonian sense: it is a moment.

Suppose that an object to be accelerated into circular motion is attached to one end of a rod, the other end of which is attached to an immobile pivot or axle, as in Figure 4.2. Further suppose that the rod, though rigid, is very light, so that we can ignore its mass. In order to produce circular motion we would apply a linear force to the object at right angles to the rod. The rod will then constrain the motion of the object so that it moves in a circle. The tension produced in the rod will provide the necessary centripetal force.

The amount of turning produced by the force applied to the rod will depend both on the magnitude of the force and the length of the rod. The product of these two factors gives the magnitude of the moment of the force, or the torque, τ:

$$\tau = Fd \tag{4.1}$$

Torque is measured in units of N m. The torque provides us with a useful way to measure the turning effect (i.e., the tendency to cause rotation) of a force applied to a rod, which we call a moment arm or **lever**.

Notice that the definition of torque at the beginning of this section and in Eq. (4.1(does not specify any particular spatial point as the pivot point or fulcrum about which torques must be calculated. The axis of rotation is determined by the physical system, and the *fulcrum* is selected by us and may be different from the axis of rotation. Changing the location of the fulcrum in a calculation will change the values of the torques used in that calculation, but will have no effect on the physical behaviour of the system and the model will still correctly predict this behaviour. While we are free to chose whichever point we like to be the fulcrum in our calculation, it is good practice to select the point which most simplifies the calculation.

4.4 The Principle of Moments

A torque may tend to turn a system in the clockwise (cw) or the counter-clockwise (ccw) direction, depending on the direction of the applied force. For a system in static equilibrium, all torques are balanced so that there is no net torque (see Figure 4.3). Since there is no net 'tendency to rotate', the system will remain motionless. (A system that is already rotating will not increase its rate of rotation if the torques are balanced.) This condition is called the **principle of moments**:

Key concept:
At equilibrium, the sum of the clockwise moments equals the sum of the counter-clockwise moments .

Therefore, at equilibrium we can write (using Σ to mean 'sum of')

$$\sum F_{\text{cw}} d_{\text{cw}} = \sum F_{\text{ccw}} d_{\text{ccw}} \tag{4.2}$$

4.5 Centre of Gravity/Centre of Mass

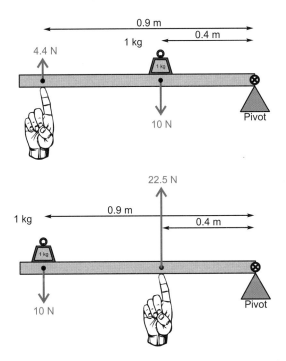

Figure 4.3 Two systems in which the torque (or moment) about a pivot is balanced.

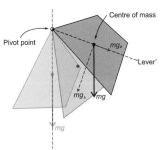

Figure 4.4 An irregular shape is hung from one of its corners. It will always hang such that it's centre of gravity is directly below the point at which it is suspended. If it is at any other position, there is always a torque produced by the component of the weight force that is perpendicular to the line joining the centre of gravity and the pivot point.

In order to make use of the full power of the principle of moments, we will make use of another concept: the centre of mass, also called the centre of gravity when dealing with a uniform gravitational field (the only case we will consider here). In the previous discussions, we were careful to stipulate that the mass of the rigid rod connecting an object to the pivot was very light so that we could ignore its mass. The reason for this condition is that we would like to be able to ignore the weight force of the rod itself. This force will produce a torque about the pivot, and in many cases this force is not negligible. In reality, any rod will have mass, so how should we include this mass in calculations? The gravitational force will act on each part of the rod, and we will have to add up the moments due to the mass of each part of our lever arm.

When a long ruler is carefully balanced on a pivot, the pivot point is located at the centre of gravity. When the ruler is balanced like this, the clockwise and anticlockwise moments balance. We can treat this situation as one in which there are equal but opposite moments about the pivot, or we could treat the ruler as though all its mass was located at the centre of mass at the pivot. The distance from the line of action of the weight force (which acts through the centre of mass) to the pivot is zero, so the total moment (or torque) is zero. If we were to try the same exercise with an asymmetric object we would still be able to find the balance point (see Figure 4.4), but in this case the centre of gravity would not necessarily coincide with the geometric centre.

The centre of gravity is the point in an object at which the force of gravity may be taken to act. The trajectory calculated for an object moving under the influence of gravity is actually the trajectory of the centre of gravity. An example is shown in Figure 4.5. In this example, a spinning juggling pin is thrown upward at a slight angle. The trajectory of the centre of gravity is shown as a line, which follows the parabolic arc characteristic of two-dimensional motion with constant acceleration. Notice also that the rotation of the juggling pin is about the centre of gravity. In general, any rotation of an object during such motion where the only force of any significance is gravitational will be about its centre of gravity. If the shape of an object changes during flight, as in the case of a diver performing somersaults, the centre of gravity may not always be located at the same point within the body, but our statement about the trajectory of the centre of gravity still stands.

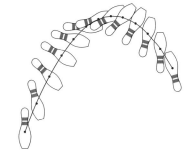

Figure 4.5 The ballistic motion of a complex projectile, a juggling pin. The centre of gravity is shown as a black dot on the pin and the trajectory of the centre of gravity is shown as a blue curve.

Example 4.1 *Principle of moments (I)*

Problem: Three people are sitting on a seesaw, as in the diagram. The masses of the two people on the right are as shown, as are their distances from the pivot. The third person is seated on the left and is 1 m from the pivot. Given that the seesaw is stationary, what is the mass of the third person?

Solution:
Since the seesaw is stationary, it is in equilibrium and the principle of moments will apply. This means that the sum of all of the counter-clockwise moments will equal the sum of all of the clockwise moments. The forces which produce the moments are the weights of the people on the seesaw. The people on the right will tend to turn the seesaw clockwise and their total moment is the sum of their individual moments:

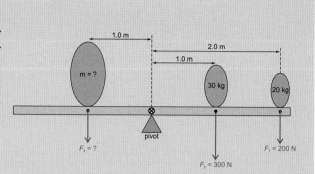

$$\tau_{\text{clockwise}} = F_1 d_1 + F_2 d_2 = m_1 g d_1 + m_2 g d_2$$
$$= (200\,\text{N} \times 2.0\,\text{m}) + (300\,\text{N} \times 1.0\,\text{m})$$
$$= 700\,\text{N m}$$

Figure 4.6 Three people on a seesaw. What is the mass of the person on the left?

There is only one counter-clockwise moment due to the weight force of the person on the left. This must be equal to the total clockwise moment provided by the people on the right, so

$$F_3 \times 1\,\text{m} = 700\,\text{N m}$$

Thus the weight force on the left is 700 N, meaning that the mass of the person on the left is 70 kg.

Example 4.2 *Principle of moments (II)*

Problem: A heavy ruler is placed on a pivot as shown in Figure 4.7. A weight is hung from the right end of the ruler and the mass of the weight is adjusted until the ruler is balanced and horizontal on the pivot. The ruler is 1 m long and symmetric and the fulcrum (pivot point) is positioned 20 cm from the right end of the ruler. A mass of 2 kg will exactly balance the ruler. What is the mass of the ruler?

Solution: We treat the ruler as having a centre of mass exactly at its centre, i.e., 50 cm from each end and 30 cm from the fulcrum. The weight of the ruler may be treated as acting through this point, thus forming an anticlockwise moment about the fulcrum of $m_{\text{ruler}} g d_2$. This moment is balanced by a clockwise moment produced by the action of gravity on the mass of the weight.

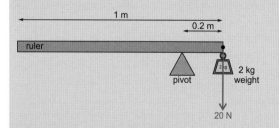

Figure 4.7 A ruler is exactly balanced by a 2 kg weight. What is the mass of the ruler?

We do not need to include the moment due to the length of ruler to the left of the fulcrum as this has been accounted for by the centre of mass.

Thus the principle of moments gives us

$$2\,\text{kg} \times 10\,\text{m s}^{-2} \times 0.2\,\text{m} = m_{\text{ruler}} \times 10\,\text{m s}^{-2} \times 0.3\,\text{m}$$

Rearranging this gives $m_{\text{ruler}} = 1.3$ kg.

Example 4.3 *Principle of moments (III)*

Problem: The elbow joint is essentially a hinge. The forearm rotates about the elbow joint, and the muscles of the upper arm, in particular the biceps muscle, attach to the bones of the forearm just below the joint. In this example we will calculate the force which the biceps muscle must apply to the forearm to hold a 4 kg weight horizontally in the hand.

Solution:

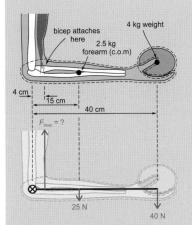

Figure 4.8 The elbow joint as a lever. How much force must the biceps muscle apply to the forearm to hold a 4 kg weight stationary in the hand?

To solve this problem we must calculate the moments about the elbow joint. Once we have calculated the clockwise and anti-clockwise moments, we use the principle of moments to equate the two and solve for the force applied by the biceps. The forearm in the model we are analysing is 40 cm long, it weighs 2.5 kg, its centre of mass is located 15 cm from the elbow joint, and the biceps muscle attaches to the forearm 4 cm from the elbow joint. These are all reasonable figures for a moderately sized person.

The mass of the forearm and the mass held in the hand both contribute to the clockwise moment, giving a total clockwise moment of

$$(2.5 \text{ kg} \times 10 \text{ m s}^{-2} \times 0.15 \text{ m}) + (4 \text{ kg} \times 10 \text{ m s}^{-2} \times 0.40 \text{ m}) = 19.75 \text{ N m}$$

The only counter-clockwise moment is provided by the biceps muscle, thus if the weight is held stationary the clockwise and counter-clockwise moments must be equal so that

$$F_{\text{biceps}} \times 0.04 \text{ m} = 19.75 \text{ N m}$$

We may now solve this equation for F_{biceps} to obtain

$$F_{\text{biceps}} = \frac{19.75}{0.04} = 490 \text{ N}$$

Thus holding a 4 kg weight in the hand (i.e., at the end of a lever arm) requires that the biceps muscle supply enough force to hold a 49 kg weight if that weight were not at the end of a lever.

The arm is optimised for speed rather than strength. The lever arrangement by which we lift objects in our hands does not maximise lifting strength. However this organisation of muscles and limbs does greatly increase the speed with which the hand is able to move. The human arm is better at throwing spears than it is at lifting rocks.

Note that this analysis also explains why it is easier to lift heavy weights if they are tucked into the arm rather than held in the hand.

Example 4.4 *Principle of moments (IV)*

Problem: In preparation for a back flip a diver weighing 80 kg stands on the very edge of a diving board, momentarily supporting all of their weight on a single foot, as shown in Figure 4.9. The diver's foot is 20 cm long, and the Achilles tendon attaches at the heel and provides an upward tension force at this point. The foot articulates about the joint located 4 cm from the heel. Assume that the only contact between the foot and the diving board occurs at the very end of the foot 16 cm from the ankle joint. What is the tension force in the Achilles tendon, and what is the force exerted downward on the ankle joint?

Solution:

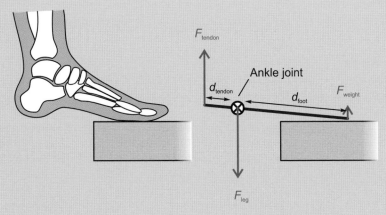

Figure 4.9 The ankle joint. How much tension is required in the Achilles tendon?

We may choose any point as the fulcrum, but the problem is considerably simplified by choosing either the point at which the foot contacts the diving board or the ankle joint. If we choose the point at which the foot contacts the board, then the torques about this fulcrum are produced by the Achilles tendon and the downward force exerted on the ankle joint. We do not know the magnitude of either of these forces yet and so we will choose to use the ankle joint as the fulcrum.

Note that the downward force on the ankle joint is *not* the weight force of the individual. The downward weight force acts at the point of contact between the foot and the floor. The force on the ankle joint is substantially greater than the weight force.

The clockwise and anti-clockwise moments about this fulcrum are shown in Figure 4.9. The normal reaction force F_{weight} is equal in magnitude to the downward weight force but is exerted upward by the floor on the foot and produces an anti-clockwise moment about the ankle. The total anti-clockwise moments about the ankle are therefore

$$F_{weight} d_{foot} = 800 \text{ N} \times 0.16 \text{ m} = 128 \text{ N m}$$

This anticlockwise moment is balanced by an equal clockwise moment about the ankle. This moment is produced by the force exerted on the heel by the Achilles tendon.

$$F_{tendon} d_{tendon} = F_{tendon} \times 0.04 \text{ m} = 128 \text{ N m}$$

We rearrange this expression to find the force exerted by the Achilles tendon:

$$F_{tendon} = \frac{128 \text{ N m}}{0.04 \text{ m}} = 3\,200 \text{ N}$$

This force is produced by the tension in the Achilles tendon. It is equivalent to the tension in a cable from which a 320 kg weight is suspended!

We can now calculate the downward force exerted on the ankle joint itself. The ankle is stationary, thus the upward and downward forces are balanced. There are two upward forces, the force exerted by the Achilles tendon at the heel and the upward normal contact reaction force at the toe. The total upward force is therefore 4 000 N. Since the ankle is stationary, this must be equal to the downward force exerted on the ankle joint. Note that this is equivalent to the weight of a 400 kg mass!

This example indicates the magnitude of the forces which may be brought to bear on the joints of the human body. It is certainly not at all surprising that these joints are often injured and have a tendency to wear out over time.

4.6 Stability

We now consider one final important application of the principle of moments and the centre of gravity. This is the analysis of stability. An object is stable if it will either remain in stable equilibrium indefinitely or will tend to move back to stable equilibrium when displaced. Whether or not a system is stable can be determined using the following rule:

Key concept:
In general, static stability occurs when the vertical line through the object's centre of gravity passes through its base of support.

The base of support of an object is the area in contact with the supporting surface. Figure 4.10 illustrates this principle. The object on the left is stable, and the middle object will fall back to this stable position as its centre of gravity lies over its base, but the object on the right will tip over as its centre of gravity falls past its base.

4.7 Summary

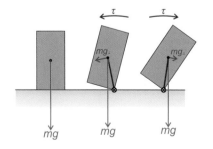

Figure 4.10 Illustrations of objects displaying static stability.

Key Concepts

equilibrium A system is in equilibrium when the net forces and torques on that system are zero, i.e., $\Sigma F = 0$ and $\Sigma \tau = 0$.

static equilibrium A static equilibrium occurs when a system is in equilibrium and stationary.

dynamic equilibrium A dynamic equilibrium occurs when a system is in equilibrium and also in motion. Dynamic equilibrium implies that the system has constant velocity and rate of rotation.

stable equilibrium An equilibrium is stable if the system will return to equilibrium if it is subject to a small displacement.

unstable equilibrium An equilibrium is unstable if the system will not return to this equilibrium if it is subject to a small displacement.

moment The product of a quantity and its perpendicular distance (or some power of the distance) from a given point. For example, torque is the moment of the force.

centre of mass The point at which the total mass of a body may be considered to be concentrated (for many purposes) in analysing its motion.

centre of gravity The point where the total weight of a material body may be thought to be concentrated. In a uniform gravitational field, this coincides with the centre of mass, but the centre of mass does not require a gravitational field.

torque (τ) The moment of force. The magnitude of the torque is the product of distance along a line from the axis of rotation to the point of application of the force and the magnitude of the component of the force perpendicular to this line. The tendency to cause rotation about an axis or point.

Equations

$$\tau = Fd$$
$$\sum F_{cw} d_{cw} = \sum F_{ccw} d_{ccw}$$

4.8 Problems

4.1 A waiter holds two plates of food in one hand. His forearm has a mass of 2.2 kg and the centre of mass of his forearm is located 13 cm from his elbow joint. The centre of mass of the two plates is located 37 cm from his elbow joint and the total mass of the two plates is 1.1 kg. His bicep is attached to the bones of his forearm 3.5 cm from his elbow joint.

(a) What is the torque produced by the mass of the two plates?

(b) What is the torque produced by the mass of the waiter's forearm?

(c) What force must be exerted by the waiter's biceps muscle to ensure that the plates and forearm are motionless?

4.2 A second waiter works at the same restaurant as the waiter in Problem 4.1. She is exactly the same size as the first waiter except the distance between her elbow and the point at which her bicep is attached is 0.5 cm shorter than his. By what percentage must the forces her bicep muscle exerts be larger than that of the first waiter?

4.3 A toddler weighs 10 kg and raises herself onto tiptoe (on both feet). Her feet are 8 cm long with each ankle joint being located 4.5 cm from the point at which her feet contact the floor. While standing on tip toe:

(a) what is the upward normal force exerted by the floor at the point at which one of the toddler's feet contacts the floor?

(b) what is the tension force in one of her Achilles tendons?

(c) what is the downward force exerted on one of the toddler's ankle joints?

4.4 The waiter in Problem 4.1 has an argument with the sous-chef over his paycheck and proceeds to throw the plates of food he is carrying straight up into the air. If the plates are accelerating at a rate of $2.5 \, \mathrm{m \, s^{-2}}$, and the center of mass of his arm is accelerating upwards at $0.9 \, \mathrm{m \, s^{-2}}$ what is the force being applied by the waiter's bicep? (Assume that the waiter's forearm also accelerates upwards at the same rate).

4.5 A diagram of a hypothetical 40 cm long arm is shown in Figure 4.11.

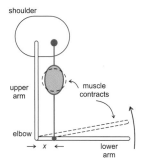

Figure 4.11 A simplified diagram of an arm showing the articulation of the lower arm by the biceps muscle.

For the purposes of answering this question assume that the arm itself is weightless. If the muscle attached to the arm can contract at a rate of $7.0 \, \mathrm{cm \, s^{-1}}$ with a force of 15 000 N then:

(a) What is the maximum angular velocity of the arm if the muscle is attached 1 cm from the elbow?

(b) What is the maximum weight that can held in the hand if the muscle is attached 1 cm from the elbow?

(c) What is the maximum angular velocity of the arm if the muscle is attached 3 cm from the elbow?

(d) What is the maximum weight that can held in the hand if the muscle is attached 3 cm from the elbow?

4.6 A 4 kg vase is placed on a shelf at the position shown in Figure 4.12. If the shelf itself has a negligible weight, what is the force F_{nail} with which the upper nail must hold to prevent being pulled out of the wall?

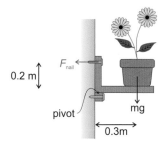

Figure 4.12 A vase sits on a shelf. The lower block acts like a pivot.

4.7 A uniform lever arm is used in conjunction with a pivot to weigh an object A. The mass of the lever arm is not negligible but is unknown. The pivot point may be moved relative to the lever arm.

(a) Where should the pivot be placed along the lever arm so that the mass of the lever arm does not appear in the calculation?

(b) If the unknown mass is balanced by a 0.05 kg mass hung from the lever arm 0.15 m from the pivot point, how much does the unknown mass weigh, given that it is hung from the lever arm at a point 0.3 m on the other side of the pivot point?

4.8 Which of the wheelbarrows in Figure 4.13 will require the smallest *upwards* force on the handles in order to lift?

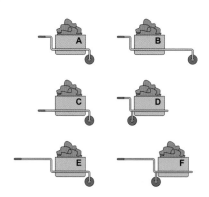

Figure 4.13 Six different ways of arranging a wheelbarrow.

4.9 Two match officials carry an injured rugby player from a rugby field on a stretcher. The rugby player weighs 95 kg and the stretcher is 2.5 m long. One of the stretcher bearers (bearer A) is able to lift a weight of 60 kg; the other (bearer B) is unsure how much weight he is able to lift.

(a) How much weight must bearer B support?

(b) How far from bearer A must the injured rugby players center of mass be in order for the stretcher bearers to carry the load without tipping the injured player out?

5 ENERGY

5.1 Introduction

When a force is exerted to accelerate an object, some quantity gets used up; the source of the force gets depleted in some way. If I drop a ball from a height of 2 m, the gravity field of the Earth exerts a force on the ball and makes it accelerate. However, once the ball reaches the ground, the ability of the Earth's gravitational field to accelerate the ball is used up. What exactly is it that is used up and how is it used? The quantity is potential energy, in this case gravitational potential energy. The potential energy is converted into another type of energy, kinetic energy. In this chapter we will investigate what energy is, how it behaves and how it can be used to solve problems in physics.

Key Objectives

- To understand the concepts of work and energy.

- To understand kinetic and potential energy.

- To understand conservative and dissipative forces.

- To be able to use energy conservation to solve kinematics problems.

- To understand power and mechanical efficiency.

5.2 What is Energy?

The idea of **energy** is slightly different to the various concepts we have introduced so far, such as mass, displacement, velocity, and acceleration. Energy is a more abstract concept. An object which has a certain mass and velocity is described as having a particular **kinetic energy** by virtue of its mass and velocity. Similarly, an object which has a certain mass and is located at a certain point in a conservative force field is described as having a particular **potential energy**.

Energy has several important properties. The single most important property of energy is that it is conserved.

> **Key concept:**
> The total energy in a closed system is constant over time.

5.3 Work

The word **work** is used frequently in everyday life and has a number of meanings in this context. However, this word has a very specific meaning in physics. Work is one process by which energy is transferred from one form to another. Work describes how much energy has changed from one form to another, and what sort of process was involved (as compared to chemical processes, or changes in energy due to heating processes). For example, a falling object gains kinetic energy and loses potential energy, and this transformation occurs because the gravitational field of the Earth does work on the object to accelerate it. In the absence of other forces such as air resistance acting on the object, the amount of work done by the gravitational force field is the numerical increase in the kinetic energy of the object.

When a force F with magnitude F acts on a body, and it moves a distance d *in the direction of the force*, the work done is

$$W = Fd \qquad (5.1)$$

The SI unit of work is the joule, symbol J, and 1 joule = 1 newton × 1 metre.

This definition connects the force exerted on an object with the change in the energy of that body. The Earth exerts a gravitational force F on an object. This force acts over a distance d – the distance that the object falls. The result is that the gravitational force does work on the object, and the object has work done on it. The amount of work is the same for the object and the Earth's gravitational field, but one gains energy and the other loses it.

The concept of work also applies when force and displacement are *not* in same direction. When the force and displacement are not in the same direction, the work is calculated using the component of force in the direction of displacement.

> **Work**
>
> Work is not a form of energy in the same sense as kinetic energy or potential energy. 'Work' instead describes the amount of energy being changed from one form into another by a force.

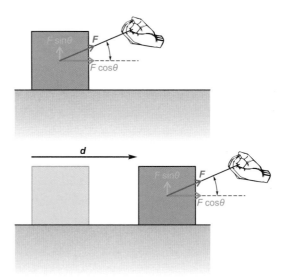

Figure 5.1 The work done by a force which produces a displacement which is not in the same direction as the force.

As illustrated in Figure 5.1, the component of the force F in the direction of the displacement d is given by $F\cos\theta$. The work expended by the applied force in this case is therefore given by

$$W = Fd\cos\theta \qquad (5.2)$$

In this equation remember that F and d are the magnitude of the applied force F and the displacement d.

5.4 Kinetic Energy

Work is the quantity of energy transferred, so if an object is able to do work then it must have energy to spend in the first place. The **kinetic energy** of an object is a measure of the work that object can do because it is moving. Consider a cricket player hitting a cricket ball with a bat. The bat exerts a force on the ball due to the speed with which it is moving. The faster the bat is moving when it hits the ball, the greater the force applied and the acceleration of the ball. As another example, compare the amount of effort involved in throwing a tennis ball and a bowling ball. Throwing a tennis ball weighing about 55 g is relatively easy but throwing a bowling ball with a mass of just over 7 kg so that it travels at the same speed is substantially harder.

These examples indicate that the kinetic energy possessed by an object is somehow dependent on both its mass and its velocity. To determine the exact form of this dependence we will consider a single force, F, acting on some object. (We will keep this as general as possible so that the expression we end up with will apply to all objects).

The amount of kinetic energy gained by an object which is initially at rest will be equal to the amount of work done on that object by an external force (ignoring friction for the time being). We will consider the 1D system illustrated in Figure 5.2. We start with a mass m which is initially at rest so that the initial speed is $v_i = 0$ and the object has no kinetic energy. We then apply a force F for a time t and the object moves a distance d while the force is being applied. At the end of this time, the final speed of the object is $v_f = v$. The object no longer has zero kinetic energy since the applied force has done work on the object. We can calculate how much work has been done on the object by the force, and so calculate the kinetic energy of the object.

The initial velocity is zero, so the average speed is just $v_{av} = \frac{1}{2}v_f = \frac{1}{2}v$. The distance travelled by the object while the force is applied is given by $d = v_{av}t = \frac{1}{2}vt$. Furthermore, the final velocity is the result of the acceleration produced by the applied force (remember that $F = ma$) and is given by $v = at$. We are now able to combine all of these facts to find an expression for the work done on the mass to accelerate it to v, and thus the kinetic energy of the mass when it is travelling at this velocity.

The work done on the mass by the external force is

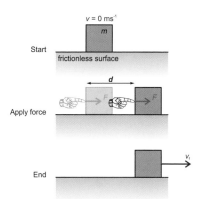

Figure 5.2 System used for derivation of kinetic energy formula.

$$
\begin{aligned}
W &= Fd \\
&= ma \times d \quad (\text{since } F = ma) \\
&= m(\frac{v}{t}) \times d \quad \left(\text{since } a = \frac{v}{t}\right) \\
&= mv \times \frac{d}{t} \\
&= mv \times \frac{1}{2}v \quad \left(\text{since } \frac{d}{t} = \frac{1}{2}v\right)
\end{aligned}
$$

When the mass reaches a velocity of v, it has had energy equal to $\frac{1}{2}mv^2$ transferred to it by the external force. This energy is the kinetic energy of the mass and is given by

$$
\text{KE} = \frac{1}{2}mv^2 \tag{5.3}
$$

In the derivation of this expression we have used speed instead of velocity and distance instead of displacement, i.e., we have used scalar quantities rather than vector quantities. However the expression we have arrived at would still be a scalar equation if we had used vector quantities – although it is associated with motion which has a direction, the kinetic energy itself is a scalar quantity.

Example 5.1 *Work done by a force*

Problem: A man pushes on a car with a force of 300 N and moves it 10 m (in the direction of the force). This is shown in Figure 5.3 below. How much work is done by the man on the car?

Solution:

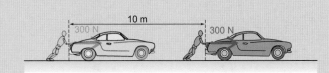

Figure 5.3 Work is done pushing a car.

The applied force and the displacement that is produced by this force are in the same direction. We are therefore able to apply Eq. (5.1)

$$Fd = 300 \text{ N} \times 10 \text{ m} = 3000 \text{ J}$$

This technique may be used to calculate the work done by the force whether or not there is friction and whether or not the car accelerates.

Example 5.2 *Work, force, and kinetic energy*

Problem: A woman pushes a toy car (initially at rest, and of mass 0.1 kg) toward a child by exerting a constant force of 5 N through a distance of 0.4 m. What is (a) the work done on the car? (b) the final kinetic energy of the car? (c) the final velocity of the car?

Solution: Work is done on the car by the force exerted by the woman i.e. by the 5 N force. Thus the work is calculated using Eq. (5.1) with $F = 5$ N and $d = 0.4$ m:

$$W = Fd = 5 \text{ N} \times 0.4 \text{ m} = 2 \text{ J}$$

This work is energy transferred by the force to the car and is then present in the car as kinetic energy. The final kinetic energy of the car is just the initial kinetic energy of the car plus the work done on the car.

$$
\begin{aligned}
\text{KE}_f &= \text{KE}_i + W \\
&= 0 \text{ J} + 2 \text{ J} = 2 \text{ J}
\end{aligned}
$$

The final velocity of the car could be found using the kinematic equation Eq. (1.9)

$$d = v_i t + \frac{1}{2} a t^2$$

as acceleration of the car may be calculated using the applied force and the mass of the car, and the initial velocity of the car is zero. Since we know the distance travelled by the car we are able to calculate the time over which the force is applied. Using this time we are then able to calculate the final velocity using the relationship between constant acceleration and the velocity Eq. (1.5).

However, we may also use a method based on energy considerations to solve this problem. The method outlined above will work well but this method becomes convoluted in more complicated situations. The energy method will work just as easily in complicated situations and also those where the acceleration is not constant.

We calculate the final kinetic energy of the car and use its mass to calculate its velocity:

$$\text{KE}_f = \frac{1}{2} m v_f^2 = 2 \text{ J}$$

(Note that the kinetic energy of the car, 2 J, has same units as work as you would expect.) Now the velocity of the car may be calculated using the mass of the car:

$$v_f^2 = \frac{2 \times 2 \text{ J}}{m} = \frac{4 \text{ J}}{0.1 \text{ kg}} = 40 \text{ m}^2\text{s}^{-2}$$

Taking the square root of this result gives the final velocity of the car:

$$v_f = 6.3 \text{ m s}^{-1}$$

5.5   Potential Energy

In many cases, an external force does work on an object, but that object does not end up in motion. Lifting a box of books onto a shelf does work on the box and its contents. A force is exerted on the box, and the box is displaced in the direction of the force. However, after the force has been applied, the box is stationary; it is sitting on a shelf. What has happened to the work done? This work has been stored in the box as **potential energy**. The box may do work on some object if it is free to move downward.

> **Key concept:**
> Potential energy is the energy an object has because of its position.

It is not necessarily gravity that supplies the force. We could place an object next to one end of a compressed spring. In this case, the object again has potential energy because of its position.

Gravitational Potential Energy

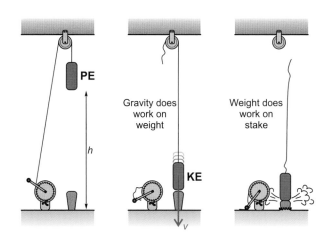

Figure 5.4 The conversion of work into gravitational potential energy in a pile driver.

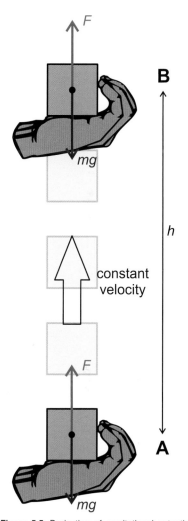

Figure 5.5 Derivation of gravitational potential energy

When an object has potential energy due to its position in a gravitational field, then it has **gravitational potential energy**. Figure 5.4 shows the conversion of work into gravitational potential energy in a pile driver. The pile driver is raised by a force to overcome gravity until the ram has been lifted the required distance. Work is done on the ram to raise it, and the ram gains gravitational potential energy. When the ram is released, the gravitational potential energy is converted into kinetic energy, and when the ram strikes the object to be driven into the ground this kinetic energy is converted into thermal energy and sound, and work is done on moving ground out of the way.

How much gravitational potential energy does the ram acquire? Clearly, the heavier the ram and the higher it is raised the more work it is able to do. Consider raising an object a distance h as shown in Figure 5.5. We would like to find an expression for the increase in gravitational potential energy of the object, so we will assume that the object is raised at constant velocity, so that none of the work done in raising it goes into increasing its kinetic energy.

A force must be applied to the object to balance the downward force of gravity so that the object is able to travel at constant velocity. This force is equal in magnitude to the gravitational force but is directed upward. The work done by this force F on the object is

$$W = Fd = Fh = mgh$$

since the magnitude of the force is $F = mg$. The increase in the gravitational potential energy is provided by this input of work from the external force. This means that the gain in potential energy is given by

$$PE_{\text{gravitational}} = mgh \qquad (5.4)$$

Note that here we have been concerned only with changes in potential energy. Calculations using energy will generally require determination of the amount of energy converted from one form to another – e.g., the amount of potential energy that is converted into kinetic energy – so that the only quantity which will be of interest will be the *change* in potential energy. This means that instead of measuring heights from the centre of the Earth, we may call a convenient point in the system under consideration the reference point, and measure heights from this point. The difference between this reference point and the centre of the Earth will be irrelevant to calculations that require only the height difference.

5.6   Conservative Forces

As we have seen, energy is intimately related to applied force. When performing energy calculations, it is important to include the initial and final kinetic and potential energies. All of the applied forces must be included when determining the potential energy of the system. There are four fundamental forces in nature and there is a potential-energy term corresponding to each of them. In the current chapter we are concerned only with gravity, and we will later consider the electrical force and its corresponding potential energy. We will not have cause to discuss the potential energy corresponding to the strong and weak nuclear forces.

As may be seen in the worked examples, potential energy can be converted into kinetic energy, and vice versa. But what about friction? Friction does not remove energy from the larger system, but it does convert it into forms which depend on the microscopic behaviour of the system. For example, when a block slides across a surface in the presence of friction the block and the surface heat up. Some of the kinetic energy of the block has been transforemd into thermal energy, i.e., friction has converted kinetic energy into thermal energy. It is very hard to keep track of this microscopic behaviour and so friction is normally treated as a special kind of force – a **dissipative force**. A dissipative force removes mechanical energy from the system under consideration.

Forces which do not do this are called **conservative**. Conservative forces are those which do not change the amount of **mechanical energy** in the system. 'Mechanical energy' is just another name for the kinetic plus the potential energy in mechanical systems such as mass–spring, pendulum, or gravitational systems. The forces acting in a mass–spring or pendulum system are conservative and we are able to derive expressions for the potential energy of these systems (and we will do this in a later chapter on Simple Harmonic Motion). The electrostatic force is also conservative.

Non-conservative forces are normally dissipative; they are forces which channel energy out of the system under investigation into microscopic or molecular motion. Work done by non-conservative forces such as friction will result in a reduction of the mechanical energy of the system. This will appear as a lower velocity than there would be in the absence of friction, or an object gaining less potential energy than it would in the absence of friction.

5.7   Conservation of Total Energy

The principle of the conservation of energy is one of the fundamental principles of physics.

> **Key concept:**
> **The Principle of the Conservation of Energy**: The total amount of energy in a closed system does not increase or decrease. A closed system is one which does not exchange energy with its surroundings.

In all experiments performed to date, it has been found that energy is never lost, and new energy is never made. All observations demonstrate that energy merely changes from one form to another. Kinetic energy may become potential energy of various forms and vice versa. The principle of conservation of energy applies to the total energy of a system. It does not necessarily apply to the mechanical energy of a system alone, but if the energy dissipated by friction and other non-conservative forces is included then the total amount of energy has never been observed to change.

Table 5.1 lists the energy produced or used by a number of well-known processes. Note that the first entry, the energy released by the Big Bang, is the total amount of energy available in the universe.

Event	Energy (J)
The Big Bang	10^{68}
Energy released in a supernova explosion	10^{44}
Solar energy incident on the Earth annually	5×10^{24}
Energy release during eruption of Krakatoa	10^{18}
Annual electrical output of power plant	10^{16}
Energy released in burning 1000 kg of coal	3×10^{10}
Kinetic energy of a large jet aircraft	10^{9}
Energy released in burning 1 L of gasoline	3×10^{7}
Daily food intake of a human adult	2×10^{7}
Kinetic energy of cricket ball hit for six	10^{3}
Work done by a human heart per beat	0.5
Work done turning a page in a book	10^{-3}
Energy in discharge of a single neuron	10^{-10}
Typical energy of an electron in an atom	10^{-18}
Energy to break one bond in DNA	10^{-20}

Table 5.1 The Energy Scale – the energy released by (mostly) common events, from the largest to some of the smallest (in joules).

5.8 Power

The definition of work given previously in Eq. (5.2) does not refer in any way to the time taken for the work to be done. The same amount of work is done by a runner who sprints up a hill and by a pedestrian who walks slowly up the hill, stopping regularly for rests. Clearly there is an important difference between these two cases which is not captured by the concept of work alone. The *rate* at which work is done is also an important quantity.

The rate at which work is done is the **power**. The power is defined to be the amount of work done divided by the time it takes to do this work:

$$\text{Power} = \frac{\text{work done}}{\text{time taken}}$$

or using mathematical symbols

$$P = \frac{W}{\Delta t} \qquad (5.5)$$

We have used Δt to indicate that we are specifically interested in the time taken to do the work.

The SI unit of power is the watt, which has the symbol W. (Take care not to confuse this with the variable W which is work, measured in J.)

$$1 \text{ watt} = 1 \text{ joule s}^{-1}$$

i.e., 1 W is produced when work of 1 J is done every second. Since the power is the work done per second, we are able to rewrite the definition so that it is a little more useful for solving many problems. We note that the work is the force multiplied by the distance, and use the fact that the velocity is the distance divided by the time:

$$P = \frac{W}{\Delta t} = \frac{Fd}{\Delta t} = F\frac{d}{\Delta t}$$

so that we end up with

$$P = Fv \qquad (5.6)$$

If an object is moving at constant velocity while a force is being applied to it (such as when there is an opposing force) then Eq. (5.6) will be useful.

Table 5.2 shows the power output of the human body while undergoing a variety of activities.

Description	Power (W)
Sleeping	83
Sitting at rest	120
Sitting in a lecture (Awake!!)	210
Walking slowly	265
Cycling at 15 km/h	400
Playing basketball	800

Table 5.2 Rate of energy expenditure in various activities.

Example 5.3 *Conservation of total energy I*

Problem: Suppose that a 0.3 kg ball is dropped a vertical distance of 7 m. What is its final speed?

Solution: We could solve this problem using the kinematic equations, first by calculating the time taken to fall this distance under the influence of the acceleration due to gravity and then using this time to calculate the final velocity of an object with this acceleration. However, we will give a solution to this calculation using energy methods, which is typically easier for more complex problems, particularly those in which the acceleration is not necessarily constant.

To begin with we set the reference point (the level of zero gravitational potential energy) at the bottom of the ball's fall. All we need to know is how much potential energy is converted into kinetic energy and to do this we need to know only the change in height, how far the ball falls. The total energy of the object is the same before and after it falls through 7 m.

We begin by writing down the balance of the final and initial energies as an equation,

$$KE_f + PE_f = KE_i + PE_i$$

Here the subscript 'f' represents an energy after the downward motion, while the subscript 'i' represents an energy before the downward motion.

The initial kinetic energy is zero and the initial potential energy is given by Eq(5.4), $PE_{gravitational} = mgh$. The final kinetic energy is given by Eq(5.3), $KE = \frac{1}{2}mv^2$ and the final potential energy is zero. We substitute these expressions into the energy balance equation along with the known numerical values to get

$$\frac{1}{2}\,0.3\,\text{kg} \times v^2 + 0\,\text{J} = 0\,\text{J} + 0.3 \times 10 \times 7\,\text{J}$$

Finally we rearrange to find v^2 and then take the square root,

$$v^2 = \frac{0.3 \times 10 \times 7\,\text{J}}{0.3\,\text{kg}} \times 2 = 140\,\text{m}^2\text{s}^{-2}$$
$$v = \sqrt{140\,\text{m}^2\text{s}^{-2}} = 11.8\,\text{m s}^{-1}$$

Example 5.4 *Conservation of Total Energy II*

Problem: Calculate the speed of the totally *frictionless* rollercoaster shown in Figure 5.6 when it reaches the ground if it starts at rest 20 m above the ground.

Solution:

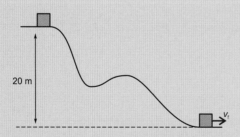

Figure 5.6 The velocity of a frictionless roller coaster calculated using gravitational potential energy.

This is an example of a problem which would be very difficult to solve using kinematic methods. We would need to know all of the details of the rollercoaster's path as it moved from the high point to the ground. To use the energy method, however, all we require is the initial and final heights and the initial velocity. We do not even need to know the mass of the rollercoaster as this will cancel out of the energy balance equation.

The first step is, as usual, to set the reference level to some appropriate value and then to write down the energy balance equation. In this case the appropriate level is the bottom of the rollercoaster's ride and the balance equation is then

$$KE_f + PE_f = KE_i + PE_i + W_{applied}$$

Note also that we have include a term W_{applied}. This term is necessary to include any energy changes due to forces which are applied to the rollercoaster apart from gravity. In this example we have stated that the rollercoaster is entirely frictionless so that this term may safely be set to zero.

Having written down our energy balance equation we find that the mass of the roller coaster appears on both sides of the equation and so may be cancelled. We then proceed as before, solving the equation for v^2, taking the square root and finding the required speed.

$$
\begin{aligned}
\frac{1}{2}mv^2 + 0 &= 0 + mgh + 0 \\
v^2 = 2gh &= 2 \times 10 \text{ m s}^{-2} \times 20 \text{ m} = 400 \text{ m}^2\text{s}^{-2} \\
v &= 20 \text{ m s}^{-1}
\end{aligned}
$$

Example 5.5 *Energy conservation and dissipative forces*

Problem: A very small bus, with a mass of 500 kg, travels down a slope as shown in Figure 5.7 and arrives at the bottom of the slope travelling at a speed of $v = 15$ m s^{-1}. The slope is 87.5 m long and starts at 20 m above the end point. Calculate the average force of friction on the bus.

Solution:

We begin by setting the reference point at the base of the slope. This allows us to calculate the initial potential energy:

$$\text{PE}_i = mgh = 500 \text{ kg} \times 10 \text{ m s}^{-2} \times 20 \text{ m} = 100\,000 \text{ J}$$

At the base of the slope this potential energy will have been completely converted into the kinetic energy of the bus, but some of it will have been dissipated as friction. The kinetic energy of the bus at the base of the slope is

$$\text{KE}_f = \frac{1}{2}mv^2 = 250 \times 15^2 \text{ J} = 56\,250 \text{ J}$$

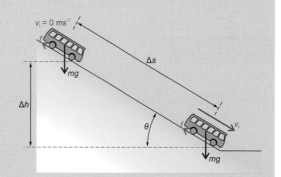

Figure 5.7 The effect of friction in energy calculations.

This is somewhat less than the original potential energy of the bus: the difference is the (negative) work done on the bus by friction. The work done on the bus is negative since the displacement of the bus is in the opposite direction to the friction force on it, and thus the product of the force and the displacement (i.e., the work) is negative.

$$W_{\text{friction}} = \text{KE}_f - \text{PE}_i = 56\,250 \text{ J} - 100\,000 \text{ J} = -43\,750 \text{ J}$$

Since we know that work is given by the force multiplied by the distance travelled in the direction of the force we are now able to calculate the magnitude of the friction force on the bus:

$$
\begin{aligned}
W_{\text{friction}} &= f_{\text{average}}d \\
f_{\text{average}} &= \frac{43\,750 \text{ J}}{87.5 \text{ m}} = 500 \text{ N}
\end{aligned}
$$

Note that friction is probably not constant along the whole length of the slope so that the quantity we have calculated is the average of the force, and it is directed up the slope.

Example 5.6 *Force and power*

Problem: A force of 500 N is required to overcome the frictional forces acting on a medium-size car travelling at 65 km h^{-1}. How much power must the car produce to maintain a speed of 65 km h^{-1}?

Solution: First we must convert the speed from 65 km h^{-1} into the appropriate SI units, i.e., m s^{-1}. Since there are 3600 seconds in an hour we have

$$65 \text{ km h}^{-1} = 65 \text{ km h}^{-1} \times \frac{1000 \text{ m}}{1 \text{ km}} \times \frac{1 \text{ hr}}{3600 \text{ s}} = 18 \text{ m s}^{-1}$$

Thus in 1 s the car travels 18 m. We are now able to calculate the amount of work that the car does in 1 s:

$$W = Fd = 500 \text{ N} \times 18 \text{ m} = 9000 \text{ J}$$

The car does this much work every second, so the power expended by the car is

$$P = \frac{W}{\Delta t} = \frac{9000 \text{ J}}{1 \text{ s}} = 9000 \text{ W} = 9 \text{ kW}$$

Alternatively we can use Eq. (5.6) in this problem, since the speed of the car is constant:

$$P = Fv = 500 \text{ N} \times 18 \text{ m s}^{-1} = 9000 \text{ W}$$

as before.

Example 5.7 *Power*

Problem: A 70 kg man runs up a flight of stairs 3 m high in 2 s. What average power does he produce in order to achieve this?

Solution:

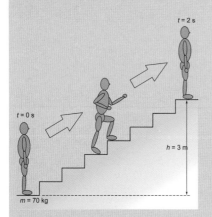

The man must do work against gravity in order to raise his centre of gravity the required 3 m.

$$
\begin{aligned}
W_{\text{applied}} &= \Delta\text{KE} + \Delta\text{PE} \\
&= 0 \text{ J} + \Delta\text{PE} \\
&= mgh \\
&= 70 \text{ kg} \times 10 \text{ m s}^{-2} \times 3 \text{ m} = 2100 \text{ J}
\end{aligned}
$$

The total work done in climbing the stairs is 2100 J. Now we calculate the power output by dividing the work done by time taken to do the work:

$$\text{Average power} = \frac{W}{\Delta t} = \frac{2100 \text{ J}}{2 \text{ s}} = 1050 \text{ W}$$

Figure 5.8 A man runs up a flight of stairs and gains 3 m of elevation in 2 s.

The total power output is 1050 W. This is a substantial power output for a human, given that the baseline metabolic rate is around 100 W and playing basketball requires a power output of about 800 W (see Table 5.2).

Mechanical Efficiency

In this discussion of power output, it is assumed that all of the energy provided to a machine is effectively utilised as work. Seen in this way, the human body is a machine that is powered by the body's metabolism. No real machine conforms exactly to this ideal. In real-world cases, some of the energy input into the machine is wasted by the machine; the input energy is not output as work, since at least some will be lost to dissipative forces. There will always be some waste heat or sound generated by a real machine. To quantify this idea, we will define the **mechanical efficiency**, η, of a machine

Task	Approximate Efficiency (%)
Cycling (bicycle producing 370 W)	20
Swimming on surface	less than 2
Swimming underwater	4
Shovelling	3

Table 5.3 Mechanical efficiency of some everyday activities

as

$$\eta = \text{Efficiency} = \frac{\text{Work out}}{\text{Work in}} = \frac{\text{Work output}}{\text{Energy used}} \qquad (5.7)$$

Below is a table of the mechanical efficiency of a number of energy-intensive everyday activities. For comparison, the efficiency of a typical petrol engine is about 40%.

Example 5.8 *Mechanical efficiency*

Problem: A certain block and tackle is 40% efficient. If an individual does 180 J of work pulling on the pulley rope, how much work could be done by the other pulley rope?

Solution:

$$\text{Work output} = 180 \text{ J} \times \frac{40}{100} = 72 \text{ J}$$

The rest of the input energy, 180 J – 72 J = 108 J has probably been converted into heat energy, i.e., the pulley will warm up as it is used.

Example 5.9 *Power and efficiency*

Problem: At what rate is energy being produced by a cyclist when cycling with a power output of 370 W and a bicycle mechanical efficiency of 20%?

Solution: We solve this problem by applying Eq. (5.7)

$$20\% = \frac{20}{100} = 0.2 = \text{efficiency} = \frac{\text{power out}}{\text{power in}} = \frac{370 \text{ W}}{P_{\text{in}}}$$

Therefore the cyclist produces energy at a rate of

$$P_{\text{in}} = \frac{370 \text{ W}}{0.2} = 1850 \text{ W!}$$

Again compare this energy expenditure with a value of about 800 W which is required to play basketball.

5.9 Summary

Key Concepts

energy The capability of a physical system to do work on another system. The SI unit of energy is the joule (J). $1 \text{ J} = 1 \text{ N m} = 1 \text{ kg m}^2 \text{ s}^{-2}$.

work The transfer of energy from one system to another, in particular in the case where a force causes a body to move in the direction of the force.

kinetic energy (KE) Energy that an object has by virtue of its motion.

potential energy (PE or U) Energy that an object has by virtue of its position.

mechanical energy The sum of the kinetic and potential energy of a system.

dissipative (non-conservative) force A force which removes mechanical energy from a system, e.g., friction forces which convert mechanical energy into thermal energy.

conservative force A force which does not remove mechanical energy from a system, e. g. gravity.

conservation of total energy The total amount of energy in a system does not change when there are no dissipative forces acting. Energy is never created or destroyed it is only transformed between the different types of energy. Dissipative forces do not destroy energy, but they remove it from the system being considered.

power (P) The power expended is the amount of work done divided by the time it takes to do this work. It is the rate at which work is done.

mechanical efficiency The mechanical efficiency of a machine is the proportion of the input work that is provided by the machine as output work.

Equations

$$W = Fd$$

$$KE = \frac{1}{2}mv^2$$

$$PE_{gravitational} = mgh$$

$$P = \frac{W}{t} = Fv$$

$$\eta = \frac{\text{Work out}}{\text{Work in}} = \frac{\text{Work output}}{\text{Energy used}}$$

5.10 Problems

5.1 In Figure 1.5, the ball had an initial vertical velocity of 30 m s^{-1}. Prove that the maximum height reached is 45 m using energy arguments instead of kinematic equations.

5.2 A pulley system is used to raise a 30 kg load upwards at a steady rate. The energy input to the system is 1000 J every second and the efficiency of the system is 50% (that is, half the energy input is used to do work on the load). How fast is the load being raised?

5.3 Two 30 kg children in a 20 kg cart are stationary at the top of a hill. They start rolling down the 80 m tall hill and they are travelling at 30 km h^{-1} when they reach the bottom. (The cart had brakes!) How much work was done on the cart by friction during its travel down the hill? (Note the use of the word *on* and remember to specify the sign of the work done.)

5.4 A car travelling at 50 km h^{-1} brakes as hard as it can and stops in a distance of 15 m. Suppose that the maximum braking force is not dependent on speed i.e. the coefficient of kinetic friction is constant. What is the shortest stopping distance when the car is travelling at 75 km h^{-1}?

5.5 A 20 kg box slides 45 m from the top down to the bottom of an incline which is at an angle of 10° to the horizontal. The box is stationary at the top of the slope and is accelerating down the slope at a rate of 1.0 m s^{-2}.

  (a) How much work is done by the force of gravity on the box?

  (b) How much work on the box is done by the frictional force?

  (c) What is the kinetic energy of the box at the bottom of the slope?

  (d) How much gravitational potential energy has the box lost while traveling down the slope?

5.6 A 70 kg physicist is running up the stairs of the physics building and makes it up 48 m vertically in 1 minute.

  (a) What is the physicist's power output while running up the stairs?

  (b) If the physicist's work efficiency is just 3% at what rate were they using metabolic energy?

  (b) If the physicist's metabolism can provide 5.6×10^6 J of energy before they need a rest, how long could they continue running up the stairs?

5.7 An 85 kg sky diver is falling through the air at a constant speed of 195 km h^{-1}. At what rate does air resistance remove energy from the sky diver?

5.8 The human body loses heat at a rate of 120 W when sitting quietly at rest. If a 65 kg student takes 100 hours to read *War and Peace* by Leo Tolstoy, how high could they have been lifted if all of the heat energy lost was utilised to lift them against gravity (assuming that the force of gravity on the student remains constant)?

5.9 An 8 g bullet leaves a gun at 700 m s^{-1}.

  (a) What is the maximum height that this bullet could reach (ignoring air resistance)?

  (b) If the gun is aimed at an angle of 30° above the horizontal what height will the bullet reach (again ignoring air resistance)?

5.10 A crane is lifting a 500 kg payload straight up at a constant speed of 0.7 m s^{-1}.

  (a) What is the power output of the crane (ignoring losses)?

  (b) If it takes the crane 2 minutes to raise the payload to its final height, how far above ground is this?

  (c) If the cable were to break as the payload reaches its final height, how fast would it be travelling as it hit the ground?

Momentum

6

6.1   Introduction

In this chapter we will discuss another conserved mechanical quantity: the linear momentum. Just as the energy can be used to solve kinematics problems, we will use the momentum to solve problems involving the interaction of objects.

Key Objectives

- To understand the concept of linear momentum and its relationship to Newton's three laws.

- To be able to identify situations in which momentum is conserved and situations in which it is not.

- To understand the difference between elastic and inelastic collisions.

- To be able to solve kinematics problems using the conservation of momentum.

6.2   Linear Momentum

The linear momentum is a vector quantity defined as follows:

> **Key concept:**
> The **linear momentum** of an object is the **mass** of that object multiplied by its **velocity**.

The symbol commonly used for momentum is p. The mathematical expression for the linear momentum is

$$p = mv \tag{6.1}$$

There is a relationship between momentum and Newton's three laws. The reformulation of Newton's laws in terms of momentum changes will show that the total momentum of a system is conserved in the absence of external forces.

6.3   Newton's Laws and Momentum

Newton originally wrote his three laws in terms of the momentum. To see how this would be done, remember that Newton's second law relates the force on an object to its mass and acceleration. The acceleration can be written as the change in velocity divided by the time over which that change occurs, so

$$
\begin{aligned}
F &= ma \\
&= m\frac{\Delta v}{\Delta t} \\
&= \frac{\Delta(mv)}{\Delta t} \\
&= \frac{\Delta p}{\Delta t}
\end{aligned}
$$

Colliding Objects	Mass (kg)	Collision Duration, Δt (ms)
Golf ball (collision with club)	0.047	1.0
Cricket ball (with bat)	0.156	2.0
Tennis ball (with racquet)	0.058	4.0
Soccer ball (with boot)	0.425	8.0
Basketball (with floor)	0.55	20

Table 6.1 Approximate interaction times of common events

This is an indication of the importance of the momentum idea; in words, the derivation above says

Key concept:
The rate of change of momentum is equal to the net external force.

In other words, the greater the force, the greater the change of momentum per unit time.

On the basis of the connection between force and momentum, we will define a new quantity: the **impulse**. The impulse is the name given to the change in the momentum, and is thus given by the expression

$$\Delta p = F \Delta t \qquad (6.2)$$

for a fixed-magnitude force. The longer the interaction time, the greater the impulse, i.e., the greater the change in momentum. If the force is varying with time, then the change in momentum over the time interval is the average value of the force. In Table 6.1 we list some common events and the approximate time intervals over which these interactions occur.

We now consider Newton's third law – 'for every force there is an equal and opposite force' – in the light of the relationship between forces and momenta. A way of stating Newton's third law is that forces always appear in third-law force pairs (action–reaction pairs). These third-law paired forces are equal and opposite at all times. This means that whenever a force changes there will be a reaction force with the same magnitude somewhere that is changing in the same way, but which is pointing in the opposite direction, and is acting on the other body in the interaction. In other words, when the momentum of one object changes, we can be sure that the momentum of another object will be changing *by exactly the opposite amount*. This point may be stated as a conservation law:

Key concept:
Newton's third law requires that momentum be conserved.

6.4 Collisions

The conservation of momentum is extremely useful, particularly in the solution of problems which involve two or more objects that interact with each other. The simplest such interaction is the 'collision', in which two objects run into each other, and their motion changes due to the contact forces acting during the collision. However, it should be understood that these methods apply equally to any interaction between two or more objects, regardless of the nature of the forces which mediate the interaction.

As a simple example, consider the collision between a blue ball and a red ball. The blue ball coming from the left collides with the (slow) red ball, as illustrated in Figure 6.1. During the collision (i.e., for the short time, Δt), the blue ball exerts a force, F_B, on the red ball, and the red ball exerts a force, F_R, on the blue ball.

Newton's third law states that action and reaction forces are equal and opposite, so that $F_R = -F_B$; therefore

$$\begin{aligned} F_R \Delta t &= -F_B \Delta t \\ \text{So} \quad \Delta p_R &= -\Delta p_B \end{aligned}$$

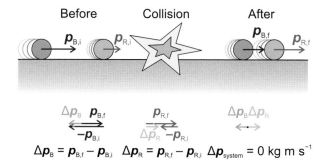

$$\Delta \boldsymbol{p}_B = \boldsymbol{p}_{B,f} - \boldsymbol{p}_{B,i} \quad \Delta \boldsymbol{p}_R = \boldsymbol{p}_{R,f} - \boldsymbol{p}_{R,i} \quad \Delta \boldsymbol{p}_{system} = 0 \text{ kg m s}^{-1}$$

Figure 6.1 A simple example of the use momentum conservation in a linear collision.

We have thus shown that the total momentum does not change during this collision, i.e.,

$$\Delta \boldsymbol{p}_{total} = \Delta \boldsymbol{p}_R + \Delta \boldsymbol{p}_B = -\Delta \boldsymbol{p}_B + \Delta \boldsymbol{p}_B = 0$$

Note that an important part of this calculation was the fact that there were no external forces present; the only forces involved were an action–reaction pair. We would find that the momentum was not conserved if there had been external forces acting on the system, as these would not have been part of action–reaction pairs in the system, so that we would not have been able to apply Newton's third law as we did. Recall also that the rate of change of momentum is equal to the net external force; an external force on our two-ball system would change their combined momentum.

Collisions are usually thought of as being between two or more objects whose properties are very simple. So simple, in fact, that we can broadly categorise all collisions into two types: **elastic** and **inelastic** collisions.

Elastic Collisions: These are collisions in which:

- momentum *is* conserved (if no external forces act on the bodies), and

- kinetic energy *is also* conserved.

In these collisions, kinetic energy is conserved, since no energy is dissipated by friction, or by heating up the colliding bodies. Often, collisions between very hard objects may be treated as elastic, since these objects do not deform.

Inelastic Collisions: These are collisions in which:

- momentum *is* conserved (if no external forces act on the bodies), and

- kinetic energy *is not* conserved – some of the kinetic energy is converted into thermal energy or sound etc. by dissipative forces.

Inelastic collisions often involve soft, deformable objects. On collision, these objects change their shape in response to the contact forces produced by the collision. This deformation may warm up the objects—this means that kinetic energy has been removed during the collision and converted into thermal energy in the objects. Objects may also emit loud sounds on collision. Noise produced by the collision will also carry off energy.

Sticky Inelastic Collisions: A subgroup of the inelastic collision category are those collisions where the objects stick together after they have collided (as in Figure 6.2). These collisions are sometimes called 'totally' or 'sticky' inelastic collisions. Sticky collisions are easier to analyse than other inelastic collisions because the objects share a common velocity after the collision.

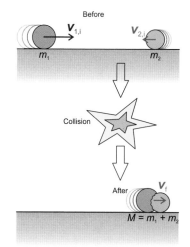

Figure 6.2 An inelastic collision in which the objects stick together (a 'sticky' inelastic collision).

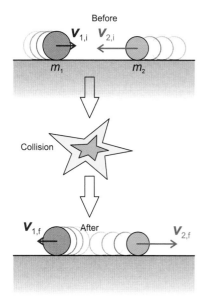

Before

$v_{1,i}$ $v_{2,i}$

m_1 m_2

Collision

$v_{1,f}$ After $v_{2,f}$

Figure 6.3 An illustrative example of an elastic collision.

6.5 Elastic Collisions

Elastic collisions are often easier to analyse than inelastic collisions for the simple reason that we have more information at our disposal. We are able to use the fact that kinetic energy, as well as momentum, is conserved. This additional information allows for significant simplifications to the solution of elastic collision problems.

We will begin by analysing one-dimensional collisions, i.e., collisions which are 'head-on'. Since we are dealing with one-dimensional problems, we are able to treat the velocity, and hence the momentum, as scalar. We can do this since the velocity is only able to point in two directions, so can only be positive or negative with respect to a chosen direction.

Collision problems such as those illustrated in Figure 6.3 may be solved using fundamental principles such as energy and momentum conservation. However, it is often possible to solve problems like these using the following key concept:

> **Key concept:**
> The relative speed of the objects before the collision equals the negative of their relative speed after the collision.

To show this, we use the principle of momentum conservation

$$m_1 \boldsymbol{v}_{1i} + m_2 \boldsymbol{v}_{2i} = m_1 \boldsymbol{v}_{1f} + m_2 \boldsymbol{v}_{2f} \tag{6.3}$$

and the conservation of kinetic energy

$$\frac{1}{2} m_1 \boldsymbol{v}_{1i}^2 + \frac{1}{2} m_2 \boldsymbol{v}_{2i}^2 = \frac{1}{2} m_1 \boldsymbol{v}_{1f}^2 + \frac{1}{2} m_2 \boldsymbol{v}_{2f}^2 \tag{6.4}$$

Combining these two equations, and employing a fair bit of algebra, gives

$$\boldsymbol{v}_{1i} - \boldsymbol{v}_{2i} = -(\boldsymbol{v}_{1f} - \boldsymbol{v}_{2f}) \tag{6.5}$$

This expression states that the difference between the initial velocities is equal to the difference between the final velocities. The difference between the initial velocities is the velocity of one object as seen by the other object.

For example, suppose two cars are heading toward each other, each travelling at 50 km h^{-1}. The car travelling in the positive-x direction has a velocity of +50 km h^{-1}, and the car travelling in the negative-x direction has a velocity of −50 km h^{-1}. The relative velocity of these two cars is (+50 km h^{-1}) − (−50 km h^{-1}) = 100 km h^{-1}. This is the velocity of the oncoming car as observed by the occupants of the other car. After an elastic collision, the car which was originally oncoming must be receding at 100 km h^{-1} for kinetic energy to be conserved.

Example 6.1 *Elastic collision*

Problem: Consider the elastic head-on collision illustrated in Figure 6.4. If the collision is totally elastic, what are the final speeds of the two balls?

Solution: This can be treated as a one-dimensional momentum-conservation problem. Using the rule that the approach and recoil relative velocities are equal for elastic collisions in the absence of external forces, we may write

$$
\begin{aligned}
v_{1i} - v_{2i} &= -(v_{1f} - v_{2f}) \\
10 - (-20) &= v_{2f} - v_{1f}
\end{aligned}
$$

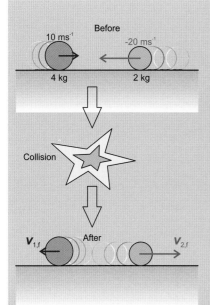

Figure 6.4 An example of an elastic collision solved using approach and recoil velocities.

so that the final velocities of masses one and two are related by

$$v_{2f} = 30 + v_{1f}$$

We are still not able to determine the velocities of the two masses after the collision. We need more information. Fortunately there is information that we have not used yet – the conservation of momentum

$$\begin{aligned} m_1 v_{1i} + m_2 v_{2i} &= m_1 v_{1f} + m_2 v_{2f} \\ 4 \times 10 - 2 \times 20 &= 4 v_{1f} + 2 v_{2f} \end{aligned}$$

Substituting our expression for v_{2f} into the latter equation gives

$$\begin{aligned} 0 &= 4 v_{1f} + 2(30 + v_{1f}) \\ 6 v_{1f} &= -60 \\ v_{1f} &= -10 \, \mathrm{m \, s^{-1}} \end{aligned}$$

Using this value for v_{1f} we are able to find v_{2f} from our earlier expression.

$$v_{2f} = 30 + (-10) = 20 \, \mathrm{m \, s^{-1}}$$

For this particular case, each of the two masses has just reversed its direction, so it is clear that we have successfully solved for the elastic collision case where the total kinetic energy is conserved.

Example 6.2 *Totally inelastic collision*

Problem: Consider the inelastic collision illustrated in Figure 6.5. In this example, the left ball has a mass of 3 kg and is initially travelling to the right with a velocity of 9 m s^{-1}, and the right ball has a mass of 2 kg and is travelling to the right with a velocity of 4 m s^{-1}. What is the velocity of the balls when they are stuck together after the collision?

Solution:

Figure 6.5 An example of a sticky inelastic collision.

First calculate the initial momentum of the system,

$$\begin{aligned} \boldsymbol{p}_i &= m_1 \boldsymbol{v}_{1i} + m_2 \boldsymbol{v}_{2i} \\ &= 3 \times 9 + 2 \times 4 \\ &= 35 \, \mathrm{kg \, m \, s^{-1}} \end{aligned}$$

Then, calculate the final momentum of the system. Since the balls are now stuck together they have the same velocity

$$\begin{aligned} \boldsymbol{p}_f &= (m_1 + m_2) \boldsymbol{v}_f \\ &= (3 + 2) \boldsymbol{v}_f \\ &= 5 \boldsymbol{v}_f \end{aligned}$$

But $\boldsymbol{p}_f = \boldsymbol{p}_i$ so that,

$$5 \boldsymbol{v}_f = 35 \quad \Rightarrow \quad \boldsymbol{v}_f = 7 \mathrm{m \, s^{-1}}$$

Example 6.3 *Totally inelastic collision*

Problem: Suppose that the red ball in Example 6.2 is moving in the opposite direction. What effect does this have on the final velocity?

Solution: First calculate the initial momentum of the system.

$$
\begin{aligned}
\boldsymbol{p}_i &= m_1 \boldsymbol{v}_{1i} + m_2 \boldsymbol{v}_{2i} \\
&= 3 \times 9 - 2 \times 4 \\
&= 19 \text{ kg m s}^{-1}
\end{aligned}
$$

Then calculate the final momentum, which is

$$
\begin{aligned}
\boldsymbol{p}_f &= (m_1 + m_2) \boldsymbol{v}_f \\
&= (3 + 2) \boldsymbol{v}_f \\
&= 5 \boldsymbol{v}_f
\end{aligned}
$$

But $\boldsymbol{p}_f = \boldsymbol{p}_i$, so that

$$
5\boldsymbol{v}_f = 19 \quad \Rightarrow \quad \boldsymbol{v}_f = 3.8 \text{ m s}^{-1}
$$

6.6 Summary

Key Concepts

momentum (p) A vector quantity defined as the product of mass and velocity. The SI units of momentum are kg m s^{-1}.

impulse The product of the average value of a force multiplied by the time interval for which it acts. The impulse equals the change in momentum produced by the force. The units of impulse are the same as for momentum.

elastic collision A collision of bodies in which the total kinetic energy is the same before and after the collision.

inelastic collision A collision of bodies in which total kinetic energy is not conserved.

sticky inelastic collision An inelastic collision where the two objects are stuck together after the collision. Sometimes known as a totally (or perfectly) inelastic collision.

Equations

$$
\boldsymbol{p} = m\boldsymbol{v}
$$

$$
\boldsymbol{F} = \frac{\Delta \boldsymbol{p}}{\Delta t}
$$

$$
m_1 \boldsymbol{v}_{1i} + m_2 \boldsymbol{v}_{2i} = m_1 \boldsymbol{v}_{1f} + m_2 \boldsymbol{v}_{2f}
$$

6.7 Problems

6.1 How fast do the following need to be travelling in order to have a momentum of $10\,\mathrm{kg\,m\,s^{-1}}$?

(a) An Airbus A380 aeroplane carrying 853 belly dancers, total weight 560×10^3 kg.

(b) An ordinary land-based bus carrying 40 belly dancers, total weight 6000 kg.

(c) A single 50 kg belly dancer.

(d) A 3.5 kg pet cat (perhaps owned by a belly dancer).

(e) A 5 g moth, being chased around by a belly dancer.

6.2 Car manufacturers conduct crash tests on their cars in order to improve crash safety. In the event of a crash the head of any child travelling in the front seat can strike the glove compartment at considerable (relative) speed, even if the child is wearing a seatbelt.

(a) The manufacturers of a particular brand of car conduct head-on collision tests and find that in the absence of a passenger side air bag, a child's head (which has a mass of 3.5 kg) goes from a speed of $40\,\mathrm{km\,h^{-1}}$ relative to the dashboard just before its collision to rebounding from the dash board at $15\,\mathrm{km\,h^{-1}}$ just after the collision. This collision lasts just 0.08 seconds. What is the average force exerted on the child's head during this collision?

(b) The manufacturer wishes to reduce the average force involved in such a collision to 200 N. In order to achieve this they install a passenger airbag on the front of the glove compartment which quickly inflates in the event of a crash and deflates as the child's head pushes into it, effectively increasing the amount of time it takes to slow the child's head (i.e., the collision lasts longer). How long would the collision between the child's head and the airbag need to last to reduce the speed of the head relative to the dashboard from $40\,\mathrm{km\,h^{-1}}$ to $0\,\mathrm{km\,h^{-1}}$ without exceeding the average force quoted above?

6.3 A 3500 kg car hits an 80 kg pedestrian who is standing in the middle of the road. Before the collision the car was travelling at $30\,\mathrm{km\,h^{-1}}$ and the pedestrian was stationary. After the collision the car was travelling at $28.5\,\mathrm{km\,h^{-1}}$. At what speed will the pedestrian be flung down the road (in $\mathrm{km\,h^{-1}}$)?

6.4 Two basketball players collide head-on. Player A weighs 80 kg and is travelling $2.5\,\mathrm{m\,s^{-1}}$ to the right while Player B weighs 68 kg and is travelling $1.2\,\mathrm{m\,s^{-1}}$ to the left. After the collision Player A is travelling at $1.0\,\mathrm{m\,s^{-1}}$ to the right.

(a) What is the change in momentum of Player A?

(b) If the collision lasted 0.1 s, what is the average force Player B must have exerted on Player A during the collision?

(c) What is the average force that Player A must have exerted on Player B during the collision?

(d) What is the change in momentum of Player B?

(e) What is the final velocity of Player B?

6.5 An enthusiastic kitten collides with a ball of string that is rolled towards it. The 0.5 kg kitten is travelling at $0.5\,\mathrm{m\,s^{-1}}$ due east before the collision and the 0.6 kg ball of string was travelling at $0.7\,\mathrm{m\,s^{-1}}$ due west. If the kitten grabs the string during the collision and does not let go, what is the final speed of the kitten+string?

6.6 An 85 kg runner is accelerating at a rate of $2\,\mathrm{m\,s^{-2}}$ and a 65 kg runner at a rate of $3\,\mathrm{m\,s^{-2}}$. If the heavier runner started at a speed of $1\,\mathrm{m\,s^{-1}}$ while the lighter runner started off stationary then how long is it before the runners have the same momentum?

6.7 A 40×10^3 kg train is travelling at $8.3\,\mathrm{m\,s^{-1}}$ when the engineer sees a sheep on the tracks. She throws the emergency brakes on. The emergency brakes can apply a maximum force of 11×10^3 N.

(a) How long is it before the train will come to a complete stop?

(b) The sheep, startled by the sparks and noise of the emergency brakes, trots off the tracks. The engineer is able to release the brakes. If the brakes had been applied for 11 s, how fast is the train travelling now?

6.8 A kererū (New Zealand wood pigeon) travelling at $1.5\,\mathrm{m\,s^{-1}}$ due west has an elastic collision with an Airbus A380 aeroplane. Fortunately for the kererū, the A380 was only travelling at $0.1\,\mathrm{m\,s^{-1}}$ in an easterly direction at the time. With what speed is the kererū travelling after the collision?

6.9 The kererū from Problem 6.8, having proven itself somewhat careless, has a totally inelastic collision with a cat. If, just before the collision, the 0.65 kg kererū was travelling $2.0\,\mathrm{m\,s^{-1}}$ north and the 3.0 kg cat was traveling $5.5\,\mathrm{m\,s^{-1}}$ south, what is the velocity of the ball of fur and feathers just after the collision?

SIMPLE HARMONIC MOTION

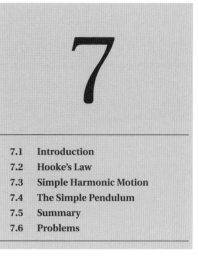

7.1   Introduction

In Chapter 3 we noted that circular motion repeated once the object had completed an entire circuit around the circle. In this respect, circular motion is periodic, and we were able to define the period of the motion. An object does not have to travel in a circle for its motion to be periodic. We will call any motion that repeats after a given period **oscillatory motion**. In this chapter we will investigate the general rules that govern the behaviour of oscillatory motion.

A very large number of physical systems are oscillatory and display behaviours which are characteristic of oscillations. There are many examples in the physical environment, from vibrations caused by trucks going over bridges to musical instruments. In the biological sciences there are a very large number of examples: many organisms detect the approach of predators by the vibrations produced by their approach, the inner ear works by transmitting sound-wave energy into a series of vibrating bones, and so on. The theory of oscillations is the base from which a large amount of physics develops.

Key Objectives

- To develop an understanding of simple harmonic motion (SHM).

- To understand the relationship between Hooke's law and simple harmonic motion.

- To be able to calculate the period and frequency of a mass–spring oscillator.

- To be able to calculate the period and frequency of a simple pendulum.

7.2   Hooke's Law

We will begin our discussion of oscillations with an investigation of the behaviour of springs. We begin at this point since the ideal massless spring is a relatively simple system and is a good model for the vibrations seen in a large number of more complex systems.

A spring is characterised by the fact that the further you stretch it, the harder it pulls back. Alternatively, the more you compress a spring, the harder it pushes back (see Figure 7.1). In mathematical language, the magnitude of the force exerted by the spring on the object attached to the spring is proportional to the 'stretch' of the spring. The 'stretch' of the spring is not how long the spring is – it indicates how far you have pulled the spring from its 'natural' position, i.e., the position that the spring will sit in if no force is applied to it. The force exerted by the spring is called the **restoring force**, and the natural resting position of the spring is the **equilibrium position** of the spring.

With these definitions we can now state the defining relationship between force and displacement for springs:

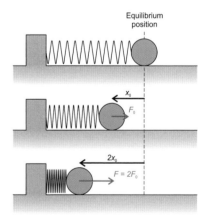

Figure 7.1 The force exerted by a spring increases with increasing displacement from equilibrium.

> **Key concept:**
> For springs (and many other elastic materials) the magnitude of the restoring force is proportional to the displacement of the spring from its equilibrium position.

This statement, known as **Hooke's law** may be put in mathematical terms as follows:

$$F = -kx \qquad (7.1)$$

The constant k is called the spring constant (and is a property of a given spring) and the distance from the equilibrium position is x. The larger the spring constant, the larger the force you would need to apply to the spring to stretch it. The negative sign indicates that the direction of the restoring force is always towards the equilibrium position.

The same discussion applies to compressing springs. This discussion will apply equally to systems which are not springs, but which respond to deforming forces in a similar way. For example, we could analyse a situation such as that shown in Figure 7.2 in which we bend a plastic ruler by applying a sideways force to its free end. The ruler will return to its equilibrium position (straight) when we no longer apply a bending force to it; Hooke's law applies to the ruler and we could determine the spring constant of the ruler if we so desired.

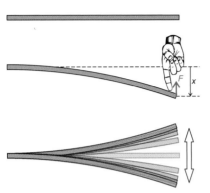

Figure 7.2 Bending a plastic ruler or some other flexible rod will result in a restoring force that may obey Hooke's law.

Rubber Bands and Hooke's Law

Some materials that we might expect to obey Hooke's law, like rubber bands, do not necessarily do so. We will see in Chapter 10 that Hooke's law requires the strain produced to be proportional to the stress applied. Some materials, while 'stretchy' have a molecular structure that stretches in a non-linear fashion for even small stresses.

Energy in Hooke's Law Deformations

In order to stretch or compress a spring, an external force (i.e., external to the spring) must have initially been applied to the object to overcome the spring's restoring force. Work must be done by this external force to stretch or compress the spring. The amount of work done is given by the product of the external force doing the work and the distance over which that force operates. The force increases linearly with distance, so the average is half the maximum value. So the work done to compress the spring by a distance x is

$$
\begin{aligned}
W &= F_{\text{average}} \times d \\
&= \frac{1}{2}kx \cdot x
\end{aligned}
$$

Thus the work done stretching or compressing a spring is given by

$$W = \frac{1}{2}kx^2 \qquad (7.2)$$

The work done on the spring is recoverable, that is, once we have done work on a spring, that spring will then be able to do work on something itself. In other words, we have stored energy in the spring. The energy stored in this way is potential energy and is equal to the amount of work we have done on the spring. The **potential energy stored in a spring** is therefore given by

$$\text{PE} = \frac{1}{2}kx^2 \qquad (7.3)$$

In the above we have used only Hooke's Law. This means that the expression we are left with is true for any system that obeys Hooke's Law: energy may be stored in all such systems.

7.3 Simple Harmonic Motion

Suppose that we have a mass attached to a spring and that we stretch this spring. We have done work on the spring, and the spring has stored this work as potential energy. The spring exerts a restoring force on the object. What will happen if we release the mass? We will consider the case where there is no friction between any of the components of the system.

To begin with, the mass will accelerate back toward the equilibrium point under the influence of the restoring force. As it approaches the equilibrium point, the restoring force will decrease as the displacement from the equilibrium point decreases, and

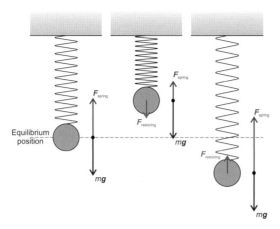

Figure 7.3 The force exerted by the spring in this case always points upwards but the net force (restoring force) on the mass always points towards the equilibrium point, thus this system will exhibit periodic motion. If the restoring force can be described using Hooke's law then it will exhibit simple harmonic motion.

when the mass reaches the equilibrium point, the restoring force will have reduced to zero. This means that at the equilibrium point, the *acceleration* of the mass has reduced to zero.

The velocity of the mass is not zero, however. If we consider the potential energy of the spring at the point at which we release the mass, we can see that the potential energy decreases from this point as the spring does work to accelerate the mass. When the spring is at its greatest stretch, the potential energy is at its maximum value, and when the mass reaches the equilibrium point, the potential energy of the spring drops to zero. The potential energy stored in the spring has been returned to the mass as kinetic energy.

The potential energy of the spring was greatest at maximum stretch, at which point the kinetic energy of the mass was zero. Similarly, at the equilibrium point, the potential energy of the spring is zero and the kinetic energy of the mass is at its greatest value. This means that the velocity of the mass is at its maximum value when the mass reaches the equilibrium point.

The mass will overshoot the equilibrium point and continue moving. The restoring force of the spring begins to grow once more, but the force is pointing in the opposite direction to the velocity of the mass. This force will therefore cause the mass to decelerate, and its kinetic energy will decrease. There is a point when the kinetic energy of the mass has been entirely converted into potential energy of the spring. However, since the kinetic energy of the mass originally came from the initial stretch of the spring, we will have that same quantity of potential energy stored once again when the spring is compressed. Thus the mass has gone from one side of the equilibrium point to the other side, compressing the spring to the same distance from the equilibrium point as its initial stretch.

The mass will now begin to accelerate back toward the equilibrium point, and again it will overshoot, and again it will stop when it gets as far on the other side of the equilibrium point as it was to begin with. This process will continue over and over again without stopping. This is what we mean by an oscillation: a restoring force with the form of Hooke's Law will produce an oscillation about an equilibrium position which will continue indefinitely. In a real situation, friction forces will always oppose the motion of the mass regardless of the direction in which it is heading, and so over time will dissipate the kinetic energy. As a result, the mass will to come to rest at the equilibrium point after a time which is dependent on the size of the friction forces.

Oscillations of this kind are very common in nature and have a particularly simple form. Such an oscillation is often called a **simple harmonic motion** or **SHM**. An understanding of SHM is particularly useful, as even oscillations which are much more complex than simple harmonic motion can always be modelled as a sum of SHMs with different periods. (This is known as the Fourier theorem.) In this way, we are able to use the physics of SHM to analyse vibrations as diverse as those which produce the sounds from musical instruments, or the vibrations of the vocal cords which produce the human voice.

Simple harmonic motion is periodic; after a set time the system will return to its initial state. The system will then repeat itself and again return to its initial state after the same amount of time. Each oscillation is called a cycle, and the time taken for the system to return to its initial state, i.e., for one full cycle of oscillation, is called the **period**, T. Since the system has a period, we are able to define a frequency for the system: the number of cycles per time interval. The **frequency**, f, in SI units of hertz (Hz), is the number of cycles per second. For example, if it takes a system 0.1 s to go through one complete cycle, then its period is $T = 0.1$ s, and there will then be 10 such cycles every second, so the frequency of the oscillation is $f = 10$ Hz. The relationship between period and frequency is

$$f = \frac{1}{T} \tag{7.4}$$

which is equivalent to

$$T = \frac{1}{f} \tag{7.5}$$

Example 7.1 *Potential energy in a spring*

Problem: The spring in a toy gun has $k = 50$ N m^{-1}, and is compressed 0.15 m to fire a 2 g plastic bullet. With what speed will the bullet leave the gun?

Solution: All of the potential energy stored in the spring will be delivered to the plastic bullet as kinetic energy. Thus we begin by calculating the potential energy stored in the spring, which is

$$\text{PE of spring} = \frac{1}{2}kx^2 = \frac{1}{2} \times 50 \text{ N m}^{-1} \times 0.15^2 \text{ m}^2 = 0.563 \text{ J}$$

We now use the fact that all of this potential energy is converted into the kinetic energy of the bullet to write

$$\frac{1}{2}mv^2 = 0.563 \text{ J}$$
$$\frac{1}{2} \times 0.002 \text{ kg} \times v^2 = 0.563 \text{ J}$$

We are then able to calculate the speed of the bullet. Since

$$v^2 = 563 \text{ m}^2 \text{ s}^2$$

thus the speed of the bullet is

$$v = 24 \text{ m s}^{-1}$$

The Relationship Between Circular Motion and SHM

Consider the following example. A black ball is constrained to move in a circle, as shown in Figure 7.4. The ball moves at constant speed around the circle. Since the motion is two dimensional, we may resolve the motion into its components along the horizontal and vertical axes. We will consider only the component of the motion in the x-axis and represent the motion in the x-axis by a blue ball.

As the black ball moves around the circle, the blue ball (the x-component of its position) will oscillate back and forth along the x-axis. If the angle between the x-axis and the black ball is θ, the distance from the centre of the circle to the blue ball is

$$x = r\cos\theta \tag{7.6}$$

Next, consider the centripetal force required to keep the black ball moving around the circle (see Figure 7.5). This force is always directed from the black ball toward the

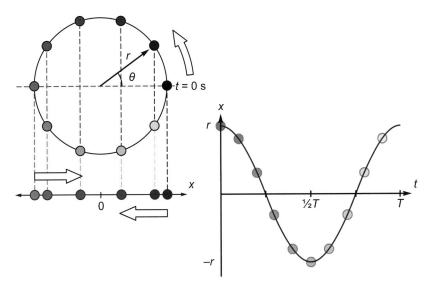

Figure 7.4 Plotting the x-component of the position of an object moving in a circle traces a sine graph.

centre of the circle so that its direction is continually changing. This force will also have vertical and horizontal components. We can find the horizontal component of this force in the same way as we found the horizontal component of the displacement. The angle between the centripetal force vector and the x-axis is again θ and thus the x-component of the centripetal force will be given by

$$F_x = -F_c \cos\theta$$

Note the minus sign! The force is pointing toward the centre of the circle, and thus the x-component of this force will point in the negative-x direction while the black ball is in the positive-x sector of the circle, and in the positive-x direction when the black ball is in the negative-x sector of the circle.

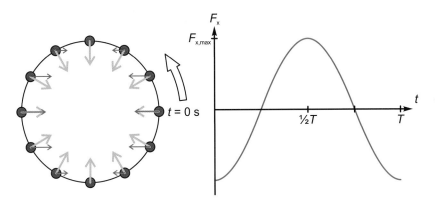

Figure 7.5 Plotting the x-component of the force (red vector) on an object moving in a circle traces a sine graph that is 180° out of phase with the position graph (see Figure 7.4).

If we combine the expression for the x component of the centripetal force and the x-component of the displacement, we get

$$\begin{aligned} F_x &= -F_c \frac{x}{r} \\ &= -\left(\frac{F_c}{r}\right)x \\ &= -kx \end{aligned}$$

where in the last step we have set $k = \frac{F_c}{r}$ (this means that if we know k and the maximum restoring force we can work out the radius of this circle). This derivation has shown that the blue ball, the x-component of the black ball's motion, obeys Hooke's Law. This means that the blue ball is undergoing simple harmonic motion along the x-axis. It is worth noting that the same analysis could have been done on the vertical component of the black ball's motion and we would have found exactly the same thing.

> **Key concept:**
> The horizontal and vertical components of the motion of an object in circular motion at constant speed are examples of simple harmonic motion.

Now we are able to apply the insights we obtained for circular motion to simple harmonic motion. We will find it useful in future to write the displacement of an object undergoing simple harmonic motion using the equation found above (Eq. (7.6)). We know from our discussion of circular motion that we may replace the angle θ in this expression using Eq. (3.3) to get

$$x = A\cos(\omega t) \tag{7.7}$$

In this expression we have used the symbol A for the maximum displacement from the equilibrium position, commonly called the **amplitude** of the oscillation. Eq. (7.7) is preferable to Eq. (7.6) since the time dependence of the displacement can be clearly seen.

We can write similar expressions for the velocity and acceleration of an object undergoing simple harmonic motion:

$$v \quad = v_x \quad = -v_{\mathrm{max}}\sin(\omega t) \tag{7.8}$$
$$a \quad = a_x \quad = -a_{\mathrm{max}}\cos(\omega t) \tag{7.9}$$

The coefficients v_{max} and a_{max} are the maximum velocity and acceleration, which occur at $x = 0$ and $x = A$ respectively. Note that the velocity and acceleration are both negative. First consider the acceleration. The negative sign indicates that when the displacement of the spring is positive, then acceleration is in the negative direction i.e., back toward the equilibrium point. Similarly, when the displacement is negative, the acceleration is positive, again pointing back toward the equilibrium position. The velocity is slightly more complicated as it is a sine function, unlike the displacement and accleration which are cosine functions. When the displacement is at its maximum value at $t = 0$, velocity is zero. After this the displacement decreases and the velocity is negative, i.e., pointing toward the equilibrium position. As the oscillator passes the equilibrium point, the displacement (and acceleration) are zero, but the velocity is at its maximum value. After this the displacement increases in the negative direction until it reaches its negative maximum, while the velocity decreases to zero (but still points in the negative direction). Then the displacement begins to decrease and the velocity turns around and points in the positive direction, toward the equilibrium again. These changes may be seen by comparing Figures 7.5 and 7.6.

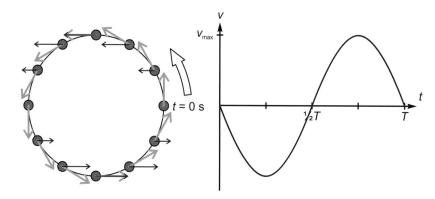

Figure 7.6 Plotting the x-component of the velocity (blue vector) on an object moving in a circle traces a sine graph that is 90° out of phase with the position graph (see Fig. 7.4). In other words the velocity is at the maximum when the displacement from equilibrium is zero.

Maximum Velocity in SHM

We will now calculate the maximum velocity of a mass attached to a spring and undergoing SHM. The total energy of a system undergoing simple harmonic motion is conserved, so the total energy at any point in the cycle of the oscillator is just the potential energy stored in the spring plus the kinetic energy due to the velocity and mass of the object attached to the spring

$$E = \frac{1}{2}kx^2 + \frac{1}{2}mv^2 \tag{7.10}$$

At the endpoints of the oscillation when the mass is as far from the equilibrium point as it gets, the distance from the equilibrium point is just the amplitude of the oscillation, i.e., $x = \pm A$. For an instant at each of these points the mass is stationary. This means that the kinetic energy at this instant is zero and the potential energy of the spring reaches its maximum value i.e., PE $= \frac{1}{2}kA^2$. Since the energy of this system is conserved (we are ignoring friction) we observe that the total energy of the system is thus equal to this maximum potential energy $E_{total} = \frac{1}{2}kA^2$.

At the centre of the oscillation, i.e., at the instant when the mass passes the equilibrium point, the position is $x = 0$ and the potential energy of the system is zero. This means that all of the energy of the system is contained in the kinetic energy of the oscillating mass, thus $E_{total} = \frac{1}{2}mv_{max}^2$; this is also the point at which the velocity is a maximum so we have used a subscript 'max'. This maximum kinetic energy is equal to the total energy of the system, and is also equal to the maximum potential energy. We are able to equate the two:

$$E_{total} = \frac{1}{2}kA^2 = \frac{1}{2}mv_{max}^2$$

We are now able to solve this equation for the maximum velocity of the oscillator. A small amount of algebra gives

$$v_{max} = \sqrt{\frac{k}{m}}A \tag{7.11}$$

Period and Frequency of SHM

Now that we have found an expression for the maximum velocity of the mass on a spring oscillator we are in a position to find an expression for the period and frequency of this oscillator. To do this we use the relationship between simple harmonic motion and circular motion. In the discussion of the relationship between circular motion and SHM, we noted that the horizontal component of a black ball moving around the circumference of a circle at constant speed undergoes simple harmonic motion. When the ball passes the highest point in the circle (i.e., directly above the centre point), the horizontal component of its velocity is at its maximum value. At this point, the vertical component of the ball's velocity is zero and the horizontal component is equal to the velocity of the ball. This is also the point at which we have calculated the maximum velocity of our object undergoing SHM.

We also know that the time it took the ball to travel around the circle (the period) is given by the distance travelled divided by the velocity, i.e.,

$$T = \frac{2\pi r}{v_{max}} = \frac{2\pi A}{v_{max}}$$

We have used the fact that the radius of the circle is equal to the amplitude of the SHM in the last equality. We now substitute into this our expression for the maximum velocity, Eq. (7.11), then a little algebra gives us an expression for the period of this oscillator as

$$T = 2\pi\sqrt{\frac{m}{k}} \tag{7.12}$$

The angular frequency of the oscillator is then found using Eq. (7.4) and is given by

$$f = \frac{1}{2\pi}\sqrt{\frac{k}{m}} \tag{7.13}$$

Alternatively, we can calculate the angular frequency in radian s^{-1}

$$\omega = \sqrt{\frac{k}{m}} \tag{7.14}$$

since $\omega = 2\pi f$, and thus $v_{\text{max}} = \omega A$.

7.4 The Simple Pendulum

We have so far considered simple harmonic motion using a mass on a spring as our model. This is not the only example of simple harmonic motion. We will now consider another important example – the simple pendulum. The pendulum consists of a mass hanging from a light cord and swinging from side to side. In this case the equilibrium point is the point at which the pendulum is hanging straight down, and the force which tends to restore the pendulum to this equilibrium position is provided by gravity. In the following we will assume that the pendulum is only slightly displaced from its equilibrium position.

As can be seen in Figure 7.7, the gravitational force may be resolved into components along the pendulum cord and perpendicular to it. It can be shown that the restoring force is given by

$$F_{\text{perp}} \approx -kx$$

with

$$k = \frac{mg}{L} \tag{7.15}$$

The force is a restoring force with the same form as Hooke's Law and the motion of the simple pendulum is simple harmonic motion. We can also calculate the period of the pendulum using Eq. (7.15) and Eq. (7.12):

$$T = 2\pi\sqrt{\frac{L}{g}} \tag{7.16}$$

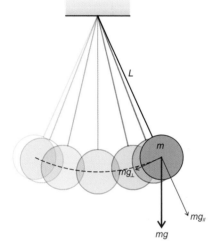

Figure 7.7 The simple pendulum

where L is the length of the pendulum. Note that the period of the pendulum's swing is independent of the mass of the object hanging from the pendulum, and varies only with the length of the pendulum and the gravitational acceleration (a pendulum will swing more slowly on the Moon than it does on Earth).

Example 7.2 *Period of SHM*

Problem: How long is a simple pendulum with a period of exactly 1 s?

Solution: To solve this problem we rearrange Eq. (7.16) to get

$$L = g\left(\frac{T}{2\pi}\right)^2$$

We now just enter the known values and calculate the required length, which is

$$
\begin{aligned}
L &= g\left(\frac{T}{2\pi}\right)^2 \\
&= 9.8\,\text{m s}^{-2} \times \left(\frac{1\,\text{s}}{2\pi}\right)^2 \\
&= 0.25\,\text{m}
\end{aligned}
$$

7.5 Summary

Key Concepts

frequency (f) The number of repetitions of a complete oscillation per unit time. Measured in cycles per second, or hertz (Hz).

period (T) The time interval between the successive occurrences of a particular phase of an oscillation.

simple harmonic motion, SHM A type of periodic motion in which the restoring force is proportional to the displacement from the equilibrium position.

Hooke's law Within the elastic limits of a material, the strain is proportional to the stress i.e., the amount of deformation is proportional to the distorting force.

spring constant (k) The ratio of force applied to a spring to the resulting change in length.

Equations

$$F = -k\boldsymbol{x}$$

$$PE = \frac{1}{2}kx^2$$

$$KE_{max} = \frac{1}{2}kA^2$$

$$f = \frac{1}{T}$$

$$T = \frac{1}{f}$$

$$v_{max} = \sqrt{\frac{k}{m}}A = \omega A$$

$$T = 2\pi\sqrt{\frac{m}{k}}$$

$$\omega = \sqrt{\frac{k}{m}}$$

$$T = 2\pi\sqrt{\frac{L}{g}}$$

7.6 Problems

7.1 A spring is pressed against a wall so that it is compressed by 0.25 m (i.e. it is 0.25 m shorter than its equilibrium length). The spring is then released. The spring constant is $k = 35\,\text{N m}^{-1}$, and the spring weighs 50 g. What is the speed at which the spring leaves the wall?

7.2 How long is a simple pendulum with a period of 5 seconds? How long would this pendulum have to be if it were to operate on the moon with the same period? ($g_{\text{moon}} = 1.62\,\text{m s}^{-2}$)

7.3 During an earthquake a skyscraper is designed to sway back and forth with simple harmonic motion with a period of 8 seconds. The amplitude at the top floor for a particular earthquake is 70 cm. With respect to the simple harmonic motion of the top floor, calculate the following quantities:
    (a) The radius of the circle used to represent the SHM.
    (b) The speed of the object moving round the circle.
    (c) The angular velocity.
    (d) The maximum speed of the top floor.

7.4 When a sound wave with a certain intensity is detected by the tympanic membrane (eardrum) the amplitude of the resultant motion is 1.0 nm (1.0×10^{-9} m). If the frequency of the sound is 600 Hz what is the maximum speed of the membrane oscillation?

7.5 A fly beats its wings at a frequency of 1200 Hz. If the expansion and contraction of the wing muscles of the fly exhibits simple harmonic motion, the angular displacement of each wing of the fly also exhibits simple harmonic motion, the length of the wing muscles varies from 750 μm to 600 μm over the course of one beat of its 1.00 cm long wings, and the wing tips move through an arc of 150° then:
    (a) What is the time period of a single beat of the fly's wings?
    (b) What is the maximum velocity of the mobile end of the wing muscle?
    (c) What is the maximum angular velocity of the wing?
    (d) What is the maximum speed of the wingtip?

7.6 Four identical springs used as part of a car's suspension system, one on each wheel. The springs compress by 6 cm when the weight of the 1900 kg car is applied to them.
    (a) What is the spring constant of each of the springs?
    (b) With what frequency will the suspension bounce if given a jolt?
    (c) If the springs are chopped in half (which will double the spring constant) with what frequency will the suspension bounce now?

7.7 A 5000 kg floating pier is moved up and down by the changing tide. If the period of this motion is 12 hours and the amplitude is 2.5 m and we treat the motion as being simple harmonic then:
    (a) What is the frequency of the motion?
    (b) What is the maximum vertical velocity of the pier?
    (c) What is the maximum vertical acceleration of the pier (hint: use circular motion as an analogue)?

7.8 A small 5 g fly is buzzing along and hits a spider web. The spider web catches the fly and proceeds to oscillate with a time period of 0.09 s and an initial amplitude of 1.9 cm.
    (a) With what velocity was the fly flying before it hit the web?
    (b) What is the maximum force that the web exerts on the fly (hint: use circular motion as an analogue to find the maximum acceleration of the fly)?
    (c) What is the spring constant of the web?

7.9 A particular person's lower leg is 56 cm long and weighs 9.5 kg. If allowed to swing freely, with what time period would the leg swing? Is this close to what you would estimate such a person's walking pace is? (Treat the leg as a simple pendulum with length equal to half the leg length and all the leg's mass concentrated at this point.)

WAVES

8

8.1   Introduction

Waves of many kinds, from ocean waves to sound and light waves, are a common feature of everyday life. Information about the environment is carried by light and sound waves, so these are of critical importance to biological organisms. Energy, in the form of heat or light, can travel as an electromagnetic wave. Furthermore, in health-science technology, a large class of diagnostic instruments rely on the propagation and reflection of various kinds of wave, for example ultrasound, or various kinds of microscopy.

In this chapter we will investigate the nature of oscillations and waves. Normally, but not always, waves travel as a disturbance in some medium. (Electromagnetic waves can travel through a vacuum, and will be discussed in Optics.) We will show that wave motion is deeply connected to oscillations in the medium in which the wave propagates.

Key Objectives

- To understand the connection between simple harmonic motion (SHM) and wave motion.

- To understand the concepts of phase, frequency, wavelength and wave velocity.

- The understand superposition and interference of waves.

- To understand the phenomenon of beats.

- To understand reflection of waves by barriers and the production of standing waves.

- To understand the transmission of energy and power by wave motion.

8.2   SHM and Waves

When a wave propagates, each spatial point on the wave is oscillating in simple harmonic motion. The wave is the result of these oscillations, and the fact that each oscillator is in a strict phase relationship with every other point, i.e., if an oscillator is one quarter of a cycle ahead of another oscillator at a particular time, this is true at all times.

In Figure 8.1, we represent a wave by a row of mass–spring oscillators. Each of these oscillators has the same spring constant, the same mass, and oscillates with the same amplitude. This means that they will all oscillate with the same period and frequency. However, each of these springs is oscillating slightly ahead or slightly behind each of its neighbours; each spring is at a different point in its cycle. This lead or lag is progressive, so moving from left to right each spring is slightly further behind the spring on the extreme left. Notice that the spring at point Q has fallen so far behind the spring at point P that is now oscillating in time with the spring at point P, i.e., it is exactly one cycle behind the spring at point P. It is because of this lead or lag between successive springs that this set of oscillators forms a wave. Each evenly spaced spring must be the same proportion of a cycle ahead of or behind its neighbours as every other spring. The distance from spring P to spring Q is called the **wavelength**, usually represented by the Greek letter lambda (λ).

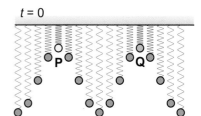

$t = 0$

Figure 8.1 The relationship between SHM and wave propagation. The oscillators at P and Q are spaced one wavelength apart and are exactly in phase.

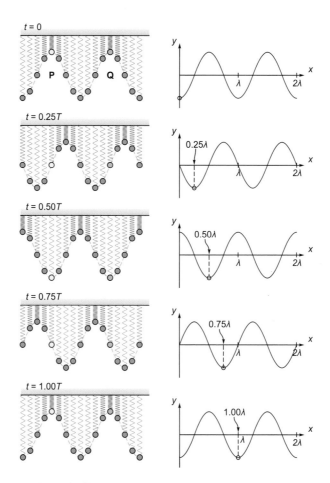

Figure 8.2 The propagation of a wave in a mass–spring system.

8.3 Frequency, Wavelength and Speed

In Figure 8.2, we have reproduced Figure 8.1, but at a sequence of successive times. In this figure, the crests of the wave (which were at points P and Q at t=0) are moving to the right. In the top diagram, the oscillators at P and Q are at the maximum vertical position, and the oscillator midway between is at the minimum vertical position. After a time equal to the period of these oscillators (the bottom of Figure 8.2), the oscillators at P and Q are again at the maximum vertical position. A continuous film made up of snapshots of the oscillator positions at successive instants (a set of which are illustrated in Figure 8.2), shows that during one period (of the oscillators), the wave-crest initially at point P has moved one wavelength to the right to point Q. We can calculate speed velocity with which the wave-crests are travelling:

$$v_{\text{wave}} = \frac{|\Delta x|}{t} = \frac{\lambda}{T} = f\lambda$$

In the last step, we used the fact that the period of an oscillator undergoing simple harmonic motion is related to its frequency by $f = 1/T$. The velocity, frequency and wavelength of a wave are related by

$$v = f\lambda \tag{8.1}$$

8.4 The Form of the Wave

The waveform that we have illustrated in Figure 8.1, using a line of mass–spring oscillators, is the familiar sine or cosine wave. We choose the cosine function to represent a propagating wave:

$$y = A\cos\left(2\pi\left(\frac{x}{\lambda} \pm \frac{t}{T}\right)\right) \tag{8.2}$$

The plus/minus symbol indicates the direction of travel as described below. Here A is the amplitude of the wave. In this expression we have used the wavelength λ and the period T. We can rewrite this in terms of some alternative parameters: the **wavenumber** k and the angular frequency ω. The wavenumber is related to the wavelength by the expression,

$$k = \frac{2\pi}{\lambda} \tag{8.3}$$

and it is related to how many cycles per metre (specifically, it is the angle in radians travelled through by that number of cycles), rather than the number of metres per wave cycle (which is what the wavelength tells us). The angular frequency is related to the frequency in hertz by

$$\omega = 2\pi f \tag{8.4}$$

and converts the frequency in hertz (i.e., in cycles per second) into the angular frequency in radians per second.

In terms of the wavenumber and the angular frequency the expression for a wave becomes somewhat simpler:

$$y = A\cos(kx \pm \omega t) \tag{8.5}$$

Compare this expression with Eq. (7.7), which describes the displacement of a simple harmonic oscillator as a function of time. The expression for the wave, Eq. (8.5) is a generalisation of the expression for the displacement of a simple harmonic oscillator and describes the displacement of simple harmonic oscillators at each point in the x direction.

The section of Eq. (8.5) between the brackets, the argument of the cosine function, is known as the **phase** of the wave. Varying the phase in this expression can be achieved by changing the point under observation (i.e., changing x) or by changing the time at which a particular point is observed (by changing t).

Note that $\frac{x}{\lambda}$ in Eq. (8.2) is just the number of wavelengths we have changed in position. The ratio $\frac{t}{T}$ is the number of periods that have elapsed.

Finally note that the phase of the wave in Eqs. (8.2) and (8.5) has a $\pm$ operation. This means that at this point in the expression we may substitute either a '+' or a '−'. The difference between these is that if we substitute a '−' sign then the expression will represent a wave travelling in the positive x direction (customarily shown as moving to the right) and if we substitute a '+' sign then the expression will represent a wave travelling in the negative x direction.

> **Choice of Cosine**
>
> Mathematically, we could equally well choose a sine function; think about the way the film of a mass on a spring is the same no matter which frame we decide to label as $t = 0$. The choice of cosine here is so that the wave described has a peak at $x = 0$ when $t = 0$.

8.5 Types of Wave

Up until this point we have considered waves as a collection of mass–spring oscillators undergoing simple harmonic motion. We have assumed in Figures 8.1 and 8.2 that these mass–spring systems are oscillating up and down and the wave is propagating horizontally. This has been useful to illustrate the connection between wave motion and simple harmonic motion, but it is by no means the only possible arrangement of the oscillators underlying wave motion. These oscillators may be any system that undergoes simple harmonic motion (such as a collection of pendula) and the oscillation need not be perpendicular to the direction of the wave's propagation. We will henceforth dispense with the underlying oscillators and concentrate on the behaviour of waves as they propagate and combine. We will first define two distinct kinds of wave: **transverse** waves, and **longitudinal** waves.

Transverse Waves

A transverse wave is one in which the medium in which the wave is travelling is oscillating in a direction which is *perpendicular* (i.e., transverse) to the direction in which the wave is propagating. The waves shown in Figures 8.1 and 8.2 are transverse waves. The waves produced in a shaken string or cable are transverse. A transverse wave is also shown in the Figure 8.3.

Electromagnetic waves, which will be covered later in this book, are another example of a transverse wave.

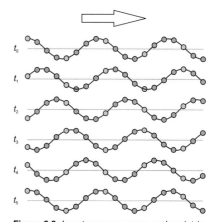

Figure 8.3 In a transverse wave each point in the wave moves perpendicular to the direction of motion of the wave.

Longitudinal Waves

As well as transverse waves, there is another class of waves in which each spatial point in the medium in which the wave is propagating is oscillating in the *same* direction as the wave is propagating. Such as wave is known as a **longitudinal** wave. Such a wave is shown in Figure 8.4.

An important example of a longitudinal wave is the pressure wave in air that we experience as sound.

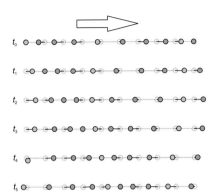

Figure 8.4 In a longitudinal wave each point in the wave moves parallel to the direction of motion of the wave.

8.6 Superposition and Interference

What happens when two waves of the same type meet and overlap? They **superpose**, creating a new oscillation. The addition of waves is achieved by adding their displacements (that is, their instantaneous displacement caused in the medium, not the maximum) at every point. These displacements add like scalar quantities (simple positive or negative numbers) at every point of the two waves, so that if at an instant in time, the displacement of one wave at point x is 1 m and the displacement of the other wave at the same point x is 2 m, then the displacement of the wave which results from the superposition of these two waves has an displacement of 3 m at x at this instant.

The primary phenomenon which results from superposition is **interference**. Interference, at its extremes, can be either a completely **constructive** or completely **destructive** process.

Constructive interference occurs when the two waves which overlap have the same wavelength and frequency and are 'lined up' so that crests of one wave are in the same place as the crests of the other wave. When this happens, the two waves are said to be **in phase**. This is illustrated in Figure 8.5. The displacements at each crest will add to give a resultant wave with an amplitude which is twice as high as the crest of each individual wave. At each trough, the displacements will add to give a trough which is twice as deep as the trough of each individual wave. Thus the wave which results from the purely constructive interference of these two identical waves will have *twice* the amplitude of each of the component waves.

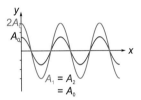

Figure 8.5 The completely constructive interference of two identical waves.

Destructive interference occurs when the two component waves are completely misaligned. This means that the crests of one wave are lined up with the troughs of the other and vice versa, and when this happens the waves are said to be completely **out of phase**. The misalignment needed for completely destructive interference is illustrated in Figure 8.6. One wave has a crest at point x and the other wave has a trough. The amplitude of these two waves is identical so the crest of one wave will exactly cancel out the trough of the other wave. Thus the amplitude of the resultant wave will be exactly zero at x. The same argument applies everywhere along both waves: at each point the displacement of one wave is exactly opposite to the displacement of the other wave, so that the two waves will *exactly cancel* each other out. Generally, the resultant displacement of the disturbance at each point in the medium is a time-dependent quantity; sometimes waves add, sometimes they cancel one another.

For the health sciences, one of the most important consequences of the interference of waves is the diffraction of light. This will be treated carefully in the Optics chapters, and limits the resolution of optical imaging devices such as microscopes and the human eye.

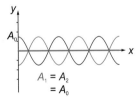

Figure 8.6 The completely destructive interference of two identical waves.

8.7 Beats

In the previous section we considered constructive and destructive interference between two identical waves, i.e., waves that had the same frequency and wavelength, and the same amplitude. In this section, we will consider an interference phenomenon which occurs when the two superposing waves are not identical. When two waves have different frequencies, the phenomenon of **beats** will be observed.

Consider a single position in space where these two waves with similar, but not equal, frequencies superpose. Suppose that initially both waves have a crest at this point. After a time corresponding to the period of the higher-frequency wave, this higher-frequency wave will have returned to a crest, but the other wave will not. This means that the amplitude of the superposition disturbance will be less than it was initially. After another cycle of the high-frequency wave, the two waves will be even further out of sync, and the amplitude of the superposition wave will have decreased further. This process will continue until the two waves are more-or-less completely out of phase and then the process will begin to reverse. Eventually the crests of the low frequency wave will again occur at the same time as the crests of the high frequency wave and the superposition wave will be back to maximum amplitude.

With time, the amplitude of the superposition wave decreases to a minimum and then increases again to its initial value. In the case of sound waves this is heard as a 'beat' in the loudness of the sound (this may be heard, for example, when tuning guitar strings). Beats are an oscillation in the amplitude of a superposition wave. In Figure 8.7, the amplitude of the superposition wave is shown as a solid line and we have drawn an envelope over this wave by connecting adjacent maxima with a dotted line. This envelope shows the beats in the maximum amplitude of the superposition wave.

As may be seen in Figure 8.7, the wave formed by the superposition of two waves with slightly different frequencies has two important frequencies. The first is the actual frequency of the resulting disturbance (the carrier wave), which is shown as a solid line in Figure 8.7. The second is the frequency with which the maximum amplitude of the superposition wave changes, the frequency of oscillation of the envelope shown as a dotted line. Some algebra and trigonometry reveals that the **beat frequency**, i.e., the frequency of the envelope is given by,

$$f_B = |f_1 - f_2| \qquad (8.6)$$

Here f_1 and f_2 are the frequencies in hertz of the two component waves. In the case of sound waves, the result is a pressure disturbance that the ear hears as the average frequency, changing 'loudness' at the beat frequency. The smaller the difference between the frequencies of the two superposing waves, the slower the beat frequency.

8.8 Reflection

Another important wave phenomenon is **reflection**. This is observed when water waves hit a wall of some sort, or light waves hit a mirrored surface. In this section we will briefly discuss two different ways in which waves reflect from barriers. These two kinds of reflection are determined by the kind of barrier that the wave encounters. To illustrate this, we will use the example of a pulse travelling along a piece of string and reflecting from a post to which the string is attached.

In Figure 8.8, the string is not able to move at the point of attachment. When this happens, the reflected wave is inverted and travels back along the string 'upside down'. In the case of a pulse travelling along a piece of string this is a π (radians) phase change. Adding π to the phase of the wave given in Eqs. (8.2) and (8.5) will change the sign of the operation, and the result will be a wave which is half a wave ahead of or behind the original wave and travelling in the opposite direction. Such a wave will be 'upside down' with respect to the original wave.

For a fixed attachment a pulse on a string will undergo a π phase change on reflection. The fixed attachment point forces the amplitude of the wave to be zero at the attachment point. The only way that the wave is able to achieve this is if the reflected wave at that point superposes with the incoming wave to give zero total amplitude. This means that the reflected wave has to be equal in magnitude but in the opposite direction at all times at the attachment point. This is achieved by shifting the phase of the reflected wave by π.

In the second case (Figure 8.9) the attachment point is not fixed and thus the phase of the reflected wave does not have to shift by π with respect to the incoming wave. This is modelled by a string attached to a post with a moveable ring. The reflected pulse travels back along the wave with the same orientation as the incoming wave. The incoming and reflected waves will superpose at the attachment point to give a wave which has twice the amplitude of the incoming wave on its own.

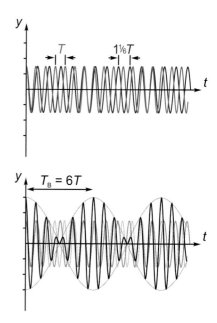

Figure 8.7 Oscillations in the amplitude of the superposition of two waves of period T and $1\frac{1}{6}T$. Note that these graphs are plotted as functions of time, not as functions of position.

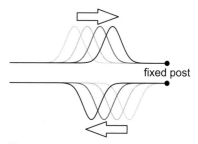

Figure 8.8 Reflection from a post with a fixed attachment point.

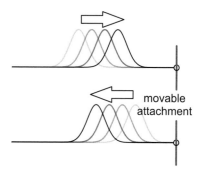

Figure 8.9 Reflection from a post with a moveable attachment point.

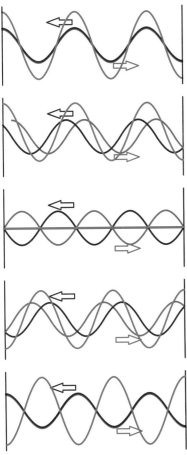

Figure 8.10 A standing wave formed by multiple reflections between two barriers.

8.9 Standing Waves

When we have two waves which have the same wavelength and speed (and so the same frequency), but are travelling in opposite directions through the same medium, we get an interesting consequence: a standing wave. As we saw in the previous section, a wave incoming to a barrier and the reflected wave superpose, with the reflected and incoming waves propagating in opposite directions, so reflection is a case where such counterpropagating waves are formed.

In Figure 8.10, the blue line represents a wave which propagates to the left, and red a wave heading to the right. The superposition of these two waves is represented by the green line. The green waveform is the standing wave which results from such counterpropagating waves. In each of the diagrams in Figure 8.10 we see the red and blue waves propagate in their respective direction and the construction of the green wave by the superposition of these two waves. The displacement of the resultant wave varies with time, but wavelength of the green superposition wave does not change, and neither does the location of the positions which have zero displacement at all times (called the nodes) and those which have the maximum movement. Note that the peaks of the green wave do not propagate to the left or right, they simply oscillate up and down. This is the reason the green superposition wave is known as a standing wave.

8.10 Waves and Energy

Energy

Waves have two extremely useful characteristics: they may be used to transmit energy from one place to another and they may be used to transmit information from one place to another. The second of these is apparent in everyday life: radio waves are used to transmit voice messages, text messages and data between cell phone and computers; light waves transmit information about the world about us into our eyes; sound waves transmit information to our ears, and so on.

The transmission of energy by waves is also experienced in everyday life. It is well known that very loud sounds can damage ears and even break the ear drum. Lasers are used to burn off tattoos and may be used to cauterise wounds. Wave energy may be even more destructive; earthquakes are an instance of the transmission of energy by waves, in this case seismic waves. Water waves may cause significant damage to coastal areas during storms, and tsunami are a particularly catastrophic example of the transmission of energy by wave motion. Wave energy has also recently begun to be harnessed for the production of electricity, in solar cells and ocean wave generators.

In an earlier section we saw that a wave is a sequence of simple harmonic oscillators. Each of these oscillators has the same amplitude and frequency. We have also seen that the potential energy stored in a mass–spring oscillator is given by

$$\mathrm{PE} = \frac{1}{2} k A^2$$

where A is the amplitude of the oscillation of the mass–spring system. When a wave transmits energy from one place to another it does so by causing the underlying medium (if there is one) to oscillate at some remote point. If the underlying medium was a sequence of mass–spring oscillators, this would mean that the energy arriving at the remote point would be proportional to the square of the amplitude of the incoming wave.

This is the case in general, not just for the case where the underlying medium is a sequence of mass–spring oscillators.

Key concept:
The energy transmitted by a wave is proportional to the square of the amplitude of that wave.

Power and Intensity

Two quantities related to the energy are the power (which we have encountered previously) and the intensity. Power is a measure of how much energy is transmitted per unit time, $P = \frac{W}{t}$. Intensity is a measure of how much power per unit area and is measured in watts per square metre.

$$I = \frac{P}{A}$$

(8.7)

where P is the power and A is the area over which it is spread.

Intensity is not a very useful measurable quantity in many of the situations we have previously discussed in Mechanics. However, when we are interested in the transmission of energy by waves, the wave motion is not likely to be confined to a single path through a material, but rather large areas of the medium will be oscillating, so how the energy is spread out may also be of interest. (Intensity may also be relevant when dealing with large numbers of moving objects, such as a beam of particles.)

The Sun is the most important source of energy in our natural environment on Earth. There are very few organisms that do not ultimately derive the energy that they need to live from the Sun. Energy produced by the nuclear fusion processes operating in the Sun is transmitted to the Earth as electromagnetic waves, i.e., 'sunlight'. On a clear day in summer the Sun provides about 1 kW m^{-2}. In other words, approximately 1000 joules of energy arrive on every square metre exposed to the Sun every second. In winter, the Sun is low in the sky and even on a clear day sunlight will deliver only about 0.3 kW m^{-2} to the same horizontal surface.

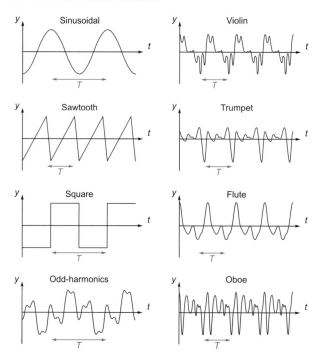

Figure 8.11 (Left) Examples of common regular periodic waveforms. (Right) Waveforms produced by a variety of different musical instruments. All of these waveforms represent the same note played on different instruments.

8.11 Complex Waveforms

The waves which we have so far encountered have all been sine waves, that is they are all waves which are well described by Eq. (8.5) or Eq. (8.2). Very few waves in nature are this simple. However all waves which repeat may be constructed from the pure sine waves we have discussed here. Figure 8.11 illustrates some of the more regular periodic waves which tend to occur most often in technological applications. Periodic waves are not necessarily this regular and some less-regular periodic waves are also shown in Figure 8.11.

Musical (and Vocal) Tone)

Musical tones are the result of the interaction of the human auditory system with periodic pressure waves produced by a wide variety of musical instruments. Several of the waveforms produced by common instruments are shown in Figure 8.11. These waveforms are significantly different but all represent the same note. The difference in the waveform results in the different sound produced by the trumpet and the violin (for example). Clearly then, a change in the shape of a periodic waveform can radically change the sound that is perceived by the human ear. An unintended change in the shape of a waveform can thus have significant consequences to the way an instrument or a piece of music sounds.

As an example, consider a pure 1 kHz sine wave produced by an electronic signal generator. Suppose this signal becomes distorted due to a fault in the amplifier used prior to sending this signal to a set of loudspeakers. This distorted signal sounds very different to the original pure 1 kHz sine wave. The reason for the change in the sound is that the new periodic wave is now the result of a number of higher frequency sine waves and thus we will hear not just a 1 kHz signal, but also some signals at 2 kHz, 3 kHz and higher. These signals are called harmonics and are all signals with frequencies that are multiples of the frequency (1 kHz) of the original signal. Distortion is thus undesirable, as it will result in poor reproduction of sound. Recorded speech can become almost incomprehensible with even moderate distortion.

8.12   Summary

Key Concepts

wavelength (λ) The length of one complete waveform, the distance from one peak to the next in a wavetrain. Measured in metres (m).

wave speed (v) The speed at which a crest or trough of a wave travels.

phase The argument of the cosine function in the mathematical expression for a wave.

transverse wave A wave in which the medium, in which the wave is propagating, oscillates in a direction which is perpendicular to the direction of propagation of the wave.

longitudinal wave A wave in which the medium, in which the wave is propagating, oscillates in a direction which is parallel to the direction of propagation of the wave.

superposition The addition of two waves; the displacements of the waves add like scalars at each point in space.

constructive interference The interference of two waves which have the same wavelength and frequency and which are in phase. Peaks and troughs of the two waves coincide everywhere so that the amplitude of the superposition is greater than that of each of the individual waves. If the two interfering waves have the same amplitude then the amplitude of the superposition wave is double that of the constituent waves.

destructive interference The interference of two waves which have the same wavelength and frequency and which are exactly out of phase. Peaks of one wave coincide with troughs of the other wave so that the amplitude of the superposition wave is less than the amplitude of either of the component waves. If the two interfering waves have the same amplitude then the superposition wave has zero amplitude.

beats When two waves superpose, but have different frequencies, the superposition wave has two distinct frequencies. The carrier frequency is the frequency with which the underlying medium oscillates, the beat frequency is the frequency at which the maximum amplitude of superposition wave oscillates.

Equations

$$v = f\lambda$$

$$y = A\cos\left(2\pi\left(\frac{x}{\lambda} \pm \frac{t}{T}\right)\right) = A\cos(kx \pm \omega t)$$

$$k = \frac{2\pi}{\lambda}$$

$$\omega = 2\pi f$$

$$f_B = |f_1 - f_2|$$

$$PE = \frac{1}{2}kA^2$$

$$I = \frac{P}{A}$$

8.13 Problems

8.1 A transverse wave propagates in a system of mass-springs as shown in Figure 8.2. The masses and springs are all identical: the masses are all 50 g weights and the springs all have the same spring constant. The separation of points P and Q (see Figure 8.2) is 0.5 m and the wave propagates through the mass-spring system with a velocity of 30 m s^{-1}.

    (a) What is the frequency of the oscillator at point P?

    (b) Suppose that the oscillator at point P has an amplitude of 25 cm. What is the spring constant of this oscillator?

    (c) Suppose that we halve the amplitude with which each of the masses is oscillating. How does the velocity of propagation of the wave change?

    (d) Suppose that we double the spring constant of each oscillator in the system (without changing anything else). How would this change the velocity of propagation of the wave?

    (e) Given the number of oscillators shown in Figure 8.2, what is the phase shift (in degrees) between two adjacent oscillators?

    (f) If this phase shift is halved (and the oscillator positions and frequency remain the same), what is the velocity of propagation of the wave?

8.2 A piano tuner wishes to use the beat frequency generated when two different notes are sounded together to tune one of the keys on a piano keyboard. She uses a tuning fork to tune the note named 'A' to 440 Hz. The next note higher than this should have a frequency of (approximately) 466 Hz.

    (a) If the speed of sound in air is 340 m s^{-1}, what is the wavelength of the note named 'A'?

    (b) When these two notes are played together how many beats per second will she hear when these two notes have the frequencies indicated?

8.3 A transverse wave propagates in the positive x-direction with a wavelength of 0.3 m and a period of 10^{-3} s.

    (a) Use Eq.(8.2) to write down an expression for this wave.

    (b) What are the frequency and angular frequency of this wave?

    (c) What is the propagation velocity of this wave?

8.4 A transverse wave propagates in the negative x-direction with a frequency of 20 Hz and a propagation velocity of 25 m s^{-1}.

    (a) What are the wavelength, period and angular frequency of this wave?

    (b) Use Eq.(8.2) to write down an expression for this wave.

8.5 A transverse wave is described by the expression

$$y = 0.05 \cos(4.19x - 1260t)$$

. You may assume all measurements are in the correct SI units.

  (a) What is the amplitude of this wave?

  (b) What is the wavelength of this wave?

  (c) What is the frequency of this wave?

  (d) How fast is this wave traveling?

  (e) What is the maximum transverse velocity of this wave?

8.6 A boat is bobbing up and down on the water as waves pass underneath it. The depth of the water under the boat oscillates between 3 m and 4 m. The boat is stationary with respect to the shore and it is 2.9 s between the crests of successive waves. A person on shore sees the crests of the wave passing by at 2 m s^{-1}. What is the distance between crests of this wave?

8.7 Two ducks are floating close together on the water near the boat in Problem 8.6. When the first duck is on the peak of a passing wave the second duck is 30 cm below it and moving upwards. How far apart are the ducks?

8.8 A violin is playing a note at 1200 Hz when a second violin starts playing. There is a distinct pulse in the resultant mix which repeats 16 times over the course of 5 seconds. What are the possible frequencies of the second violin?

SOUND AND HEARING

9.1   Introduction

If you dive into a swimming pool, one thing you will notice is a sudden decrease in how loud everything sounds. This is because the sound waves originating in the air do not get transmitted into the water all that well. The human body has a similar problem; we live surrounded by air, through which sound travels to reach us, but our internal biology is water-based, so the sound waves do not get efficiently transmitted into the fluid-filled inner ear without a huge loss of intensity which must somehow be compensated for. In this chapter we will investigate the properties of sound waves, how the human ear detects sounds, and give an overview of the function of the human vocal organs. Lastly, a brief overview of the Doppler effect is given. The uses of sound waves and Doppler techniques in diagnostic medicine will be covered alongside other medical imaging processes at the end of the book.

In discussing the transmission of sound waves we will use some concepts not yet covered in this text, such as pressure, strain, density and bulk modulus. These are covered in the next two chapters in Bulk Materials.

Key Objectives

- To understand the production and transmission of sound waves.

- To understand how the frequency and amplitude of a sound wave influence the pitch and loudness experienced by a listener.

- To get an overview of the anatomy and function of the human vocal organs and ear.

- To see how the motion of a sound source or observer changes the observed frequency.

9.2   Sound Waves in Media

In the previous chapter, we investigated waves and noted that a wave is a disturbance propagated through some medium. The term **sound waves** is generally used to refer to compression or strain waves transmitted through a medium that have sufficient intensity and are of a frequency that can be detected as sound. The term is also used more loosely to apply to similar waves outside this range, such as waves with a frequency too high (ultrasonic) and too low (infrasonic) for the human ear to hear. Sounds inside the range that the human ear can detect are known as *sonic*.

Pressure Waves in Gases

Imagine you have a tube full of air with a moveable piston at one end. If the piston is moved in, air molecules are displaced forward, causing the molecules to get bunched up. As the concentration of molecules is higher, the pressure is also higher, and this region of higher pressure will tend to move away from the piston as the molecules bump into one another. (Pressure will be covered in detail in a later chapter.) When

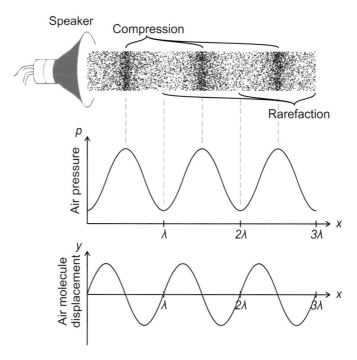

Figure 9.1 A speaker produces an acoustic wave in which the air molecules are variously compressed and rarefied. Also shown are two plots of the pressure, and displacement of the air molecules at each point.

the piston is moved back out, the molecules will get more spread out, leaving a region of lower pressure, which will also move outwards away from the piston. This is what a sound wave is: a moving disturbance of the molecules of the air. The areas where the molecules are closer together are called **compressions** and the area where the molecules become more sparse are called **rarefactions**.

The direction in which the air molecules are displaced is the same direction as that in which the wave pulse propagates, so the sound wave in air is **longitudinal**.

Sound waves in air can be described two ways: by looking at the displacement of the molecules at each position in the medium, or by looking at the pressure at each position in the medium. Figure 9.1 shows how these are related. The position of maximum compression is a position toward which gas molecules from either side have been pushed, so this is a position with the least displacement. Because of the phase difference between the two graphs describing the same wave, it can get somewhat confusing, so we will try to be quite clear about what is being plotted in our graphs.

Waves in Solids and Liquids

Sound waves can also be transmitted though liquids and solids by oscillation of the molecules of the medium. In solids, the waves can be either longitudinal (as compressive and tensile strain waves) or transverse (as shear strain waves).

Wave Speed

The speed of sound is not fixed and depends on the material through which it is travelling. In dry air at 20 °C, the speed is around 343 m s^{-1}. This varies with local ambient conditions such as temperature. When the temperature is higher, so is the speed of sound. For gases, the speed also has some dependence on molecular weight and composition, though it is not very dependent on pressure.

In general, the **speed of sound**, c_{sound}, can be predicted from the elastic properties of the medium, described by the bulk modulus, B, and the density, ρ:

$$c_{\text{sound}} = \sqrt{\frac{B}{\rho}}$$

(9.1)

Notation

It is common to use c for wave speeds rather than v, to distinguish the speed of moving objects emitting and receiving the sound waves from the speed of the sound waves. Most of the time, c stands for the speed of light in a vacuum, which has a constant value, but in this chapter it will be used to refer to the speed of any wave.

Substance	ρ (kg m^{-3})	c_{sound} (m s^{-1})	Z (kg m^{-2} s^{-1})
Air	1.204	343	413
Water	1.00×10^3	1440	1 440 000
Fat	0.92×10^3	1450	1 330 000

Table 9.1 Typical material values for sound speed and acoustic impedance at around 20 °C.

The speed of sound increases with increasing stiffness of the medium, and decreases with increasing density. This equation agrees well with the measured speed of sound in water (1482 m s^{-1} at 20 °C):

$$c_{sound, water} = \sqrt{\frac{2.2 \times 10^9 \text{ N m}^{-2}}{1000 \text{ kgm}^{-3}}} \approx 1480 \text{ m s}^{-1} \tag{9.2}$$

It is often the case in crystalline solids that the speed is dependent on direction through the material, so this adds complexity.

Example 9.1 *Speed of sound in air*

Problem: To put the speed of sound in air in context, calculate how many times a sound wave will travel across a 5 m by 5 m room during the average time it takes to utter one syllable.

Solution: In normal human speech, it is common to speak at a rate of 250 to 300 syllables per minute, so a single syllable averages about 60 s/300 syllables = 0.2 s per syllable. In this time, sound can travel

$$d = ct = 343 \text{ m/s} \times 0.2 \text{ s} \approx 70 \text{ m}$$

So the sound will have crossed the room about 14 times. This contributes to the richness of the voice indoors, and explains why we often have to speak louder outside.

Acoustic Impedance

A property of a medium which determines many of its acoustic properties is its **acoustic impedance**. This is given the symbol Z and is defined as

$$Z = \rho c_{sound} \tag{9.3}$$

where ρ is the density and c_{sound} is the wave speed. Some typical values are shown in Table 9.1.

The acoustic impedance is a crucial parameter in determining how waves will reflect and transmit at a boundary between media. If we call the acoustic impedances for the two media Z_1 and Z_2, such that the incident wave propagates in medium 1 and the transmitted wave in medium 2, then the proportions of the wave intensity reflected and transmitted depends on the ratio, $r = Z_1/Z_2$:

$$\text{proportion reflected} = \frac{(1-r)^2}{(1+r)^2} = \frac{(Z_1 - Z_2)^2}{(Z_1 + Z_2)^2} \tag{9.4}$$

The total intensity of the incoming wave is divided between reflection and transmission. Thus the transmitted intensity is just

$$\text{proportion transmitted} = 1 - \text{proportion reflected} \tag{9.5}$$

9.3   Pitch and Loudness

A pure tone is a sound wave which has an amplitude that varies sinusoidally. The two main qualities that we associate with a pure tone on hearing it are its *pitch* and its *loudness*. Here we will look at the wave properties that determine how we perceive the pitch and loudness of pure tones. We will discuss more complex sound waves later.

Frequency and Pitch

The **pitch**, how 'high' or 'low' it sounds, or where the sound lies on a musical scale, is determined by the wave frequency. The higher the frequency, the higher the pitch. They are not the same thing: frequency refers to an objectively measurable wave property, whereas pitch describes the subjective psychological impression. As for all waves, the frequency has units of cycles per unit time, and so the pitch of any pure tone can be specified in the usual SI units of Hz. For example, the musical note 'middle C', which is in roughly the middle of a piano keyboard, is usually tuned to about 262–4 Hz.

The range of frequencies audible to the human ear ranges from about 20 Hz to 20 kHz. The upper end of the range typically decreases with age, and may be as high as 40 kHz in children. The ear is most sensitive to frequencies between about 100 and 4000 Hz.

Some tones have a very distinct similarity. If you hit two keys on a piano that are both labelled as note 'C', they clearly have a common feel even though they are different in frequency. There is a simple mathematical relationship between the notes, though: dividing one frequency by the other will give an integer result. Two such notes that are separated by one *octave* have a 2:1 frequency ratio. A commonly-used system of tuning instruments in Western society is the *twelve-tone equal-tempered* system, where each octave is divided into 12 steps. Adjacent notes (those with a semitone step between them) have a frequency ratio of $2^{1/12}$. Other tuning systems are also in use and so that a particular note may not have the same frequency in different systems.

Some notes tend to sound pleasant when played together, and this property is called *consonance*. The opposite property, where the combination of notes is unpleasant, is called *dissonance*. Often, notes that have frequencies that form ratios with low numbers in the numerator and denominator ($\frac{3}{2}$, $\frac{5}{4}$, ...) sound nice together. An often mentioned theory of consonance is based on this idea, but it does not seem to hold true in every case. There are many other theories to explain which musical notes sound pleasing together, some based on the properties of the sound signal, others on the psychophysiological aspects of the human auditory system, and some on other factors such as learning and culture.

Hearing Range

The range of frequencies to which the human ear is most sensitive is similar to that of a seven-octave piano keyboard, which ranges from 27.5 Hz to 3 520 Hz.

Galileo on Frequency Ratios

'Those Pairs of Sounds shall be Consonances, and will be heard with Pleasure, which strike the *Tympanum* in some Order; which Order requires, in the first Place, that the percussions made in the same Time be commensurable in Number, that the Cartilage of the *Tympanum* or Drum may not be subject to a perpetual Torment of bending itself two different Ways, in Submission to the ever disagreeing Percussion.' From *Discorsi e dimostrazioni matematiche interno à due nuove scienze attenenti alla mecanica ed i movimenti locali*, 1638.[?]

> **Example 9.2 *Reflection and transmission of sound waves at a water/air boundary***
>
> **Problem: Calculate the proportion of a sound wave's energy transmitted at an air/water boundary.**
>
> **Solution:** $Z_1 = 413 \text{ kg m}^{-2}\text{s}^{-1}$ and $Z_2 = 1.44 \times 10^6 \text{ kg m}^{-2}\text{s}^{-1}$, so $r = 2.87 \times 10^{-4}$.
>
> $$\text{proportion transmitted} = 1 - \text{proportion reflected} = 1 - \frac{(1-r)^2}{(1+r)^2} \approx 0.001$$
>
> Only about one thousandth of the intensity of the sound wave in the air is transmitted into the water.

Amplitude and Intensity

The loudness of a sound is largely but not entirely determined by the amplitude of the pressure fluctuations, which is directly related to the amplitude of the molecular displacements. The amplitude of any wave is related to the wave's **intensity**, a measure of how much energy is transported through a unit area every second, measured in watts per square metre. In terms of the amplitude of the displacement of the molecules, A,

Sound source	Sound level (dB)
Jet aeroplane	140
Jackhammer	130
Threshold of pain	120
Busy traffic	80
Vacuum cleaner	70
Normal conversation	50
Whisper	30

Table 9.2 Approximate sound levels of various noises from loud to quiet.

the intensity is

$$I = \rho c A^2 (2\pi f)^2 / 2 \tag{9.6}$$

where ρ is the density, f is the frequency and c is the speed. An alternate expression for the intensity in terms of the amplitude of the pressure fluctuations p rather than displacement is

$$I = p^2 / (2Z) \tag{9.7}$$

where, as before, Z is the acoustic impedance.

Intensity, Loudness and the Decibel Scale

The human ear can detect sound waves that vary in intensity by a remarkable amount – 12 orders of magnitude. In terms of the amplitude of the pressure changes that can be detected, the smallest detectable amount is about 20 µPa, which corresponds to an intensity of 10^{-12} W m^{-2}. The sound of a nearby jet engine is over one million million times higher in intensity. The apparent loudness does not scale linearly with intensity, but instead a factor of 10 increase in intensity is perceived as an approximate doubling in volume, so a non-linear scale is more useful to describe the intensity of sounds instead of simply specifying the number of watts per square metre. Due to the logarithmic nature of the apparent loudness, the **decibel** is used to compare sound intensities. The decibel is not so much a unit as a way of specifying a ratio, and is used in other areas of physics, such as electronics and communications, for comparing voltages, intensities, powers and other quantities. For two intensities, I_1 and I_2, the *intensity ratio* in decibels (dB) is defined as:

$$\text{intensity ratio in dB} = 10 \log_{10} \frac{I_2}{I_1} \tag{9.8}$$

To give the **sound intensity level** in dB, we need to specify a reference intensity, which is taken to be 10^{-12} W m^{-2}. The sound intensity level in dB is therefore

$$L_I = 10 \log_{10} \frac{I}{10^{-12} \text{ W m}^{-2}} \tag{9.9}$$

We can similarly define the **sound pressure level** in dB. The reference pressure is 2×10^{-5} Pa, which corresponds approximately to an sound intensity of 10^{-12} W m^{-2}. The sound pressure level in dB is

$$L_p = 20 \log_{10} \frac{p}{2 \times 10^{-5} \text{ Pa}} \tag{9.10}$$

The 20 in the last equation is due to the relationship between pressure and intensity described in Eq. (9.7), which introduces an extra factor of 2, as $\log a^2 = 2 \log a$.

The human ear does not respond equally well to all sounds – the response is strongly dependent on frequency. We can plot on a graph lines that represent the number of dB a sound signal needs to be in order for a person to think they sound equally loud. The **phon** is a unit that is related to the psychophysically measured response of a typical human ear. At 1 kHz, the number of phons and the dB reading are by definition the same. At other frequencies, the number of dB required to appear to be equally loud will usually be larger. There is one place where the ear is more sensitive than at 1 kHz, due to the effect of resonance in the ear canal, which we will cover in the next section.

dB and phons

Sound level is a measurable quantity, while the loudness is something that is perceived. The level in dB is a physically measurable quantity. Scales relating this to perceived loudness were created by asking volunteers to adjust volumes until they thought they were equally loud.

dBA

The human ear does not respond equally to all frequencies, so some sound meters are fitted with filters to mimic the response of the human ear. If the 'A weighting filter' is used, then the sound pressure level is given in units of dBA.

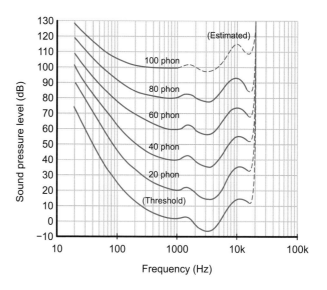

Figure 9.2 Equal loudness contours from the ISO 226:2003 revision.

9.4 Resonance and Sound Generation

To generate sound, the only requirement is that something moves sufficiently to create an air pressure wave of enough intensity. If you drop a rock on a hard surface, you will hear a sharp noise made up of many different frequencies at once, and this sound will quickly cease as the vibrations in the rock and ground are quickly damped. If instead you strike a metal rod, or a hollow plastic tube, the sound will tend to have distinct frequencies in the mix that are determined by the length, from constructive interference of reflected sound waves. Understanding how the dimensions of a rod or pipe determine the standing-wave frequencies allows us to see how we can alter the pitch of the sound produced by our vocal chords and nasopharyngeal cavity, and the increased sensitivity of our ears to certain frequencies from resonance in the ear canal.

Modes of Vibration of a String

A wave which is repeatedly reflected between two parallel reflecting boundaries will form a standing wave. We can achieve such a situation by sending waves along a string that is fixed at both ends. However, because the string has fixed places that must have zero vibrational amplitude, there are limits placed on the standing waves that can be formed, and so only wavelengths that have nodes at the fixed ends will keep bouncing back and forth. Figure 9.3 shows some of the standing-wave patterns that are possible in a string.

The longest wavelength pattern has a wavelength twice the length of the string. This is called the **fundamental mode of vibration** and the frequency is known as the **fundamental frequency**, or **first harmonic**. The next possibility has a wavelength equal to the string length and so has a frequency that is twice that of the fundamental. This is called either the **first overtone** or the **second harmonic**. The next possibility has a wavelength that is 2/3 of the length, which is 1/3 of the wavelength of the fundamental, and so it is three times its frequency. The overtones all have frequencies which are integer multiples of the fundamental.

$$f_n = n\frac{c}{2L} = nf_1, \; n = 1,2,3,\ldots \tag{9.11}$$

The speed at which a transverse oscillation travels along a wire will affect the frequencies with which a taut string will vibrate. A useful estimate of the speed of a wave

www.wiley.com/go/biological_physics

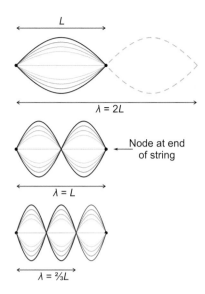

Figure 9.3 The first three modes of vibration for a string or wire.

along a wire such as a guitar string is

$$v = \sqrt{\frac{T}{\mu}}$$

(9.12)

where T is the tension in the wire and μ is the mass per unit length. This means that the wave speed is lowered in thicker wires and increased as the wire is placed under more tension. As $v = f\lambda$, a lower speed will correspond to a lowered frequency, and a reduced pitch. The thicker guitar strings sound lower in pitch, and tightening the tuning pegs raises the pitch.

> **Overtones and Harmonics**
>
> In addition to the numbering difference, harmonic and overtone are customarily used in slightly different ways: an overtone is resonance of a system with any frequency that is higher than the fundamental frequency, whereas a harmonic is a frequency of vibration of a system which is an integer multiple of the fundamental frequency.

Modes of Vibration of an Open Pipe

An open pipe acts in a very similar way to a string with both ends fixed. In this situation, the sound waves in the pipe reflect off the open ends and form a standing wave pattern like that shown in Figure 9.4. The constraint this time is not that the wave has a fixed zero displacement at each end, but that the pressure at each end must be that of outside the pipe, so the ends are nodes in terms of pressure. Other than that, the same maths applies, so tubes with two open ends have a fundamental frequency that is determined by the length, which is half the fundamental wavelength. The overtones all have frequencies which are integer multiples of the fundamental.

$$f_n = n\frac{c}{2L} = nf_1, \; n = 1, 2, 3, \ldots$$

(9.13)

The position of the effective pressure node is slightly outside the exact end of the pipe, and a correction factor is needed to explain the small differences between the frequencies observed in practice, and those predicted based purely on the tube length.

Modes of Vibration of a Half-Open Pipe

In the case of a pipe with one end open and one closed, the open end is a pressure node and the closed end is a displacement node. The maximum displacement will occur at a pressure node, so the displacement of the air will have a form like that shown in Figure 9.4. The minimum displacement will occur at a pressure antinode, again as shown in Figure 9.4.

The standing-wave pattern with the longest wavelength is has λ equal to four times the pipe length. For the next, the tube length is 3/4 of a wavelength, so λ is 4/3 the

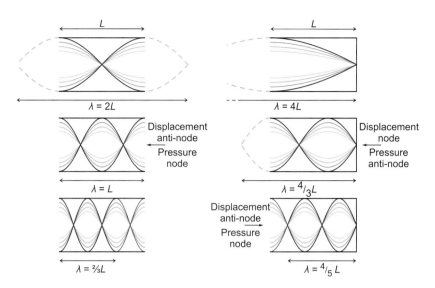

Figure 9.4 The first three modes of vibration for an open (left) and a half-open (right) pipe. The waveforms shown are for the *displacement* of the air molecules.

length. This will give us a factor of three change in frequency. The frequencies will be $f, 3f, 5f$ etc.

$$f_n = n\frac{c}{4L} = nf_1, \quad n = 1, 3, 5, \ldots \tag{9.14}$$

Example 9.3 *Ear canal*

Problem: The ear canal extending from the open outer ear to the closed surface of the eardrum is around 26 mm in length for a particular person. What are the frequencies of the first three resonant modes in this person's ear canal? Do these modes fall within the 20 Hz to 20 kHz range that a normal human can hear?

Solution: The ear canal is open at one end (the outer end) and closed at the eardrum and so the eardrum is a half-open pipe. The resonant frequencies for a half-open pipe are given by

$$f_n = n\frac{c}{4L}, \quad n = 1, 3, 5, \ldots$$

and so for the first three modes we will use $n = 1$, $n = 3$ and $n = 5$. The speed of sound in air is $c = 343 \text{ m s}^{-1}$ and the length used is $L = 0.026 \text{ m}$

$$f_1 = \frac{c}{4L} = 3300 \text{ Hz}$$

$$f_3 = \frac{3c}{4L} = 9900 \text{ Hz}$$

$$f_5 = \frac{5c}{4L} = 16\,500 \text{ Hz}$$

The frequencies of the first three resonant modes in this person's ear canal are 3.3 kHz, 9.9 kHz and 16.5 kHz and all of these are within the normal range of human hearing.

Complex Waveforms

In general, even for a simple case like a wave on the string fixed at both ends, the resulting wave will be a combination of the fundamental frequency and various amounts of the overtones: the wave will no longer be a pure sinusoid. One way of analysing

www.wiley.com/go/biological_physics

such complex waves is **Fourier analysis**. This is a mathematical technique for taking a time-dependent function like the pressure in a sound wave, and calculating the corresponding frequency-dependent function. It is like asking how much of frequency 'x' is in the wave signal, how much of frequency 'y' and so on. This is basically what the graphic equaliser display on a stereo shows – how much of the energy of the sound is in the low-frequency range, how much is in the middle, etc.

The Human Vocal Organs

Figure 9.5 shows some of the parts of the human mouth, nose and throat involved in sound production. The cavities of the mouth, nasal cavity and pharynx are known as the **vocal tract**, and changes in the shape of these areas will alter the resonant characteristics of the cavities and produce different sounds. It is the huge flexibility of the vocal tract that allows humans to generate such a range of sounds. Compared to other primates, the human larynx is much lower and we have a much longer and more flexible pharynx.

To create sound, an energy source is needed, and most of human speech and communication is carried out using air from the lungs while exhaling as the source. This air needs to cause vibrations somehow, and the most important way of producing these is in the larynx, where the **vocal folds** or **cords** are located. The vocal cords are two bands of muscular tissue that have variable and controllable dimensions, tension and elasticity, and the space between them is called the **glottis**. When making a voiced sound such as a vowel sound, or some consonants like m or b, the vibration of these cords can be felt by placing the fingers on the throat. In adult males, the average vibration frequency is 120 Hz, while for females it is 220 Hz.

The vocal organs can also be used in other ways to generate sounds used in speech. Clicking the tongue against the teeth, palate and cheeks is a key part of many languages (mostly southern African in origin). The sound generated when the vocal cords are opened and air moves through the glottis (much like the start of a cough) can be heard in many dialects. We can even make a cartoon character-style noise by compressing air in the cheeks.

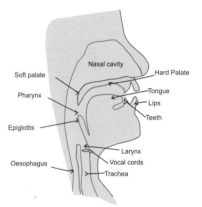

Figure 9.5 The human vocal organs.

9.5   The Ear

Anatomy

The ear, the human version of which is shown in Figure 9.6, has three distinct parts: the inner, middle and outer ear.

The outer ear consists of the **pinna**, that part of the ear that sits outside the skull, the **auditory canal**, and the **tympanic membrane**, also known as the eardrum. The sound waves are channelled and slightly modified by the pinna and canal, and these waves cause the tympanic membrane to vibrate.

In the middle ear, shown in more detail in Figure 9.7, are a series of small bones, suspended by ligaments, known collectively as the **ossicles**. These sit in a bony enclosure that is connected to the oral cavity by the **eustachian tube**. One of the ossicles is attached to the tympanic membrane: the **malleus**, more commonly called the *hammer*. This connects to the **incus** (*anvil*), and this in turn connects to the **stapes** (pronounced 'stay-peez' and often called the *stirrup*). The stapes is connected to the **oval window**, through which movement of the ossicles causes movement in the fluid inside the cochlea.

The inner ear contains the **cochlea**, a cavity encased in bone, and filled with a sea-water-like fluid. The cochlea is coiled up rather like a snail shell, but would have a length of 3.5 cm stretched out. In addition to the oval window, where the stapes is anchored, there is another membrane-covered entry in to the cochlea nearby (the round window), which moves in and out in response to the pressure fluctuations in the cochlea. The cochlea is internally divided into two halves by a membrane, which has a small hole in the farthest end. It is on this **basilar membrane** that the sound waves are turned into signals.

The speed of the pressure waves in the fluid-filled cochlea is much greater than that of air, and a quick calculation will confirm that the pressure is basically the same all over the basilar membrane at any time. However, the membrane acts like a series of oscillators coupled together, each of which only reacts to certain input frequencies. This will create a kind of travelling wave in the basilar membrane. Along the interior side of the basilar membrane runs a structure called the **organ of Corti**, to which are attached four rows of **hair cells**. From each of these, up to 100 **cilia** protrude, touching a membrane above. Movement in the basilar membrane beneath a cell will cause the hairs to bend and cause the release of a chemical neurotransmitter, resulting in electrical discharge through the neurons. These signals reach the brain along the **auditory nerve**.

Example 9.4 *Pressure and wavelength in the cochlea*

Problem: Calculate the wavelength of a 4 kHz sound wave in water and compare this to the length of the cochlea.

Solution: The speed of sound in water was given earlier as 1482 m s^{-1}. For $f = 4000$ Hz,

$$\lambda = \frac{c}{f} = \frac{1482 \text{ m s}^{-1}}{4000 \text{ Hz}} = 0.37 \text{ m}$$

The length of the cochlea is about 3.5 cm, so the wavelength is much greater than this, and the pressure is roughly the same over the whole of the basilar membrane at each instance.

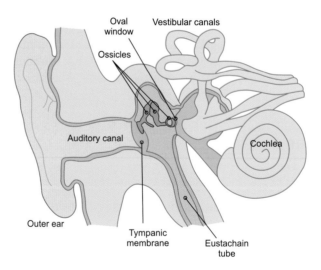

Figure 9.6 The anatomy of the human ear.

Effects of Resonance in the Ear Canal

The ear canal is shaped almost like a half-open pipe with a length of 2.5 cm. This gives a fundamental resonant frequency which is easily calculated:

$$\frac{\lambda}{4} = 25 \text{ mm}$$

$$\lambda = 0.1 \text{ m}$$

$$f = \frac{344 \text{ m s}^{-1}}{0.1 \text{ m}} \approx 3.4 \text{ kHz} \tag{9.15}$$

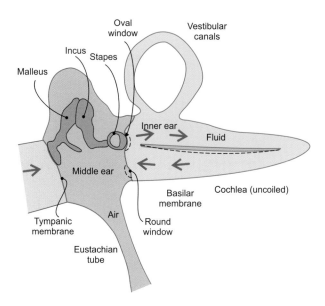

Figure 9.7 The (simplified) anatomy of the human ear. The cochlea is shown uncoiled (which is not possible in reality).

This is the reason for the curves showing the dB sound levels that have the same loudness to the ear have a dip around 3.4 kHz.

The Ear and the Problem of Impedance

As mentioned in the introduction, there is a problem getting the pressure waves in air, which are really quite small, into the body's liquid interior. There are several features of the ear's anatomy that help. Firstly, resonance in the ear canal and some focussing by the pinna helps. Secondly, the ossicles act like a system of levers to slightly amplify the vibrations (by a factor of around 1.3). However, the difference in size between the ear drum and the oval window has a more dramatic effect. The force applied to the ossicles is the product of the area of the ear drum and the pressure exerted on it. This force is amplified by the bones and then applied to the much smaller oval window, producing a larger force per unit area and hence a larger pressure. The ear drum is about 55 mm^2, whereas the area of the oval window is around 3.2 mm^2.

Example 9.5 *Amplification and loss in the ear*

Problem: In a previous example, the loss in intensity in sound waves being transmitted from air to water was found to be a factor of 1000. Calculate the loss in dB and compare this to the amplification of pressure waves in the ear.

Solution: An intensity drop to 1/1000 corresponds to

$$10\log_{10}\frac{1}{1000} = -30 \text{ dB}$$

This is a loss of 30 dB.

The ratio of the area of the tympanic membrane to the oval window is 55/3.2 = 17. If the force on the ossicles from the tympanic membrane was transmitted unclanged, this would give a factor of 17 increase in the force per unit area, i.e., the pressure. The pressure is actually further increased by the lever action of the ossicles to 1.3 × 17=22.1. This is an increase of

$$20\log_{10}22.1 = 26.9 \text{ dB}$$

This is close to the 30 dB loss we calculated.

The ear is able to reduce the transmission of the sound into the cochlea also, to reduce the risk of damage. The ossicles are suspended by ligaments that can reduce their movement. This produces a 0.6 dB decrease in apparent sound level for every

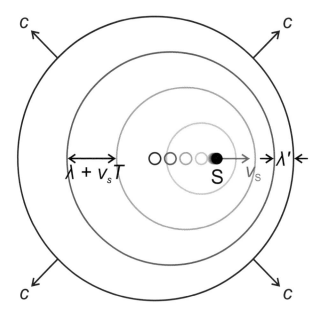

Figure 9.8 When the source is moving, successive wave peaks from the source are emitted at different locations. They all travel to the observer at the same speed, so because the wave peaks are squeezed closer together or spread further apart, this is observed as a change in wavelength and frequency, and hence pitch.

1 dB increase over 80 dB. However, there is a time lag of a few milliseconds, so this provides less protection against loud percussive noises. So, if you are exposed to loud, abrupt noise, this is more likely to be painful and damaging.

9.6   The Doppler Effect

The **Doppler effect** is a well-known and frequently observed phenomenon in which the apparent pitch of a sound is changed by the relative motion between the sound source and the observer. This is most frequently noticed in the modern world in traffic, where the pitch of an approaching car engine or siren appears to have a higher pitch when approaching and a reduced pitch when moving away.

The Doppler effect has proven to be a useful tool in many areas of scientific exploration. As the effect applies for electromagnetic waves as well as sound, the Doppler shift in the frequency of radiation from distant galaxies was used to show that most of them are moving away from us, indicating that the universe is expanding. The effect is also used in diagnostic medical procedures to measure flow velocities of fluids, and in meteorology to measure wind speeds and map air flows with Doppler radar, to name a few applications.

Moving Source, Fixed Observer

We can show how a source moving relative to an observer results in a change in pitch in Figure 9.8.

If the source, S, is moving at speed v_S towards the observer, each wave peak is emitted a distance $d = v_S T$ closer, where T is the wave period. This leads to a wavelength decrease of $v_S T$, from the original wavelength of $\lambda = cT$ for a sound wave travelling at speed c. The new frequency will be observed to be

$$f' = \frac{c}{\lambda'} \tag{9.16}$$

$$= \frac{c}{cT - v_S T} = \frac{c}{\frac{c}{f} - \frac{v_S}{f}}$$

$$= f \frac{c}{c - v_S} \tag{9.17}$$

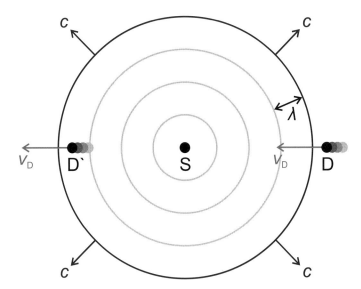

Figure 9.9 When the source is stationary, successive wave peaks from the source are all evenly spaced. The observer, D, moving towards the source encounters the peaks more frequently, though, due to their motion relative to the air.

The situation is similar for a sound source moving away from the observer, but the wavelength is increased, so the minus sign becomes a plus sign. To summarize

$$f' = f \frac{c}{c \pm v_S}$$

(9.18)

where a minus sign is used for an approaching sound source and a plus sign for a retreating source.

Fixed Source, Moving Observer

If a shift in frequency occurs when the source is moving relative to observer, it seems reasonable to expect a shift in frequency when the observer is moving and the source is not, and this is the case. However, the two cases are not completely symmetric, as the sound waves travel at a fixed rate through the still air, and so we need to derive a different formula. Figure 9.9 shows the effect of an observer moving towards a source that is emitting sound waves.

The situation is rather like being in a moving car, observing a string of evenly spaced cars moving at a fixed speed in the other lane. To the moving observer, the oncoming cars appear to move at a higher speed than if the observer's car was stopped. But the spacing of the cars is the same, so a new car is passed more often. For a moving observer, the time between successive peaks in the sound waves will be decreased when moving towards the source at speed v_D. The new period, T', will be shortened

$$T = \frac{\lambda}{c}$$

$$T' = \frac{\lambda}{c + v_D}$$

(9.19)

$$f' = \frac{1}{T'} = \frac{c + v_D}{\lambda}$$

$$= f \frac{c + v_D}{c}$$

(9.20)

Once again, if the observer is moving in the other direction (away from the source) the sign is reversed, so we can write the more general formula

$$f' = f \frac{c \pm v_D}{c}$$

(9.21)

where a plus sign indicates an approaching observer and a minus sign indicates retreating.

General Case

The two cases can be combined to provide a general formula that applies to any situation where the observer and/or source are moving:

$$f' = f\frac{c \pm v_{\mathrm{D}}}{c \pm v_{\mathrm{S}}}$$

(9.22)

Key concept:
To remember which sign to use in the general formula, remember that if the detector or source is moving towards the other, then the sign on its speed must give an increase in f', and if they are moving away, the corresponding speed has whichever sign would decrease f'.

9.7 Summary

Key Concepts

sound Mechanical waves transmitted as compression or strain waves through a medium which are the objective cause of hearing.

acoustic impedance A material property which is useful for describing acoustic properties.

pitch The apparent highness or lowness of a sound which is determined by its frequency.

loudness The magnitude of the auditory sensation produced by sound waves, which is determined in large part by the amplitude, but is also frequency dependent.

decibel (dB) The decibel is a logarithmic unit used to compare ratios, whether that be sound pressure, power, intensity etc.

sound intensity level (L_{I}) The sound intensity level is measured in decibels (dB) on a logarithmic scale against a reference intensity of 10^{-12} watts per square metre.

sound pressure level (L_{p}) The sound pressure level is measured in decibels (dB) on a logarithmic scale against a reference pressure of 20 μPa.

phon The phon is a unit of measurement that is related to the psychophysically measured response of a typical human ear. At 1 kHz, the number of phons and the dB reading are by definition the same. At other frequencies, a sound with the same loudness in phons will vary in dB as the ear responds differently to differing frequencies.

fundamental mode of vibration The mode of vibration of a resonating system with the lowest possible frequency.

overtone A resonance of a system with a frequency that is higher than the fundamental frequency.

harmonic A frequency of vibration of a system which is an integer multiple of the fundamental frequency.

Doppler effect The apparent shift in frequency (and hence pitch) of a sound when the source and observer are in relative motion.

Equations

$$c_{\text{sound}} = \sqrt{\frac{B}{\rho}}$$

$$Z = \rho c_{\text{sound}}$$

$$\text{intensity ratio in dB} = 10 \log_{10} \frac{I_2}{I_1}$$

$$L_I = 10 \log_{10} \frac{I}{10^{-12}\ \text{W}\,\text{m}^{-2}}$$

$$f_n = n\frac{c}{2L} = nf_1,\ n = 1, 2, 3, \ldots$$

$$f' = f\frac{c}{c \pm v_S}$$

$$c_{\text{sound}} = \sqrt{\frac{T}{\mu}}$$

$$I = \rho c A^2 (2\pi f)^2 / 2$$

$$L_p = 20 \log_{10} \frac{p}{2 \times 10^{-5}\ \text{Pa}}$$

$$f_n = n\frac{c}{4L} = nf_1,\ n = 1, 3, 5, \ldots$$

9.8   Problems

9.1 What is the speed of sound through ice? ($B_{ice} = 8.8 \times 10^9$ Pa, $\rho_{ice} = 920\,\text{kg}\,\text{m}^{-3}$)

9.2 A 0.500 m guitar string is placed under a tension of 270 N and the fundamental mode of vibration is at 150 Hz.

(a) What is the weight of the string per unit length (μ in $\text{kg}\,\text{m}^{-1}$)?

(b) What tension would need to applied to the string such that its fundamental mode of vibration is a middle C (440 Hz)?

9.3 A wave travels from air ($Z_{air} = 413\,\text{kg}\,\text{m}^{-2}\,\text{s}^{-1}$) into a liquid and with a density of $950\,\text{kg}\,\text{m}^{-3}$ and in which the speed of sound is $750\,\text{m}\,\text{s}^{-1}$.

(a) What proportion of the wave is reflected from the boundary?

(b) What proportion of the wave is transmitted through the boundary?

9.4 During an extremely loud sound the amplitude of the pressure in the sound wave in air is 1.0 kPa ($Z_{air} = 413\,\text{kg}\,\text{m}^{-2}\,\text{s}^{-1}$).

(a) What is the intensity of the sound (in $\text{W}\,\text{m}^{-2}$)?

(b) What is the intensity level of the sound (in dB)?

(c) What is the sound pressure level (in dB)?

9.5 A quiet whisper is measured at 30 dB and a loud shout at 110 dB. ($A_{ear\,canal} = 1.54 \times 10^{-4}\,\text{m}^2$, $Z_{air} = 413\,\text{kg}\,\text{m}^{-2}\,\text{s}^{-1}$)

(a) What power, in watts, is delivered to the opening of the ear canal during the whisper?

(b) How many times larger is the power (in watts) delivered to the ear canal by the shout?

(c) What is the pressure variation during the whisper?

(d) How many times larger is the pressure variation during the shout?

9.6 An oboe and a double bass are playing the same note, a 'G' at 392 Hz. The speed of sound in air is $343\,\text{m}\,\text{s}^{-1}$ and the speed of propagation of a wave on the string of the double bass is $500\,\text{m}\,\text{s}^{-1}$. The oboe can be modeled as a pipe, open at one end, and the bass as a string fixed at both ends. What are the lengths of the bass string and the oboe cavity if the note being produced is a result of the fundamental mode in each case?

9.7 After giving an intense performance, a confused and disoriented flautist has wandered onto the motorway! They are playing a constant 300 Hz tone on their flute and are essentially stationary. If you are driving along the motorway at $100\,\text{km}\,\text{h}^{-1}$ ($27.8\,\text{m}\,\text{s}^{-1}$), what is the frequency you hear from the flautist's instrument before you pass them, and after you pass them? ($c_{air} = 343\,\text{m}\,\text{s}^{-1}$.)

9.8 Coincidentally the horn on your car, which you sound as you narrowly miss the flautist in Problem 9.7, also gives a constant 300 Hz tone. What frequency does the flautist hear before and after you pass them?

II

Bulk Materials

In the Mechanics section, we saw how forces determine the motion of rigid objects. We assumed that the forces acting on the objects moved or rotated them as a whole, and did not deform them. In real-world cases, forces exerted on objects often cause the molecules in the object to move relative to each other. In this section we will look at how forces deform solids, and cause liquids and gases to flow. Unlike gases where intermolecular forces do not play a significant role, intermolecular forces are important in the behaviour of liquids and solids.

In solids, the molecules are tightly packed and bound in place by intermolecular bonds that do not allow the molecules to change position, though the molecules are not completely motionless – they are always jiggling about their fixed positions. However, the energy required to break one of the intermolecular bonds is greater than the thermal energy associated with the jiggling atoms. Because the molecules in a solid cannot readily move relative to each other, a solid has fixed volume, density and shape. Solids do not expand to fill a container, as a gas does. Solids also do not conform to the shape of the container as liquids do. We will look at what happens when forces deform solids in Chapter 10.

Molecules are also tightly packed in a liquid giving it a fixed volume and density. Unlike solids, there is significant movement possible between the molecules of a liquid, meaning that liquids do not have a fixed shape. The thermal kinetic energy of the molecules is enough to break intermolecular bonds, but not for a molecule to completely escape from its neighbouring molecules. This allows movement between molecules whilst maintaining a fixed volume. When a liquid is transferred between containers it maintains the same volume but conforms to the shape of the container. Some of the properties of liquids will be discussed in Chapters 11, 12 and 13.

When paired sideways forces (creating shear stress) are applied to a liquid, the molecules slide over each other and the liquid is said to flow. Liquids are defined by their continuous deformation under shear stress. Due to differences in intermolecular bonding, liquids vary in how easily the molecules can slide over each other. Liquids in which the molecules can slide over each other readily have low viscosity, and those where the molecules can only move past each other with more difficulty have high viscosity. A low-viscosity liquid will flow faster than a high-viscosity liquid when the same shear stress is applied to it. We will look into the movement of liquids in Chapters 14 and 15.

We will examine the behaviour of materials and systems from an energy and temperature perspective in the next topic on Thermodynamics.

ELASTICITY: STRESS AND STRAIN

10

10.1 Introduction

When studying mechanics, we were concerned with the overall motion of an object and how this was related to the forces applied to it. Often we examined only the motion of the object's centre of mass. In this chapter we will examine the effects of applied forces on the *shape* of an object.

In order to make meaningful statements about the strength of a material, we will need clear definitions of the various ways that the shape of an object may change, as materials may deform differently when pushed, pulled, twisted, flexed and struck sharply. The behaviour of a material under stress depends on the types and strengths of intermolecular bonds, and also depends a great deal on defects and impurities in the material.

We will begin by looking at the effects of different kinds of stressing forces: compressive, tensile and shear.

Key Objectives

- To be able to calculate tensile, compressive, shear and bulk stresses.

- To be able to relate stress and strain.

- To develop an understanding of elastic and plastic deformation, and how the behaviour of materials under load depends on their stress–strain curves.

10.2 Tension and Compression

Stress and Strain

> **Key concept:**
> **Stress** is a measure of the force per unit area applied to an object, and the size of the internal forces acting within the object as a reaction to the externally applied forces.
> **Strain** measures the change of shape of an object subject to a stress.

The stress tells us the force on imaginary internal surfaces within an object; the strain, how much it changes shape along some axis as a result. In this chapter we will carefully define stress and strain for several cases: tension, compression, shear and bulk.

Tensile Stress and Strain

When an object is being subjected to stretching forces so that its length will increase, it is said to be under **tensile stress**. This is achieved by applying forces to the opposite ends of an object, directed away from one another, as in Figure 10.1. A tensile stress stretches intermolecular bonds, and if it is sufficiently high it will break these intermolecular bonds, causing the material to rupture. The force exerted on each intermolecular bond acting to stretch it depends on both the total force applied to stretching the object, and what area (in other words how many intermolecular bonds) that force is

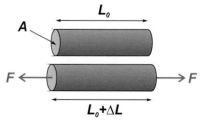

Figure 10.1 Tensile stress and strain. A 'stretching' force is applied to an object that causes an increase in the length of the object. This increase in length is usually small for rigid or 'solid' objects.

distributed over. This suggests that the force per unit area is important in determining how much an object will stretch. We define the tensile stress, σ, as the stretching force per unit area:

$$\text{Tensile stress} = \sigma = \frac{F}{A} \tag{10.1}$$

where F is the (equal) force applied to each end, and A is the cross-sectional area of the object at right angles to the direction of the stretching force.

A tensile stress applied to an object increases its length. The amount by which its length increases depends on several things:

- the tensile stress applied to it,

- the material it is made of (as, clearly, the strengths of the intermolecular bonds matter),

- the length of the object (because each segment of the object stretches in proportion to its length).

For a given material and a given tensile stress, the amount by which an object stretches is determined by its length. If we double its length, we double the amount by which it stretches. We define the **tensile strain**, ε, by the amount of stretch per unit length:

$$\text{Tensile strain} = \varepsilon = \frac{\Delta L}{L_0} \tag{10.2}$$

where ΔL is the amount by which the object is stretched and L_0 is the original length of the object. (This is the original length *in the direction of stretch.*)

For small tensile stresses, Robert Hooke (1635–1703) found that the change in length of an object was proportional to the applied force; we know this fact as **Hooke's law**, which states that the amount of stretch is proportional to the applied force:

$$F \propto \Delta L$$

where F is the applied force, and ΔL is the change in length. For a particular object, the cross-sectional area and original length are fixed, so this is like saying strain is proportional to stress – if we double the stress, we double the strain.

$$\frac{F}{A} \propto \frac{\Delta L}{L_0}$$

The proportionality constant, which tells us how much stress is required to generate a given strain in a material, is the **Young's modulus**, which we will represent with the Greek letter gamma, γ. As the strain is dimensionless, the Young's modulus and the stress both have units of force per unit area (N m^{-2} or Pa).

$$\gamma_{\text{tension}} = \frac{\text{tensile stress}}{\text{tensile strain}} = \frac{\frac{F}{A}}{\frac{\Delta L}{L_0}} \tag{10.3}$$

which is the same as saying

$$\text{stress} = \gamma \times \text{strain} \tag{10.4}$$

Young's modulus is a measure of a material's resistance to stretching. It does not depend on the size or shape of the object, but only on the material from which the object is formed. A bar or rod with a large Young's modulus will need a larger tensile stress applied to it to stretch it the same amount as a bar of the same length with a smaller Young's modulus.

Isotropic materials have the same Young's modulus in all directions so it does not matter in which direction we are stretching the object. Other materials are **anisotropic** and have different Young's moduli in different directions. This is usually due to an

www.wiley.com/go/biological_physics

asymmetry in the microstructure of the material. The Young's modulus of an anisotropic material depends on the direction of the stress relative to the crystal lattice. Bone is an example of an anisotropic material. In long bones, the bone is more resistant to stretching (and hence has a larger Young's modulus) in the longitudinal direction than in the transverse direction.

Compressive Stress and Strain

A **compressive stress** is one that tries to compress an object, that is, to reduce the length of the object. A compressive force is produced by applying forces directed towards one another on either end of an object. This pushes the molecules together, shortening the intermolecular bonds. The deformation of the object will be proportional to the amount by which these bonds are shortened, which in turn depends on the compressing force per intermolecular bond and hence on the compressing force per unit area. As with tension, it is the force per unit area, or compressive stress, that determines by how much a given material will be shortened.

The compressive stress is $\frac{F}{A}$ where F is the size of the force applied to each end of the object, and A is the cross-sectional area of the object at right angles to the direction of compression. A compressive force causes a **compressive strain** in the object, $\frac{\Delta L}{L_0}$, where ΔL is the amount by which the object has been shortened in the direction of compression, and L_0 is the original length of the object in the same direction. For small compressive forces, the compressive stress is proportional to the compressive strain.

As for tensile stress, we define a Young's modulus, γ, for a material under compression:

$$\gamma_{\text{compression}} = \frac{\text{compressive stress}}{\text{compressive strain}} = \frac{\frac{F}{A}}{\frac{\Delta L}{L_0}} \tag{10.5}$$

Many materials resist compression and stretching equally, so the ratio of compressive stress to compressive strain is the same as the ratio of tensile stress to tensile strain. In other words, for most materials the Young's modulus for tension and the Young's modulus for compression are the same. However, there are some notable exceptions, one of which is human bone, which is more resistant to tension than compression. A compressive stress will deform a bone more than the same size tensile stress.

Note that under compression the force is taken to be negative in value, and ΔL will be negative, so the stress and strain have negative values.

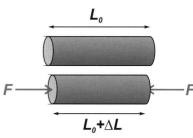

Figure 10.2 Compressive stress and strain. A 'squeezing' force is applied to an object, which causes a decrease in the length of the object. ΔL is negative.

Substance	γ (10^9 N m^{-2})
Aluminium	70
Copper	120
Bone, Tensile	16
Bone, Compressive	9
Iron, Wrought	190
Steel	200
Fused Quartz	70
Brick	20

Table 10.1 Some representative values for common materials. Individual samples may vary widely. [Data reprinted from *Physics For The Life Sciences*, A. H. Cromer, McGraw-Hill.]

Example 10.1 *Young's modulus (compression)*

Problem: A 45 kg child whose femur has a length of 0.3 m and a radius of 1.43 cm jumps off of a low wall and lands on one leg. The child is travelling at a speed of 9.0 m s^{-1} as they hit the ground. It takes 0.10 s for the child to be brought to a stop and during this period the femur is compressed by 0.18 mm. After the child has landed they continue to stand on one leg. How much is their femur compressed under these circumstances? (Note: you can assume the whole weight of the child acts on the femur.)

Solution: We can use the impulse required to stop the child upon landing to find the force on the femur during landing:

$$F_{\text{net}} = \frac{\Delta p}{\Delta t} = \frac{45 \text{ kg} \times 9 \text{ m s}^{-1}}{0.1 \text{ s}} = 4050 \text{ N}$$

This is the *net* force on the child. To find the force acting on the femur we need to take into account the 450 N weight force of the child as well. The force on the femur must be 4500 N. From this we can find the Young's modulus of the femur:

$$\gamma = \frac{\frac{F}{A}}{\frac{\Delta L}{L_0}} = \frac{\frac{4500 \text{ N}}{\pi \times (0.0143 \text{ m})^2}}{\frac{0.00018 \text{ m}}{0.3 \text{ m}}} = 11.7 \times 10^9 \text{ N m}^{-2}$$

Thus when the child is merely standing still ($F = mg = 450$ N), the compression of the femur will be:

$$\Delta L = L_0 \frac{\frac{F}{A}}{\gamma} = 0.3 \text{ m} \times \frac{\frac{450 \text{ N}}{\pi \times (0.0143 \text{ m})^2}}{11.7 \times 10^9 \text{ N m}^{-2}} = 0.018 \text{ mm}$$

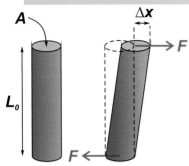

Figure 10.3 Shear stress and strain arise from a force that acts 'sideways'. Such a force would spin the object if it were allowed to move freely.

A Note on Symbols

A number of different symbols are in common use for the quantities we have covered in the last few pages, and because of that, and to avoid excessive use of confusing Greek letters, we'll try to write out the words most of the time. It is now most common to see Young's modulus given the symbol E, a convention we haven't used here to avoid confusion with energy. However, the symbol we have chosen, γ, is used in many places as the symbol for shear strain, so be cautious when browsing through other texts.

10.3   Shear Stress and Strain

A **shear stress** (often given the symbol τ, though we will tend to refer to it by name to avoid confusion) is applied to an object when we apply a sideways force parallel to a surface that is held fixed. Think of pushing sideways on the top surface of a thick book (but not enough to make it slide along the table). There is a force on the top of the book in one direction and an equal force on the bottom of the book in the opposite direction (supplied by static friction between the book and table).   The pages slide relative to each other deforming the book. The same thing happens between layers of molecules in a solid. As the shear forces are equal and opposite there is no net acceleration of the object as a whole but there is relative movement between layers of molecules in the material, resulting in a macroscopic deformation. The greater the force we apply, the greater the deformation.

The shear stress is again F/A, the force applied to the surface divided by the area of the surface that the force is applied to. Unlike tensile and compressive stresses, the area is *parallel* to the applied force.

We can quantify the degree of deformation by defining a **shear strain**:

$$\text{shear strain} = \frac{\Delta x}{L_0} \tag{10.6}$$

where Δx is the amount by which the top surface moves, and L_0 is the distance from the top surface, which moved a distance Δx, to the bottom surface which didn't move at all.

Some materials deform more readily than others under shear stress, and we quantify this with the **shear modulus**, G. For small stresses, the shear stress is proportional to the shear strain and we define the shear modulus of the material by

$$\text{shear modulus} = G = \frac{\frac{F}{A}}{\frac{\Delta x}{L_0}} \tag{10.7}$$

Materials with small shear moduli are easily deformed, whereas materials with large shear moduli are resistant to deformation. A common way of defining a fluid is 'a material that deforms continuously when subjected to a shear stress'.

Example 10.2 *Shear modulus*

Problem: A standard 40 cm plastic ruler is placed so that 30 cm of its 40 cm length protrudes from a desk. A 50 g weight is hung from the end and results in a displacement of 1 cm (as in Figure 10.4). Treat the ruler as a simple rectangular solid of dimensions 30 cm × 4 cm × 0.3 cm and assume that the portion of the ruler held against the desk cannot bend. What is the shear modulus of this ruler?

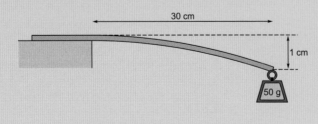

Figure 10.4 Shear displacement in a long, thin ruler.

Solution:
We can use Equation 10.7 to solve this question where $\Delta x = 0.01$ m, $L_0 = 0.3$ m, $F = 0.05$ kg $\times 10$ m s$^{-2} = 0.5$ N, and $A = 0.04$ m $\times 0.003$ m $= 1.2 \times 10^{-4}$ m^2.

$$G = \frac{\frac{F}{A}}{\frac{\Delta x}{L_0}} = \frac{\frac{0.5 \text{ N}}{1.2 \times 10^{-4} \text{ m}^2}}{\frac{0.01 \text{ m}}{0.3 \text{ m}}} = 1.25 \times 10^5 \text{ N m}^{-2}$$

10.4 Bulk Stress and Strain

If we apply a compressive force over all of an object's surface, then it will decrease in size in all directions simultaneously, causing a volume decrease. We use the term **bulk stress** to describe this situation. As the force isn't being applied in a single direction here, it makes more sense to refer to the force per unit area as the pressure increase. Pressure is force per unit area, and has units of newtons per square metre, also known as pascals (Pa). (The next chapter will cover pressure in detail.) Similarly, the change in length is not along a single axis, so instead of change in length and initial length, we need to concern ourselves with the change in volume as a fraction of the initial volume. Therefore the bulk stress depends on the pressure change, ΔP:

$$\text{bulk stress} = -\Delta P \qquad (10.8)$$

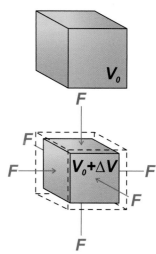

Figure 10.5 A compressive (or tensile) force acting in three dimensions can cause a change in volume of an object. In this example, a positive pressure change produces a negative bulk stress, and a negative volume strain, hence the decrease in volume (ΔV is negative).

The minus sign indicates that an increase in pressure on the surface is compressive, so must give a negative stress to be consistent with our earlier definitions. This produces a volume strain:

$$\text{volume strain} = \frac{\Delta V}{V_0} \qquad (10.9)$$

The *change* in volume, ΔV, is the final volume minus the initial volume (V_0), so this is negative when the object gets smaller.

As with the other moduli, the **bulk modulus**, B, is defined as the ratio of the bulk stress to the volume strain:

$$B = -\frac{\Delta P}{\frac{\Delta V}{V_0}} \qquad (10.10)$$

Substance	B (10^9 N m^{-2})
Ethanol	0.9
Water	2.2
Aluminium	70
Copper	120
Steel	158

Table 10.2 Some representative values for the bulk modulus of common materials. Individual samples may vary widely.

Some materials are easily compressed, and others change volume only a small amount when large additional pressures are applied to them. These differences are quantified by the bulk modulus. Materials with large bulk moduli are resistant to compression. Most liquids fall into this category. Materials with low bulk moduli, e.g., gases, are easily compressed, with a small additional pressure resulting in a large change in volume.

Example 10.3 *Bulk modulus*

Problem: At the start of an incredible journey you fill a water bottle with exactly 1.00 L of water. You start your journey from Dunedin and the atmospheric pressure at the time you filled the bottle was 102.1 kPa. [$B_{\text{water}} = 2.20 \times 10^9$ N m^{-2}.]

(a) The first stop on your journey is the top of Mount Everest. The atmospheric pressure at the top of Mount Everest is 35.1 kPa. By how much does the volume of water in your container *change*?

(b) The next leg of your amazing journey takes you down under the ocean in the Mariana Trench to a depth of 10.8 km. You notice that the volume of the water in your container is now 0.951 L. What is the pressure at this depth?

Solution: (a) To solve these problems we will need to use Eq. (10.10). The difference in pressure between the top of Mount Everest and Dunedin, where the bottle was filled, is $\Delta P = 35.1\text{ kPa} - 102.1\text{ kPa} = -67.0\text{ kPa}$. i.e. the pressure is lower at the top of mount Everest. This means that the volume of the water in the bottle should increase. Using Eq. (10.10) we can find the change in volume of the water.

$$B = \frac{-\Delta P}{\frac{\Delta V}{V_0}}$$

$$\Delta V = -\frac{\Delta P V_0}{B} = -\frac{-67.0 \times 10^3\text{ Pa} \times 1.00\text{ L}}{2.20 \times 10^9\text{ Pa}} = 3.05 \times 10^{-5}\text{ L}$$

This is a very small amount, and you would likely not notice the change at all.

(b) When taking your bottle to the bottom of the Mariana Trench it reduces in volume. This makes sense as the pressure at the bottom of the ocean is larger than that at the top, resulting in compression of the water ($\Delta V = 0.951\text{ L} - 1.00\text{ L} = -0.049\text{ L}$). Using Eqn. (10.10) again:

$$B = \frac{-\Delta P}{\frac{\Delta V}{V_0}}$$

$$\Delta P = -\frac{B \Delta V}{V_0} = -\frac{2.20 \times 10^9\text{ Pa} \times -0.049\text{ L}}{1.00\text{ L}} = 107.8 \times 10^6\text{ Pa}$$

This is a very large increase in pressure indeed! The question asks what the pressure is at the bottom of the ocean and so we still need to add our original pressure to this result ($P = P_0 + \Delta P$) but because ΔP is so large, after the addition our answer is only slightly revised. $P = 102.1 \times 10^3\text{ Pa} + 107.8 \times 10^6\text{ Pa} = 107.9 \times 10^6\text{ Pa}$.

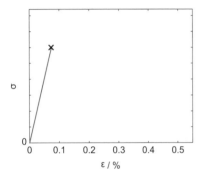

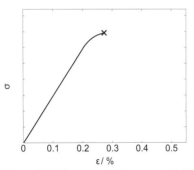

Figure 10.6 The stress–strain curves of brittle (top) and slightly ductile (bottom) materials.

10.5 Elasticity

Stress–Strain Curves

The ratio of stress to strain for real materials is constant only over a certain range, which depends on the type of material. It is often of interest to show where the linear region lies, and to convey other information (such as the stress that causes a material to break) on a **stress–strain plot**.

For many materials, the length of a rod-shaped sample changes in proportion to the stress. This region, where the stress/strain relationship is linear, is known as the **elastic region**. For this range, the material will return to its original shape once the stressing load is removed; any shape changes are *reversible*.

For *brittle* materials, such as ceramics, brittle metals, and materials that have been chilled sufficiently, the stress is directly proportional to the strain until the point at which the material fractures. Examples of brittle materials are glass and (non-reinforced) concrete under tension. The stress–strain curve for a brittle material is shown in Figure 10.6. A slightly *ductile* solid shows a linear stress–strain relationship for most of the range, but deforms more readily before fracturing.

A ductile metal is shown in Figure 10.7. This metal will not return to its original shape once it is stressed past a certain point, known as the *yield strength* or *elastic limit*. Some degree of deformation will remain after the stress is removed, and this is known as **plastic deformation**.

The decrease in stress before the point of rupture that the curve shows is due to the way we defined the stress – the plot shows engineering stress. In this region, the cross-sectional area is decreasing, so the true stress is still increasing, though the engineering stress is not. (A decrease in stress just after the yield point is also often a feature on such a stress–strain plot, and is a complicated effect of dislocations in the material's molecular structure.)

Some materials can tolerate large strains reversibly. Materials (such as rubber) that can be elastically stretched to twice their length or more are called *elastomers*. Over the range of interest, many of the tissues that make up our bodies have stress–strain curves that show behaviour similar to rubber. Ligaments, which connect bones to other bones, and tendons, which connect muscles to bones, have just such a J-shaped curve (see Figure 10.7). Their resistance to stretching is low at low stress loads, but increases with higher stress loads. This happens because initially the collagen fibres, which are a major component of ligaments, are not stretched straight, so small stresses easily straighten out the fibres giving significant elongation. Once the collagen fibres are straight, further stretching is more difficult. Other areas of soft tissue in the body that contain the protein elastin, such as the skin, arterial walls (particularly in the large arteries like the aorta), the bladder and the skin behave similarly.

Hair (based on the protein keratin) is an example of a material that has a rather S-shaped curve. For tensile strains of up to about 5%, hair behaves elastically. After this, the keratin molecules can unwind, and the hair can elongate by 25% with relative ease, and may stretch to as much as twice its length before breaking. Hair is very permeable and absorbs water readily, changing its properties and making it more fragile when wet.

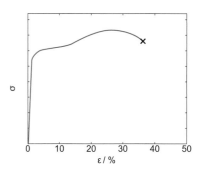

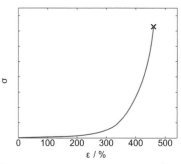

Figure 10.7 Stress–strain curves for a ductile metal (top) and an elastic substance (bottom), such as a ligament.

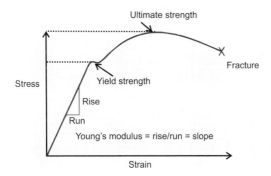

Figure 10.8 A general stress–strain curve showing the initial elastic region, which ends when the material reaches its yield strength, and the point of fracture.

The *stiffness* of a material, shown by the initial slope of the stress–strain curve, is not always the most important property for a particular use. Often *toughness*, which is basically the amount of energy that can be absorbed before the material fractures, is more important. The toughness can be estimated by looking at the area under the stress–strain curve. Another useful material property is the *resilience*, the ability to absorb energy during elastic deformation, which can be estimated from the area under the elastic region of the curve.

Change in Cross-Sectional Area

The elastic properties of solids result from the properties of their intermolecular bonding. Any form of deformation results from either the stretching or compressing of bonds within the material. Consider isotropic materials, where the bond strengths and the spatial organisation of bonds is not direction-dependent. In these materials, if you stretch the material by applying a tensile stress to it, this will lengthen all bonds in the same direction as the tensile stress, but it will also lengthen those bonds which have a component in the direction of the tensile stress. The angles of these bonds will change, resulting in compression of the material at right angles to the direction of tensile stress. Just think of how when you stretch something, it also tends to get thinner. The amount of compression at right angles to the tensile stress depends on the amount of tensile strain and on the nature of the intermolecular bonding. The ratio of the compressive strain to the tensile strain is known as **Poisson's ratio** and is in the range 0.25 to 0.35 for most materials.

10.6 Summary

Key Concepts

stress The force per unit area applied to a material. It is the stress that causes deformation of the material. It has units of N m^{-2} or Pa.

tensile stress A stress that causes molecular bonds to lengthen. The stretching force per unit area.

compressive stress A stress that causes the shortening of intermolecular bonds. The compressing force per unit area.

shear stress Stress that causes two parts of a material to slide across each other, caused by application of a force parallel to the plane dividing the parts.

bulk stress The force per unit area applied perpendicular to the surface of an object in three dimensions. It is equal in magnitude to the applied surface pressure, and is negative for a compressive force.

strain A measure of the deformation of the material, normalised to the size of the sample. For example, tensile strain is the fractional change in length $\frac{\Delta L}{L_0}$ of a stretched rod. Strain is dimensionless.

deformation Alteration of shape or form.

elastic deformation Deformation which is reversible when the stress is removed.

plastic deformation Non-reversible deformation.

Young's modulus (γ) A measure of a material's resistance to stretching or compressing. It has units of N m^{-2}.

shear modulus (G) A measure of a material's resistance to shearing stress. It has units of N m^{-2} or Pa.

bulk modulus (B) A measure of a material's resistance to bulk stress. It has units of N m^{-2} or Pa.

Equations

$$\text{stress} = \sigma = \frac{F}{A} \qquad \text{tensile or compressive strain} = \frac{\Delta L}{L_0}$$

$$\text{tensile or compressive stress} = \gamma \times \text{tensile or compressive strain}$$

$$\text{shear strain} = \frac{\Delta x}{L_0} \qquad \text{shear stress} = \text{shear modulus} \times \text{shear strain}$$

$$\Delta P = -B \frac{\Delta V}{V_0}$$

10.7 Problems

10.1 The elasticity of a cylindrical sample of an unknown material is to be tested. The sample is 40 cm long and has a cross sectional area of 2.5 cm^2. The sample is hung vertically and a 50 kg weight is attached to its free end. It is found that the sample stretches to a length of 40.1 cm.

(a) What is the tensile stress on the sample?
(b) What is the tensile strain?
(c) What is Young's modulus for this material?

10.2 (a) Bone has a tensile Young's modulus of $\gamma_{\text{tensile}} = 16 \times 10^9$ N m^{-2}. If the sample of unknown material in Problem 10.1 was replaced with an identical sample of bone, what would the length increase of the bone be?

(b) Bone has a compressive Young's modulus of $\gamma_{\text{compressive}} = 9 \times 10^9$ N m^{-2}. If the 50 kg weight was used to compress the sample rather than stretch it, how much would the length of the sample change?

10.3 Which of the following is most likely to be the Young's modulus of a rubber band? (Hint: What effect will a force of 1 N might have on a piece of rubber band 1 mm square. What is the strain found for such an applied stress for each value of the Young's modulus) (a) 5×10^{10} Pa, (b) 5×10^5 Pa, (c) 5×10^{-5} Pa.

10.4 A brand new type of rubber is discovered that can be manufactured from a combination of air and wishful thinking. A solid cylindrical rod made of this new type of rubber is fixed to the ceiling and a 1 kg weight is hung from the lower end. The rod was originally 20 cm long, and when it is hung it is 35.9 cm long. The radius of the rod is 0.37 cm. What is the Young's modulus of the rubber used to make the rod?

10.5 The type of rubber featured in Problem 10.4 is understandably very cheap to produce and as a result the manufacturers of all sorts of devices wish to incorporate it into their products. The makers of a car suspension system wish to see if a 0.3 m solid cylindrical length of this new rubber can be used in a car's suspension system. The rubber cylinder must be compressed by just 2 cm under a load of 5500 N. What radius must the cylinder be? (Is this practical?)

10.6 The unknown material in Problem 10.1 is now tested for resistance to shear. One end of the cylinder is clamped to a heavy table and a horizontal force of 300 N is applied to the free end. The top of the sample is found to deflect by 5 cm. What is the shear modulus of the sample?

10.7 The bulk modulus of water is 2.2×10^9 N m^{-2}. What increase in pressure needs to be applied to the surfaces of a cube of water 1 m on a side to reduce its volume by 1%?

10.8 Refer to Figure 10.9 to answer the following questions:

(a) As shown in Figure 10.9 a cylinder is completely filled with water and a force F applied to the airtight movable piston which forms one end of the cylinder. What force will need to be applied to the piston for the volume of the cylinder to *increase* by 1 cubic millimetre (which equals 10^{-9} m^3)? ($B_{\text{water}} = 2.2 \times 10^9$ Pa)

(b) If the cylinder was filled with air instead ($B_{\text{air}} = 1.42 \times 10^5$ Pa) what force would be required to increase the volume of air by 1 cubic millimetre?

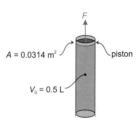

Figure 10.9 A force is applied to a piston attached to a cylinder filled with water. This causes the volume of the water to change.

PRESSURE

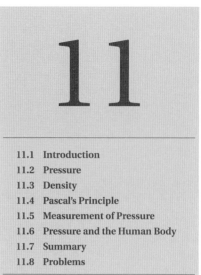

11.1 Introduction

How a force is applied is as important as the magnitude of the force itself – just think of how much more painful it would be to have your foot trodden on by a person wearing stiletto heels as opposed to flat-soled shoes. The same force (from the person's weight) is distributed over a smaller area, so the force per unit area, which we call the *pressure*, is greater.

The pressure exerted by fluids in biological systems is often of interest, with some examples being the pressure exerted on us by the atmosphere, or the pressure of our blood or of our cerebrospinal fluid. The pressure from the atmosphere varies with altitude, which is responsible for the way our ears 'pop' when we go up or down a hill, and has serious implications for the human body when at high altitude climbing mountains or flying in planes. The pressure increases rapidly with depth in the ocean, placing limits on how deep we can go without needing to encase ourselves inside the protective shell of a submarine. The measurement of blood pressure is routinely used in medical diagnosis, as it is both a symptom of, and a cause of, health problems.

In this chapter, we will investigate pressure, looking at how it is defined, how it is measured, and how it varies.

Key Objectives

- To be familiar with the concept of density.

- To understand that pressure in a liquid of uniform density is the same at all points at the same level.

- To understand that pressure at a point in a liquid is the same in all directions.

- To be able to determine the pressure at a given depth in a fluid of a given density.

- To be able to distinguish between, and convert between, absolute and gauge pressure.

- To understand how pressure is measured.

11.2 Pressure

Pressure is simply a measure of the force exerted per unit area (see Figure 11.1), that is

$$P = \frac{F}{A} \tag{11.1}$$

where P is the pressure, and F is the force applied normal to the area A. Pressure is a scalar quantity. The SI unit of pressure is the pascal, which has the symbol Pa. One pascal corresponds to a pressure of one newton of force per square metre. To put this number in perspective, normal atmospheric pressure is around 100 kPa at sea level.

Note that throughout this textbook we have chosen to use a capital P to represent pressure so that it is easier to distinguish between it and ρ, the symbol used for density. Many other books choose to use a lowercase p instead, reserving the capital to indicate power.

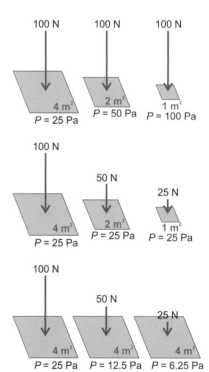

Figure 11.1 (Top) Applying the same force to half the area doubles the pressure. (Middle) Applying half the force to half the area results in the same pressure. (Bottom) Applying half the force to the same area results in half the pressure.

Introduction to Biological Physics for the Health and Life Sciences   Franklin, Muir, Scott, Wilcocks and Yates
©2010 John Wiley & Sons, Ltd

Solids

There are differences in how solids, liquids, and gases respond to and transmit pressure. A solid exerts pressure on whatever it rests on. This pressure depends on its weight and the surface area of contact. A solid doesn't exert a sideways pressure on something that it is sitting beside, but not leaning on or compressing. Consider a non-accelerating solid sitting on a surface; there must be a support force acting upon the solid object equal to its weight. The solid object exerts a force equal to its weight downwards on the surface it is sitting on, so for a non-accelerating solid, the downwards pressure is equal to the weight divided by the surface area of contact.

Gases

Things get more interesting with gases and liquids. Gases consist of large numbers of fast-moving molecules. These molecules collide with the walls of the container the gas is confined to. These collisions exert forces and hence pressure on the container walls. The pressure depends on the average magnitude of these collision forces and on the number of collisions happening per second. The average force per collision depends on how fast the molecules are moving on average (which is determined by the temperature), and the density and speed will determine how often collisions occur. In Chapter 18 we will see how the pressure of a gas depends on density and temperature. The pressure exerted by a gas is the same in all directions, because the gas molecules move randomly in all directions.

For a small volume of gas, the additional pressure at the bottom of the sample due to the weight of the sample above is negligible (as it is very small compared to the pressure due to random collisions), and we can often treat the pressure as being the same throughout the container. For large samples, like a section through the atmosphere, the change in pressure with elevation due to the additional weight of the gas is significant, and it is necessary to take elevation into account in determining the pressure.

Liquids

In a liquid, the molecules are also moving randomly (though more slowly than in a gas) and are able to slide past each other. The higher density of liquids means that the additional pressure at the bottom of a liquid sample due to the weight of liquid above that point is significant. The slow speed and high density of molecules in a liquid mean that the average forces involved in collisions are much smaller than in a gas. These two facts mean the dominant term in determining the pressure set up by a liquid is the weight of the liquid above a certain point rather than the random motion of the molecules. The pressure in a non-flowing liquid varies with elevation in the liquid, but is the same at all points at the same elevation.

A liquid exerts a downward force due to its weight in the same way a solid does, but unlike a solid, a liquid in a container will exert pressure sideways on the sides of the container, as shown in Figure 11.2. This happens because the weight of the liquid above a certain point tends to compress the fluid below it. As the molecules get squashed closer together, they resist being pushed closer and tend to move sideways unless there is a horizontal force preventing this.

Imagine a glass full of water placed upside down in a larger empty container. The weight of the water pushes down on the water molecules near the bottom. As the water molecules can slide past each other they would flow out sideways if it were not for the sideways force from the walls of the glass constraining them. If we now lift the glass slightly, there is no sideways force constraining the bottom water molecules and they flow out sideways until they come into contact with the walls of the larger container. This is why liquids flow and conform to the shape of their container.

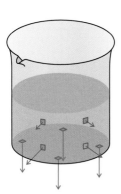

Figure 11.2 Fluid in a beaker exerts both sideways force on the vertical sides of the beaker and downwards force on the bottom of the beaker. The deeper the level of fluid, the larger the force exerted on adjacent small elements of the beaker. The force exerted on each small element is perpendicular to the surface of the beaker.

11.3 Density

When dealing with continuous media like liquids and gases, we don't have an object with a well-defined shape and composition, and hence mass. Of more use to us is the fluid **density**. The density, ρ, is defined as

$$\rho = \frac{m}{V} \qquad (11.2)$$

where m is the mass, and V is the volume that the mass occupies. The density depends on the composition of a liquid or solid, but not on how much of it there is. For example, the density of water at room temperature is about 1000 kg m^{-3}, whether we have a tiny droplet, or a full bathtub. Seawater has a different composition, and has a slightly higher density than fresh water.

Solids and liquids have distinct densities which depend only on the average lengths of the intermolecular bonds within the material, so for a specific solid or liquid at a specific temperature, there is a well-defined density. As the intermolecular bonds in gases are weak, gases expand to fill their container, so the density of a gas depends on the size of the container.

> **P and ρ**
>
> The symbol used for density, ρ, is the lowercase form of the letter rho from the Greek alphabet. It is also commonly used in electricity to denote resistivity. It is also often confused with p, especially when written by hand, so we have chosen to use uppercase P for pressure.

11.4 Pascal's Principle

> **Key concept:**
> **Pascal's principle** states that pressure applied to an enclosed fluid is transmitted undiminished to every part of the fluid as well as the walls of the container.

As a consequence, in a static (non-moving) liquid the pressure, P_h, at depth h depends on the pressure applied to the surface and how much liquid there is above that point exerting additional pressure:

$$P_h = P_{\text{surface}} + \rho g h \qquad (11.3)$$

where P_{surface} is the pressure exerted on the surface of the liquid, ρ is the density of the liquid, and g is the acceleration due to gravity.

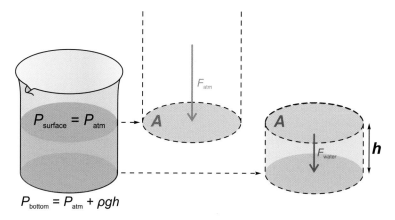

Figure 11.3 The pressure at the bottom of a beaker full of water can be found by using Eq. (11.3). This pressure has contributions from the weight of the water acting over the bottom surface of the beaker, and also from the weight of air above that, acting over the surface of the water.

To see why the pressure varies with the height of the liquid above a particular point, consider a horizontal plane of area A at depth h in the liquid, such as the bottom of the beaker illustrated in Figure 11.3. The force pushing down on this surface is equal to the force being exerted on the surface of the liquid (by the atmosphere, for example), plus the weight of the liquid above the plane. In terms of pressure this gives us

$$P_h = \frac{\text{weight}}{A} + P_{\text{surface}}$$
$$= \frac{mg}{A} + P_{\text{surface}}$$

We previously defined the density of a substance as the ratio of mass to volume, so the density of the water is $\rho = \frac{m}{V}$. The volume of the cylinder of liquid above the plane we are considering is the product of the area and the height of the liquid, $V = Ah$. So, substituting ρV for m, and Ah for V

$$= \frac{\rho V g}{A} + P_{\text{surface}}$$
$$= \frac{\rho A h g}{A} + P_{\text{surface}}$$
$$= \rho g h + P_{\text{surface}}$$

This also gives the relationship between the pressures at any two arbitrary points in the liquid

$$\Delta P = \rho g \Delta h \qquad (11.4)$$

where ΔP is the pressure difference and Δh is the difference in elevation (see Figure 11.4). The pressure is greater at the lower point.

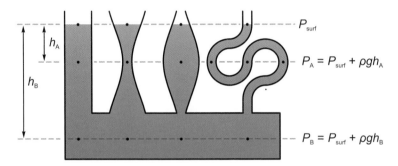

Figure 11.4 The pressure at any point in a stationary fluid is a function *only* of the height of the fluid above, and the pressure at the surface (Eq. (11.3)). The pressure does not depend on the shape or size of the container, and so all of those points at height h_A are at the same pressure. Applying Eq. (11.4), we can also see that $P_B = P_A + (h_B - h_A)\rho \, g$.

If we apply a force F_{applied} to the surface of the liquid, then the pressure at the liquid surface P_{surface} increases by $\Delta P = \frac{F_{\text{applied}}}{A}$, and the pressure at all other points in the liquid increases by the same amount because of Pascal's principle. This is the basis of **hydraulic systems,** used for a myriad of applications such as car lifts, dentists' chairs and hydraulic brake systems.

In a hydraulic system, like that in Figure 11.5, the increase in pressure in the fluid is a result of a force applied to piston X, $\Delta P_{\text{fluid}} = \frac{F_X}{A_X}$. This results in an increase in the force applied to piston Y by the fluid:

$$F_Y = \Delta P_{\text{fluid}} A_Y = \frac{A_Y}{A_X} F_X \qquad (11.5)$$

The ratio of the magnitude of the forces on pistons X and Y is the same as the ratio of the areas. Note that the pistons will move through different distances, as for an incompressible liquid the volume of liquid that is displaced by piston X is the same as the volume of liquid that displaces piston Y.

$$V = A_X d_X = A_Y d_Y \implies d_Y = \frac{A_X}{A_Y} d_X \qquad (11.6)$$

As work done equals force times distance, this is equivalent to saying that the work done by piston X is the work done on piston Y.

The pressure at a point in a liquid is *the same in all directions.* This means that a small, flat plate placed at a particular height within a liquid will experience the same force per unit area regardless of its orientation. The force is always exerted perpendicular to the area. We can see that this must be the case by looking at a tiny wedge-shaped

www.wiley.com/go/biological_physics

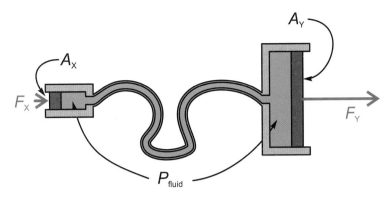

Figure 11.5 A very simple hydraulic system. A small force applied to piston X causes an increase in pressure in the hydraulic fluid. This causes a large force to act on piston Y.

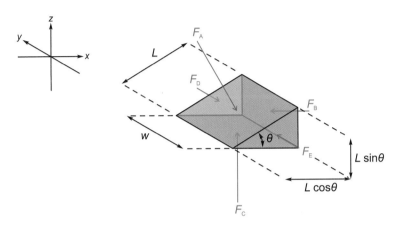

Figure 11.6 The net force on a small wedge-shaped segment of a fluid is zero. It is easy to see that the two forces F_D and F_E acting on the ends of the shape cancel each other out. Figure (11.7) shows that the other three forces, F_A, F_B and F_C also cancel out.

section of the liquid. Choosing a wedge sufficiently small, we can make the assumption that the pressure difference between the top and bottom of the wedge approaches zero. Figure 11.6 illustrates our small wedge of fluid. The net force acting on our wedge must be zero if the fluid is in equilibrium.

Consider the vertical components of the forces. The upward force on the bottom surface is the pressure times the area

$$F_{\text{up}} = F_C = P_C A_C$$
$$= P_C wL\cos\theta$$

The force on the upper, tilted surface is

$$F_A = P_A A_A$$
$$= P_A wL$$

The downward component of this force is

$$F_{\text{down}} = F_A \cos\theta$$
$$= P_A wL\cos\theta$$

The vertical forces are balanced, so $P_C = P_A$. Clearly, by symmetry, the same arguments work for the horizontal forces and so $P_B = P_A$, and this extends to all directions. In order for the wedge of liquid to be in equilibrium (that is, with zero net force acting on it), the pressure must be the same in all directions.

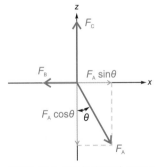

Figure 11.7 The horizontal component of the large force, F_A, acting on the largest surface of the wedge is equal and opposite to the horizontal force, F_B, acting on the vertical surface of the wedge. Similarly, the vertical component of F_A is equal and opposite to the vertical force, F_C, acting on the horizontal surface of the wedge.

Example 11.1 **Pressure and force**

Problem: Given that at sea level air is usually at a pressure of around 100 kPa, what force is the atmosphere exerting on any given 1 cm^2 of our body?

Solution: The relationship between force and pressure is given by Eq. (11.1):

$$P = \frac{F}{A}$$
$$F = PA$$

We can use this to find out what force is required to produce a given pressure over a given area. To produce a pressure of 100×10^3 Pa over an area of 1 cm$^2 = 1 \times 10^{-4}$ m^2 a force of $F = 100 \times 10^3$ Pa $\times 1 \times 10^{-4}$ m$^2 = 10$ N is required. This is equivalent to the weight force of a 1 kg object. This is quite a large a force for such a small area, if we were to look at 1 m^2 we would find the force is equivalent to a 10000 kg weight force!

If these forces are so large why are we not crushed? The answer is because our internal pressure is much the same as the external atmospheric pressure. Our insides push out with the same force as the atmosphere pushes in.

11.5 Measurement of Pressure

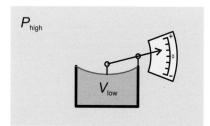

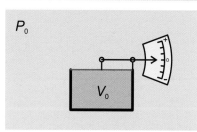

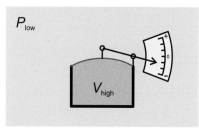

Figure 11.8 A simple mechanical diaphragm pressure meter. A *sealed* container holds some fixed amount of a gas. The container has at least one flexible surface, the diaphragm, which means that the pressure inside is always the same as that outside the container. As the pressure changes the volume of the gas inside the sealed container also changes (see the section on the Ideal Gas Law in Chapter 18 for an explanation of this) and the diaphragm bends. The degree and direction of the bending of the diaphragm indicate changes in pressure.

Pressure has been defined as force per unit area. It follows that, conceptually at least, the simplest way to measure a pressure in a fluid would be to measure the force that the fluid applies to a known area, using Eq. (11.1) to calculate the pressure. In practice using such an appealingly straightforward method is somewhat complicated by the difficulty of accurately measuring the force acting on any given surface. Any such measurement usually has to be indirect, as any surface has two sides, each of which is generally exposed to a fluid of some kind. (It is extremely hard to produce and maintain a vacuum under most conditions.)

Any property which changes with pressure can, in principle, be used to measure it, and there are a great many kinds of devices in use today that rely on a wide variety of pressure-sensitive phenomena to do so. The cheapest, and hence most common, rely on measuring the expansion/contraction of a fixed *mass* (not volume) of gas in a sealed compartment (see Figure 11.8) and are called diaphragm pressure meters.

The Manometer

A **manometer** is a particular type of pressure measurement apparatus that historically has seen wide use as a simple, reliable and potentially very accurate way to measure pressure and pressure differences. This role has only recently been taken over by other kinds of sensors as materials technology and fabrication techniques have improved. A manometer is a *hydrostatic* device, which means that it relies on the physics of stationary fluids.

A manometer relies on Pascal's principle (Eq. (11.3)) which states that, for a stationary fluid, the pressure at some depth below the surface can be found in terms of the pressure at the surface, the density of the fluid and the depth. The manometer itself is a U-shaped tube filled with a fluid, as in Figure 11.9.

The pressure at the surface of the fluid in each arm of the manometer is the same as the pressure of the gas above it. Because any two points at the same level in a stationary fluid must be at the same pressure, we can use Pascal's principle to express the difference in pressure in one arm of the manometer relative to the pressure in the other.

Using Eq. (11.3), the pressure at point 1 is equal to P_A plus a contribution due to the height, h, of fluid above

$$P_1 = P_A + \rho g h$$

Because point 2 is on the surface of the fluid, the pressure at point 2 is the same as P_B

$$P_2 = P_B$$

Point 1 and point 2 are at the same height in a continuous fluid, and so must be at the same pressure.

$$P_1 = P_2$$
$$P_A + \rho g h = P_B$$

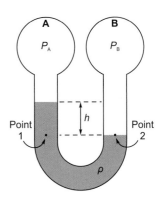

Figure 11.9 A very simple manometer. Two bulbs, A and B, each contain a gas at different pressures. They are connected by a U-shaped tube, which is partially filled with a fluid of density ρ. P_B must be higher than P_A, as the level of fluid in the right arm of the manometer is lower. The difference in pressure can be found from Eq. (11.4).

This is essentially a restatement of Eq. (11.3). If the pressure in one arm of the manometer is known, then the pressure in the other can be found. The difference in pressure between the two arms of the manometer can be calculated:

$$\Delta P = P_B - P_A = \rho g h \tag{11.7}$$

In most cases the two arms of the manometer are not connected to two closed bulbs as in Figure 11.9, but instead one arm is connected to a system in which the pressure is to be measured, and the other is connected to a system in which the pressure is known, most commonly the open atmosphere (Figure 11.10).

The Barometer

We inhabit a world at the bottom of an ocean of air, and the combined weight of all of this air acting upon surfaces at ground level creates the ubiquitous phenomenon we call **atmospheric pressure**. An Italian scientist, Evangelista Torricelli (1608–1647), is credited with making this discovery while helping silver miners who were trying to keep water out of their mines. The suction pumps they were using were only capable of raising the water about 10 m. This is because the weight of the atmosphere only produces sufficient pressure to support a column of water of this height.

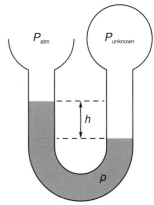

Figure 11.10 A manometer used to measure an unknown pressure with reference to atmospheric pressure.

Example 11.2 *Manometer*

Problem: A manometer connecting two closed chambers is filled with two different fluids (which are prevented from mixing by a thin membrane), water and an unknown fluid, as shown in Figure 11.11. What is the density of the unknown fluid? [$\rho_{water} = 1000$ kg m^{-3}.]

Solution:

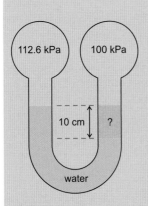

Figure 11.11 A closed manometer.

The pressure at the interface of the unknown fluid and the water in the right of the manometer is the same as the pressure at the same depth (10 cm) in the water at the left of the manometer, the pressure in each arm is given by $P = P_0 + \rho g h$.

$$P_{left} + \rho_{water} g h = P_{right} + \rho_{unknown} g h$$
$$\rho_{unknown} = \frac{P_{left} - P_{right}}{g h} + \rho_{water}$$
$$= 13\,600 \text{ kg m}^{-3}$$

The unknown substance is most likely mercury.

Example 11.3 *Barometer*

Problem: A mercury barometer ρ_{Hg} = 13 600 kg m^{-3}) shows a pressure of 767 mmHg at 5 pm and a pressure of 759 mmHg at 7:30 pm. What is the rate of change of the pressure in Pa min^{-1}?

Solution: A pressure of 760 mmHg is equivalent to the standard atmospheric pressure of 101.3 kPa. Thus a change of −8 mmHg (note the negative sign) is the same as $\frac{-8\,\text{mmHg}}{760\,\text{mmHg}} \times 101.3 \times 10^3$ Pa = −1066 Pa. This is a rate of change of $\frac{-1066\,\text{Pa}}{150\,\text{min}} = -7.11$ Pa min^{-1}.

An alternative way of converting to Pa is to use $\rho g h$

$$\rho g h = -8\,\text{mmHg} = -13600\,\text{kg m}^{-3} \times 9.8\,\text{m s}^{-2} \times 8 \times 10^{-3}\,\text{m} = 1066\,\text{Pa}$$

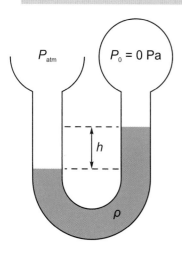

Figure 11.12 The right bulb of this manometer is evacuated and so the pressure difference between the arms is the same as the current atmospheric pressure. This pressure difference, and hence the atmospheric pressure, can be found using the difference in height of the fluid and its density. See Eq. (11.8)

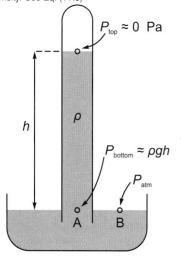

Figure 11.13 A typical barometer. The two points A and B are at the same height in a stationary continuous fluid and so must also be at the same pressure. The pressure at the bottom of the column of fluid inside the tube is entirely due to the weight of the fluid and given by $\rho g h$.

The air pressure varies with altitude and weather patterns, so the value of atmospheric pressure is not constant with time or location. When the exact value is not known, or high accuracy is not needed, we use the *mean atmospheric pressure at sea level*. In SI units, this is 1.013×10^5 Pa (101.3 kPa).

The atmospheric pressure at a particular geographic location varies with altitude. Even driving up quite a small hill quickly enough can cause the ears to 'pop' as the internal pressure in the auditory canal equalises with the lowered external atmospheric pressure. Atmospheric pressure at any particular altitude also varies by a few percent (several kPa) over time, as it is influenced by changing temperature, local wind-speed, and the large-scale movement of air, among other things.

A **barometer** is a pressure-measuring device used to measure local atmospheric pressure. A specialised manometer is one such device. In this kind of barometer, one arm of the manometer is evacuated and used as a known pressure reference (i.e. 0 Pa, see Figure 11.12). The difference in height between the level of the fluid in each arm of this manometer is now directly related to the current atmospheric pressure.

$$P_{atm} = P_0 + \rho g h = \rho g h \tag{11.8}$$

A barometer is not usually constructed with a U-tube, as shown in Figure 11.10, but instead consists of a straight tube filled with some fluid that is inverted into a reservoir, such that the closed top of the tube remains evacuated and the open bottom end lies below the level of fluid in the reservoir (See Figure 11.13). The space that forms at the top of tube has no air in it, and so is at nearly zero absolute pressure. The pressure at the surface of the reservoir is equal to the local atmospheric pressure. As the pressure in a liquid of uniform density is equal at all points at the same elevation, the pressure in the column of fluid at the same height as the fluid surface is also equal to atmospheric pressure.

In principle, *any* fluid could be used inside the barometer, but in practice mercury is most often used, because of its very high density. If we were to construct a barometer using a tube filled with water (ρ_{water} = 1000 kg m^{-3}), then at normal atmospheric pressures we would need a tube that could contain the resulting 10 m column of water. In contrast, the height of the column of mercury in a mercury barometer is only 0.76 m tall at 101 kPa, because mercury has a density of 13 600 kg m^{-3}.

In addition to its greater density, mercury also has a very low vapour pressure, which means that the pressure in the evacuated region above the mercury is closer to zero than if another liquid were used. For example, if a vacuum pump is used to remove the air above a column of water, the water will actually boil at room temperature, and the resulting pressure will be limited by the vapour pressure of water at the room's temperature (see Section 19.2).

Gauge Pressure

Most pressure gauges work by measuring a difference in pressure between two points or two systems as in Figure 11.14. Commonly, the reference pressure is the local atmospheric pressure, and the pressure gauge measures the value of the pressure at the place of interest in terms of how much higher or lower it is than the atmosphere. The reading on a gauge of this type, when the reference point is atmospheric pressure, is known as the **gauge pressure**.

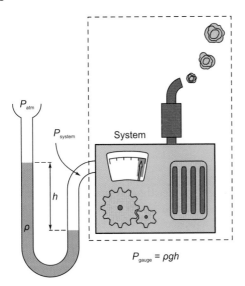

Figure 11.14 A manometer being used to measure the pressure in some system relative to atmospheric pressure. The *gauge* pressure is $P_{system} - P_{atm} = \rho\, g\, h$.

A gauge pressure can be either positive or negative. A positive gauge pressure indicates that the pressure being measured is higher than the reference pressure. A negative gauge pressure indicates that the pressure being measured is lower than the reference pressure.

Absolute Pressure

The **absolute pressure** is the pressure measured on a scale that has a perfect vacuum, which will be zero absolute pressure, as its reference point. The relationship between the two scales, gauge and absolute, is simple:

$$\text{absolute pressure} = \text{gauge pressure} + \text{absolute atmospheric pressure} \qquad (11.9)$$

(Remember that when you are using this for calculations, you'll need to check that you are using the same units for all the pressure values, as there are so many in use. It is only possible to add numbers that are in the same units. See the section on units below.)

While a gauge pressure can be either positive or negative, an absolute pressure must always be positive. Remember that pressure is a scalar quantity. An absolute pressure of zero indicates that the magnitude of the force acting on a given area is also zero. Since there can never be a force with a negative *magnitude*, there cannot be a negative absolute pressure.

The relationship between absolute pressure and gauge pressure is similar to the relationship between absolute temperature, in kelvin, and temperature measured in degrees Celsius. A temperature of 25°C means that the temperature is 25 degrees *above the freezing point of water* and corresponds to an absolute temperature of 298 kelvin. Similarly, a gauge pressure of 1000 Pa corresponds to an absolute pressure atmospheric pressure plus 1000 Pa (so something in the vicinity of 102 kPa at sea level).

Units

Pressure is the force per unit area, and as there are many non-SI units that are still in use for length, area and force, many non-SI units are still widely used for pressure also. The SI unit of pressure is the pascal, which has the symbol Pa. One pascal is equal to one newton per square metre. In SI units, the sea-level atmospheric pressure on Earth is close to 1×10^5 Pa, so writing the pressure in pascals can have 'readability' problems, and the prefix *hecto*, meaning 10^2, is sometimes used (particularly on weather maps). A related unit is the *bar*. One bar is the same as 100 kPa, so a millibar (mbar) is the same as a hectopascal.

$$1 \times 10^5 \, \text{Pa} = 1000 \, \text{hPa} = 100 \, \text{kPa} = 0.1 \, \text{MPa} = 1 \, \text{bar}$$
$$100 \, \text{Pa} = 1 \, \text{hPa} = 0.1 \, \text{kPa} = 1 \, \text{mbar}$$

Example 11.4 *Absolute pressure and gauge pressure*

Problem: What is the gauge pressure at points A, B and C in Figure 11.15 if $\rho_{\text{water}} = 1000$ kg m^{-3}, $\rho_{\text{oil}} = 1500$ kg m^{-3}, and $P_{\text{atm}} = 101.3$ kPa?

Solution:

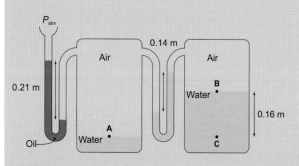

Figure 11.15 Two chambers containing different amounts of water are connected via a water-filled tube which acts as a manometer. Chamber A has a mercury manometer attached to it. The pressure at the three points A, B and C can be found from the information given.

P_A is higher than atmospheric pressure as the level of oil in the left manometer is lower on the chamber side. The absolute pressure at A can be found using the equation

$$P_A = P_{\text{atm}} + \rho_{\text{oil}} g h_{\text{oil}}$$

but we are asked for the gauge pressure, which is the difference between the absolute pressure and atmospheric pressure, or simply $\rho_{\text{oil}} g \Delta h_{\text{oil}}$. This gives a gauge pressure at A of

$$P_{\text{A,gauge}} = 1500 \text{ kg m}^{-3} \times 10 \text{ m s}^{-2} \times 0.21 \text{ m} = 3150 \text{ Pa}$$

The gauge pressure at point A is 3.15 kPa.

The difference in pressures at points A and B is given by $\rho_{\text{water}} g \Delta h_{\text{water}} = 1000 \text{ kg m}^{-3} \times 10 \text{ m s}^{-2} \times 0.14 \text{ m} = 1400 \text{ Pa}$. This is true for both gauge and absolute pressures and further P_B must be smaller than P_A. The gauge pressure at point B is thus $P_{\text{B,gauge}} = P_{\text{A,gauge}} - \rho_{\text{water}} g \Delta h_{\text{water}} = 3.15 \text{ kPa} - 1.4 \text{ kPa} = 1.75 \text{ kPa}$.

The pressure at point C is simply $P_C = P_B + \rho_{\text{water}} g \Delta d_{\text{water}}$ where $\rho_{\text{water}} g \Delta d_{\text{water}} = 1.6 \text{ kPa}$ and so $P_{\text{C,gauge}} = 3.35 \text{ kPa}$.

Other units of pressure still commonly found in use include psi (pounds per square inch), torr, mmHg (millimetres of mercury) and atmospheres (atm), so we will mention these briefly.

- 1 atm is the standard atmospheric pressure, 101 325 Pa.

- 1 torr is 1/760 atm.

- 1 mmHg is the pressure exerted at the base of a column of fluid exactly 1 mm high, when the density of the fluid is exactly 13.5951 g cm^{-3}, and where the acceleration of gravity is exactly 9.80665 m s^{-2}. The density chosen in the definition is the density of mercury (chemical symbol Hg) at 0 °C. One mmHg is very close to one torr.

- 1 psi is the pressure exerted by gravitational force from one pound of mass over an area of one square inch. 1 psi is equivalent to 6894.76 Pa.

Blood-pressure readings are still given in units of **mmHg**, and this is unlikely to change as the old-fashioned mercury blood-pressure meters do not go out of calibration, and are used to check the calibration of other types. Mercury *sphygmomanometers* (from the Greek sphygmós, meaning pulse, and manometer, meaning pressure meter) are still required in some clinical situations where accuracy is particularly important.

11.6 Pressure and the Human Body

Blood Pressure

High or low blood pressure can result from a number of medical conditions and can cause a number of health problems. As a result, blood pressure is routinely measured in medical practice. One method of measurement would be to connect a manometer directly to an artery, but this clearly far too invasive!

The standard method uses a sphygmomanometer (see Figure 11.16). An inflatable cuff is used to apply a pressure to the outside of the artery. The blood pressure in the artery varies with time throughout the cardiac cycle, and is maximum just after the left ventricle of the heart contracts, sending a surge of blood into the arteries. To start with, the external pressure applied to the artery is beyond the expected maximum blood pressure. At this point, the artery collapses. The applied pressure is slowly reduced as the physician listens. When sounds (known as Korotkoff sounds) are first heard, it indicates that the applied pressure has dropped below the maximum pressure in the artery. When the pressure in the artery is maximum, the artery opens briefly allowing a spurt of blood through, and the sound results from the brief turbulent flow of blood through the artery. The pressure at which sounds are first heard is known as the *systolic pressure*. It is the maximum pressure in the artery at the location of measurement. This ranges from 95 to 140 mmHg for a normal adult, with the average being 120 mmHg, when the pressure is measured level with the heart, as is customary.

As the external pressure on the artery is reduced further, the artery remains open for longer fractions of the cardiac cycle. When the pressure is reduced to a level where no further sounds are heard, the artery remains open throughout the cardiac cycle. This is known as the *diastolic pressure*, and is the lowest pressure in the artery at the location of measurement. For a normal adult, this number is in the range from 60 to 85 mmHg. Blood pressures are normally reported as two numbers, both in mm of Hg; the first number is the systolic pressure and the second number is the diastolic pressure. These pressures are gauge pressures, so they are the pressure in the artery relative to atmospheric pressure.

As we have seen, pressure varies with elevation in a liquid, so the height at which we make the measurements is important. When using a sphygmomanometer, the blood pressure is measured in the brachial artery (the main artery supplying blood to the arm). Readings will be different depending on whether the patient is lying or sitting, and how their arm is held relative to their heart. Measurements are conventionally taken with the patient sitting with their arm by their side so the point of measurement is level with their heart. If the upper arm was raised to head level, the pressure readings would be about 35 mmHg lower than with the arm by the side.

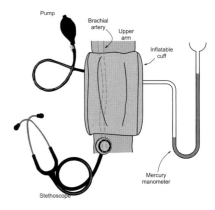

Figure 11.16 A sphygmomanometer being used to measure blood pressure.

11.7 Summary

Key Concepts

solid A state of matter that does not flow in response to a shearing force.

liquid A state of matter which flows in response to a shearing force.

gas A state of matter where intermolecular bonding is negligible and its properties are determined by molecular collisions.

pressure (P) The force per unit area. The SI unit of pressure is the pascal (Pa) which is equivalent to 1 newton per square metre (1 N m^{-2}), although many units are in common use such as *bar, torr, atm, mmHg* and *psi*. Solids, liquids and gases all exert pressure.

gauge pressure The pressure relative to the local atmospheric pressure.

absolute pressure The pressure measured relative to a perfect vacuum.

atmospheric pressure The pressure at a given point in the Earth's atmosphere. The atmospheric pressure varies with geographic location, altitude and time. The pressure at sea level is on the order of 1×10^5 Pa.

density (ρ) The mass per unit volume. The SI unit of density is the kilogram per cubic metre (kg m^{-3}).

manometer A pressure-measuring device used to measure relative pressures. The name is usually applied to a liquid-column hydrostatic instrument.

barometer An instrument that measures atmospheric pressure.

Equations

$$P_h = P_{\text{surface}} + \rho g h$$

$$\Delta P = \rho g h$$

absolute pressure = gauge pressure + atmospheric pressure.

$$\rho = \frac{m}{V}$$

11.8 Problems

11.1 Steve has a mass of 85 kg and wears size 11 shoes. The bottom of each of Steve's shoes has an area of 0.03 m² (approximately 30 cm by 10 cm). What pressure do the soles of Steve's shoes exert on the ground when

(a) he is standing still?

(b) he is standing on one leg?

(c) he is in the process of jumping (with both feet on the ground), and thus accelerating upwards at a rate of 5 m s^{-2}?

11.2 Water has a density of $1 \times 10^3 \text{ kg m}^{-3}$. How many litres of water would weigh as much as an 80 kg man?

11.3 Blood has a density of 1060 kg m^{-3} whereas the density of the cerebrospinal fluid is 1007 kg m^{-3}. What *mass* of cerebrospinal fluid will have the same *volume* as 50.0 g of blood?

11.4 Given that the density of water is $1 \times 10^3 \text{ kg m}^{-3}$, how deep would you have to dive to experience an absolute pressure of 2 atm? How deep would you have to dive to experience an absolute pressure of 5 atm? (Note that 1 atm = 101.3 kPa.)

11.5 What would you estimate that the difference in average blood pressure between the top of 1.7 m tall person's head and the bottom of their feet is? ($\rho_{\text{blood}} = 1060 \text{ kg m}^{-3}$)

11.6 Blood pressure is generally quoted in mmHg and is a *gauge pressure*. In this question we will convert this into SI units. Suppose that the systolic pressure of a particular patient is 120 mmHg and the diastolic pressure is 80 mmHg. Given that 760 mmHg is equivalent to 101.3 kPa, what is the systolic and diastolic blood pressure of this patient in SI units (Pa, or kPa)?

11.7 You wish to measure the blood pressure at the top of a patient's head but the patient is unable to lie flat. You measure the blood pressure at the bicep (i.e. level with the heart) and find that it is 140 mmHg systolic and 80 mmHg diastolic. What would you expect the blood pressure to be at the top of the head given that this point is 45 cm above the measurement point? Remember this is a gauge pressure and give your answers in both mmHg and kPa. (The density of blood is about 1060 kg m^{-3}.)

11.8 The aortic valve is located at the base of the aorta and controls the flow of blood from the left ventricle of the heart into the aorta. The valve has an area of about 3.5 cm² when closed (there are variations in this area from person to person). The aortic valve closes just before the diastolic phase of the cardiac cycle at which point the blood pressure is at about 90 mmHg. Using the figure just quoted calculate the force exerted on the aortic valve by the blood in the left ventricle (remember, the blood pressure shown above is a gauge pressure, not an absolute pressure.).

11.9 The systolic pressure in a major artery is measured at 115 mmHg. What is the net force on a 1 cm² section of the arterial wall if the (absolute) pressure in the tissue outside the arterial wall is 109 kPa? ($P_{\text{atm}} = 101.3 \text{ kPa}$)

11.10 What is the density of the unknown fluid in Figure 11.17? (You many assume that the unknown liquid is prevented from mixing with the water or displacing it in the event it is the more dense. $\rho_{\text{water}} = 1000 \text{ kg m}^{-3}$.)

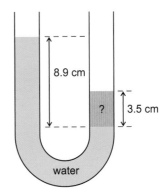

Figure 11.17 Water and an unknown fluid are placed in an open u-tube.

BUOYANCY

12

12.1 Introduction

When an object is submerged in a fluid, there is an upward force; we call this the buoyant force. This force is responsible for the apparent change in weight of submerged objects. This is of vital importance to animals which live in aquatic environments, and many are able to change their apparent weight by adjusting their volume by use of a swim bladder, an organ evolved for this purpose.

In this chapter, we will introduce the buoyant force and show its dependence on the volume of displaced fluid.

Key Objectives

- To be able to calculate the buoyant force acting on an object.

- To be able to calculate the fraction of an object submerged when it is floating in a known fluid.

12.2 The Buoyant Force

Archimedes' Principle

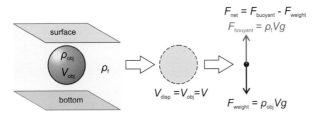

Figure 12.1 An object in a fluid experiences an upwards buoyant force. The magnitude of this force is equal to the weight force on *the volume of fluid displaced*. If the buoyant force is larger than the weight force, the object will rise; if less, it will sink. In this case, where the object displaces a volume of fluid equal to its own volume, the net force will depend on the object's density relative to the fluid.

Archimedes' principle, named after Archimedes of Syracuse, states that the buoyant force on an object is equal to the weight of the fluid it displaces.

> **Key concept:**
> **Buoyancy** is the upward force exerted on an object that is fully or partially submerged in a fluid, resulting from the increase in pressure with depth.

This force allows some objects to float on a liquid, while others sink. Mathematically, we can express this as

$$F_{\text{buoyant}} = m_{\text{f}}g = \rho_{\text{f}}V_{\text{f}}g \qquad (12.1)$$

where m is the mass of the fluid displaced, V is the volume of the fluid displaced, ρ is the fluid density and g is the acceleration due to gravity on Earth's surface. To avoid confusion, it is a good idea to use subscripts liberally to indicate which densities and volumes are being referred to. Here we have used 'f' to indicate that the density is the *fluid* density and the volume is the volume of fluid that has been displaced, not the volume of the object. *These will not be the same unless the object is fully submerged.* The *net* downwards force on the object in the case of only the buoyant force and gravity acting is

$$F_{net} = mg - \rho_f V_f g = \rho_{obj} V_{obj} g - \rho_f V_f g \qquad (12.2)$$

Examining Eq. (12.2) and referring to Figure 12.1, it can be seen that for an object fully submerged in a liquid, where $V_{obj} = V_f$, there will be a net upwards force only if ρ_f is larger than ρ_{obj}. The object will rise until it reaches the surface of the liquid. As it rises out of the liquid, the volume displaced, and hence the buoyant force, decreases. The object will likely bob up and down a few times, but will eventually come to rest at a position where the buoyant force is equal and opposite to the weight force, and the net force is zero.

If the object is more dense than the liquid, there will be a net force downwards acting upon the object, which will cause it to sink towards the bottom of the container. The object will end up sitting on the bottom of the container such that the combination of the upwards contact (normal) force and the upwards buoyant force is equal and opposite to the weight force of the object. See Figure 12.2.

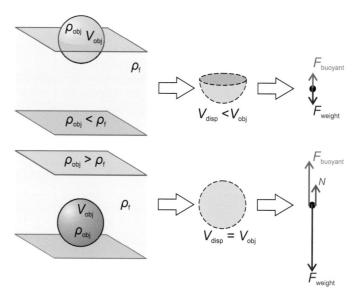

Figure 12.2 (Top) Objects with lower densities than the fluid in which they are placed will float on the surface of the fluid. When doing so they displace a *volume* of fluid that is less than the volume of the object, but the *weight* of the fluid displaced is equal to the weight of the object. (Bottom) Objects more dense than the fluid in which they are placed will sink to the bottom where a combination of the upwards buoyant force and an upwards contact (normal) force is equal and opposite to the object's weight force.

An object less dense than the fluid in which it is placed floats only partially submerged, so does not displace a volume of fluid equal to its own volume (Figure 12.3). Because the floating object is stationary, the net force acting upon it must be zero, and we can rewrite Eq. (12.2) to find the ratio of the submerged volume and the object's volume:

$$F_{net} = 0 = \rho_{obj} V_{obj} g - \rho_f V_{disp} g$$
$$\rho_f V_{disp} g = \rho_{obj} V_{obj} g$$
$$\frac{V_{disp}}{V_{obj}} = \frac{\rho_{obj}}{\rho_f} \qquad (12.3)$$

If the object in question has an irregular shape or composition, determining whether or not it will float is slightly more complex that than a mere comparison of densities. A boat can be made out of materials that are much denser than water, and it is capable of both floating and sinking. If the boat is oriented so that as the weight of the boat and

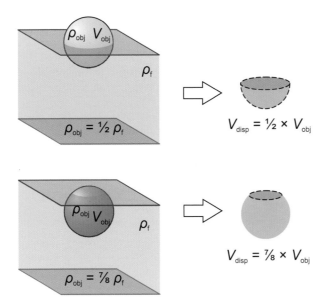

Figure 12.3 For a floating object, the fraction of the volume submerged is the same as the ratio of the densities of the object and the fluid (Eq. (12.3)).

its cargo is increased, more liquid is displaced also, then it may float. The same boat may sink if it fills with water and is not able to displace a sufficient volume of water to achieve a buoyant force equal to its weight.

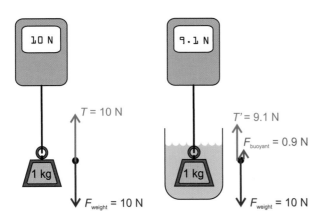

Figure 12.4 A 1 kg lead weight ($\rho_{\text{lead}} =11\,300$ kg m^{-3}) is suspended on a set of hanging scales. When the lead weight is submerged in water ($\rho_{\text{water}} = 1000$ kg m^{-3}) the buoyant force applied to the lead weight by the water partially supports it and so the reading on the scales is reduced.

In the preceding paragraphs it was noted that there is an upwards buoyant force on a submerged object. This means that any submerged object has an *apparent weight* that is not the same as its actual weight, so it requires less additional force to support it than usual. Figure 12.4 shows this effect. A measurement of the apparent weight of a submerged object can be used to determine its density relative to the liquid:

$$\text{apparent weight} = \text{actual weight} - \text{buoyant force}$$

$$\text{actual weight} - \text{apparent weight} = \rho_{\text{f}} V g$$

$$\Rightarrow \frac{\text{actual weight} - \text{apparent weight}}{\text{actual weight}} = \frac{\rho_{\text{f}} V g}{mg} = \frac{\rho_{\text{f}} V g}{\rho_{\text{obj}} V g} = \frac{\rho_{\text{f}}}{\rho_{\text{obj}}}$$

Example 12.1 *Salvage diver*

Problem: A salvage diver is attaching a cable to a box of treasure she has found on the bottom of the sea. Once her job is done she will be lifted at constant speed back onto the salvage ship by a rope. If the combined weight of her body and all her equipment is 125 kg and her diving suit has a volume of 0.09 m³, what will the tension in the rope be as she is lifted up to the surface?

Solution: As she is lifted, at a constant speed, the net force on the diver must be zero. The weight force of the diver is acting downwards and must be balanced by the upwards buoyant force and tension force. The weight force acting on the diver (and her equipment) is 1250 N. The buoyant force acting on the diver is

$$F_{\mathrm{B}} = \rho_{\mathrm{water}} g V = 1000 \text{ kg m}^{-3} \times 10 \text{ m s}^{-2} \times 0.09 \text{ m}^3 = 900 \text{ N}$$

The tension force in the rope must be the difference between the two, $T = 1250 \text{ N} - 900 \text{ N} = 350 \text{ N}$.

Example 12.2 *Lead sphere*

Problem: A sphere of lead is hung on a rope and submerged in water. The tension force in the rope is 160 N less when the sphere is submerged than when it is not. What is the radius of the sphere? ($\rho_{\mathrm{lead}} = 11\,300 \text{ kg m}^{-3}$)

Solution: The buoyant force acting on the sphere must be 160 N. We can find the volume of the sphere directly by using Eq. (12.1).

$$F_{\mathrm{B}} = \rho_{\mathrm{fluid}} g V_{\mathrm{displaced}}$$

$$V_{\mathrm{displaced}} = \frac{F_{\mathrm{B}}}{\rho_{\mathrm{fluid}} g} = \frac{160 \text{ N}}{10^3 \text{ kg m}^{-3} \times 10 \text{ N kg}^{-1}} = 0.016 \text{ m}^3$$

which gives a radius of

$$V = \frac{4}{3} \pi r^3$$

$$r = \sqrt[3]{\frac{3V}{4\pi}} = 0.156 \text{ m}$$

Example 12.3 *Synchronised swimmer*

Problem: A person will typically float with just 4% of their volume above the surface of the water. If a 55 kg synchronized swimmer is performing a manoeuver in which they raise 30% of their volume out of the water and hold themselves there, what 'thrust' force must they generate by kicking their legs?

Solution: If a person floats with just 4% of their volume above the surface of the water then the density of a person must be $0.96 \rho_{\mathrm{water}} = 960 \text{ kg m}^{-3}$. Thus a 55 kg person will have a volume of $\frac{55 \text{ kg}}{960 \text{ kg m}^{-3}} = 0.057 \text{ m}^3$.

When performing the manoeuver specified in the question 30% of the swimmer's volume is out of the water leaving just 0.040 m³ submerged. This means that the buoyant force on the swimmer is $F_{\mathrm{B}} = \rho_{\mathrm{water}} g V_{\mathrm{displaced}} = 1000 \text{ kg m}^{-3} \times 10 \text{ m s}^{-2} \times 0.040 \text{ m}^3 = 400 \text{ N}$.

While performing the manoeuver the net vertical force on the swimmer must be zero (they are not rising or sinking) and therefore the total upwards force on the swimmer (400 N buoyant force + X N thrust force) must be equal in magnitude to the downwards weight force (550 N). This means that the swimmer is required to produce a thrust force of 150 N to maintain the manoeuver.

Example 12.4 *Is it gold?*

Problem: A person is paid for a job with a 0.100 kg 'gold' coin. Suspicious, the person decides to check to see if the coin is really gold (ρ_{gold} = 19 300 kg m^{-3}). They hang the coin on a piece of string and submerge it in water. The apparent mass while the coin is submerged is 0.0912 kg. Is the coin gold?

Solution: The buoyant force acting on the coin is $1 \, \text{kg} - 0.0912 \, \text{kg} \, g = 0.0088 \, \text{kg} \, g$. From this we can find the volume of the coin using $F_B = \rho_{fluid} g V$ which gives $V = \frac{0.0088 \, \text{kg}}{1000 \, \text{kg m}^{-3}} = 8.8 \times 10^{-6} \, \text{m}^3$. This indicates that the density of the coin is $\frac{0.100 \, \text{kg}}{8.8 \times 10^{-6} \, \text{m}^3} = 11\,400 \, \text{kg m}^{-3}$. The density of the coin is too low to be gold, it is however quite close to the density of lead ($\rho_{lead} = 11\,300 \, \text{kg m}^{-3}$).

12.3 Summary

Key Concepts

buoyant force The upward force from the fluid acting on an object wholly or partially submerged in the fluid.

Archimedes' principle A floating object displaces a volume of liquid with the same weight as the object.

Equations

$$\rho = \frac{m}{V}$$

$$F_{buoyant} = \text{weight of displaced fluid} = \rho_f V_{disp} g$$

12.4 Problems

12.1 A swimmer finds that she just floats in water. If she weighs 70 kg what is her volume ($\rho_{water} = 1 \times 10^3 \text{ kg m}^{-3}$)?

12.2 Law 2 of the game of soccer specifies that the ball is an air-filled sphere with a circumference of 68–70 cm, and a mass of 410–450 g. A particular ball has a circumference of 69 cm, and a mass of 430 g. Calculate the fraction of the volume of this ball that floats above the surface of water.

12.3 A product designer for a range of nautical safety devices conceives of a floating container in which emergency supplies can be stored, which would automatically become detached from a boat if it sunk. The container is to be a cylinder with a radius of 20 cm and a height of 1 m. When in the water 20% of the volume of the container must be above the surface. The container is attached to the deck of the boat by a cord which should break as the boat sinks and the container is submerged and pushed upwards by it's own buoyancy.

(a) What is the maximum mass of the container and it's contents?

(b) At what tension should the cord attaching the container to the boat break?

12.4 A small 0.5 m radius weather balloon is filled with helium ($\rho_{He} = 0.164 \text{ kg m}^{-3}$). What is the maximum payload (including the balloon mass) of this weather balloon? ($\rho_{air} = 1.18 \text{ kg m}^{-3}$)

12.5 A helium shortage forces some under-funded meteorologists to investigate alternative gases to use in their weather balloons. They settle on methane ($\rho_{CH_4} = 0.657 \text{ kg m}^{-3}$). What is the minimum radius of a methane filled weather balloon that will allow the same minimum payload as the helium filled balloon in Problem 12.4?

12.6 A piece of polystyrene packaging material (density = 25 kg m^{-3}) that has a mass of 0.2 kg is tethered to the bottom of a container of water (density = 1×10^3 kg m^{-3}) with a piece of string. What is the tension in the string?

12.7 In an experiment to determine the density of an unknown material, its apparent weight when fully submerged in water is measured. The apparent weight in water is 17.5 N and the weight in air is 27.5 N. What is the density of the material?

12.8 Some salvage divers are raising a rectangular box of treasure measuring 1 m × 1.3 m × 1.9 m from the bottom of the sea. They are using a cable which can handle a maximum force of 10 000 N without breaking. They raise the treasure from the bottom of the ocean at a constant speed. All goes well while the treasure is rising through the water but much to the despair of the salvors as soon as the treasure is clear of the water the cable breaks, dropping the treasure back into the briny deep. What are the maximum and minimum possible weights of the lost treasure?

12.9 A large air-filled rubber ball is tethered to the bottom of a swimming pool. The tension in the tether is 100 N. The mass of the rubber in the ball itself is 2 kg while $\rho_{water} = 1000$ kg m^{-3} and $\rho_{air} = 1.2$ kg m^{-3}. What is the volume of the ball?

12.10 A wooden cube 3 cm on a side floats level on water with just 1.5 mm of the cube showing above the surface. What is the density of the wood? ($\rho_{water} = 1000$ kg m^{-3})

SURFACE TENSION AND CAPILLARITY

13.1 Introduction

The attractive force between substances that are alike is known as **cohesion**, while the attraction between unlike substances is called **adhesion**. Cohesive forces are responsible for **surface tension**; adhesive forces between the surfaces of a liquid and solid can cause the edges of the liquid surface to be distorted, pulling the liquid up or down, an effect known as **capillary action** or **capillarity**.

Key Objectives

- To understand the nature of surface tension.

- To understand capillarity.

- To be able to calculate the height to which capillary action is able to raise a fluid in a pipette.

13.2 Surface Tension

Surface tension is a property of liquid surfaces resulting from intermolecular bonding which causes the liquid to minimise its surface area and resist deformation of its surface. It causes liquids to act rather like they have a thin, elastic skin. This is not true, but is a useful analogy to visualise the behaviour of liquids.

Molecules in a liquid are attracted to their neighbours by cohesive forces. Inside the bulk liquid, a molecule is attracted equally in all directions by its neighbours. A gas has a lower density, and hence fewer molecules in a volume, so at the liquid–gas surface the molecules have fewer neighbours on one side. Consequently, they experience a stronger attraction to the molecules in the liquid than to the molecules of the other medium in contact (such as air). This results in an inwards force due to cohesive forces, which causes a liquid to reduce its surface area (see Figure 13.1). The inward force is eventually balanced by the liquid's resistance to compression.

The size of the surface tension can be measured by determining the force required to hold in place a wire that is being used to stretch a film of liquid as in Figure 13.2. The surface tension is defined as the force per unit length along a line where the force is parallel to the surface and perpendicular to the line:

$$\gamma = \frac{F}{L} \tag{13.1}$$

The length over which the force is being applied in the diagram shown is *twice* the length of the wire, as the liquid film has two surfaces. The surface tension γ has the units of N m^{-1}.

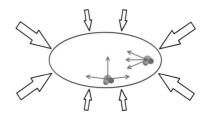

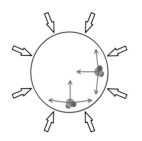

Figure 13.1 Surface tension acts to reduce the surface area of a body of liquid. In general, the intermolecular forces on a molecule of liquid from its neighbours in the liquid are stronger than those from any neighbouring gas molecules. This means that the net force on surface molecules is directed into the liquid, and is stronger for more curved surfaces.

Alternate Description

An alternative description of surface tension is often used in thermodynamics. It can also be thought of as an energy per unit area (N m^{-1} is equivalent to J m^{-2}) and the shape is formed which minimises the energy.

Introduction to Biological Physics for the Health and Life Sciences   Franklin, Muir, Scott, Wilcocks and Yates
©2010 John Wiley & Sons, Ltd

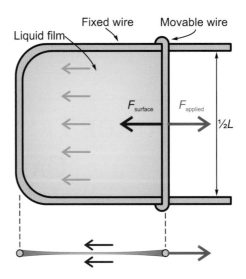

Figure 13.2 A thin film of fluid is held in a metal loop, one side of which is movable. Surface tension on *both* the fluid surfaces acts to reduce its surface area, so *L*, the length along which the surface tension is pulling, is twice the wire length. In this situation pictured, this will pull the movable wire to the left, unless some external force to the right is applied. Measurement of the force required to keep the movable wire stationary can give the surface tension, γ. Note: It is necessary to ensure that frictional forces are very low in such an experiment.

Example 13.1 *Surface tension*

Problem: A thin film of a mystery fluid is formed on a device like that shown in Figure 13.2. If the width of the apparatus is 3 cm and the force required to hold the movable wire steady is 4.8 mN, what is the surface tension of the fluid?

Solution: The surface tension can be found using Eq. (13.1):

$$\gamma = \frac{F}{L}$$

where $F = 4.8 \times 10^{-3}$ N and $L = 0.06$ m. It is important to remember that L is twice the width of the apparatus as there are two surfaces to the fluid.

This gives a surface tension of:

$$\gamma = \frac{4.8 \times 10^{-3}}{0.06 \text{ m}} = 0.08 \text{ N m}^{-1}$$

Pressure in Bubbles

Surface tension is important in the functioning of the lung. To see why, we will start by looking at the pressure in a spherical bubble. The pressure inside the bubble must be higher than the outside pressure to stop the surface tension from collapsing the bubble.

$$\Delta P = P_{\text{int}} - P_{\text{ext}} = \frac{4\gamma}{r} \tag{13.2}$$

To see why this should be the case, consider a cross section through the bubble, as in Figure 13.3. The force applied along this circular edge from the surface tension is the circumference doubled (because there are two surfaces) times the surface tension γ, so it is $4\pi r\gamma$. The force that balances this is the pressure difference times the cross-sectional area, $\Delta P\pi r^2$. This gives us the pressure difference of $4\gamma/r$ in Eq. (13.2). A consequence of this is that the (gauge) pressure inside grows larger as the bubble decreases in size. More pressure difference is needed to inflate a bubble than to keep it inflated.

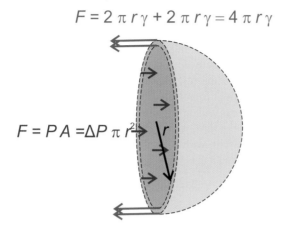

$$F = 2\pi r\gamma + 2\pi r\gamma = 4\pi r\gamma$$

$$F = PA = \Delta P\,\pi\,r^2$$

Figure 13.3 The pressure inside a hollow bubble. The total force pulling on the surfaces is twice the circumference times the surface tension. The net force over the hemispherical surface is the pressure times the cross-sectional area.

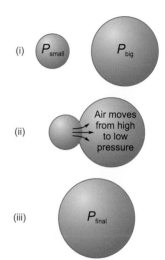

(i) P_{small} P_{big}

(ii) Air moves from high to low pressure

(iii) P_{final}

Figure 13.4 When two bubbles collide, air always flows from the smaller to the larger. This is because the pressure in the small bubble P_{small}, is larger than that in the bigger bubble, P_{big}. Eq. (13.2) shows that the pressure inside a bubble is inversely dependent on the radius of the bubble: the larger the bubble, the smaller the pressure. It should also be noted that the pressure P_{final} inside the new, even larger bubble is lower still.

Surfactants

A substance that, when added to a liquid, reduces the liquid's surface tension is called a **surfactant**. This is a shortened form of 'surface-active agent'. The surfactant molecules tend to concentrate near the surface. An example of a surfactant is soap in water. A needle that can be supported by the surface tension of water will break through the surface and sink when soap or detergent is added to the water. Detergents and soap are surfactants because they have one hydrophilic ('water-loving') and one hydrophobic ('water-hating') side, so the lowest-energy position for them is at the surface, with the hydrophobic end farther from the water molecules. Surfactants are of major importance to lung function. We will examine some aspects of the respiratory system in more detail in Section 13.4.

Example 13.2 *Bubbles*

Problem: A bubble of water ($\gamma = 0.068$ N m^{-1}) forms such that it has an internal gauge pressure of 13.6 Pa.

   (a) How large is this bubble?

A surfactant is added to the water reducing the surface tension to 0.021 N m^{-1}.

   (b) If the gauge pressure inside a new bubble is the same, what would the radius be in this case?

Solution: (a) The difference in pressure (i.e. the gauge pressure) between the inside of a bubble and the outside is given by

$$\Delta P = \frac{4\gamma}{r}$$

$$r = \frac{4\gamma}{\Delta P} = \frac{4 \times 0.068\ \text{N m}^{-1}}{13.6\ \text{Pa}} = 0.020\ \text{m}$$

So the bubble has a radius of 2 cm. After the surfactant has been added the bubble will *decrease* in size!

$$r = \frac{4\gamma}{\Delta P} = \frac{4 \times 0.021\ \text{N m}^{-1}}{13.6\ \text{Pa}} = 0.0062\ \text{m}$$

13.3   Capillarity

Interfacial Tension

In examining surface tension, the adhesive forces between the liquid and any molecules of gas near the surface could be largely ignored. In the case of liquid in contact with other immiscible (non-mixing) liquids, or solids, such as the walls of the container that the liquid is held in, these forces are no longer negligible. The size of the adhesive forces between the materials will determine the **interfacial tension** (i.e. the tension at the interface). The interfacial tension of a water droplet in contact with glass is different to that of water in contact with wax. This difference results in 'beading' or 'wetting' of surfaces, as shown in Figure 13.5.

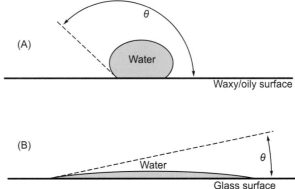

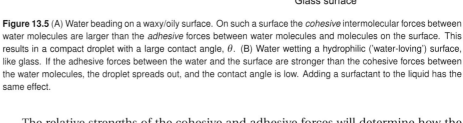

Figure 13.5 (A) Water beading on a waxy/oily surface. On such a surface the *cohesive* intermolecular forces between water molecules are larger than the *adhesive* forces between water molecules and molecules on the surface. This results in a compact droplet with a large contact angle, θ. (B) Water wetting a hydrophilic ('water-loving') surface, like glass. If the adhesive forces between the water and the surface are stronger than the cohesive forces between the water molecules, the droplet spreads out, and the contact angle is low. Adding a surfactant to the liquid has the same effect.

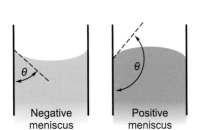

Figure 13.6 Strong adhesive forces result in a small contact angle, θ, and a negative meniscus. Stronger cohesive forces result in a large contact angle and a positive meniscus. Note: In all but the smallest diameter tubes the meniscus will not be as prominent as that shown here and will be noticeable only at the fluid/tube boundary, with the bulk of the fluid surface being near flat.

The relative strengths of the cohesive and adhesive forces will determine how the liquid behaves. If the cohesive forces in a liquid are stronger than the adhesive forces between it and an adjacent substance, the liquid will tend to 'bead', minimizing the contact area between the two substances. An example of this is the beading of water on a waxed car. The liquid molecules are attracted to each other more than to the neighbouring substance, so the liquid arranges itself to keep the distances between liquid molecules as small as possible.

If the cohesive forces in a liquid are weaker than the adhesive forces between the two substances, the result is 'wetting' – the liquid will spread out across the surface, as water does on a glass surface.

A quantitative measure of the tendency to bead is the **contact angle**. This is the angle that the edge of the liquid–air surface makes with the liquid–solid surface (see Figure 13.5). Contact angles more than 90° are indicative of beading, and angles less than 90° show wetting.

The angle of contact at the surface between a liquid and its container will depend on the relative strengths of cohesive and adhesive forces also. For a fluid in a vertical tube, if the cohesive forces are weak in comparison to the adhesive forces, the contact angle will be small and the liquid will be pulled slightly up at the edges of the tube, giving a **negative meniscus**. This is the case with water in a glass. If the cohesive forces are strong in comparison to the adhesive forces, the contact angle will be large and a **positive meniscus** is formed, as with mercury in a glass tube. Figure 13.6 shows both a positive and a negative meniscus.

Capillary Action

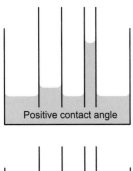

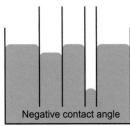

Figure 13.7 The distance the liquid travels up or down the tube is inversely proportional to the radius. In the case where the contact angle is greater than 90°, the liquid surface is depressed.

When a thin glass tube is placed in a liquid such as water, the liquid often rises up the tube. This is known as **capillary action**, or **capillarity**. Capillary action is important in many biological systems – it contributes to the rising of sap in trees and to the blood flow into our capillaries.

The thinner the tube, the more the liquid rises (Figure 13.7). A smaller radius means more contact with the surface for a particular volume of liquid, and hence a greater mass of liquid that can be supported by the contact force. In fact, the height depends on the inverse of the radius of the tube:

$$h = \frac{2\gamma\cos\theta}{\rho g r} \qquad (13.3)$$

where h is the height the liquid travels up the tube above the level of the surrounding liquid, γ is the surface tension, θ is the angle the liquid surface makes with the tube surface, ρ is the liquid density, r is the tube radius and g is the acceleration due to gravity.

Figure 13.8 illustrates the forces involved. At the contact points between the liquid and the surface of the tube, the liquid is being pulled in the directions shown by the surface tension. The horizontal components of these forces will cancel out, but the upward component of the force will be $F_{st}\cos\theta$. To find the upwards force, we note that surface tension is the force per unit length, so the upwards component of the force will be $\gamma \times \text{length} \times \cos\theta = \gamma 2\pi r\cos\theta$. This will pull the liquid upwards until the downwards force on the mass due to gravity is equal. The downward force is $F_{weight} = mg = \rho V g = \rho(\pi r^2 h)g$, and this rearranges to give Eq. (13.3)

$$\rho\pi r^2 hg = \gamma 2\pi r\cos\theta$$
$$\Rightarrow h = \frac{\gamma 2\pi r\cos\theta}{\rho\pi r^2 g}$$
$$= \frac{2\gamma\cos\theta}{\rho g r}$$

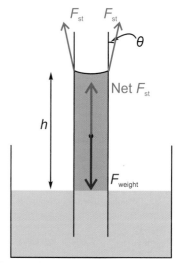

Figure 13.8 Surface tension pulls the fluid up until the weight force of the fluid supported is the same as the net force due to surface tension.

In the cases where the angle θ is larger than 90°, the force is actually downwards, and the surface of the liquid in the tube is *lower* than the surrounding liquid, as seen in Figure 13.7. This is the case with mercury and glass.

Example 13.3 *Capillary Action*

Problem: An engineer is designing an experimental dialysis machine. A vital component of this machine is a small hollow tube made from the element unknownium with a 100 µm inner radius which will dip down into a chamber filled with blood. The surface tension of blood is 0.058 N m^{-1} and the density of blood is 1050 kg m^{-3}.

(a) The engineer does not know what the adhesive forces, and hence contact angle, between blood and unknownium is. What is the maximum possible range of heights to which blood will be raised/lowered in the tube?

(b) If the contact angle between unknownium and blood is measured at 78°, to what height is blood drawn up the tube of unknownium?

(c) Blood needs to be drawn up the small tube at least 3 cm. What is the simplest change that can be made to the tube of unknownium to enable this to happen?

Solution:

(a) The height to which a liquid will be drawn up a capillary tube is given by Eq. (13.3).

$$h = \frac{2\gamma\cos\theta}{\rho g r}$$

The cosine term could have any value from +1 for $\theta = 0°$ (fully wetting the surface) to −1 for $\theta = 180°$ (fully beaded upon the surface). Given this the possible range of heights to which blood will be drawn up a tube of unknownium are

$$h = \pm \frac{2\gamma}{\rho g r} = \pm \frac{2 \times 0.058 \text{ N m}^{-1}}{1050 \text{ kg m}^{-3} \times 10 \text{ m s}^{-2} \times 100 \times 10^{-6} \text{ m}}$$

$$= \pm 0.110 \text{ m}$$

(b) With a specific contact angle, $\theta = 78°$, we can calculate a specific height to which the blood will be drawn into the capillary. Given that the angle is less than 90° we can say with certainty that the blood will be drawn *up* into the capillary ($h > 0$) Given that $\cos(78°) = 0.208$

$$h_{78°} = \pm \frac{2\gamma}{\rho g r} \cos(78°) = 0.110 \text{ m} \times 0.208 = 0.0229 \text{ m}$$

(c) If the blood needs to be drawn up the tube 3 cm then the capillary as it stands is inadequate. In order increase the height to which the blood is drawn into the capillary we can: increase the viscosity of blood; use a different material that has a higher contact angle (lower adhesive forces between the material and blood); or decrease the radius of the tube.

Of these three options decreasing the radius of the tube is likely the easiest. Changing the viscosity of blood will require adding some sort of surfactant to it which is quite the opposite of the general purpose of a dialysis machine and using a different material may require other aspects of the machine to be redesigned.

We can rearrange Eq. (13.3) to calculate what radius tube will give us the required height of 0.03 m

$$h = \frac{2\gamma \cos\theta}{\rho g r}$$

$$r = \frac{2\gamma \cos\theta}{\rho g h} = \frac{2 \times 0.058 \text{ N m}^{-1} \times 0.208}{1050 \text{ kg m}^{-3} \times 10 \text{ m s}^{-2} \times 0.03 \text{ m}} = 76.6 \times 10^{-6} \text{ m}$$

13.4 Surfactants and the Lung

The alveoli in our lungs are similar in many respects to a collection of air bubbles surrounded by water, with the air being free to move about between them. This is an unstable situation for bubbles – the small bubbles have the highest internal air pressure (see Eq. (13.2)), which would tend to force the air to move to lower pressure regions in the larger bubbles, causing large bubbles to get larger and small bubbles to get smaller. This would be undesirable behaviour for our lungs – we need to have a large surface area to allow the most diffusion of oxygen into the bloodstream. Surfactants can stabilise bubbles by making their surface tension size dependent.

Adding surfactant has the desired effect because the concentration of the surfactant decreases as the surface expands and increases as the surface contracts. An increase in concentration of surfactant from a decreasing surface area will reduce the surface tension, and will reduce the pressure at which the bubble becomes stable.

Pulmonary (lung) surfactant has the same effect. It allows alveoli with slightly different sizes to have the same internal pressure, giving a stable arrangement. Also, in the absence of surfactant, the pressure required to change the size of an alveolus would be greatest when the alveolus was smallest. This would make the initial inflation difficult, rather like the first part of blowing up a balloon being the hardest. Without pulmonary surfactant, the pressure difference required to inflate the alveolus would be greater than that generated by chest expansion during inhalation.

Premature infants often lack sufficient pulmonary surfactants, causing breathing difficulties. Understanding this has led to improved treatment of premature infants and a great increase in survival rates. The role of surfactants is also important in the case of drowning. In instances of 'secondary drowning', a person initially appears fine, but later deteriorates. This is caused by the interaction of small quantities of water with the surfactant in the lungs. Fresh water denatures the surfactant and salt water dilutes it, with both cases possibly resulting in late-developing breathing difficulties.

13.5 Summary

Key Concepts

cohesion The intermolecular attraction between like molecules.

adhesion The intermolecular attraction between unlike molecules.

surface tension The property of a liquid surface that causes it to behave like an elastic sheet, as the result of cohesive forces.

capillarity The distortion of a liquid surface due to adhesive forces between the surface of the liquid and an adjacent solid surface. This can result in the liquid being pulled up or down a narrow tube.

surfactant A substance that, when added to a liquid, reduces the liquid's surface tension.

Equations

$$\gamma = \frac{F}{L}$$

$$\Delta P = P_{int} - P_{ext} = \frac{4\gamma}{r}$$

$$h = \frac{2\gamma \cos\theta}{\rho g r}$$

13.6 Problems

13.1 A device such as that shown in Figure 13.2 is used to measure the surface tension, γ, of glycerol. The length of the moveable wire is 0.5 cm and a force of 0.63 N must be applied to this wire to maintain a constant area fluid film. what is the surface tension of glycerol?

13.2 A soap bubble is formed using a mixture of detergent and water. The surface tension of the mixture is 0.030 N m^{-1}. If the bubble has a radius of 2 cm and atmospheric pressure is 101.2 kPa, what is the gauge pressure inside the bubble?

13.3 A water bubble has radius 1 mm in air. Atmospheric pressure is 101.3 kPa and the surface tension of water at room temperature is 0.073 N m^{-1}.

  (a) What is the gauge pressure inside the bubble?

  (b) If atmospheric pressure had been 101.2 kPa and the absolute internal pressure of the bubble was the same as in part (a), what would the radius of the bubble have been?

  (c) If the surface tension of the water was lowered to 0.037 N m^{-1} (due to the addition of a surfactant for example), what would the radius of the bubble in part (b) be?

13.4 An entertainer, when performing a bubble trick, forms one bubble inside another. The surface tension coefficient of the bubble liquid used was $\gamma = 0.04$ N m^{-1}. If the outer bubble has a radius of 4.5 cm and the inner bubble has a radius of 2 cm, what is the gauge pressure in the inner bubble?

13.5 The interface between blood and stainless steel makes a angle of 110°. Would you expect that capillary action would draw blood into a stainless steel needle, or expel it?

13.6 Fluid A is found to have a surface tension of 0.080 N m^{-1}, a density of 1.2×10^3 kg m^{-3} and a contact angle of 70° with dry glass.

Fluid B is found to have a surface tension of 0.100 N m^{-1}, a density of 3.1×10^3 kg m^{-3} and a contact angle of 110° with dry glass.

A glass capillary tube with inner radius 1 mm is lowered into a container of fluid A and an identical capillary tube is lowered into a flask of fluid B. To what height above (or below) the fluid surface will fluids A and B rise in their respective capillary tubes?

13.7 Drops of two liquids are placed onto a glass slide. Liquid A remains a small rounded drop, sitting on the glass. Liquid B spreads out to form a thin film on the glass. If a narrow glass capillary tube is placed in a container of each liquid, would you expect the level of liquid in the capillary to be higher or lower than that in the container for each liquid?

13.8 What is the minimum surface tension of a fluid that can sustain a gauge pressure of 0.1 kPa in a 0.5 cm radius bubble?

FLUID DYNAMICS OF NON-VISCOUS FLUIDS

14.1 Introduction

An understanding of the physics of fluid flow is vital to an understanding of biological systems as diverse as the human circulatory system and the distribution of nutrients in plants. In this chapter we will introduce the physical foundations of fluid flow in the absence of viscosity. We will discuss viscosity in Chapter 15.

Key Objectives

- To be able to relate volume flow rate to fluid velocity and cross-sectional area.

- To understand how mass conservation leads to the continuity equation.

- To understand how energy conservation leads to Bernoulli's equation.

- To be able to use the continuity equation and Bernoulli's equation to calculate fluid velocity and pressure at various points in a flowing fluid.

14.2 Definitions of Some Key Terms

There are a number of potentially unfamiliar terms that we will use throughout this chapter and the next, so we will start off with a few definitions:

incompressible fluid The fluid has a constant density throughout.

viscosity The resistance of a fluid to flow.

laminar flow A situation in which layers of fluid slide smoothly past each other. Laminar flow is characteristic of lower fluid velocities.

turbulent flow Non-laminar flow. The flow is irregular and complex, with mixing and eddies. This occurs at higher velocities or where there are objects in the flow producing large changes in velocity.

streamlines A family of curved lines that are tangential to the velocity vector of the flow (i.e. always in the same direction as the flow). They provide a kind of snapshot of flow throughout the fluid at an instant of time.

14.3 The Equation of Continuity

Volume Flow Rate

The **volume flow rate**, $\mathscr{F}$, tells us how much fluid is flowing across some surface, such as a pipe's cross section, in a given time. It is usually measured in cubic metres per second. Imagine that all the fluid that flowed though some cross section of a pipe was collected in a bucket – the volume flow rate would tell you the how many cubic metres of fluid would be collected in one second.

For an incompressible fluid, the volume flow rate is equal to the product of the cross-sectional area and the velocity; if the cross-sectional area of the pipe or the velocity of the fluid in it were increased, then more fluid would collect in the bucket and the volume flow rate would be higher. This can be seen by considering a cross-sectional surface, area A, in a pipe such as the right-hand end of the shaded cylinder shown in Figure 14.1. If the fluid velocity is v, then in time Δt, all the fluid from distance Δx to the left of this surface will cross through it, where $\Delta x = v\Delta t$. The volume of liquid that crosses the surface must be $A\Delta x$, and so the volume flow rate is

$$\mathcal{F} = \frac{\Delta V}{\Delta t} = \frac{A\Delta x}{\Delta t} = Av \tag{14.1}$$

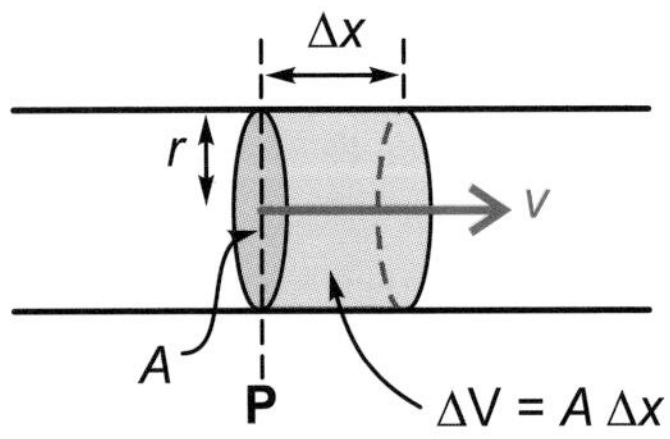

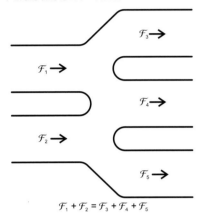

Figure 14.1 Fluid flowing through a pipe at a velocity v past a point P will travel a distance Δx in some small time Δt. The volume of fluid passing P in this time is $\Delta V = A \Delta x$.

Continuity of Flow

Under certain conditions, when the fluid is incompressible and there is no fluid gained or lost, the volume flow rate is constant along a pipe or channel. This is due to the conservation of mass – the amount of material entering one end of the pipe must be the same as the amount coming out the other end. There must also be the same amount *per unit time*, and as we are talking about an incompressible fluid, a fixed mass implies a fixed volume, hence a constant volume flow rate. We've established the relationship between volume flow rate, area and velocity, so this brings us to the **continuity equation**:

$$A_1 v_1 = A_2 v_2 \tag{14.2}$$

Figure 14.2 The continuity equation states that the flow rate $\mathcal{F}$ in a single pipe must be constant. This can be generalised to pipes that branch multiple times. In the case of multiple inflows or outflows, the sum of all incoming flow rates must equal the sum of all outgoing flow rates.

$$\mathcal{F}_1 + \mathcal{F}_2 = \mathcal{F}_3 + \mathcal{F}_4 + \mathcal{F}_5$$

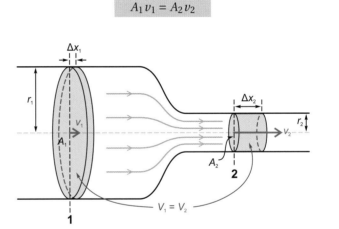

Figure 14.3 The volume flow rate of an incompressible fluid flowing through a pipe is constant. The pipe is wide at point 1, but narrow at point 2. Eq. (14.2) shows that in such a case the velocity of the fluid must be higher at point 2.

The continuity equation implies that when a fluid enters a more constricted section, as in Figure 14.3, it will speed up. We can apply the continuity equation even when we have multiple pipes joining; the volume of liquid flowing into a given location per second is equal to the volume flowing out. If there are multiple pipes, as in Figure 14.2, then the sum of the volume flow rates into the junction is equal to the sum out.

Example 14.1 *Continuity of flow*

Problem: A water pipe carries 1000 L of water past a certain point every minute.

(a) If the speed of the water in this pipe is 2 m s^{-1}, what is the radius of the pipe?

(b) The radius of the pipe narrows by 10% as it passes from one suburb to another. What is the velocity of the water in the pipe now?

(c) The pipe then splits up into two pipes, each of which of has an area equal to the area of the pipe just before it splits. What is the speed of the water in each pipe now?

Solution: (a) Given that $1\text{ L} = \frac{1}{1000}\text{ m}^3$, the volume flow rate through the pipe is $\mathscr{F} = \frac{1\text{ m}^3}{60\text{ s}} = 0.0167\text{ m}^3\text{s}^{-1}$. The area of the pipe must be

$$\mathscr{F} = Av$$

$$A = \frac{\mathscr{F}}{v} = \frac{0.0167\text{ m}^3\text{s}^{-1}}{2\text{ m s}^{-1}} = 8.33 \times 10^{-3}\text{ m}^2$$

Which corresponds to a radius of 5.15 cm.

(b) Because the volume flow rate must be conserved, if the radius of the pipe decreases the speed of the water in the pipe must increase, $A_1 v_1 = A_2 v_2$. If $r_2 = 0.9 r_1$ then by $A = \pi r^2$, $A_2 = 0.9^2 A_1 = 0.81 A_1$. So it follows that $v_2 = \frac{A_1}{A_2} v_1 = \frac{A_1}{0.81 A_1} v_1 = \frac{v_1}{0.81} = 1.23 v_1$ or 2.47 m s^{-1}.

(c) The *total* volume flow rate must be the same as before the pipe split. Thus each pipe will have half the volume flow rate of the pipe before it splits. If each pipe also has the same area as the pipe before it splits it follows that the speed of the water in the pipe must half, $\mathscr{F} = Av$, $v' = \frac{\mathscr{F}'}{A'} = \frac{\frac{1}{2}\mathscr{F}}{A} = \frac{1}{2}\frac{\mathscr{F}}{A} = \frac{1}{2}v$.

14.4   Bernoulli's Equation

Bernoulli's Principle and Incompressible Fluid Flow

Bernoulli's principle is named after Daniel Bernoulli (1700–1782), one of several famous men from his family, and is in essence a statement of the law of energy conservation for fluids. When viscosity can be neglected, an increase in fluid velocity is accompanied by a decrease in pressure and/or a decrease in gravitational potential energy (see Figure 14.4).

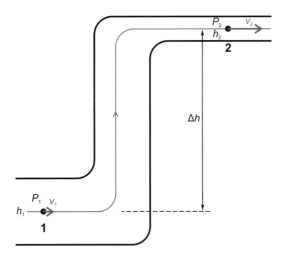

Figure 14.4 Bernoulli's principle allows the combination of pressure, speed, and height of a fluid at one point to be compared to the same three properties at a different point in the fluid.

We can use this to write an equation relating together pressure, speed and elevation for the case of an incompressible fluid. This will be valid for most liquids, and for gases when no expansion or compression is happening. In addition, if the fluid flow is laminar, steady (i.e., independent of time), and we can ignore the effects of friction, then we have **Bernoulli's equation**:

$$P + \frac{1}{2}\rho v^2 + \rho g h = \text{constant} \tag{14.3}$$

In the above equation, P is the pressure at a chosen point, g is the acceleration due to gravity, v is the fluid velocity along a streamline at the point, h is the height of the point above a selected reference level, and ρ is the density of the fluid. By constant, we mean that the sum is constant along a streamline.

Energy Density

Previously, pressure has been defined as the force per unit area. The SI unit of pressure, the pascal, corresponds to 1 newton per meter squared ($N\,m^{-2}$). As work must have units of force times distance, we see that from $W = Fd$, which can be expressed as $F = \frac{W}{d}$ 1 N $=1\,J\,m^{-1}$, and so

$$1\,Pa = 1\,N\,m^{-2} = \frac{1\,J\,m^{-1}}{m^2} = 1\,J\,m^{-3} \tag{14.4}$$

so pressure can also be thought of as energy per unit volume.

Perhaps a better way to understand this is if we consider a gas under pressure in a cylinder with a movable piston. Suppose the gas exerts a force F on the piston. If the piston is allowed to move a very small distance, Δx, then some work, $W = F\Delta x$, has been done on the piston by the gas. Thus a fluid under pressure can do work on some other system, and in order to increase the pressure of a fluid, work must be done on it.

Each of the contributing terms in Eq. (14.3) has units of $J\,m^{-3}$ or energy per unit volume. As we have seen previously, density is defined as the mass per unit volume, $\rho = \frac{m}{V}$, and so it is easy to see how the second two terms, $\frac{1}{2}\rho v^2$ and $\rho g h$, are the kinetic energy per unit volume and gravitational potential energy per unit volume, respectively.

Pressure and Velocity

Another way of writing Bernoulli's Equation relates the parameter values at two points on a streamline, labelled 1 and 2

$$P_1 + \frac{1}{2}\rho v_1^2 + \rho g h_1 = P_2 + \frac{1}{2}\rho v_2^2 + \rho g h_2 \tag{14.5}$$

Consider a case where there is no change in height, as in Figure 14.3 where h is fixed, so the gravitational potential energy is not changing. Rearranging Bernoulli's equation

$$\frac{1}{2}\rho v_2^2 - \frac{1}{2}\rho v_1^2 = P_1 - P_2 \tag{14.6}$$

So, the change in pressure gives us the change in kinetic energy per unit volume. If, as in Figure 14.3, the velocity at 2 is higher, then the pressure at 2 is lower.

The signs in the equations above can seem a bit counterintuitive at first sight. Why would higher velocity mean *lower* pressure? We know from the continuity equation that the fluid speeds up as it moves from point 1 to 2, so it gains kinetic energy. As the pipe is horizontal, there is no change in the fluid's gravitational potential energy. Positive work must have been done on the mass of fluid in order to speed it up. The only force available to do this work is the pressure difference between points 1 and 2. To do positive work on a mass of fluid moving from point 1 to 2, the pressure must be higher at point 1 than at point 2, so the pressure is indeed lower in the more constricted region where the velocity is higher.

Example 14.2 *Bernoulli's equation*

Problem: A pipe in an industrial plant is designed to carry a fluid of density 1500 kg m^{-3} at a speed of 3 m s^{-1}. Any faster than this and the flow could become turbulent, with undesirable results. Any slower than this and the fluid could start to congeal on the sides of the pipe. The fluid is to be carried from a holding tank which is at a pressure of P_t to a manufacturing line at atmospheric pressure (P_{atm} = 100 kPa) which is 2.5 m below the holding tank. At what pressure must the tank be maintained?

Solution: This pipe is designed to keep the fluid velocity constant despite lowering the height of the fluid. We can easily see that as the fluid gets lower the height term in Bernoulli's equation will reduce, and we already know that the velocity of the fluid remains constant. We can write down Bernoulli's equation and solve for the pressure in the tank:

$$P_t + \frac{1}{2}\rho v_t^2 + \rho g h_t = P_l + \frac{1}{2}\rho v_l^2 + \rho g h_l$$

$$P_t + \rho g h_t = P_l + \rho g h_l$$

$$P_t = P_l + \rho g (h_l - h_t)$$

This indicates that the pressure in the tank must be lower than the pressure at the manufacturing line at the other end of the pipe. This make sense; as the fluid gets lower it loses gravitational potential energy ($h_l - h_t = -2.5$ m) and as it is travelling at a constant velocity it must gain energy in the form of increased pressure.

$$P_t = P_l + \rho g (h_l - h_t)$$
$$= 100 \times 10^3 \text{ Pa} + 1500 \text{ kg m}^{-3} \times 10 \text{ m s}^{-1} \times (-2.5 \text{ m}) = 62.5 \times 10^3 \text{ Pa}$$

The pressure in the holding tank must be 62.5 kPa.

Applications of Bernoulli's Equation

Fluid Flow Out of a Tank

How fast will water flow from the outlet pipe of a tank and what does it depend on? We can apply Bernoulli's equation to show that it depends on the height of water above the outlet, provided the surface area of the tank, A_s, is significantly greater than the cross-sectional area of the outlet, A_o. Figure 14.6 shows such a case.

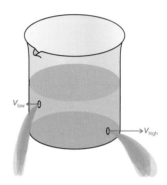

Figure 14.5 Two holes in a beaker full of fluid. The difference in pressure between surface and outlet creates a force that accelerates the fluid out of the hole. The greater the pressure difference, the higher the resultant velocity of the stream of fluid.

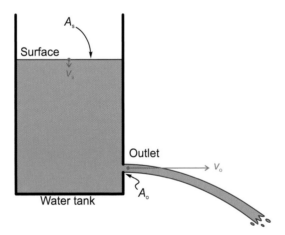

Figure 14.6 Speed of water flowing out a hole in a tank depends only upon the height of liquid in the tank and that liquid's density

Applying Bernoulli's equation to water at the surface (subscript 's') of the tank and at the outlet ('o') of the tank we have

$$P_s + \frac{1}{2}\rho v_s^2 + \rho g h_s = P_o + \frac{1}{2}\rho v_o^2 + \rho g h_o$$

Now we assume that both the surface of the tank and the tank outlet are at atmospheric pressure, so

$$\frac{1}{2}\rho v_s^2 + \rho g h_s = \frac{1}{2}\rho v_o^2 + \rho g h_o$$

If the surface area of the tank is much larger than the cross-sectional area of the outlet, then the continuity equation suggests

$$v_s \ll v_o$$

In other words, the speed at which the surface of water in the tank drops is much less than the speed at which water leaves the tank outlet pipe. In this case, we have that

$$\rho g h_s \approx \frac{1}{2}\rho v_o^2 + \rho g h_o$$

and we can neglect v_s. Solving for v_o gives

$$v_o = \sqrt{2g(h_s - h_o)}. \tag{14.7}$$

This relationship between the speed of outflow (sometimes referred to as the speed of efflux) and the distance from the liquid surface (called the head height) is known as **Torricelli's theorem**.

Plaque Deposits and Aneurysms

We have already looked at the change in pressure in a liquid when the height is unchanged and the liquid speeds up or slows down. This has consequences for blood vessels that are narrowed by plaque deposits, or widened at the site of an aneurysm.

In the case of narrowing (called *stenosis*) by plaque deposits, the blood velocity must be increased, which decreases the pressure, and may result in further narrowing, leading the artery to close entirely. When the artery is narrowed, the flow will also become more turbulent, possibly damaging the arterial wall. To make matters worse, the decrease in elasticity changes the wall's vibrational characteristics, which can lead to resonant vibrations which can dislodge the deposits.

An aneurysm is a localised, balloon-like bulge in an artery. As the radius increases and velocity decreases, the pressure increases. As the wall is already likely to be weakened, this further increases the chances of a rupture.

Example 14.3 *Aneurysm*

Problem: An aneurysm forms in a small blood vessel through which blood travels at 3 m s^{-1}. The diameter of the blood vessel increases by 20%. What is the increase in pressure inside this aneurysm? ($\rho_{\text{blood}} = 1060$ kg m^{-3}.)

Solution: We will assume that the blood vessel is nearly horizontal and so we can drop the height terms in Bernoulli's equation. This leaves us with

$$P_1 + \frac{1}{2}\rho v_1^2 = P_2 + \frac{1}{2}\rho v_2^2$$

$$\Delta P = P_2 - P_1 = \frac{1}{2}\rho\left(v_1^2 - v_2^2\right)$$

We will need to use the continuity equation to find how v_1 and v_2 are related.

$$A_1 v_1 = A_2 v_2$$

$$v_2 = \frac{A_1}{A_2}v_1 = \frac{1}{1.2^2}v_1 = 0.69 v_1$$

this gives us

$$\Delta P = \frac{1}{2}\rho\left(v_1^2 - v_2^2\right)$$

$$= \frac{1}{2}\rho\left(0.69^2 v_1^2 - v_1^2\right) = 0.52\frac{1}{2}\rho v_1^2$$

$$= 0.52 \times \frac{1}{2} \times 1060 \text{ kg m}^{-3} \times \left(3 \text{ m s}^{-1}\right)^2 = 2.48 \times 10^3 \text{ Pa}$$

14.5 Summary

Key Concepts

continuity equation A statement of the conservation of some quantity, which in the this case is mass. The rate at which mass enters a system is equal to the rate at which mass leaves the system, which, for an incompressible fluid in the absence of any sources or sinks, results in a constant volume flow rate along a closed pipe or set of pipes.

Bernoulli's law A statement of conservation of energy for fluids. The sum of the pressure, the gravitational potential energy per unit volume, and the kinetic energy per unit volume is conserved along a streamline.

Torricelli's theorem The speed of efflux through an outlet pipe is proportional to the square root of the head height: $v = \sqrt{2gh}$.

Equations

$$\mathscr{F} = \frac{\Delta V}{\Delta t} = Av$$

$$A_1 v_1 = A_2 v_2$$

$$P + \frac{1}{2}\rho v^2 + \rho g h = \text{constant}$$

14.6 Problems

14.1 A large artery has a diameter of 7 mm. This artery divides into two identical smaller arteries, the velocity of the blood in the smaller arteries is the same as the velocity of the blood in the larger artery. What is the diameter of the smaller arteries?

14.2 The diameter of a blood vessel narrows by 70% due to the presence of a plaque on the blood vessel walls.

   (a) By what factor does the blood velocity increase?

   (b) If the blood velocity in a normal blood vessel is 0.15 m s^{-1} and the systolic blood pressure is 130 mmHg, what is the systolic blood pressure in the narrowed vein (in mmHg)?

14.3 A small plastic pipe carries water horizontally at a speed of 10 m s^{-1}. A section of the pipe bulges out so that the radius is twice that of the rest of the pipe. If the gauge pressure in the pipe is ordinarily +90 kPa what is the gauge pressure in the bulge (in kPa) (the density of water is 1000 kg m^{-3})?

14.4 In this example we will construct a simple model of the circulatory system to investigate the rate at which cuts bleed. In this model we will assume that blood is a Newtonian fluid at all length scales so that the equations of fluid flow which we have been studying will apply. We will also assume that the effects of viscosity may be ignored. These will NOT be a good approximations for real blood in capillaries as the diameter of capillaries is about the same as the size of the red corpuscles and this has a major effect on blood flow in capillaries. However, our model will serve as an indication of the effects of Newtonian fluid flow in circulatory systems.

A large artery has a diameter of 7 mm and carries blood which flows with a peak velocity of 0.15 m s^{-1}. This vessel eventually feeds a network of capillaries which together have an area approximately 400 times that of the large artery which feeds into them. In this model, the capillaries are identical to each other and have a diameter of 7.5 μm.

   (a) Suppose that the diastolic blood pressure is 130 mmHg at the level of the heart and the blood velocity in the large artery at the heart is 0.15 m s^{-1}. What is the blood velocity in the artery at a point 1 m below the heart? (The density of blood is 1050 kg m^{-3})

   (b) If the artery is severed at a at a point 1 m blow the heart, what is the maximum velocity of blood flow from the artery?

   (c) What is the blood velocity in a capillary in the capillary net at this point (1 m below the heart)?

   (d) What is the blood pressure in the capillary net at this point (in mmHg and with the assumptions discussed above)?

   (e) If a capillary is severed, what is the blood velocity leaving the wound? (again on the basis of this model)

14.5 At what rate (Pa m^{-1}) does the pressure need to change in a vertical pipe filled with water to keep the velocity of the water flowing through it constant?

14.6 A hole is punched in the side of a tank below the surface of the fluid in it. The fluid is coming out at a speed of 7 m s^{-1}.

   (a) How far below the surface of the fluid was the hole punched?

   (b) If the volume flow rate of the fluid coming out of the tank is 0.5 L min^{-1}, what is the radius of the hole?

14.7 A hydroelectric power plant draws water from a lake whose surface is 55 m above the turbines. It draws the water through a pipe with radius 1.2 m. (P_{atm} = 100 kPa, ρ_{water} = 1000 kg m^{-3})

   (a) If, at point A, 55 m below the surface of the lake, the pipe is horizontal and the water is flowing through it at a rate of 9 m s^{-1}, at what pressure is the water? (Hint: compare this point to the surface of the lake using Bernoulli's Law.)

   (b) In order for the turbines to work most efficiently the water should enter them at a speed of 20 m s^{-1}. In order to achieve this the pipe narrows to what radius just before it enters the turbines?

   (c) What is the pressure just before the water enters the turbines?

   (d) After passing through the turbines the water is now open to the atmosphere again. At what speed is it traveling (assuming that only a negligible fraction of the energy contained in the flow is removed by the turbines).

FLUID DYNAMICS OF VISCOUS FLUIDS

15

15.1 Introduction

In the last chapters we ignored friction within the fluid, and between the fluid and the material it was flowing past. Ignoring this friction is not a valid assumption in many cases. When blood or other fluids are injected into a person's vein, they need to go through a narrow-diameter needle, providing a large amount of friction and resistance to fluid flow. Work needs to be done against this friction, and there needs to be a force creating a pressure difference between the two ends of the needle to keep the fluid moving. The fluid viscosity also influences how readily the flow ceases to be smooth and laminar.

In this chapter we will see how friction affects fluids by investigating fluid viscosity and turbulence.

Key Objectives

- To understand what viscosity is.

- To understand the relationship between the viscosity, pressure and flow rate.

- To develop a qualitative understanding of turbulent fluid flow.

15.2 Viscosity

When a shear stress is applied to a solid it causes it to deform. We saw in Chapter 10 that a given shear stress will result in a certain amount of deformation or shear strain within a solid. If we apply a constant shear stress to a solid, we will have a constant shear strain. To increase the amount of deformation of a solid we need to increase the amount of shear stress. When the shear stress is removed, the solid returns to its initial shape.

Shear stresses have a different effect on fluids; in fact, this is a useful definition of a fluid. When a shear stress is applied to a fluid, it causes it to flow, that is, to deform continuously.

Imagine a layer of fluid between two plates. If we apply a force F over an area A to give a shear stress of $\frac{F}{A}$, the fluid deforms at a constant rate as long as this shear stress is applied. The longer this shear stress is applied, the greater the deformation.

Viscosity is the resistance of the fluid to flow. We define it by finding the shear stress required to generate a shear-strain rate of one per second. (If you're thinking 'one what?', remember that strain is dimensionless – it is a ratio of distances.)

Imagine a fluid with depth L between two plates of surface area A. To cause the top plate to move at the constant speed v, which will give us a constant shear-strain *rate*, we need to apply a steady force F to the top plate which just balances the kinetic friction force of the fluid on the plate. To keep the lower plate stationary, an equal and opposite force needs to be applied to it, so a shear stress of $\frac{F}{A}$ to the fluid causes the top layer of fluid to move at a velocity v, whilst the bottom layer of fluid is stationary. The

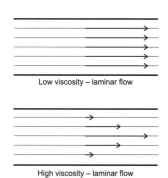

Figure 15.1 A viscous fluid experiences shear forces between layers of flow. This results in non-uniform flow speeds throughout the cross section of a flow.

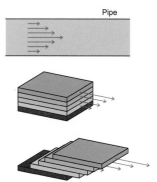

Figure 15.2 A viscous fluid experiences shear forces between layers of flow. These result in different flow rates in different parts of the fluid.

shear strain in the fluid is constantly increasing, and the rate of increase of shear strain is equal to the change in strain, divided by the time interval Δt:

$$\text{Rate of change of strain} = \frac{\Delta x / L}{\Delta t} = \frac{v}{L} \tag{15.1}$$

It is found experimentally that for some fluids the shear stress is proportional to the shear strain rate, and the proportionality constant is the **fluid viscosity**, η. (η is the Greek letter eta.)

$$\frac{F}{A} = \eta \frac{v}{L} \tag{15.2}$$

Viscosity is a property of the fluid. Fluids with high viscosity do not flow readily; a large shear stress is required to produce a given shear strain rate or flow rate. Fluids with a low viscosity, e.g. water, flow readily. Viscosity has units of N s m^{-2}, which is equivalent to Pa s. The **poise**, a non-SI unit, is sometimes used instead, where 1 N s m^{-2} = 10 poise.

Poiseuille's Law

Flow of a viscous fluid along a pipe, whether that is a water pipe, an artery or a hypodermic needle, requires a pressure difference to overcome the fluid's viscosity. Anyone who has ever had a thickshake knows that it is much harder to suck through a straw than a glass of cola is, and they often come served with a larger diameter straw. The narrower the pipe, the larger the required pressure difference, and the longer the pipe the larger the required pressure difference. The higher the viscosity of the fluid the larger the required pressure difference. All of these are related quantitatively by **Poiseuille's law**. This states that the volume flow rate, $\mathscr{F}$, for a fluid of viscosity η through a cylindrical pipe of length l and radius r, when the pressure difference between the ends is ΔP is

$$\mathscr{F} = \frac{\Delta P \pi r^4}{8 \eta l} \tag{15.3}$$

Example 15.1 *Drug delivery*

Problem: A drug is being delivered into a patient's arm at a rate of 10 mL min^{-1}. The drug is being delivered from a syringe through a 5 cm long needle with an internal diameter of 1 mm. If patient's blood pressure is 110 mmHg (+15.0 kPa gauge pressure, and ignoring the variation from systolic to diastolic), what must the pressure in the syringe be? ($\eta_{\text{drug}} = 8.90 \times 10^{-4}$ Pa s)

Solution: The volume flow rate is given in non-standard units so we will need to convert into SI units.

$$\mathscr{F} = 10 \times 10^{-3} \text{ L min}^{-1} \times \frac{1 \times 10^{-3} \text{ m}^3 \text{L}^{-1}}{60 \text{ s min}^{-1}} = 1.67 \times 10^{-7} \text{ m}^3 \text{s}^{-1}$$

To solve this problem we use Poiseuille's law:

$$\mathscr{F} = \frac{\Delta P \pi r^4}{8 v l}$$

$$\Delta P = \frac{8 \mathscr{F} v l}{\pi r^4} = \frac{8 \times 1.67 \times 10^{-7} \text{ m}^3 \text{s}^{-1} \times 8.90 \times 10^{-4} \text{ Pa s} \times 0.05 \text{ m}}{\pi \times \left(0.5 \times 10^{-3}\right)^4} = 300 \text{ Pa}$$

Which means the pressure in the syringe must be 300 Pa larger than that in the vein, or at 15.3 kPa gauge pressure.

Blood Viscosity

Blood is a heterogeneous mixture, not a simple liquid, as the blood cells are physically separate from the plasma. As a result of this, its behaviour is more complicated than a simple proportional relationship between shear stress and strain rate. Its viscosity is not constant, but depends on a number of factors.

The *hematocrit* is the volume fraction of the blood composed of red blood cells. A higher hematocrit leads to a higher viscosity. In general this will lead to higher blood pressures, as a greater blood pressure is required to push the blood through the circulatory system. Males in general have a higher hematocrit (47% versus 42% for women) and hence have higher blood viscosity. This is a possible factor in their higher rates of hypertension (high blood pressure) and consequently greater risk of heart disease and strokes.

At high altitude, the number of red blood cells is increased – this is one of the body's responses to *hypoxia*, an inadequate supply of oxygen. This leads to higher blood viscosity, higher blood pressure and a greater risk of complications arising from raised pressure and reduced flow velocity.

The heterogenous nature of blood means that, unlike a simple liquid, its viscosity also depends on the velocity of flow. There is a positive-feedback loop between blood velocity and viscosity at low blood speeds. At high blood speeds, the blood cells do not group together, and the blood behaves like a low-viscosity mixture of two liquids. At low blood speeds, there is a greater risk of red blood cells stacking, causing the blood to behave like solid particles suspended in a liquid, giving a higher viscosity. The higher viscosity leads to slower flow speeds and more stacking in a positive feedback loop. This can happen in anaphylactic shock, when release of histamine into the blood vessels causes the vessels to dilate; the increased cross-sectional area causes a reduction in flow speed, stacking of red blood cells, and an increase in viscosity. The increase in viscosity further slows the blood flow, resulting in more red-blood-cell stacking and a further increase of viscosity.

15.3 Turbulence

So far, we have been concerned only with **laminar flow**, where the fluid flows in smooth layers without mixing. When the velocity of a fluid is increased, the flow becomes more complex, with mixing between layers and *eddies*, where the flow is in a different direction to the net fluid flow (see Figure 15.3). The speed at which the flow becomes turbulent depends on the viscosity of the fluid, the density of the fluid and the dimensions and shape of the pipe it is flowing through. We can define a single, dimensionless number which takes account of all of these properties and determines whether the fluid flow is laminar or turbulent. This number is known as the **Reynolds number**, Re, and is defined as

$$Re = \frac{\rho v L}{\eta} \tag{15.4}$$

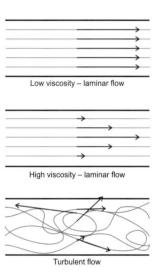

Low viscosity – laminar flow

High viscosity – laminar flow

Turbulent flow

Figure 15.3 Laminar flow is characterised by straight or smoothly curved flow lines. Given the right conditions flow can become turbulent, in which flow is complex and mixing occurs

where ρ is the fluid density, η is the fluid viscosity and L is a characteristic length, which is chosen by convention and depends on the shape. For flow through a pipe with a circular cross section, this may be the diameter or the radius, so frequently a subscript is used to indicate the dimension that was chosen. Any examples in this book will use *diameter* of the pipe the fluid is flowing through for L.

For fluid flow in a closed pipe, if Re is less than 2000, the flow is laminar, and if Re is more than about 3000, the flow is turbulent.

Example 15.2 *Reynolds number*

Problem: A drug is being delivered into a patient's arm at a rate of 319 mL min^{-1}. The drug is being delivered from a syringe through a needle with an internal diameter of 1 mm. This drug has a density of 1050 kg m^{-3}. Is the flow laminar or turbulent? ($\eta_{drug} = 8.90 \times 10^{-4}$ Pa s)

Solution: The velocity of the fluid must be found from the flow rate and needle size before the Reynolds number can be calculated

$$\mathscr{F} = 319 \times 10^{-3} \text{ L min}^{-1} \times \frac{1 \times 10^{-3} \text{ m}^3 \text{L}^{-1}}{60 \text{ s min}^{-1}} = 5.32 \times 10^{-6} \text{ m}^3 \text{s}^{-1}$$

and so the velocity of the fluid through the needle is

$$v = \frac{\mathscr{F}}{A} = \frac{\mathscr{F}}{\pi r^2} = \frac{5.32 \times 10^{-6} \text{ m}^3 \text{s}^{-1}}{\pi \times \left(0.5 \times 10^{-3} \text{ m}\right)^2} = 6.77 \text{ m s}^{-1}$$

The Reynolds number for this fluid flow is

$$\text{Re} = \frac{\rho v L}{\eta}$$

$$= \frac{1050 \text{ kg m}^{-3} \times 6.77 \text{ m s}^{-1} \times 1 \times 10^{-3} \text{ m}}{8.90 \times 10^{-4} \text{ Pa s}} = 8000$$

This is above the cutoff point for turbulent flow.

15.4   Summary

Key Concepts

viscous fluid A viscous fluid is one where we cannot ignore the effects of friction within the fluid and between the fluid and neighbouring interfaces.

viscosity (η) A measure of the internal friction of a fluid. It is a property of a particular fluid, and is a measure of the fluid's resistance to flow. The viscosity has units of N s m^{-2}, which are the same as Pa s.

Poiseuille's law The volume flow rate of a viscous fluid along a pipe is proportional to the pressure difference and pipe radius to the power of four, and is inversely proportional to the viscosity and pipe length.

turbulent flow Non-laminar flow. The flow is irregular and complex, with mixing and eddies. This occurs at higher velocities or where there are objects in the flow producing large changes in velocity.

Reynolds number (Re) A dimensionless quantity that allows us to distinguish between laminar and turbulent flow.

Equations

$$\frac{F}{A} = \eta \frac{v}{L}$$

$$\mathscr{F} = \frac{\Delta P \pi r^4}{8 \eta l}$$

$$\text{Re} = \frac{\rho v L}{\eta}$$

15.5 Problems

15.1 Blood in large arteries and veins may be treated as a Newtonian fluid, which means we are able to ignore the effects of the cellular material in blood on blood flow. The viscosity of blood with a normal red blood cell count is 2.7×10^{-3} N s m^{-2}. Suppose we are considering blood flowing at a speed of 0.3 m s^{-1} in a large blood vessel with a radius of 2 mm.

(a) What is the pressure drop along a 2 cm length of this blood vessel?

(b) At approximately what velocity will the blood flow definitely be turbulent (take the density of blood to be 1050 kg m^{-3})?

15.2 Saline solution is delivered into a patient's vein through a needle. The saline solution has a viscosity of 0.37×10^{-3} N s m^{-2}, a density of 1060 kg m^{-3}, and is delivered through a 7 cm long needle with an internal diameter of 0.24 mm directly into the patient's vein in which the blood pressure is 130 mmHg. The saline solution must be delivered at a flow rate of 0.04×10^{-6} m^{3}s^{-1}. How high must the saline solution be suspended in order to achieve this flow rate? (Ignore variations between systolic and diastolic blood pressure when doing this question. Also assume that the viscosity of the saline solution is low enough that it does not affect the flow of the solution through the IV tube, only the needle itself.)

15.3 The viscosity of cerebrospinal fluid is 0.8×10^{-3} N s m^{-2}. What pressure difference is required to produce a cerebrospinal fluid flow rate of 0.1 m s^{-1} in 1 cm long tubes of the following diameters: 1 mm, 5 mm, 1 cm?

15.4 A small blood vessel near the skin surface has a radius of 10 μm, a length of 1 mm and the pressure drop along the blood vessel is 2.5 Pa (about 19 mmHg). The viscosity of blood is 2.7×10^{-3} N s m^{-2}.

(a) What is the volume flow rate of blood through this blood vessel? What is the velocity of blood flow?

(b) Vasodilation causes the radius of this blood vessel to increase to 12 μm, while leaving the pressure drop along the vessel unchanged. What is the volume flow rate through this blood vessel now? What is the velocity of blood flow?

15.5 Water flows via gravity from a high water tank to a point on the ground some distance away through a hose of diameter 1 cm and length 100 m. How high must the tank be for the flow rate to equal 1 L min^{-1}? ($\eta_{water} = 8.90 \times 10^{-4}$ Pa s, $\rho_{water} = 1000$ kg m^{-3})

15.6 With what minimum speed would blood need to travel through a small blood vessel with a radius of 1 mm before the flow was turbulent? ($\eta_{blood} = 2.7 \times 10^{-3}$ Pa s, $\rho_{blood} = 1050$ kg m^{-3})

15.7 If the blood vessel from Problem 15.6 had water flowing through it instead of blood, approximately what minimum speed would the flow become turbulent?

MOLECULAR TRANSPORT PHENOMENA

16.1 Introduction

Things begin to move only when they experience an unbalanced force of some kind. The transport of mass, energy, momentum or electricity requires a driving force. Throughout this book we have looked at, and will continue to look at, transport phenomena, even though we don't often call them that specifically. In this chapter, we wish to introduce one kind of transport – transport of mass in the form of individual molecules.

This process has more in common with viscous fluid flow than is apparent at first glance; the viscosity of a fluid is related the rate at which momentum 'diffuses' in response to a velocity gradient, and the diffusion constant is related to the rate at which mass is transferred in response to a concentration gradient.

Our goal here is the brief introduction of diffusion and osmosis, which are both important in biological systems, but on the way we will point out how diffusion is analogous to other processes that we cover in this book.

Key Objectives

- To understand diffusion.

- To understand osmosis and osmotic pressure.

- To be able to use Fick's Law to understand the rate of diffusion of a gas through a membrane.

16.2 Diffusion

If you were to take all the oxygen molecules in a room and crowd them into one corner, leaving the nitrogen molecules uniformly spread throughout the room and then release them, then eventually these oxygen molecules would be evenly spread out too. The process by which these oxygen molecules spread out uniformly throughout the nitrogen-filled room is known as **diffusion** – the net migration of molecules from a region of high concentration to a region of low concentration (see Figure 16.1).

Diffusion is a transport process which often occurs in conjunction with convection. Particles of liquids, solids and gases undergo spontaneous movement due to thermal motion and tend to intermingle, which results in the movement of molecules from regions of higher concentration to those with lower concentrations. Diffusion is responsible for the transport of oxygen from the air in the lungs into the bloodstream. The process of dialysis to cleanse the blood and eliminate waste products that is performed by the kidneys, or by machines for patients with impaired kidney function, is a diffusion one.

We will simply state without proof that the average distance x_{rms} travelled by a molecule of type A in a space filled with type B molecules in a time t is given by

$$x_{\text{rms}} = \sqrt{2D_{\text{AB}}t} \qquad (16.1)$$

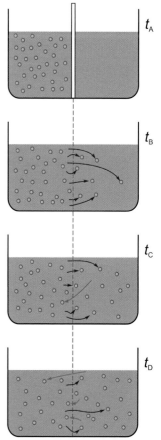

Figure 16.1 Diffusion results in a net movement of molecules from regions of high concentrations to regions of low concentrations. At time t_A a dissolved solute is confined to one half of a volume of fluid. When the barrier is removed at t_B the random movement of molecules of solute results in a net movement of these molecules across the imaginary line where the barrier used to be. As time progresses and the concentration of solute on the right increases the net rate of movement of solute across our imaginary line gets lower. At time t_D the concentrations of solute are equal in both halves of the volume and the net rate of movement across the imaginary line is zero.

The rms subscript

The subscript 'rms' used on the average distance x_{rms} stands for 'root mean square'. This is a kind of statistical measure that is often more relevant than the straight average. Consider the sine function – it has a positive value and a negative value equally often, so over an interval with a whole number of cycles, the average is zero. The rms value isn't zero. $\sin(x)$ has an rms value of $1/\sqrt{2}$ over a cycle. The mathematical definition for a collection of N values is $x_{rms} = \sqrt{(x_1^2 + x_2^2 + x_3^2 + \ldots + x_N^2)/N}$, that is, the square root of the mean of the squared quantity.

Flux

The flux is the amount of something transported per unit area per unit time. This can be used to describe the amount of anything – mass, energy, number of charges – that passes through some area each second. The diffusive flux is measured in kilograms (or moles) per square metre per second.

where D_{AB} is the diffusion constant for type A molecules diffusing through a substance B. The diffusion constant has units of $m^2\,s^{-1}$.

Fick's Law

> **Key concept:**
> **Fick's Law** states that the rate of diffusion of one material through another is proportional to the gradient of its concentration.

This is typically stated mathematically as

$$J_A = -D_{AB}\frac{\Delta c_A}{\Delta x} \tag{16.2}$$

where J_A is the diffusive flux (see the margin note), D_{AB} is the diffusion constant for A travelling through B, and the last term is the concentration gradient, that is, how rapidly the concentration changes with distance. The negative sign indicates that the direction is from high concentration to low.

This can be used to produce a form of Fick's Law useful in biological situations. For transport of gas across a membrane:

$$\text{rate of diffusion} = \frac{AD}{d}(P_1 - P_2) \tag{16.3}$$

where P is the partial pressure of the gas in question on a given side of the membrane, d is the membrane thickness, A is the surface area of the membrane and D is the diffusion constant, which will depend on the molecule and membrane. We will cover partial pressure of gases in the Thermodynamics topic.

Relationship to Other Transport Processes

The way mass in the form of molecules diffuses from high to low concentration is analogous to some other transport processes that we will cover later in the book, namely thermal conduction and electrical conduction.

The flux of thermal energy, how much energy is transported per unit area per unit time, depends on the *temperature gradient* and the *thermal conductivity, k* (see Equation (21.1)).

The movement of electrical charges is also similar in nature. Charges move in response to a gradient of electrical potential, and the proportionality constant in this case is the *electrical conductivity*. Ohm's Law, which we most often use in the form $V = IR$ (and we will encounter in Electricity and DC Circuits as Eq. (27.6)) can be written in form like Eq. (16.2).

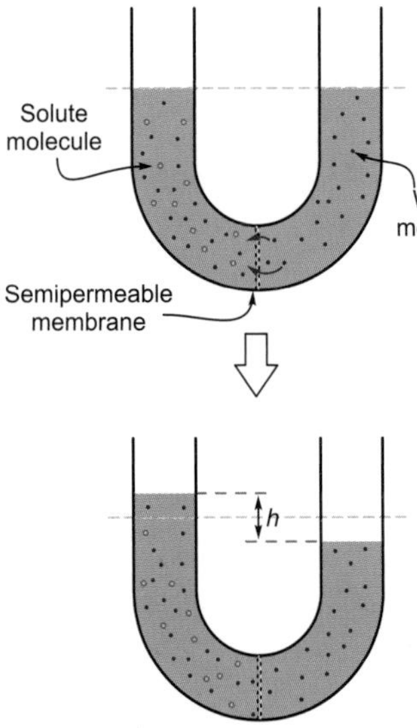

16.3 Osmosis

Osmosis is the diffusion of water through a semipermeable membrane, from high to low concentration (as in Figure 16.2. A semipermeable membrane is a one through which only some smaller molecules, such as water, can pass.

Osmosis regulates the movement of water across cell membranes. If a cell is placed in a *hypotonic* solution (one with lower solute concentrations and hence higher water concentration than the cell), water will move by osmosis into the cell causing the cell to swell and eventually rupture. If a cell is placed in a *hypertonic* solution (one with higher solute concentration and hence lower water concentration than the cell), then water will move by osmosis out of the cell, resulting in dehydration of the cell.

Osmotic Pressure

If we have a semipermeable membrane with water in higher concentration on the right, so it has less solute dissolved in it, the concentration gradient will cause water to move

Figure 16.2 Osmosis across a semipermeable membrane causes the level of water in the region containing a dissolved solute to rise

from right to left (Figure 16.2 (top)). As water moves to the left side of the membrane, the pressure in the left-hand compartment will rise above that in the right. This pressure difference will tend to cause water to move from left to right. Water molecules will continue to move in this way until the pressure difference becomes high enough that water movement to the right due to the pressure difference matches the water movement to the left due to the concentration difference. This is shown in Figure 16.2 (bottom).

The pressure difference at which there is no net movement of water across the membrane is known as the **relative osmotic pressure**. If one solution is pure water, the back pressure that stops osmosis is called the **osmotic pressure**. Also, the application of a pressure difference by some external agent can alter the diffusion rate across the membrane. Applying sufficient pressure will result in reverse osmosis. This is one technique used for the desalination of sea water. The fresh water at the Scott Base, New Zealand's Antarctic station, is supplied by a reverse osmosis system.

Example 16.1 *Diffusion*

Problem: The diffusion constant of oxygen through water is an important limiting factor in the size of biological organisms that do not have active transport mechanisms to circulate oxygen through their systems. The diffusion constant of oxygen through water is 8×10^{-10} m s^{-1}. On average, how long will it take an oxygen molecule to diffuse 1 mm through water?

Solution: The rms distance that a given molecule will diffuse in a given time is given by

$$x_{rms} = \sqrt{2D_{AB}t}$$

In this case $D_{AB} = D_{O_2H_2O} = 8 \times 10^{-10}$ m s^{-1} and $x_{rms} = 1 \times 10^{-3}$ m. We can rearrange the equation to give

$$t = \frac{x_{rms}^2}{2D_{O_2H_2O}} = \frac{\left(1 \times 10^{-3} \text{ m}\right)^2}{2 \times 8 \times 10^{-10} \text{ m s}^{-1}} = 625 \text{ s}$$

16.4 Applications to Biological Systems

Diffusion and the Lung

One example of how Eq. (16.3) is applicable to the human body is in the lung. Oxygen diffuses into the bloodstream through the alveoli walls. Any medical condition that acts to thicken the barrier and increase d (such as inflammation, or a buildup of mucus caused by cystic fibrosis) will reduce the lung's ability to oxygenate the blood. Emphysema, a condition that destroys the alveolar walls, reduces the surface area (A) available for diffusion, and causes breathing problems.

Contact Lenses and Diffusion

Unlike other cells in the human body, the cornea has no blood supply, and receives its oxygen via diffusion from the air. Contact lenses block this to a degree. Conventional contact lenses only allow oxygen to be delivered to the cornea by allowing some oxygen-rich tear fluid under the lens when it moves during blinking. At night, they need to be removed. All-day lenses are made of a material through which oxygen can diffuse, and so they can be worn at night without depriving the cornea of oxygen.

16.5   Summary

Key Concepts

diffusion The net transport of molecules from a region of higher concentration to one of lower concentration.

osmosis The diffusion of a solvent (usually water) from solution with low solute concentration to one with higher solute concentration through a semipermeable membrane.

Equations

$$x_{rms} = \sqrt{2D_{AB}t}$$

$$\text{rate of diffusion} = \frac{AD}{d}(P_1 - P_2)$$

16.6 Problems

16.1 An experiment is performed to determine the diffusion constant of ants on a smooth tabletop. A handful of ants is placed in the centre of a large flat tabletop and a photograph of the tabletop is taken 1 minute after the ants are released. The number of ants within 5 cm concentric bands are counted and the numbers are recorded below. (Hint: use the inner radius of each zone to solve this problem and look at the text box on page 146 entitled 'The rms subscript')

(a) What is the average displacement of the ants?

(b) What is the rms displacement of the ants?

(c) What is the diffusion constant for ants on a tabletop?

Number of Ants	Inner radius of circular band (in cm)
3	0
5	5
15	10
19	15
37	20
23	25
11	30
1	35
0	40
0	45
2	50

16.2 The size of spherical aerobic bacteria is limited by the rate at which oxygen diffuses through water. A bacterium with a radius greater than about 10 μm is not able to obtain enough oxygen from the surrounding water to sustain itself. Given that the diffusion constant of oxygen in water is 8×10^{-10} m^2 s^{-1}, how long does it take oxygen in water to diffuse an rms distance of 10 μm?

16.3 The diffusion constant of ATP is 3×10^{-10} m^2 s^{-1}. How long would it take for ATP to diffuse across an average cell (about 20 μm across)?

16.4 A cylinder of water contains oxygen in solution. The cross-sectional area of the cylinder is 2 cm^2 and the length of the cylinder is 5 cm. At one end of the cylinder the concentration of oxygen is maintained at 0.2 mol m^{-3}, this concentration falls linearly to 0.05 mol m^{-3} at the other end of the cylinder. The diffusion constant of oxygen in water is 8×10^{-10} m^2 s^{-1}. How many moles of oxygen pass down this cylinder every second? What mass of oxygen passes down the cylinder each second?

III

Thermodynamics

Thermodynamics is the study of thermal energy, its movement, and its transformation. In the following chapters we will develop a quantitative understanding of thermal energy, heat and temperature, and how thermal energy is exchanged between different systems, especially the body. We will also focus on how these concepts are important for understanding human metabolism and how we interact with our environment. The human body operates within only a narrow temperature range because the rates of the biochemical reactions that sustain our lives, and the conformation of the many proteins in our bodies, rely on this. In this topic we will look at the regulatory processes that keep our core body temperature constant despite the changes in our external environment.

The following chapters will cover a wide range of topics: the fundamental concepts of temperature, thermal energy and thermal equilibrium; the properties of gases; the relationship between thermal energy and the states of matter; the properties of water-vapour/air mixtures; the transfer of heat between systems and the factors that affect it; and how these things are relevant to the human body.

TEMPERATURE AND THE ZEROTH LAW

17.1 Introduction

Temperature is an important property for the body because it is a sensitive indicator of health status. In this chapter we will introduce the concept of temperature by looking at temperature scales, temperature measurement, and how materials expand and contract in response to temperature changes.

Key Objectives

- To develop an understanding of temperature and how we measure it.

- To understand the concept of thermal equilibrium.

- To be able to calculate amounts of thermal expansion.

17.2 Thermal Equilibrium

Defining Temperature

Temperature is a measure of how hot or cold something is. There are several ways to describe what it is. A good starting point for understanding temperature is from the laws of thermodynamics. The law that defines temperature was not identified as such until after the other laws of thermodynamics, but was considered in many ways to be more fundamental, so it has become known as the *zeroth law*.

First, we need to take a look at the concept of **thermal equilibrium**. When two systems are in equilibrium, they are balanced in some way; they share a property. When systems are in thermal contact, they exchange energy until an equilibrium state is reached, and no more net energy transfer occurs. The **zeroth law of thermodynamics** states it in this way: if two systems, A and B, are in thermal equilibrium and a third system, C, is in thermal equilibrium with A, then it is also in thermal equilibrium with B. The property that the systems share is called temperature. When we use a thermometer we place it in contact with an object and allow it reach thermal equilibrium so it has the same temperature as the object. We can therefore say, for all practical purposes, temperature is what a thermometer reads.

Thermal Energy, Equilibrium and Heat

On a microscopic level, the atoms and molecules of matter are in constant motion. Any gas, liquid or solid has an amount of kinetic energy associated with this random motion. There is also energy associated with the rotational and vibrational motion of atoms within molecules. All together this energy is the **thermal energy**. The thermal energy of an object depends on the number of molecules in the object and molecular composition, as well as the object's temperature. At higher temperatures, the randomly moving atoms and molecules of an object move faster and the thermal energy is higher.

In Chapter 18 we will see that the thermal energy of the atoms of an ideal gas depends entirely on its temperature.

When two objects at different temperatures are placed in contact, collisions occur between the molecules in the two objects. In these collisions, some thermal energy is transferred. The result of many collisions is that thermal energy is transferred from the hotter object to the colder object. The thermal energy that is transferred in this way from a hot object to a cold object is known as **heat**. The movement of thermal energy due to a temperature difference is known as **heat transfer**. Heat transfer between the objects continues until the objects are at the same temperature. When the objects are at the same temperature they are said to be in **thermal equilibrium**.

Thermal equilibrium is a *dynamic* equilibrium as collisions between molecules continue to transfer thermal energy. At thermal equilibrium, equal amounts of thermal energy are being transferred in each direction.

Temperature Scales

The SI unit of temperature is the **kelvin**, symbol K, and it is one of the seven base units of the SI system, i.e., it is one of the fundamental units in terms of which other units are defined. The Kelvin scale is defined by two points: absolute zero (the temperature at which all thermal motion theoretically ceases) is defined to be 0 K, and the triple point of water (see margin note) is defined to be at 273.16 K. A 1 K temperature change is thus 1 part in 273.16 of the difference between these two points.

The **Celsius** scale (once known as the centigrade scale) is a temperature scale on which the freezing point of water is at zero degrees Celsius and the boiling point is at one hundred degrees Celsius when the environmental pressure is one atmosphere. It uses the symbol °C and was devised by Anders Celsius (1701–1744). A change of 1 °C is the same as a change of 1 K. As the triple point of water is at 0.01 °C,

$$0\,\text{K} = -273.15\,°\text{C} \tag{17.1}$$

$$273.16\,\text{K} = 0.01\,°\text{C} \tag{17.2}$$

The **Fahrenheit** scale is a temperature scale on which the freezing point of water is 32 degrees Fahrenheit and the boiling point is 212 degrees Fahrenheit, putting them 180 degrees apart. The symbol is °F. The original scale, devised by Daniel Gabriel Fahrenheit in 1724, used a mixture of ice, water and ammonium chloride to define the 0 °F temperature point. The reason for choosing this mixture, which has a temperature lower than the freezing point of water, is that it is an example of a frigorific mixture. A frigorific mixture is one which maintains a constant temperature as long as all components of the mixture are present. This temperature is independent of the starting temperature of the component of the mixture.

The human body is 70% water, and water is an important substance in the human environment. The Celsius scale, which has the freezing and boiling points of water at such easy-to-remember positions, is typically the most convenient scale for everyday purposes – measuring the temperatures of rooms, people, refrigerators, ovens and so on. (The Fahrenheit scale is rarely used outside the US.) It is also frequently the case that even when we are performing scientific calculations, we are still able to work with temperatures in Celsius, provided we are dealing with situations where only the temperature *change* matters, as the change has the same numeric value whether the units are °C or K. All the key formulae throughout this section of the book use SI units, so putting temperature in kelvin will *always work* (with the obvious exception of formulae that are for converting from one temperature scale to another).

17.3 Measuring Temperature

The Thermometer

Now that we have defined what temperature is, we will look at how we can measure it. The scales that were defined in the previous sections were created by defining the tem-

---

The triple point

The triple-point temperature of a pure substance is the unique temperature at which the three phases – solid, liquid and vapour – are in equilibrium. We will look at this in Chapter 19.

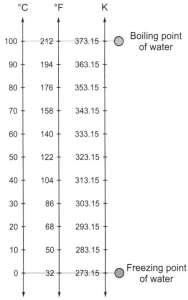

Figure 17.1 Comparison of the common temperature scales.

Temperature Conversion

T in kelvin = T in °C + 273.15 K

T in °C = T in kelvin − 273.15 °C

T in °F = $\frac{9}{5}\,T$ in °C + 32 °F

T in °C = $\frac{5}{9}(T$ in °F −32 °C$)$

Units in calculations

To save some time and effort, though, you can remember that you might not need to convert numbers from °C to K if the formula you are working with has a ΔT for change in temperature. If the formula uses only a T, you *must use kelvin*, as it almost certainly refers to *absolute* temperature.

perature at two (or more) reference points and interpolating a linear scale (i.e., equally spaced degrees). To measure temperatures in between, we use a **thermometer**, which is something that has a property that varies with temperature. By measuring this property, and interpolating between the fixed points, we can measure other temperatures.

In theory, any property of a material, system or device that has a unique value at each temperature in the range of interest can serve as a thermometer. Some examples of useful properties are the volume of a liquid, the pressure of a fixed volume of gas, the intensity of emitted radiation, the equilibrium vapour pressure of a liquid–gas mixture and electrical resistance.

Thermometers that are based on a fundamental physical law, and so do not need to be calibrated before use, are called *primary thermometers*. Most common thermometers are *secondary thermometers* – each one needs to be individually calibrated against another thermometer before it can be used to measure the true temperature.

The Constant-Volume Gas Thermometer

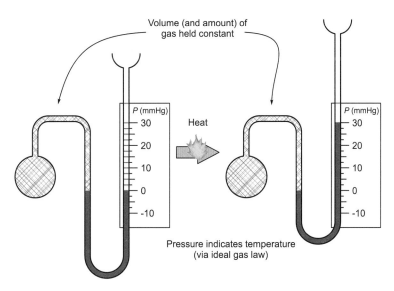

Figure 17.2 In a constant-volume gas thermometer, the volume of the air sample is held fixed by adjusting the pressure. The change in pressure gives an accurate measure of the change in temperature.

One type of primary thermometer is the constant-volume gas thermometer, illustrated in Figure 17.2. The gas-filled bulb of the thermometer is encased in the material whose temperature is to be measured, thus bringing the gas inside to thermal equilibrium with the material. The volume of the chamber is adjusted to ensure that it is equal to a particular reference volume. The pressure inside the chamber then indicates the temperature of the gas. This will be covered in more detail in the next chapter.

The constant-volume gas thermometer is special, because all gases (at low density) behave the same way, and so it is used to define the linear Kelvin scale. Another thermometer could have been chosen, but the fact that all gases provide the same scale suggests that there is something universal and fundamental about the constant-volume gas thermometer.

However this type of thermometer is not practical to use in many circumstances, and achieving highly accurate readings is hampered by a number of systematic errors, such as absorption and desorption of the gas by the walls of the chamber and thermal expansion.

Secondary Temperature Measurements

Liquid-Filled Thermometers

Liquid-filled thermometers are still the most common, and work on the thermal expansion of a liquid, usually mercury or ethanol. These thermometers contain a fixed mass

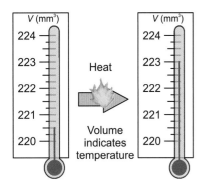

Figure 17.3 A liquid-filled thermometer works on the basis of the thermal expansion of the liquid. Commonly used liquids are alcohol and mercury.

of liquid whose volume is dependent on temperature. As the temperature increases, the liquid's volume increases and it moves further up the thin column inside the glass, as shown in Figure 17.3. A linear scale on the side is marked with temperatures. Clinical mercury thermometers used for taking oral temperatures have a kink in the tube just above the mercury reservoir which prevents the mercury returning to the reservoir until the thermometer is shaken sharply. This allows a person's temperature to be read after the thermometer is removed from the mouth.

Resistance as a Measure of Temperature

The electrical resistance of most substances changes with temperature in a reproducible fashion. A *thermistor* thermometer is a thermometer based on the temperature-dependent resistance of a special compound within a probe (the thermistor). The temperature probes used in many physics teaching laboratories are based on changes in resistance (see Figure 17.4). Modern digital thermometers used for measuring oral (mouth) temperature are also based on the temperature-dependent resistance of the probe. These are sometimes preferred over mercury thermometers today due to health concerns regarding mercury exposure from breakages.

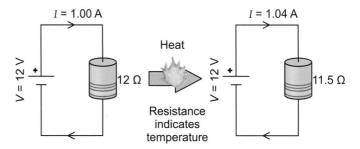

Figure 17.4 The electrical resistance of a resistor is temperature dependent. This dependence on temperature is the basis for the thermistor. Thermistors can be designed to have resistance that either increases or decreases with increasing temperature. The resistance of the thermistor illustrated here decreases with increasing temperature. In contrast, the resistance of metals generally increases with increasing temperature.

Infrared Sensors

Tympanic thermometers are designed to measure the temperature of the ear drum. This is a good place to measure body temperature because of the proximity of the ear drum to the hypothalamus, the part of the brain involved in temperature control, and their shared blood supply. Tympanic thermometers measure the infrared radiation emitted by the ear drum. The relationship between emitted radiation and temperature will be covered in Section 21.4.

Thermal Expansion of a Bimetallic Strip

A bimetallic strip consists of a 'sandwich' of two different metals which expand at different rates as their temperatures increase. This results in the bimetallic strip bending, with the degree of curvature of the strip being related to the temperature. Bimetallic strips were once commonly used as switches for devices such as heaters or air conditioners. In most modern devices, they have been replaced with thermistors and electronic control circuits, which are more accurate and adjustable. We will cover the thermal expansion of materials in Section 17.4.

Temperature and the Human Body

Many early temperature scales used human body temperature as a standard reference point – a practice that was abandoned long ago. Because our brains and internal organs function well only within a narrow range of temperatures, our bodies have highly developed systems to monitor and control our temperature. Nevertheless, the temperature of the body still varies slightly with location and time, as shown in Table 17.1. To keep

Temperature	Circumstance
Less than 33 °C	Hypothermia. Metabolic processes may be affected.
35–36 °C	Core temperature in early morning.
36–37.5 °C	Normal range for day to day activities.
38 °C	Moderate exercise.
39–40 °C	Hard exercise. Unwell with fever.
More than 42 °C	Hyperthermia. May cause irreversible damage to vital organs.

Table 17.1 Core temperature variation in the human body.

the internal organs within the required temperature range, the body uses thermoregulation processes to stabilise our core temperature, despite varying external temperature and activity levels. The surface temperature is normally lower than the core temperature, which reduces the rate of heat loss to a surrounding cooler environment. Both surface temperature and core temperature vary throughout the day. Core temperature tends to decrease at night, whereas skin surface temperature usually increases at night.

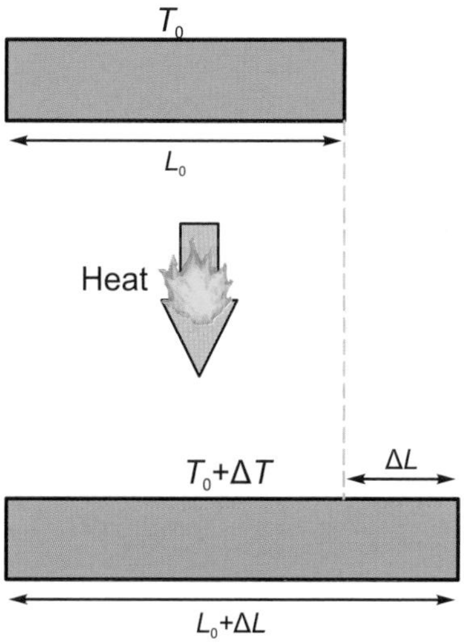

17.4 Thermal Expansion of Materials

Linear Expansion

As an object is heated, its atoms or molecules move faster, tending to move further apart, so most objects expand as they are heated (see Figure 17.5). For a solid rod of some material, the change in length, ΔL, due to heating is

$$\Delta L = L_0 \alpha \Delta T \tag{17.3}$$

where L_0 is the original length (at temperature T_0), ΔT is the change in temperature, and α is a constant associated with the particular material, known as the **linear coefficient for thermal expansion**. The expansion coefficient is a measure of the fractional change in length per degree of temperature change, and so has the units $°C^{-1}$ or K^{-1}. Table 17.2 gives some coefficients for various materials. The equation above gives the change in length, so the *final* length at temperature $T_0 + \Delta T$ is $L_0 + \Delta L$.

Figure 17.5 A rod heated uniformly will expand due to the increase in thermal energy of its constituent atoms/molecules. This results in an increase in length of the rod.

Example 17.1 *Linear thermal expansion*

Problem: The Waitaki bridge (North Otago, New Zealand) is a concrete structure, 0.9 km long. The summer-winter temperature extremes are 0 °C and 30 °C. $\alpha = 12 \times 10^{-6}\,°C^{-1}$ (the same as steel). What is the seasonal change in length?

Solution:

$$\Delta L = L_0 \alpha \Delta T$$

$$\Delta L = 900\text{ m} \times 12 \times 10^{-6}\,°C^{-1} \times 30\,°C = 0.324\text{ m}$$

To accommodate this length change, many expansion joints are needed.

Expansion in Two and Three Dimensions

When a flat plate of some material is heated, it expands in both directions at once (see Figure 17.6). At the initial temperature T_0, a square plate with sides of length L_0 has an

Thermal Expansion Coefficients, at 20 °C		
Material	*Linear (α), $\times 10^{-6}\,°C^{-1}$*	*Volume (β), $\times 10^{-6}\,°C^{-1}$*
Aluminium	23	69
Brass	19	57
Concrete	12	
Copper	17	51
Ethanol		750
Gasoline		950
Water		207
Fused quartz	0.59	
Glass	9	
Ice (at 0 °C)	51	
Lead	29	87
Mercury		182
Pyrex glass	3.2	
Steel	11–13 (varies with composition)	33–39
Gold	14.2	
Amalgam filling	15–35 (depends on alloy)	
Composite filling	19–57 (varies with composition)	
Tooth Enamel	17	
Invar	1.2	

Table 17.2 Expansion coefficients of some materials. As α is determined from $\beta(=3\alpha)$ for liquids, it is more common to quote the volume coefficient than the linear coefficient. [Values from Wikipedia.]

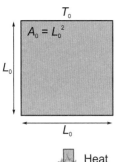

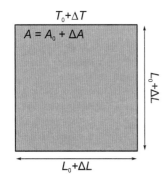

Figure 17.6 When a plate is heated uniformly, each side expands, and the total area of the plate increases.

area L_0^2. When the plate is heated to temperature $T_0 + \Delta T$, the new area will be

$$A_{\text{new}} = (L_0 + \Delta L)^2$$
$$= L_0^2 + 2L_0\Delta L + (\Delta L)^2$$
$$= A_0 + 2L_0\Delta L + (\Delta L)^2$$

so the change in area is

$$\Delta A = 2L_0\Delta L + (\Delta L)^2 = 2L_0(L_0\alpha\Delta T) + (\Delta L)^2$$

As ΔL^2 is often small enough to ignore,

$$\Delta A \approx A_0(2\alpha)\Delta T \tag{17.4}$$

This indicates that the **coefficient of surface thermal expansion** is 2α. Similarly, for isotropic materials, the **coefficient of volume thermal expansion**, given the symbol β, is 3α

$$\Delta V = V_0\beta\Delta T \tag{17.5}$$

Imagine a plate with a hole cut out of it. As the plate is heated, the hole in the plate will actually expand in exactly the same way as the piece of the material that was removed to make the gap; the removed segment would expand, so the hole *expands* also, provided the plate is heated uniformly, as in Figure 17.7.

Examples of Thermal Expansion

From ice cream to hot coffee, our teeth are exposed to temperatures ranging from below 0 °C to over 70 °C. If dental fillings had a different coefficient of thermal expansion to tooth enamel, the join between the tooth and the filling would tend to fail, with a high risk of pain and filling loss. Materials for dental fillings are chosen to have coefficients of thermal expansion similar to tooth enamel.

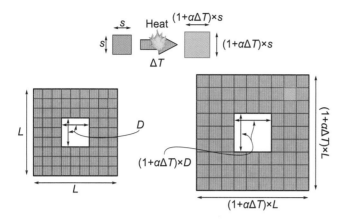

Figure 17.7 When a plate is heated uniformly, each section expands by the same percentage, so any holes expand in the same fashion.

Example 17.2 *Volume thermal expansion*

Problem: Suppose the temperature of the sea increases by 2 °C throughout. The mean depth of the ocean is 3.8 km. By how much will sea-level rise? Assume the surface area stays the same, so that all the increase in volume appears as raised surface level.

Solution: If all the increase in volume appears as raised surface level, then the fractional change in depth will be given by the volume expansion coefficient.

$$\Delta V = V_0 \beta \Delta T$$
$$V_0 = A h_0 \text{ and } \Delta V = A \Delta h$$
$$\text{so } \Delta h = h_0 \beta \Delta T$$

The volume expansion coefficient, β, is $207 \times 10^{-6}\,°\text{C}^{-1}$ at 20 °C, so

$$\Delta h = 3800\,\text{m} \times 207 \times 10^{-6}\,°\text{C}^{-1} \times 2\,°\text{C} = 1.6\,\text{m}$$

Something to think about: When would we use one-third of the volume expansion coefficient?

The coefficients of linear thermal expansion of concrete and steel are similar. Steel-reinforced concrete is a common construction material, and the similarity of the rates of thermal expansion is one of the factors that makes it so useful.

One of the major factors that will contribute to the rise in sea level if the global average temperature continues to rise is the volume expansion of the oceans.

Example 17.3 *Surface expansion*

Problem: What happens to the gap when this C-shaped piece of metal is heated uniformly all over?

Solution:

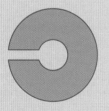

The gap will get bigger provided the C-shape is heated uniformly all over. You can envisage this by considering what would happen if you heated a single sheet of metal with the C-shape drawn on it. All the metal inside the C would expand as it is heated. So if we now cut the C-shape out of this sheet, the same results would be obtained. The gap would expand, just as if it had the original metal in it. If, on the other hand, just the arms of the 'C' were heated selectively, then the gap would tend to close.

Example 17.4 *Expansion of a fluid-filled cylinder*

Problem: A long, hollow cylinder, closed at one end, is held vertically and contains some ethyl alcohol. The height of the ethyl alcohol in the cylinder is 50 cm. How much will the height of the ethyl alcohol increase if the temperature of the ethyl alcohol increases by 10 °C? (Assume that the change in the diameter of the cylinder is negligible and that the average value for the volume coefficient of expansion of ethyl alcohol is 750×10^{-6} K^{-1})

Solution: We can use the following equation to find the change in volume of the ethyl alcohol in terms of its initial volume, the volume expansion coefficient and the change in temperature

$$\Delta V = V_0 \beta \Delta T$$

This is not, however what the question asks for. We are required to find the change in *height* of the ethyl alcohol in the cylinder. We can find the height of by using the fact that the volume of a cylinder is $V_{\text{cylinder}} = Ah$, where A is the cross-sectional area of the cylinder, and h is the height of the cylinder. Of course we don't know the crosssectional area of the cylinder, or the initial volume of the ethyl alcohol. But we can combine the two equations to generate a useful expression

$$\Delta V = A\Delta h \quad \text{and} \quad V_0 \beta \Delta T = Ah_0 \beta \Delta T$$
$$\text{so } A\Delta h = Ah_0 \beta \Delta T$$
$$\text{and } \Delta h = h_0 \beta \Delta T$$

Now that we have an expression in which only one quantity is unknown we can solve for the change in height

$$\Delta h = h_0 \beta \Delta T = 0.5 \, \text{m} \times 750 \times 10^{-6} \times 10 \, \text{K} = 0.0038 \, \text{m} = 0.38 \, \text{cm}$$

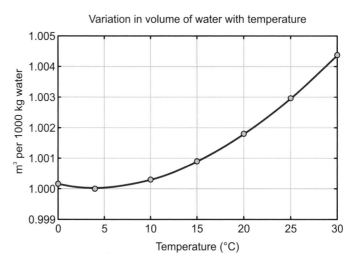

Figure 17.8 Water is most dense at about 4 °C. [Data source: *ASHRAE Fundamentals Handbook, 2001.*]

Anomalous Thermal Expansion of Water

The volume of water increases with increasing temperature above about 4 °C, but below this, the volume increases with decreasing temperature, leading to the odd situation of water being most dense at 4 °C. This can be seen in the graph in Figure 17.8 and means that the temperature of a body of water may in fact be warmer at the bottom than at the surface. It is also one of the few substances, and the only non-metallic one, that is generally less dense in its solid form than its liquid, so ice floats on water and lakes, and rivers don't freeze from the bottom up. (There are 15 different forms of solid ice known, although most occur only at very high pressure. Two of these are more dense than water; the common form, hexagonal crystalline, is less dense.)

Example 17.5 *Thermal expansion of water*

Problem: A fresh-water lake 10 m deep is covered with ice 10 cm thick. The temperatures of the top and the bottom of the water (respectively) could be: (a) 0 °C, –1 °C; (b) 0 °C, 4 °C; (c) 4 °C, 0 °C; (d) 0 °C, 10 °C; (e) none of these.

Solution:

(a) *Wrong* because the water at the bottom would be frozen if the temperature were –1 °C.

(b) *Correct.* We can have ice at 0 °C, and water at 4 °C would be more dense than water between 0 °C and 4 °C, so it would tend to sink to the bottom.

(c) *Wrong* because the ice would melt at 4 °C.

(d) *Wrong* because at 10 °C the density of the bottom water would be less than at 0 °C, so it would tend to rise.

17.5   Summary

Key Concepts

temperature (T) A quantitative measure of how hot or cold an object is. In the SI system temperature is measured in kelvin, though degrees Celsius and degrees Fahrenheit are in common use.

Celsius temperature scale The Celsius scale (once known as the centigrade scale) is a temperature scale on which the freezing point of water is at 0 degrees Celsius and the boiling point is at 100 degrees Celsius at a pressure of one atmosphere. It uses the symbol °C.

Fahrenheit temperature scale The Fahrenheit scale is a temperature scale on which the freezing point of water is at 32 degrees Fahrenheit and the boiling point is at 212 degrees Fahrenheit. It uses the symbol °F.

kelvin temperature scale The kelvin is the SI unit of temperature and is one of the seven base units of the SI system. It has the symbol K. The temperature scale is defined by absolute zero, which is 0 K, and the triple point of water, which is at 273.16 K. A 1 K temperature change is thus one part in 273.16 of the difference between these two points, and is identical to a change of 1 °C.

heat The energy transferred from a hot object to a cold object due to the temperature difference.

thermal equilibrium A state in which objects in thermal contact reach a common temperature and the net transfer of heat between them is zero.

zeroth law of thermodynamics The zeroth law states that two systems that are at the same time in thermal equilibrium with a third system are in thermal equilibrium with each other.

coefficient of linear thermal expansion (α) The fractional change in length per kelvin. i.e., the length by which a 1 m unconstrained rod will increase when its temperature is raised by 1 K.

coefficient of volume thermal expansion (β) The fractional change in volume per kelvin, i.e., the increase in volume of $1\ \mathrm{m}^{-3}$ of material when its temperature is increased by 1 K.

Equations

$$\Delta L = L_0 \alpha \Delta T$$
$$\Delta V = V_0 \beta \Delta T$$
$$\beta = 3\alpha$$

17.6 Problems

17.1 The core temperature of the human body is about 37.0 °C and a temperature of 40.0 °C is regarded as a high fever. What are 37 °C and 40 °C in kelvin and degrees Fahrenheit?

17.2 Which of the following statements are correct? (**Note:** more than one of the statements may be correct)

(a) The volume coefficient of thermal expansion of water is negative for temperatures in the range 0 to 4 °C.

(b) The triple point of water, 0.01 °C, and 611.73 Pa, is one of the primary fixed points on the Kelvin absolute temperature scale.

(c) The volume coefficient for the thermal expansion of a solid is twice the linear coefficient for thermal expansion.

(d) A temperature of 26.85 °C is the same as 300 K.

(e) The surface temperature of the sun (3142 °C) is the highest temperature found in nature.

17.3 The Rankine temperature scale has the same temperature unit intervals, °R, as the Fahrenheit scale, °F, but it is an absolute scale, so 0 °R is the same as 0 K. At what temperature does water boil on the Rankine scale (at standard atmospheric pressure)?

17.4 A titanium metal rod has been inserted into the tibia of an injured soccer player. The rod, which runs the length of the tibia, is 0.55 m long and is normally at a constant temperature of 37 °C. Suppose that the soccer player develops a severe fever and his core temperature rises to 40 °C. The linear coefficient for thermal expansion of titanium is 8.6×10^{-6} K^{-1}. By how much will the length of the titanium rod increase?

17.5 You are scheduled to implant a metal brace into a fractured femur to secure the broken ends. You are concerned that the temperature variations in the body will cause the length of the brace to change, creating a risk that the break will not heal satisfactorily. A 2 m length of the metal from which the brace is made is available. You heat this rod by 10 °C and find that the length of the rod has increased by 0.2 mm. How much (in μm) would the length of a 10 cm brace change if the body temperature of your patient were to increase by 3 °C? (during a high fever for example)

17.6 The temperature coefficient of linear expansion of steel is 12×10^{-6} K^{-1}. When the temperature increases from 5 °C to 25 °C, what is the increase in the length of a straight 25 m length of un-clamped railway track?

17.7 At 20 °C a steel ring has an inside diameter that is 0.5 mm smaller than the diameter of a steel rod, which is 0.2 m. The ring is heated until it fits over the rod, which remains at 20 °C. The temperature coefficient of linear expansion of steel is 12×10^{-6} K^{-1}. What is the temperature of the ring when it is just large enough to fit over the rod?

17.8 If you drink 0.5 L of extremely cold water at 4 °C, how much will its volume increase (in mL) once the water has reached your core body temperature of 37 °C in your stomach? (Assume the average value for the volume coefficient of expansion of water is 207×10^{-6} K^{-1}.)

17.9 A cube of aluminium measures 1 m along each side. In the centre of the cube there is a small spherical cavity with a diameter of 1 cm. When the cube is heated from 10 °C to 30 °C, what is the percentage increase in the volume of the central cavity? (The linear thermal expansion coefficient of aluminium is 23×10^{-6} K^{-1}.)

17.10 A long hollow cylinder of aluminium, closed at one end, is held vertically and contains some water. The cylinder has an internal diameter of 1.5 cm and the height of the water in the cylinder is initially 20 cm. The linear coefficient of expansion for aluminium is $\alpha_{Al} = 23 \times 10^{-6}$ K^{-1} while the volume coefficient of expansion of water is $\beta_{water} = 207 \times 10^{-6}$ K^{-1}.

(a) How much will the height of the water increase if the temperature of both the cylinder and the water increase by 10 °C?

(b) What result do you get for part (a) if you ignore the expansion of the cylinder?

IDEAL GASES

18.1 Introduction

Gases are important for biological physics, mainly because our environment is dominated by the atmosphere, which is a mixture of gases. Gases are also important because they allow us to understand the link between thermal energy and temperature.

Key Objectives

- To understand Charles' law, Boyle's law, the ideal gas law and Dalton's law.

- To understand the concept of the mole.

- To understand the principle of the constant-volume gas thermometer.

- To understand how the thermal energy of an ideal gas relates to the absolute temperature.

18.2 The Gas Laws

Charles' Law

When a fixed quantity of gas is held at constant pressure, it is found experimentally that the volume of the gas increases linearly with temperature, as shown in Figures 18.1 and 18.2:

$$V = aT \qquad (18.1)$$

where V is the volume of the gas sample, T is the absolute temperature (in kelvin), and a is the proportionality constant, which depends on the number of gas molecules and the pressure. However, a is *not* dependent on the chemical structure of the gas. This equation is equally valid for any gas, but it fails if the temperature is too low, or the density is too high. In these situations, the interactions between the molecules become significant and Eq. (18.1) no longer holds true.

Boyle's Law

When a fixed quantity of gas is held at a fixed temperature, it is found experimentally that the pressure is inversely proportional to volume (see Figure 18.3). For example, doubling the pressure will halve the volume of the sample. We can express this relationship as

$$P = \frac{b}{V} \qquad (18.2)$$

where P is the absolute pressure and V is the volume. The proportionality constant, b, depends on the absolute pressure of the gas sample and the number of molecules of gas present, but again it does not depend of the type of gas, provided the density is not too large.

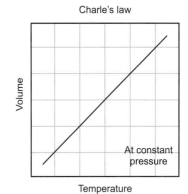

Charle's law

Figure 18.1 Charles' law. The volume and temperature of a sample of ideal gas are linearly related at constant pressure.

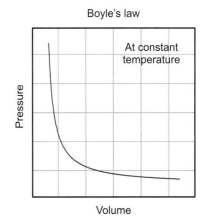

Boyle's law

Figure 18.3 Boyle's Law. The pressure and volume of a sample of ideal gas are inversely related at constant temperature.

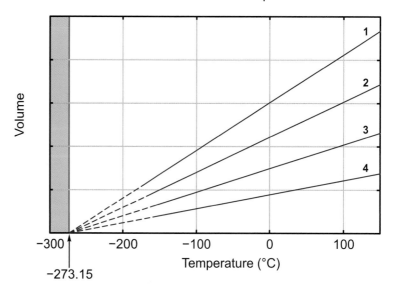

Figure 18.2 The relationship between volume and temperature for four different gas samples. When the temperature is low, the straight-line relationship fails, but, by extrapolation, the temperature at which the volume would reach zero is the same for all gases. This temperature, which is known as absolute zero, is $-273.15\,°C$.

The Ideal Gas Law

We can combine Charles' and Boyle's laws together into one equation:

$$PV = cT \tag{18.3}$$

where P is the pressure, V is the volume, T is the absolute temperature and c is a proportionality constant. Experimentally, it is found that c depends only on the *number* of gas molecules present, provided the temperature is not too low, or the density too high. It does not depend on the mass of the individual atoms or molecules, or on their structure (whether the gas is monatomic, diatomic or more complex). This happens because the molecules tend not to stick together when they collide under the conditions stated. Also, the volume occupied by the molecules is very small compared with the volume available.

Because c is proportional to the number of molecules, then we can write $c = Nk$, where N is the number of gas molecules, and k is a proportionality constant. k is universal constant called **Boltzmann's constant**. In SI units, $k = 1.381 \times 10^{-23}$ J K^{-1}. Substituting this into Eq. (18.3), we obtain the **ideal gas law**

$$PV = NkT \tag{18.4}$$

The word ideal is used here because the equation holds strictly only in the ideal limit of a gas of very low density and high temperature.

There is a useful alternate way of writing the ideal gas relationship. The SI unit for the amount of a substance is called the **mole**, for which the symbol is the abbreviation mol. One mole of a substance contains the same number of 'elementary entities' as 12 g of carbon-12. The number of carbon-12 atoms in one mole is a fixed number, known as Avogadro's number, which is usually given the symbol N_A. It is equal to 6.022×10^{23} mol^{-1} (to four significant figures). The number of molecules and the number of moles in a sample of gas are directly proportional, and are related to one another by Avogadro's number.

Key concept:

$$n = \frac{N}{N_A} \tag{18.5}$$

where n is the number of moles of gas, N is the number of molecules and N_A is Avogadro's number.

This gives us another form of the ideal gas law:

$$PV = nRT \qquad (18.6)$$

where P is the pressure, V is the volume, n is the number of moles of gas and T is the absolute temperature. R is called the **universal gas constant** or the **ideal gas constant**, and has the value $8.314 \, \mathrm{J \, K^{-1} mol^{-1}}$. The relationship between the constants R and k is

$$R = N_A k \qquad (18.7)$$

Dalton's Law of Partial Pressures

All the gas laws mentioned so far apply equally well (under the appropriate conditions) to all gases, regardless of their type or structure, or the mass of the individual atoms or molecules. This means they can also be applied to mixtures of gases, such as air. In this situation, it is useful to use the concept of **partial pressure**, which refers to the pressure exerted by one individual component of the gas. The partial pressure of that gas is the pressure it would have if *only that component of the gas* were present in the same volume (see Figure 18.4).

The sum of the partial pressures of all the component gases is the same as the total pressure of the gas mixture; this is **Dalton's law of partial pressures**.

$$P_{\text{total}} = P_1 + P_2 + P_3 + \ldots \qquad (18.8)$$

> **Moles**
>
> One mole of any substance contains Avogadro's number of elementary units. One mole of lead contains the same number of atoms as one mole of sodium. One mole of water (H_2O) contains the same number of molecules as one mole of hydrogen gas (H_2), but in this case, the numbers of *atoms* are not the same. At times, it is important to be clear which entities are being referred to, molecules or atoms.

> **Caution needed with gas mixtures**
>
> When we talk about percentages of gases we need to know whether it is a percentage of the number of molecules (which is the same as the percentage of the number of moles), or a percentage of the weight (which is different, as the different gases have different molecular weights). Unless otherwise stated, when we refer to percentages of a gas mixture we mean the percentage of *molecules*.

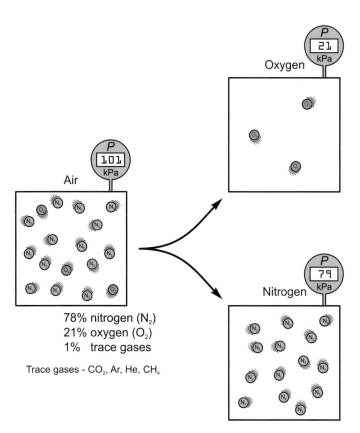

78% nitrogen (N_2)
21% oxygen (O_2)
1% trace gases

Trace gases - CO_2, Ar, He, CH_4

Figure 18.4 Dalton's law of partial pressures. The total pressure is the sum of the partial pressures of the individual gases.

For example, at sea level we have a total pressure of (on average) 1.013×10^5 Pa for dry air, which consists of 78% N_2, 21% O_2 and 1% other gases (mostly argon), expressed as mole percentages. In this case, the partial pressure of oxygen, O_2, is $0.21 \times$ the total, or 2.1×10^4 Pa. The partial pressure of nitrogen is $0.78 \times 1.013 \times 10^5$ Pa $= 7.9 \times 10^4$ Pa.

If we use the total number of molecules of all gases, then the pressure in the ideal gas law will be the total pressure exerted by all the gases. For a mixture of m gas samples

$$P_{\text{total}} = P_1 + P_2 + ... + P_m = \frac{(N_1 + N_2 + ... + N_m)kT}{V} \qquad (18.9)$$

If we are interested in only one component of the mixture, molecules of gas of type j, then the pressure in the ideal gas law is the *partial pressure* of gas j

$$P_j = \frac{N_j kT}{V} \qquad (18.10)$$

where N_j is the number of molecules of gas j.

Example 18.1 *Ideal gas law*

Problem:

(a) In 1.000 m^3 of ideal gas at atmospheric pressure (101.3 kPa), how many moles of gas are there at 20 °C?

(a) What is the mass of gas if it is all O_2?

(b) What is the mass of gas if it is all N_2?

(c) What is the mass if the gas is composed of 21% oxygen and 79% nitrogen?

(d) Suppose the gas mixture also contains some water vapour along with the O_2 and N_2, at the same total pressure (101.3 kPa). Would the density increase or decrease?

Solution: (a) To get the number of moles, use the ideal gas equation

$$PV = nRT$$

with the following values: $P = 101.3 \times 10^3$ Pa, $V = 1.000$ m^3, $R = 8.314$ J K^{-1}mol^{-1} and $T = 293.1$ K. So

$$n = \frac{101\,300 \text{ Pa} \times 1.000 \text{ m}^3}{293.15 \text{ K} \times 8.314 \text{ J K}^{-1} \text{ mol}^{-1}} = 41.56 \text{ moles}$$

(b) The mass of 1 mol of molecular oxygen is 32.00×10^{-3} kg. So if the gas is oxygen only then the mass will be

$$41.56 \text{ mol} \times 32.00 \times 10^{-3} \text{ kg mol}^{-1} = 1.330 \text{ kg}$$

(c) For molecular nitrogen, N_2, the molar mass is 28.01×10^{-3} kg, so the mass of the 1 m^3 volume will be

$$41.56 \text{ mol} \times 28.01 \times 10^{-3} \text{ kg mol}^{-1} = 1.164 \text{ kg}$$

(d) For the 21%-79% mixture, the number of moles of O_2 will be

$$0.21 \times 41.56 \text{ moles} = 8.728 \text{ moles}$$

Similarly, the number of moles of N_2 will be

$$0.79 \times 41.56 = 32.83 \text{ moles}$$

Hence the mass of 1 m^3 of air will be

$$8.728 \text{ mol} \times 32.00 \times 10^{-3} \text{ kg mol}^{-1} + 32.83 \text{ mol} \times 28.01 \times 10^{-3} \text{ kg mol}^{-1} = 1.199 \text{ kg}$$

(e) Water (H_2O) has a molar mass of 18 g mol^{-1}, which is significantly lower than either oxygen or nitrogen, so since the other gases are displaced by water, the density will be lower. This has consequences for the aviation industry – when the atmospheric humidity is higher, planes need to travel faster to get enough lift for take-off, as the lift is dependent on the air density.

Example 18.2 *Ideal gas law*

Problem: A helium-filled balloon is initially at a pressure of 101.3 kPa when the temperature is 10 °C. What is the pressure when the volume has increased by a factor of 10 and the temperature is −50 °C?

Solution: Starting with the ideal gas equation

$$PV = nRT$$

we can rearrange to get

$$\frac{PV}{T} = nR$$

As nR is constant here, then $\frac{PV}{T}$ is constant also. If we label the initial parameters with a 1 and the final parameters with a 2, then we have

$$\frac{P_1 V_1}{T_1} = \frac{P_2 V_2}{T_2}$$

$$\Rightarrow P_2 = \frac{P_1 V_1 T_2}{V_2 T_1}$$

Now we know $V_2 = 10V_1$, $T_1 = 283.15$ K, $T_2 = 223.15$ K and $P_1 = 101\,300$ Pa

$$P_2 = \frac{P_1 T_2}{10 T_1} = \frac{101\,300 \text{ Pa} \times 223.15 \text{ K}}{10 \times 283.15 \text{ K}} = 7983 \text{ Pa}$$

Example 18.3 *Ideal gas law/Boyle's law*

Problem: A gas sample with a fixed number of moles of gas is compressed at constant temperature. If it initially has a volume of 0.5 m^3 at sea level atmospheric pressure (1.013 × 10^5 Pa), what is its volume if the pressure is doubled to 2.026 × 10^5 Pa?

Solution: In this case, $P_1 = 1.013 \times 10^5$ Pa, $V_1 = 0.5$ m^3, $P_2 = 2.026 \times 10^5$ Pa$=2P_1$, $T_2 = T_1$ and $n_2 = n_1$. The quantity we want to find is the final volume V_2. We can apply the ideal gas law to both situations, so

$$P_1 V_1 = n_1 R T_1$$

and

$$P_2 V_2 = n_2 R T_2$$

Because $n_1 = n_2$ and $T_1 = T_2$ then the right-hand sides of the above two equations are equal, and therefore we can equate the left-hand sides

$$P_1 V_1 = P_2 V_2$$

$$\Rightarrow V_2 = \frac{P_1 V_1}{P_2} = \frac{1}{2} 0.5 \text{ m}^3 = 0.25 \text{ m}^3$$

An alternate way of finding the final volume is to use Boyle's law. If a gas sample is held at constant temperature, then the volume and pressure have an inverse relationship, so doubling one will *halve* the other. If the pressure is doubled, the volume is halved from 0.5 m^3 to 0.25 m^3.

Example 18.4 *Ideal gas law*

Problem: A gas sample is held at constant volume, and contains a fixed number of moles of gas. Its temperature is initially 16 °C at atmospheric pressure 1.013×10^5 Pa. If the gas is warmed up to 20 °C, what will the gas pressure be?

Solution: Here the fixed quantities are n and V. Rearranging the ideal gas equation so that all the constant quantities are on one side

$$PV = nRT$$

$$\Rightarrow \frac{P}{T} = \frac{nR}{V}$$

Because the right-hand side is constant throughout in this example, then

$$\frac{P_1}{T_1} = \frac{P_2}{T_2}$$

The unknown quantity we are trying to determine is P_2. The temperatures were given in Celsius, so these need to be converted into kelvin. $T_1 = 16\,°C = 289.15$ K, $T_2 = 20\,°C = 293.15$ K and $P_1 = 1.013 \times 10^5$ Pa

$$P_2 = \frac{P_1 T_2}{T_1} = \frac{1.013 \times 10^5 \text{ Pa} \times 293 \text{ K}}{289 \text{ K}} = 1.027 \times 10^5 \text{ Pa}$$

We should check that this makes sense. We have increased the temperature, without altering the volume or number of moles, so we would expect the pressure to increase as the molecules move faster.

18.3   Biological Applications

Breathing

We breathe (Figure 18.5) by altering the volume of our chest cavity. To inhale, we contract the intercostal muscles and diaphragm, enlarging the chest cavity. This increase in volume (which occurs at constant temperature) results in a reduction in the pressure in our lungs to below atmospheric pressure (Boyle's law). Since a gas moves from a high-pressure region to a low-pressure region, this results in air flow into our lungs. To exhale, we relax the intercostal muscles and diaphragm, reducing the lung volume, hence increasing the pressure in the lungs above atmospheric pressure and forcing air out. The amount by which the volume of our lungs changes determines the volume of air exchanged per breath.

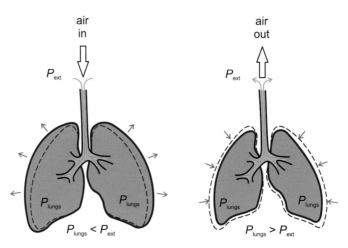

Figure 18.5 During inflation, expanding the lungs drops the internal lung pressure below atmospheric pressure, and results in the movement of air into the lungs.

Tension Pneumothorax

A tension pneumothorax is a collapse of the lung caused by air building up within the pleural space, the space between the lung and the chest wall. Normally, pressure differences prevent our lungs from collapsing.

Our lungs are not affixed to the walls of the chest cavity. When we inhale, the pressure between the lungs and chest wall is lower than the pressure within the lungs ,and this results in the lungs being held against the chest wall (Figure 18.6). In a tension pneumothorax, damage to the lung results in air entering the pleural space between the lung and chest wall. When the patient breathes they expand the chest cavity, but not the lung, and insufficient air moves into the lung. In severe cases, pressure build-up can also cause the collapse of the uninjured lung resulting in respiratory failure (Figure 18.7).

Diving

As divers descend, the pressure exerted by the water increases (at a rate of about 100 kPa for every 10 m extra depth). The pressure in the lungs must go up too. If no extra air is breathed in, this pressure increase is achieved by a decrease in lung volume. Conversely, as divers ascend, the external pressure drops, thus reducing the pressure in the divers lungs also. The lungs expand as the pressure is reduced. If divers ascend too rapidly without breathing out, their lungs may rupture.

18.4 Kinetic Theory of Gases

The ideal gas law (Eq. (18.4) and Eq. (18.6)) shows the relationship between the bulk properties of the gas: temperature, pressure, volume and quantity. However, the way the gas exerts pressure on the walls of a container is by individual molecules striking the walls with momentum. Here we will take a closer look at the speeds and energies of the molecules in order to get a better understanding of temperature.

Energy of an Ideal Gas

Temperature is a measure of how rapidly atoms or molecules are moving. The amount of movement increases with increasing temperature, so the particles must have more kinetic energy as the temperature rises. For a monatomic gas, i.e., a gas in which each molecule consists of just one atom, the thermal energy is the kinetic energy due to random translational motion of the molecules in the sample. More complex molecules can have other forms of internal energy – this will be discussed later.

The pressure of the gas is dependent on the number of collisions per square metre of wall, and the average force exerted on the wall per collision. Molecules exert a force on the wall during collisions as a result of their momentum being changed by the collision. The total pressure exerted on the wall will be the average pressure exerted on the wall by each molecule multiplied by the number of molecules in the container. If the gas contains N molecules,

$$P = N P_{\text{per molecule}} \tag{18.11}$$

Imagine our gas is contained in a cube with all sides having length a, and sides parallel to x, y and z axes. Provided the cube is not too big, the pressure in the gas is the same throughout, and so the pressure exerted on each of the walls is the same. Pressure is force per unit area, so the average pressure exerted per molecule on a wall equals the average force exerted per molecule on that wall, divided by the wall area

$$P_{\text{per molecule}} = \frac{F_{\text{per molecule}}}{A} = \frac{F_{\text{per molecule}}}{a^2} \tag{18.12}$$

In the Mechanics topic (Chapter 6), it was shown that the force exerted on a wall when an object collides with it depends on the change in momentum of the object and the duration of the collision (see Figure 18.8). To determine the *average* force exerted on the wall per molecule, we need to know by how much the molecule's momentum

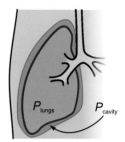

Figure 18.6 The lungs are held against the chest wall by a pressure difference, because the pleural space (highlighted in red) is a sealed volume. If the lung begins to collapse, P_{cavity} falls. As P_{cavity} falls below P_{lungs}, the higher pressure in the lung as compared to the cavity will tend to re-inflate the lung.

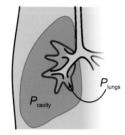

Figure 18.7 Air build-up in the pleural space between the lung and the chest wall can cause it to collapse.

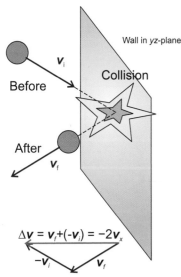

Figure 18.8 When a molecule collides with a wall in the *yz*-plane, the force is only in the *x*-direction, and the velocity component in the *x*-direction is changed.

changes per collision, and the average time between collisions. We will assume collisions between gas molecules and the wall are elastic (that is, kinetic energy is conserved) and that the walls are stationary before and after the collision. Each molecule rebounds with the component of velocity perpendicular to the wall reversed in direction, but unchanged in size.

The change in momentum of a molecule following a collision with a wall in the *yz*-plane (perpendicular to the *x*-direction) is in the *x*-direction. In other words, the molecule's momentum in the *x*-direction reverses in the collision, but its momentum in the *y*- and *z*-directions does not change

$$\Delta p = -2mv_x \qquad (18.13)$$

where Δp is the change in the molecule's momentum, m is the molecule's mass, and v_x is the molecule's initial velocity in the *x*-direction. The momentum imparted to the wall during the collision has the same magnitude

$$\Delta p_{\text{wall}} = -\Delta p_{\text{molecule}} = 2mv_x \qquad (18.14)$$

The average force exerted on this wall due to each collision is $\Delta p / \Delta t$, where Δt is the time between collisions. At points in time between consecutive collisions, our molecule exerts no force on the wall. The force we are interested in is the average force our molecule exerts on the wall *over all time*, not just the collision time, so to get the average force, we divide by the time between consecutive collisions of the same wall–molecule pair. This time is the average time for the molecule to cross the container twice in the *x*-direction, travelling a distance $2a$

$$\Delta t = \frac{2a}{v_x} \qquad (18.15)$$

Our molecule exerts, on average, a force on the wall of

$$F_{\text{molecule}} = \frac{\Delta p}{\Delta t} = \frac{\Delta p v_x}{2a} = \frac{2mv_x^2}{2a} = \frac{mv_x^2}{a} \qquad (18.16)$$

hence the average pressure on the wall due to this one molecule will be

$$P_{\text{molecule}} = \frac{F_{\text{molecule}}}{a^2} = \frac{mv_x^2}{a^3} = \frac{mv_x^2}{V} \qquad (18.17)$$

as the volume of the container $V = a^3$.

Because there are a large number of molecules colliding with the wall at random times, the forces exerted on the wall get smoothed out to give a time-independent force on the wall. The total pressure exerted on the wall will be the sum of the pressures from all the molecules:

$$P = \frac{m}{V}\left(v_{x_1}^2 + v_{x_2}^2 + \ldots + v_{x_N}^2\right) = \frac{Nm\overline{v_x^2}}{V} \qquad (18.18)$$

where $\overline{v_x^2}$ is the average of the squares of the *x* components of the velocity for each of the *N* molecules, which is

$$\overline{v_x^2} = \frac{v_{x_1}^2 + v_{x_2}^2 + \ldots + v_{x_N}^2}{N} \qquad (18.19)$$

Writing this another way

$$PV = N\left(m\overline{v_x^2}\right) \qquad (18.20)$$

As we already know that $PV = NkT$, it follows that

$$m\overline{v_x^2} = kT \qquad (18.21)$$

so

$$\frac{1}{2}m\overline{v_x^2} = \frac{1}{2}kT \qquad (18.22)$$

www.wiley.com/go/biological_physics

In other words, the average kinetic energy of the gas sample associated with random motion of gas molecules in the x-direction is $\frac{1}{2}NkT$, or $\frac{1}{2}kT$ per molecule.

As the motion is random, each direction is the same. The average velocity squared is

$$\overline{v^2} = \overline{v_x^2} + \overline{v_y^2} + \overline{v_z^2} = 3\overline{v_x^2} \tag{18.23}$$

This gives us an important relationship

$$\mathrm{KE}_{\text{average, per molecule}} = \frac{1}{2}m\overline{v^2} = \frac{3}{2}kT \tag{18.24}$$

The average kinetic energy of an atom is $\frac{3}{2}kT$, showing that absolute temperature is proportional to average kinetic energy associated with the random translational motion of atoms.

For an ideal monatomic gas, the only form of thermal energy is the kinetic energy associated with random translational motion of the gas atoms, so we can conclude that the thermal energy (U) of the gas sample is proportional to the absolute temperature of the gas and given by

$$U = N\frac{1}{2}m\overline{v^2} = \frac{3}{2}NkT = \frac{3}{2}nRT \tag{18.25}$$

In diatomic or polyatomic gases we also need to consider kinetic energy associated with rotation and/or vibration of the molecules. The thermal energy of these gases is higher for the same temperature. To see how much energy is associated with rotation and vibration, we apply the principle of *equipartition of energy*, which we will state here without proof.

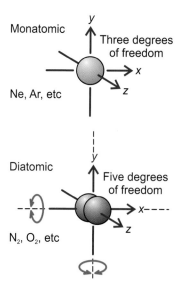

Figure 18.9 Both monatomic and diatomic gases have three translational degrees of freedom. Diatomic molecules have two more due to rotation. They have rotational symmetry about one axis, and energy is required for rotation about the other two.

Key concept:
The thermal energy of a system in equilibrium has the same value for each of the degrees of freedom, being $\frac{1}{2}kT$.

By degrees of freedom, we mean all the independent ways that a molecule can possess energy. For example, as shown in Figure 18.9, there are three directions in space, so there is $3 \times \frac{1}{2}kT$ per molecule associated with translational motion – $\frac{1}{2}kT$ per degree of freedom. For a monatomic species such as helium, the atom is rotationally symmetric, so there is no energy associated with rotation. For a diatomic molecule like oxygen, there is one axis of symmetry, and any rotation about the other two axes requires energy, and so the are two more degrees of freedom. For **diatomic gases** like nitrogen and oxygen at *room temperature*

$$U_{\text{per molecule}} = \frac{5}{2}kT \tag{18.26}$$

The atoms in diatomic molecules like these are rather like two masses attached by a spring (the bond holding them together), so the atoms can also vibrate. There are two degrees of freedom (kinetic and potential) so the average thermal energy per molecule is expected to be $\frac{7}{2}kT$. In practice, however, the vibrational motion occurs only at temperatures well above room temperature, so for most situations, Eq. (18.26) is the correct equation.

The Maxwell–Boltzmann Distribution

The molecules in the gas do not all move at the same speed – they collide with each other frequently and undergo changes in velocity. Some of the molecules will be moving faster than the average and some will be moving slower. The distribution of speeds is known as the Maxwell–Boltzmann distribution.

Figure 18.10 shows the distribution of speeds for some gases at 25 °C. The area under any portion of the curve gives the relative probability that the speed lies in that range, so the higher the curve at a particular speed, the more molecules have that

Maxwell-Boltzmann Molecular Speed Distribution for Noble Gase

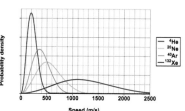

Figure 18.10 The Maxwell–Boltzmann distribution. This plot shows the distribution of speeds for some of the noble gases at 25 °C. [Public domain image from Wikipedia.]

speed. The figure also shows that lighter atoms (like He) have higher speeds than heavier ones (like Xe). The average speed obtained from Eq. (18.24) is

$$v_{\text{rms}} = \sqrt{\overline{v^2}} = \sqrt{\frac{3kT}{m}} \tag{18.27}$$

Example 18.5 *Kinetic theory of gases*

Problem: What is the mean speed of molecules of oxygen when the temperature is $T = 293$ K (20 °C)?

Solution: The kinetic energy depends on temperature

$$\frac{1}{2}mv_m^2 = \frac{3}{2}kT$$

where v_m is the mean speed of the molecules. Here m is the mass of an oxygen molecule, which is the molar mass of O_2 divided by Avogadro's number

$$m = \frac{32.0 \times 10^{-3} \text{ kg mol}^{-1}}{6.022 \times 10^{23} \text{ mol}^{-1}} = 5.316 \times 10^{-26} \text{ kg}$$

and $k = 1.381 \times 10^{-23}$ J K^{-1}. So

$$v_m^2 = \frac{3kT}{m} = \frac{3 \times 1.381 \times 10^{-23} \text{ J K}^{-1} \times 293 \text{ K}}{5.316 \times 10^{-26} \text{ kg}} = 228\,000 \text{ m}^2 \text{ s}^{-2}$$

$$\Rightarrow v_m = 478 \text{ m s}^{-1}$$

This is about 1720 km h^{-1}. Something to think about: would v_m be more or less than this for N_2?

Example 18.6 *Kinetic theory of gases*

Problem: If we have 1 kg of oxygen at $T = 293$ K, what is the energy due to the random linear motion of the molecules?

Solution: The kinetic energy per molecule is

$$\frac{3}{2}kT = 1.5 \times 1.381 \times 10^{-23} \text{ J K}^{-1} \times 293 \text{ K} = 6.07 \times 10^{-21} \text{ J}$$

We calculated the mass of an oxygen molecule in the previous example (5.316×10^{-26} kg), so the number of oxygen molecules will be

$$\frac{1 \text{ kg}}{5.316 \times 10^{-26} \text{ kg}} = 1.881 \times 10^{25}$$

So the total energy due to linear motion is

$$6.07 \times 10^{-21} \text{ J} \times 1.881 \times 10^{25} = 114 \times 10^3 \text{ J} = 114 \text{ kJ}$$

18.5   Summary

Key Concepts

Charles' law At low densities and fixed pressure, the volume of a fixed amount of gas is proportional to the absolute temperature.

Boyle's law At low densities and fixed temperature, the absolute pressure of a fixed amount of gas is inversely proportional to the volume.

Ideal gas law At low density, for a fixed quantity of gas, the value of $\frac{PV}{T}$ is constant. This is usually written as $PV = nRT$ or $PV = NkT$, where P is the pressure, V is the volume, T is the absolute temperature, n is number of moles of gas, N is the number of molecules of gas, R is the universal gas constant and k is Boltzmann's constant.

Dalton's law The sum of the partial pressures of all the component gases is the total pressure of the gas mixture.

partial pressure The partial pressure of a component of a gas mixture is the pressure it would have if only that component of the gas was present in the same volume.

universal gas constant or ideal gas constant (R) A physical constant which appears in the ideal gas law, and is closely related to Boltzmann's constant. $R = 8.314 \, \text{J K}^{-1} \, \text{mol}^{-1}$.

Boltzmann's constant (k) A physical constant which relates particle energy to temperature. $k = R/N_A$, where R is the universal gas constant and N_A is Avogadro's number. $k = 1.381 \times 10^{-23} \, \text{J K}^{-1}$.

Avogadro's number or Avogadro constant (N_A) The number of atoms in exactly 12 g of carbon-12. $N_A = 6.022 \times 10^{23} \, \text{mol}^{-1}$.

mole The SI unit of amount of substance. One mole of anything contains the same number of elementary units as 12 g of carbon-12. Its symbol is mol. See also Avogadro's number.

molar mass (M) The mass of one mole of a substance. Molar masses may be given in g mol^{-1} or kg mol^{-1}. For example, $M(\text{H})$, the molar mass of hydrogen, is $1.008 \, \text{g mol}^{-1}$ or $1.008 \times 10^{-3} \, \text{kg mol}^{-1}$. The mass of a sample of n moles is $m = nM$.

Maxwell–Boltzmann distribution A probability distribution which predicts the fraction of the molecules in a gas sample which have speeds in a particular range.

Equations

$$V = aT \text{ at constant } P$$

$$P = \frac{b}{V} \text{ at constant } T$$

$$PV = NkT = nRT$$

$$P_{\text{total}} = P_1 + P_2 + \ldots + P_m$$

$$n = \frac{N}{N_A} \quad m = Mn$$

$$\text{KE}_{\text{average, per molecule}} = \frac{1}{2}m\overline{v^2} = \frac{3}{2}kT$$

$$U_{\text{per molecule}} = \frac{5}{2}kT \text{ (diatomic, room temp)}$$

$$v_{\text{rms}} = \sqrt{\overline{v^2}} = \sqrt{\frac{3kT}{m}}$$

18.6 Problems

18.1 A sample of an unknown gas (gas A) has a volume of $3.2 \, m^3$ at a temperature of 10 °C. Another sample of an unknown gas (gas B) has a volume of $4.5 \, m^3$ at a temperature of 250 °C. Assuming that both gases obey Charles' law at all temperatures and are at the same pressure,

  (a) Could gas A and gas B be samples of the same gas at different temperatures?

  (b) What is the volume of gas A at the following temperatures: -50 °C, 0 °C, 50 °C, 100 °C ?

18.2 A sample of a hypothetical ideal gas has a volume of $0.5 \, m^3$ at a temperature of 5 °C and a pressure of 250 kPa.

  (a) How many molecules of gas are there in this sample?

  (b) How many moles of gas are there in this sample?

18.3 A novice pearl diver takes a deep breath before diving. She fills her lungs to their maximum capacity, which is 3 litres. Before diving the pressure in her lungs is 101.3 kPa and the temperature is 37 °C. She then dives to a depth of 20 m.

  (a) How many moles of air has she inhaled?

  (b) If the temperature in her lungs does not change, she does not exhale at all, and the pressure in her lungs is the same as the surrounding water pressure, what is her lung volume at the bottom of her dive (in litres)?

18.4 Two identical industrial gas cylinders (cylinder A and B) each have a volume of $2.25 \times 10^{-2} \, m^3$ and are maintained at 20 °C ($M_{O_2} = 32 \, g \, mol^{-1}$, $M_{N_2} = 28 \, g \, mol^{-1}$, and $M_{CO} = 28 \, g \, mol^{-1}$).

  Cylinder A contains: 1 kg of O_2, 1 kg of N_2, and 1 kg of CO.
  Cylinder B contains: 2 kg of O_2, 0.5 kg of N_2, and 0.5 kg of CO.

  (a) What is the partial pressure of O_2, N_2, and CO in each cylinder?

  (b) What is the total pressure in each cylinder?

  (c) What is the total thermal energy in each cylinder?

  (d) What is the rms velocity of O_2, N_2, and CO in each cylinder?

18.5 A sealed cylinder (with fixed volume) contains one mole of He gas and is slowly heated until the temperature of the gas has increased by 50 K. Which of the following statements are correct? (Note: more than one statement may be correct)

  (a) Rotation of the molecules contributes to the thermal energy of the gas as it is heated.

  (b) The average kinetic energy of the individual atoms of He increases by 4.2×10^{-19} J.

  (c) The total thermal energy of the gas increases by 620 J (to 2 s.f.).

  (d) The density of the gas decreases as the gas is heated.

  (e) The pressure of the gas decreases as the gas is heated.

18.6 Container A has 1.0 mole of O_2 gas and container B has 1.0 mole of He gas. The containers, which have different volumes, are brought into thermal contact and reach thermal equilibrium. Which of the following statements are correct?

  (a) The total thermal energy of the two gases is the same.

  (b) The average kinetic energy of the atoms of He is the same as that of the molecules of O2.

  (c) The temperatures of the two gases need not be the same.

  (d) The gases in the two containers have the same mass.

  (e) The gases in the two containers have the same pressures.

18.7 Suppose air at 20 °C contains 10 g of water vapour per cubic meter of air. Given the molar mass of water is $18 \, g \, mol^{-1}$, what is the partial pressure of water vapour (to 3 s.f.)?

18.8 Which of the following statements are correct? (Note: more than one statement may be correct)

  (a) The average kinetic energy of an atom of an ideal gas approaches zero at a temperature of 0 °C.

  (b) Two moles of helium (He) has the same total thermal energy as one mole of nitrogen (N_2) when they are both at 10 °C, because nitrogen is a diatomic gas.

  (c) The average translational kinetic energy of the molecules of air is $\frac{3}{2}kT$, where k is the Boltzmann constant and T is the temperature in kelvin.

  (d) The average thermal energy of the molecules of air is $\frac{5}{2}kT$, where k is the Boltzmann constant and T is the temperature in degrees kelvin.

  (e) The ideal gas law can be written as $PV = NkT$, where N is the number of molecules of the gas in the volume V and k is the Boltzmann constant.

18.9 A marine mammal holds its breath and dives 200 m below the sea surface, where the total pressure is 2.061 MPa. The volume of the mammal's lungs when fully inflated at sea surface (air pressure = 101.3 kPa) is 7 L. Assuming that the mammal's core temperature remains constant at 310 K, what will the volume of its lungs be when it reaches a depth of 200 m?

18.10 A gas mixture is contained in a sealed flask at atmospheric pressure, 101.33 kPa. When all the carbon dioxide is chemically removed from the sample, keeping the same temperature, the final pressure is 67.89 kPa. What percentage of the molecules of the original sample was carbon dioxide?

18.11 In the "death-zone" on Mt Everest the atmospheric pressure is typically 34 kPa. The air at this elevation contains a negligible quantity of water vapour. On a molar basis the composition of the air is 20% O_2, 79% N_2 and 1% Ar. Determine the partial pressure of O_2 in the death-zone.

PHASE AND TEMPERATURE CHANGE

19.1 Introduction

This chapter looks at phase-change phenomena and thermal energy in order to provide a framework for understanding how the body controls its temperature, energy and moisture balances.

Key Objectives

- To understand the concepts of phase change and latent heat.
- To be aware of vapour pressure and its dependence on temperature.
- To understand the concept of specific heat.
- To be able to apply these concepts to energy-conservation problems.

19.2 Phase Changes

Real Gases

For real gases, the ideal gas equation no longer holds when the interaction between the atoms or molecules becomes significant. This happens when the temperature is low and the density is high.

Consider what you know of the properties of a solid. A solid retains its shape, which indicates that there *must be attractive forces between the molecules* when they are close enough. When the density is high, the molecules are closer together. Also, when the temperature is low, the potential energy associated with the attractive force between the molecules can no longer be ignored (see Figure 19.1) compared to the kinetic energy – the attraction becomes significant and the behaviour of a gas deviates from ideal gas behaviour.

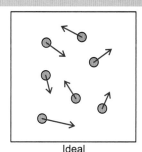

Ideal

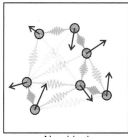

Non-ideal

Figure 19.1 In a real gas, there are short-range interactions between the molecules that cannot always be ignored.

States of Matter

Pure samples of chemically-simple substances are typically found in one of three different **physical states**: solid, liquid or gas. Other states exist under more extreme conditions, such as plasma (ionised gas) which exists at high temperature, and the Bose-Einstein condensate state, which requires extremely low temperatures.

For this section we will be considering only the simplest materials, such as samples of pure elements and simple chemicals with relatively small molecules. This excludes such things as emulsions, gels, complex biological materials and long-chained molecules like plastics and fatty acids.

A **solid** has a definite shape and volume. In a solid the particles are not free to move, only to vibrate.

A **liquid** has a fixed volume (at a particular pressure and temperature) but the shape is determined by its container. The molecules are usually further apart than in the solid state.

State and phase

These two terms are often used interchangeably, which is fine when we are discussing pure substances. The Merriam–Webster dictionary defines phase as 'a homogeneous, physically distinct, and mechanically separable portion of matter present in a nonhomogeneous physicochemical system'. An example where state and phase are not the same thing is an oil–water mixture; there are two distinct *phases*, one of which is oil-rich, and the other which is water-rich. They are both in the liquid *state*.

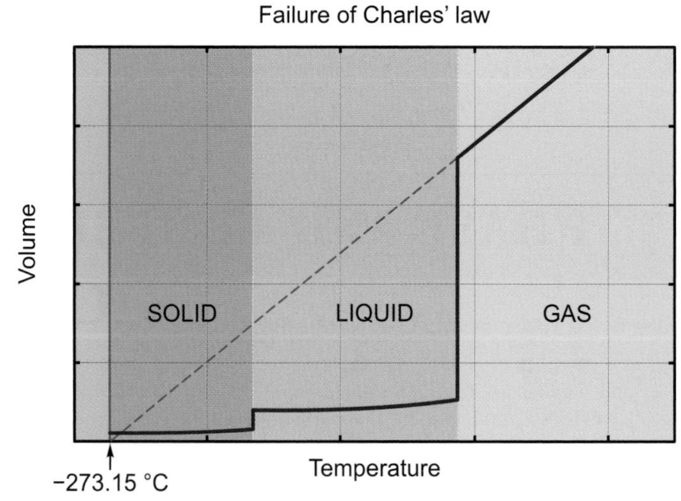

Figure 19.2 At low temperatures, Charles' law no longer holds true. On cooling at constant pressure, the gas will condense to form a liquid, and in most cases will eventually solidify at lower temperature. For water, the volume of the solid is slightly larger than the liquid, whereas for most substances it is smaller, as shown here. (The change in volume on transition from liquid to solid has been exaggerated.)

In the **gas** state, the molecules are far apart on average, and the gas has its shape and volume fixed by its container.

We are all familiar with matter changing phase: when heated, solid ice will become liquid water, and then turn into a vapour as steam. At normal atmospheric pressure, these phase changes occur at 0 °C and 100 °C. However, these temperatures are highly pressure-dependent. On top of Mt Everest, water will boil at about 69 °C, as the pressure is about 30% of the pressure at sea level.

There are different names given to the various phase transitions:

Vaporisation Liquid to gas

Condensation Gas to liquid

Melting Solid to liquid

Freezing Liquid to solid

Sublimation Solid to gas

Deposition Gas to solid

We will now take a look at how pressure, temperature, volume and the state of matter are related.

Phase Diagrams

Boiling Point

The boiling point of a liquid is the temperature at which the liquid and vapour phases are in equilibrium, which depends on the pressure. If the vapour pressure of the gas phase is lower than the saturation vapour pressure, then molecules can continue to break away from the liquid and join the gas phase. If the liquid is not in a sealed container, the gas molecules will escape, ensuring the vapour pressure will stay below saturation. This is while water bubbles continuously at 100 °C at sea level.

Consider a sample of water completely isolated from the rest of the world. The sample is inside a container with all the air removed, and sealed by a movable piston so the volume and pressure can be varied. To begin with, imagine that the volume of the container is held fixed (and is greater than the liquid volume). At around room temperature, say 20 °C, the water is mostly in the liquid state. However, some of the molecules have enough energy to break free of the liquid surface. Even though there is no air in our container, there will be some **vapour** above the liquid, and the liquid and vapour regions will be easily distinguishable.

Key concept:
The vapour pressure of a substance is the gas pressure created by the solid or liquid phases, and is a consequence of the faster molecules breaking away from the liquid or solid.

When the system reaches thermodynamic equilibrium at this temperature then the rate at which the molecules leave the liquid and the rate at which they rejoin it are the same. The pressure exerted by this saturated vapour is 2.33×10^3 Pa: this is the **saturation**, or equilibrium, **vapour pressure** of water at 20 °C. In this mixed-phase stage where liquid and vapour are in equilibrium, the pressure is not dependent on the volume. It depends only on the temperature of the mixture. This particular pressure corresponds to a water vapour density of 17.2 g m^{-3} at 20 °C. The density of the liquid phase will be close to 1000 kg m^{-3} at this temperature, which is more than 53 000 times larger than the vapour density.

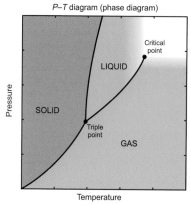

Figure 19.3 A typical *P–T* phase diagram.

> **Key concept:**
> The saturation vapour pressure depends only on temperature.

Note that as the piston is moved to increase the volume, the system is no longer in thermodynamic equilibrium. More molecules will evaporate than will condense until the saturation vapour pressure is reached.

Now consider what will happen as the temperature is raised while the volume is held fixed. More water molecules from the liquid phase will have the energy to break free and join the vapour phase, and so the pressure and density of the vapour phase will increase. The liquid water will expand as the temperature is raised and it will become slightly less dense. When the temperature reaches 100 °C, then the saturation vapour pressure will reach 1.01×10^5 Pa – standard atmospheric pressure at sea level.

As the temperature increases still further, then the vapour increases in density and the liquid density decreases until eventually they will be the same. The temperature at which this happens is called the **critical temperature**. Above this temperature, the liquid phase can't exist, and there is no distinction or phase boundary between liquid and gas, and it is called a super-critical fluid. The critical point for water occurs 647.1 K (374.0 °C), at a pressure of 22.06×10^6 Pa (219 atm).

If the container is instead cooled, then at some temperature ice begins to form. At this temperature, the solid, liquid and gas phases are all in thermodynamic equilibrium: this is called the **triple point**. The triple point occurs at a unique temperature, which is not affected by the volume of the container, or the amount of the substance in it. For this reason, it is the triple-point temperature of water, not the freezing point (which is pressure dependent) that is used to define the kelvin temperature scale. The triple-point temperature of water is defined to be at 273.16 K, so that it is at 0.01 °C, and the freezing point at standard atmospheric pressure still occurs at 0 °C. Below the triple-point temperature, liquid and vapour cannot exist together in equilibrium. Instead, the vapour co-exists with the solid phase, provided the volume is large enough.

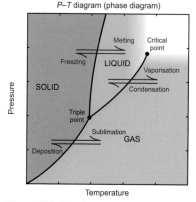

Figure 19.4 The lines show where two phases can co-exist in equilibrium, and these separate regions where each phase is stable.

If the piston in our container is moved so that there is no room for vapour, only the solid and liquid phase can co-exist and the temperature at which this happens is the melting point. The melting point typically varies only a little with changes in pressure.

The ***P–T* phase diagram** shows this behaviour, plotting the regions in which the different phases exist together as a function of temperature and pressure. Figure 19.3 shows the phase diagram for a typical pure substance. The co-existence lines show the conditions under which two phases co-exist in dynamic equilibrium. These lines separate the regions where each phase is stable. A phase change corresponds to movement across the boundary lines. The boundary line between the liquid and vapour phases ends at the critical point, and at temperatures above this, the substance cannot be liquefied at any pressure. At these temperatures, a **supercritical fluid** is formed, which has properties that vary between an incompressible liquid and a gas, depending on the pressure, but there is no distinct change of phase. However, if the pressure is high enough, further phase changes can occur. In water, for example, even the supercritical fluid will change into the solid phase, a form of ice, at pressures above 10^{10} Pa.

The slope of the boundary line between the solid and liquid phases is very steep as there is little change in temperature with pressure. For water, the slope of this line is negative, so it is in fact possible to turn ice into water by increasing the pressure. If the temperature is near the melting point and the pressure is increased sufficiently, the solid ice will become liquid water, which will refreeze when the additional pressure is removed. The boundary between the *gas* phase and a condensed phase varies strongly

P–T diagram (Water)

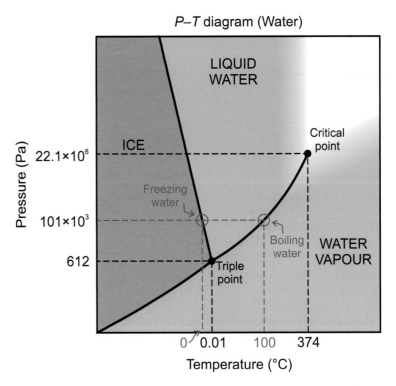

Figure 19.5 The P–T diagram for water. Water is unusual in that the slope of the melting point line is negative.

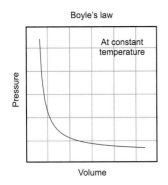

Figure 19.6 For an ideal gas at constant temperature, the product of P and textitV is fixed, giving the P–V graph a hyperbolic shape.

with pressure, as compressing a gas increases the rate of collisions between molecules and favours condensation.

Another type of phase diagram is the **P–V phase diagram**. This shows how pressure and volume are related when the temperature is held fixed. Recall from earlier that Boyle's law states that for a gas at a fixed temperature, pressure is inversely proportional to volume. Plotting this relationship gives a graph that looks like Figure 19.6. A curve with this shape is called a hyperbola. This relationship between pressure and volume breaks down at low temperatures when the attraction between molecules becomes significant and the gas begins to condense into a liquid.

P–V diagram

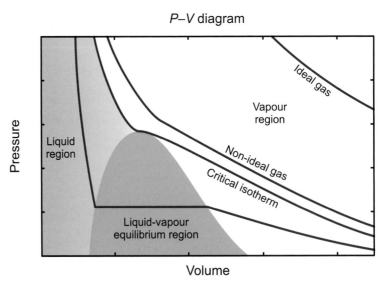

Figure 19.7 A typical P–V diagram, showing isotherms. In the region where the gas behaves like an ideal gas, the isotherms are hyperbolic in shape. At lower temperatures, the isotherms deviate from this behaviour. The critical isotherm has zero slope at the critical point.

Imagine that we again have a sealed container, with a movable piston, containing only water molecules. At around room temperature, if we make the volume large enough we can see from the P–T diagram that all the water will exist as water vapour at low pressure. As we decrease the volume, the pressure will go up. Eventually, when the pressure reaches the vapour pressure (2.33×10^3 Pa at 20 °C), the water vapour will begin to condense into a liquid. As the volume continues to decrease, the pressure will not change, as there is only one pressure at each given temperature when the gas and liquid phases co-exist. If the temperature of our sample of water is instead 100 °C, then the liquid–vapour equilibrium region happens at normal atmospheric pressure (1.013×10^5 Pa). Eventually, all the gas will condense into the liquid phase. Figure 19.7 shows the corresponding P–V diagram. The lines on the plot each correspond to one temperature, and are called **isotherms**.

At the critical temperature, there is no pressure at which the liquid phase exists – there is no liquid-vapour equilibrium region, so the isotherm has zero slope only at a single point, the critical point.

Phase Changes and Latent Heat

To change the phase of a substance requires thermal energy. To turn a solid into a liquid, the bonds that hold the molecules in place can only be broken with the addition of energy. Similarly, energy is required to turn a liquid into a gas. Conversely, to change a gas into a liquid or a liquid into a solid requires that thermal energy be removed.

The amount of energy required to transform a substance from one phase to another depends on what the substance is, as this determines the strength of the molecular forces involved, and on how much of the substance there is. The amount of heat, Q, is therefore

$$Q = mL \tag{19.1}$$

where m is the mass and L is the **latent heat of phase change**. The word latent means *hidden*, and is used because during the phase change, when the substance exists in both phases simultaneously, the temperature does not change.

The value of L differs for each type of phase change, and has some dependence on temperature. L_f is the **heat of fusion**, the energy required to change 1 kg of the substance from solid to liquid. L_v is the **heat of vaporisation**, which is the energy needed to change 1 kg of the substance from liquid to vapour. Some substances can go straight from the solid to the vapour phase, so L_s is used then – the **heat of sublimation**.

The sign of Q is positive when energy is put into a substance and negative when it is taken out. The same coefficient can be used to calculate the energy required to go from, say, solid to liquid or liquid to solid, but the sign of the latent heat transferred is different.

The latent heat of phase change is dependent on temperature. For example, for water at 100 °C, the latent heat of vaporisation is 2256.3 kJ kg^{-1}. At lower temperatures, it is higher. Between 0 °C and 100 °C (with T in Celsius)

$$L_v = 2500 - 2.269\,T - 0.00164\,T^2 \tag{19.2}$$

> **Why doesn't the temperature change during a phase change?**
>
> Consider what happens when we add energy to ice to melt it. The ice can't exist at temperatures higher than the melting point, but the water can. If the heat went into the water and increased its temperature, it would then be warmer than the ice, and so because heat goes from warmer objects to colder ones, it warm the ice, melting it. This means that the water doesn't rise in temperature until all the ice melts.

19.3 Temperature Changes

Heat and Temperature

When a substance has thermal energy transferred to it, it can produce a phase change, as discussed in the previous section. The other possible result is a change in temperature. The amount of temperature change depends on the substance being heated. Heat and temperature are not the same. *Heat* is energy transferred due to a *temperature* difference, and is measured in joules.

In everyday English, it is common to use the word heat in ways that are incorrect when used in a scientific setting, such as when people refer to heat when they really mean temperature. It is also wrong to say an object possesses heat; it possesses internal or thermal energy. An object's internal energy, and hence temperature, can be increased by doing work on it. It is not possible to tell whether the object was heated, had work done, or if there was a combination of both. (James Prescott Joule (1818–1889) performed experiments determining the mechanical equivalent of heat, helping to show that energy is conserved. His contributions were considered important enough that the SI unit of energy now bears his name.)

Specific Heat

Temperature is a measure of the average kinetic (thermal) energy of the molecules in a substance, as shown by Eq. (18.24). Therefore, it seems reasonable that putting more thermal energy into something will raise the temperature. Doubling the amount of energy input should double the increase in temperature. Also, if we double the mass of an object, we will need to put in twice the amount of energy for the same change in temperature. In other words, when there is no change of phase, Q, the heat input, is proportional to both the mass and temperature change. Hence

$$Q = mc\Delta T \tag{19.3}$$

where m is the mass, and ΔT is the temperature increase caused by the amount of heat, Q. The symbol c stands for the **specific heat capacity** – the amount of heat required to increase the temperature of 1 kg of a particular substance by 1 K. The specific heat capacity is dependent on the substance in question, its temperature and the phase of the substance. For example, the specific heat capacity of water is 4186 J kg^{-1} K^{-1} at 15 °C, and for ice it is 2072 J kg^{-1} K^{-1} at –5 °C.

The SI unit for heat is the joule, as it is a transfer of energy. The historic unit for heat, the **calorie**, is still in widespread use. A calorie is the amount of energy required to raise the temperature of 1 g of water by 1 °C. (The value of c for water is slightly temperature dependent, so care is required in defining this. The 15 °C calorie is the energy to change 1 g of water from 14.5 °C to 15.5 °C.) One calorie is equal to 4.186 J. It is still common to see the energy content of foods listed in both kilojoules (kJ) and in kilocalories (kcal), which are also called 'large calories' (written as Calorie) sometimes. This can cause some confusion, especially when food labellers don't distinguish between calories and Calories, so it is wise to stick with kcal or kJ.

19.4   Energy Conservation

When objects are placed in thermal contact, energy is exchanged until the objects reach thermal equilibrium. At this point, the *net* exchange of energy ceases, and the objects have the same temperature. We can use the conservation of energy principle to predict how the temperature of objects placed in contact will change.

The Simple Case – No Phase Change

In the case where the amount of energy transferred between objects in contact will not cause any phase changes, we can write a simple equation relating the masses, heat capacities and temperature changes. If there are only two objects in thermal contact and they are isolated from their surroundings, then the heat lost by one will be same magnitude as the heat gained by the other:

$$Q_{\text{obj 1}} = -Q_{\text{obj 2}} \tag{19.4}$$

Note the minus sign in the Eq. (19.4). One object loses energy, so Q is negative, and the other gains energy, having Q positive, so they are not equal. The sign must be taken into account, and forgetting to do so will give an unphysical solution if you are trying to solve a problem. A more general way of writing energy conservation for a system of objects placed in thermal contact is

$$\text{sum of heat inputs} = Q_{\text{obj 1}} + Q_{\text{obj 2}} + Q_{\text{obj 3}} + \ldots = 0 \tag{19.5}$$

Recall from Eq. (19.3) above that $Q = mc\Delta T$ tells us the amount of heat transferred to increase the temperature by ΔT. As ΔT is defined as the final temperature minus the initial temperature, then Q is positive for an increase in temperature and negative for a decrease.

For the case of two objects placed in contact and reaching a common final temperature, T_f, we have

$$m_{obj\,1}c_{obj\,1}(T_f - T_{obj\,1,\,i}) + m_{obj\,2}c_{obj\,2}(T_f - T_{obj\,2,\,i}) = 0 \qquad (19.6)$$

where the m values are the masses of the two objects, the c values are their specific heat capacities, and $T_{obj\,1,\,i}$ and $T_{obj\,2,\,i}$ are their initial temperatures. This can be simplified to

$$T_f = \frac{m_1 c_1 T_1 + m_2 c_2 T_2}{m_1 c_1 + m_2 c_2} \qquad (19.7)$$

(We have simplified the notation a little for clarity.)

This formula closely resembles the formula for calculating the centre of mass of a two-mass system. This is worth keeping in mind, as it can be a great help in checking that calculated temperatures seem reasonable. For example, suppose we are mixing 1 kg of water at 20 °C with 2 kg of water at 50 °C. In the same way the center of mass of a system of a 1 kg mass and a 2 kg mass will be along the line joining the masses, and will be two-thirds of the distance along this line, closer to the 2 kg mass, the final temperature of the mixture will be two-thirds of the temperature difference higher than 20 °C. So, for this example, the final temperature will be 40 °C.

Example 19.1 *Thermal equilibrium*

Problem: Two objects of known initial temperature and mass are placed in contact and allowed to come to thermal equilibrium. Derive an expression for their final temperature.

Solution: Here, energy as heat is transferred from the hotter object to the cooler object until they reach the same temperature. The initially hot object loses a total amount of energy

$$Q_{hot} = -m_{hot}c_{hot}\Delta T_{hot}$$
$$= -m_{hot}c_{hot}(T_{hot,\,final} - T_{hot,\,initial})$$

The energy gained by the initially cold object is

$$Q_{cold} = m_{cold}c_{cold}\Delta T_{cold}$$
$$= m_{cold}c_{cold}(T_{cold,\,final} - T_{cold,\,initial})$$

Assuming our experiment is isolated from its surroundings so there are no energy transfers between the objects and the surroundings, then the energy lost by the hot object must equal the energy gained by the cold object so $Q_{cold} = -Q_{hot}$. When the two objects finally reach thermal equilibrium, they must have the same final temperature:

$$T_{hot,\,final} = T_{cold,\,final} = T_{final}$$

so (with slightly simplified notation)

$$-m_h c_h(T_f - T_h) = m_c c_c(T_f - T_c)$$

We can rearrange this to find a formula for T_f:

$$-m_h c_h T_f + m_h c_h T_h = m_c c_c T_f - m_c c_c T_c$$
$$T_f(m_h c_h + m_c c_c) = m_h c_h T_h + m_c c_c T_c$$
$$T_f = \frac{m_h c_h T_h + m_c c_c T_c}{m_h c_h + m_c c_c}$$

Example 19.2 *Thermal Equilibrium*

Problem: An indoor swimming pool containing 20×10^3 L of water has a temperature of 20 °C. In an attempt to increase the pool's temperature some heated rocks are added to the pool. These rocks have an initial temperature of 80 °C and a total mass of 500 kg of hot rock is added to the pool. What is the pool's final temperature?

Solution: Energy is transferred as heat from the rocks to the water in the pool. (Note: we will assume that heat transfer between the pool and its surroundings is negligible. This is a reasonable assumption if the air is also at 20 °C.) In the previous example we derived an expression that we can apply to this situation

$$T_f = \frac{m_r c_r T_r + m_w c_w T_w}{(m_r c_r + m_w c_w)}$$

The density of water is 1000 kg m^{-3}, which gives a mass of 1 kg for every litre of water

$$\begin{aligned} m_{\text{water}} &= \rho V \\ &= 1\,\text{kg}\,\text{L}^{-1} \times 20000\,\text{L} \\ &= 20000\,\text{kg} \end{aligned}$$

The specific heat capacities are

$$c_r = 790\,\text{J}\,\text{kg}^{-1}\text{K}^{-1}\text{(for granite)} \quad \text{and} \quad c_w = 4190\,\text{J}\,\text{kg}^{-1}\text{K}^{-1}$$

so

$$T_f = \frac{500\,\text{kg} \times 790\,\text{J}\,\text{kg}^{-1}\text{K}^{-1}\,80\,°\text{C} + 20000\,\text{kg} \times 4190\,\text{J}\,\text{kg}^{-1}\text{K}^{-1} \times 20\,°\text{C}}{(500\text{kg} \times 790\,\text{J}\,\text{kg}^{-1}\text{K}^{-1} + 20000\,\text{kg} \times 4190\,\text{J}\,\text{kg}^{-1}\text{K}^{-1})} = 20.3\,°\text{C}$$

Note that we can use temperatures in °C in Eq. (19.7) as the factor of +273 K that is added to each temperature in °C to get the temperature in kelvin cancels out in Eq. (19.6).

Example 19.3 *Thermal equilibrium*

Problem: Your coffee is too hot to drink at 90 °C, so you would like to cool it to 65 °C. You are impatient and don't want to wait, so you decide to add some tap water which has come out of a Dunedin, NZ, tap in winter, and is at about 7 °C. If your hot coffee has a mass of 250 g, what mass of tap water do you need to add to it to achieve the required temperature?

Solution: Energy is transferred as heat from the hot coffee to the cold tap water until they both reach the final temperature of 65 °C. The energy lost as heat by the hot coffee must equal the energy gained as heat by the cold tap water. (We are neglecting any energy transfers between the coffee and the air. This will be a good approximation, as stirring water into coffee is quick, so not too much heat will be lost to the surroundings during this time.) We can calculate the heat transferred to or from the hot coffee from its mass, specific heat capacity and temperature change

$$\begin{aligned} Q_{\text{coffee}} &= m_{\text{coffee}} c_{\text{coffee}} \Delta T_{\text{coffee}} \\ &= -0.250\,\text{kg} \times \times 4190\,\text{J}\,\text{kg}^{-1}\,\text{K}^{-1}(65-90)\text{K} \\ &= -2.6 \times 10^4\,J \end{aligned}$$

The heat is negative, and the coffee is losing energy. Conservation of energy means that $Q_{\text{coffee}} + Q_{\text{water}} = 0$. From this we can see that $Q_{\text{water}} = 2.6 \times 10^4$ J.

The mass of tap water required is then determinable by rearranging $Q_{\text{water}} = m_{\text{water}} c_{\text{water}} \Delta T_{\text{water}}$

$$m_{\text{water}} = \frac{Q_{\text{water}}}{c_{\text{water}} \Delta T_{\text{water}}}$$

$$m_{\text{water}} = \frac{2.6 \times 10^4\,\text{J}}{4190\,\text{J}\,\text{kg}^{-1}\text{K}^{-1} \times (65-7)\text{K}} = 0.1\,\text{kg}$$

You will need a reasonably large coffee cup to do this.

Latent heat of fusion (0 °C)	334.4 kJ kg^{-1}
Latent heat of vaporisation (100 °C)	2256.3 kJ kg^{-1}
Specific heat capacity of liquid (15 °C)	4.186 kJ kg^{-1} K^{-1}
Specific heat capacity of solid (–5 °C)	2.072 kJ kg^{-1} K^{-1}

Table 19.1 Latent and specific heat values for water

A helpful hint: Many people are quite comfortable with the idea of 'magnitude of heat lost equals magnitude of heat gained' and will happily solve a problem like those here this way. This approach quickly falls apart when there are more than two objects involved, and will also give a result that is unphysical if you leave out a critical minus sign. A much more reliable approach to such problems is to remember that the *sum* of the Q values will be zero if no heat is exchanged with anything outside the system, and if there is no phase change, $Q = mc\Delta T$. By using $\Delta T = T_{\text{final}} - T_{\text{initial}}$, the sign of Q will always be correct.

Thermal Equilibrium With Phase Change

The equation we previously gave for conservation of energy where objects are in thermal contact, Eq. (19.5), is still valid for cases where a change of phase occurs. We now need to include the heat transfers that occur during the changes in phase, however. For example, suppose we have a glass of water in which we place an ice cube (at 0 °C). At least some of the ice will melt, and the resulting melt-water will then mix with the original water to come into thermal equilibrium. If the ice melts completely, and the final temperature of the mixture is T_f, then

$$m_{\text{water}} c_{\text{water}} (T_f - T_{\text{water, i}}) + m_{\text{ice}} c_{\text{water}} (T_f - 0\,°C) + m_{\text{ice}} L_f = 0 \qquad (19.8)$$

Once the ice has melted, it is still at 0 °C, so this is the initial temperature of the resulting water. It takes energy input to melt ice into water, so there is a plus sign in front of mL. If the phase change was, say, vapour to liquid instead, we would need a minus sign.

19.5 L and c Values for Water

Table 19.1 summarises the most important values for water.

These numbers show some interesting facts. It takes as much energy to melt 1 kg of ice as it does to raise the temperature of that 1 kg of water by about 80 K. It is also quite clear from the large value of the latent heat of vaporisation that the phase change from water to steam takes a lot of energy. This is responsible for the very strong cooling effect of evaporation from the skin, and for the severity of burns caused by steam. As steam condenses to water on the skin, it transfers a large amount of heat, so the effect is much worse than a burn from water at 100 °C.

The latent heat of vaporisation changes with temperature, but at normal skin temperatures it is still large ($\approx$ 2450 kJ kg^{-1}). When water or sweat evaporates from the skin, this transfers heat from the skin, so this is one of the key ways our bodies regulate temperature. Humans are one of the few species that can sweat, and many animals pant instead, cooling the mouth and lungs. Like us, horses and primates also have sweat glands in the armpits (or the equivalent part of their anatomy).

Cooling by evaporation can occur only if the air is not already carrying as much water vapour as possible, so the *humidity* plays a crucial role in determining how comfortable we are in an environment. We will look at the importance of water vapour in the air in the next chapter.

Example 19.4 *Thermal equilibrium with phase change*

Problem: You want to add just enough ice to your room-temperature (25 °C) beer to bring it to a temperature of 4 °C, which is what it would have been if you had remembered to put it in the fridge. What mass of ice (at 0 °C) is required? Assume the mass of beer is 300 g and that beer and water have the same specific heat capacity.

Solution: This problem is similar to the previous one. However this time we need to take into account that the ice will have energy put into it to change state, as well as to warm up to the final temperature.

Energy is transferred as heat from the warm beer to the ice, cooling the beer and melting the ice and then warming up the resulting water from its starting temperature of 0 °C.

For the beer

$$Q_{beer} = m_{beer} c_{beer} \Delta T_{beer} = 0.3 \, \text{kg} \times 4190 \, \text{J kg}^{-1} \text{K}^{-1} \times (4 - 25) \text{K} = -2.6 \times 10^4 \, \text{J}$$

For the ice/ice-water

$$Q_{ice} = m_{ice} L_{water} + m_{ice} c_{water} \Delta T_{icewater} = m_{ice} \times \left(334 \times 10^3 \text{J kg}^{-1} + 4190 \, \text{J kg}^{-1} \text{K}^{-1} \times (4 - 0)\text{K} \right)$$

There must be heat transferred to the ice to melt it, so the sign of mL is positive. We use the specific heat capacity of water, not of ice, because the ice has turned into water before its temperature increases.

$$Q_{ice} + Q_{beer} = 0$$

so

$$m_{ice} = \frac{2.6 \times 10^4 \, \text{J}}{334 \times 10^3 \text{J kg}^{-1} + 4190 \, \text{J kg}^{-1} \text{K}^{-1} \times (4 - 0)\text{K}} = 0.074 \, \text{kg}$$

This is quite a lot of ice, and would dilute the beer by almost 25%. This would ruin the taste, and is why nobody puts ice in their beer!

19.6   Summary

Key Concepts

phase   The term phase usually refers to a part of a sequence, and it is used differently in several areas of physics, such as wave motion and astronomy. In thermodynamics, it refers to a state of a macroscopic physical system that has uniform composition and physical properties.

state (of matter)   A form of matter. The three traditional states are solid, liquid, and gas. Other forms of matter, such as plasma, are now also referred to as states.

P–T phase diagram   A diagram showing the stable states of matter for a particular substance as pressure and temperature are varied. On the boundary lines between these regions, two phases can exist in equilibrium.

P–V phase diagram   A diagram showing how pressure and volume are related. Each line plotted on the diagram shows the P–V relationship for a fixed temperature, and these are called isotherms.

heat   Thermal energy transferred due to a temperature difference.

specific heat capacity (c)   The amount of heat required to change the temperature of 1 kg of a substance by 1 K.

latent heat coefficient (L)   The amount of heat that must be transferred to change 1 kg of a substance from one phase to another. The latent heat coefficient depends on the phase change involved and the temperature.

evaporation   A change of state from liquid to gas.

condensation   A change of state from gas to liquid.

melting   A change of state from solid to liquid.

freezing   A change of state from liquid to solid.

sublimation   A change of state from solid to gas.

deposition   A change of state from gas to solid.

vapour pressure The gas pressure created by the equilibrium between vaporisation/sublimation and condensation/deposition. It depends only upon the temperature.

vapour The term vapour is used when a substance is in its gas phase at a temperature lower than its critical temperature.

Equations

$$Q = mc\Delta T$$

$$Q = mL$$

$$\text{sum of heat inputs} = Q_1 + Q_2 + Q_3 + \cdots = 0$$

19.7 Problems

19.1 Which of the following statements are correct? (Note: more than one statement may be correct)

(a) When the liquid and vapour phases of a substance co-exist the pressure depends only on the volume, not the temperature.

(b) When the temperature of a substance is less than the critical temperature there is a distinct phase change between the liquid and solid phases involving latent heat.

(c) The latent heat for the vaporisation of water is greater at 0 °C than at 100 °C.

(d) When two objects (1 and 2) exchange heat, the corresponding heat quantities, Q_1 and Q_2, satisfy the equation $Q_1 = Q_2$.

(e) The specific heat capacity of a substance is numerically the same as the amount of thermal energy required to increase the temperature of 2 kg of the substance by 0.5 °C.

19.2 A 1 kg block of copper and a 2 kg block of wood each absorb the same amount of heat, and the temperature of the wood increases by 2 °C. The specific heat of wood is 1700 J kg^{-1} K^{-1}. The specific heat of copper is 387 J kg^{-1} K^{-1}. How much does the temperature of the block of copper increase?

19.3 To cool a hot bath containing 50 litres of water at 50 °C down to 20 °C,

(a) what volume of cold water at 5 °C would be needed (in litres)?

(b) what mass of ice at 0 °C would be needed?

19.4 How much heat must be transferred into a 50.0 kg block of ice at -10 °C to raise its temperature to 0 °C, melt it into liquid water, heat it to 100 °C and then evaporate all of the water?

19.5 By mistake I run a bath using 60 kg of water from just the hot tap and the water temperature is 65 °C. I decide to cool the water to 40 °C with snow, which is at 0 °C. The specific heat capacity of the water is 4.19 kJ kg^{-1} K^{-1} and the latent heat of fusion of ice is 333 kJ kg^{-1}. What weight of snow should I use?

19.6 A 65 kg patient suffers from hyperthermia, having a mean body temperature of 41 °C. The patient is placed in a bath containing 50 kg of water. The specific heat capacity of the water is 4.19 kJ kg^{-1} K^{-1} and the average specific heat capacity of the patient is 3.49 kJ kg^{-1} K^{-1}. In order for the final temperature of the water and patient to be 37 °C, what should the initial temperature of the water be?

19.7 A hyperthermic male, weighing 104 kg, has a mean body temperature of 42 °C. He is to be cooled to 37 °C by placing him in a water bath, which is initially at 25 °C. What is the minimum amount of bath water required to achieve this result? The specific heat of the body is 3.5 kJ kg^{-1} k^{-1}. The specific heat of water is 4.19 kJ kg^{-1} k^{-1}.

19.8 A beaker of water and a beaker of an unknown liquid are weighed and their temperatures measured. The unknown fluid has a mass of 1.2 kg and the water has a mass of 0.8 kg. They are both at an initial temperature of 20 °C. The beakers are then simultaneously heated on the same heating element for the same length of time and then they are weighed and their temperatures measured again. Their weights did not change when they were heated in this way, but the water now has a temperature of 28 °C whereas the unknown fluid has a temperature of 34 °C. What is the specific heat of the unknown fluid?

19.9 Steam condenses on a 5 kg iron plate. The plate was initially at 15 °C and it is found that 10 g of steam has condensed onto the plate. What is the temperature of the plate after the steam has condensed? (The specific heat capacity of iron is 449.4 J kg^{-1} K^{-1})

19.10 A 70 kg runner loses 0.5 kg of water each hour through evaporation of perspiration in order to maintain a stable temperature. The latent heat of water at his skin temperature is 2440 kJ kg^{-1} and the average specific heat capacity of his body is 3.5 kJ kg^{-1} K^{-1}. If he stopped perspiring, how much would his temperature rise in the following 30 minutes?

WATER VAPOUR AND THE ATMOSPHERE

20

20.1 Introduction

The air in the Earth's atmosphere is made up of a mixture of gases, mostly nitrogen and oxygen, with smaller amounts of argon, carbon dioxide and other gases. Excluding water vapour, the composition of this mixture does not change to any significant degree in most situations, so for many purposes it can be treated like a single gas. However, the quantity of water vapour mixed in with this dry air changes as environmental conditions change, which has important consequences for the body's thermoregulation. In this chapter, we will learn about the effects of water vapour in the air, and how this affects our ability to survive in our environment and regulate our body temperature.

Key Objectives

- To be able to determine the relative humidity from the partial pressure and saturated vapour pressure of water.

- To understand the various quantities used to characterise water-vapour/dry-air mixtures, such as wet-bulb temperature, dry-bulb temperature, dew-point temperature and moisture content.

- To be able to use a psychrometric chart to determine the properties of water-vapour/dry-air mixtures.

20.2 Mixtures of Water Vapour and Air

Dalton's Law

Dalton's law was covered earlier along with the other ideal gas laws. Recall that the partial pressure exerted by one gas in a fixed volume is the pressure it would exert if it was the only gas present. The total pressure is the sum of the partial pressures of all the gases present. In the atmosphere, the total atmospheric pressure is the sum of the partial pressure exerted by the dry air and the partial pressure due to water vapour. The total pressure at sea level varies with time and location, and the contribution due to water vapour varies as the amount of water vapour changes. The partial pressure due to water vapour can be as high as 4% of the total atmospheric pressure in humid tropical conditions. In cold, dry, polar environments it can be 0.1% or less.

Water Vapour in the Air

In the section on phase diagrams we considered what would happen to water in a sealed container at different temperatures. When the vapour and liquid phases are in equilibrium, i.e. evaporation and condensation are occurring at the same rate, the vapour density (and hence the pressure exerted by the vapour) has a fixed value at each temperature. If any more vapour is added, it will condense and if vapour is removed, more evaporation will occur.

If another gas is introduced into our container, this does not affect the partial pressure due to the water vapour. This is important when looking at water vapour in the atmosphere. There is a fixed maximum amount of water vapour that a volume can hold at any particular temperature, corresponding to the maximum partial pressure due to water vapour, the **saturated water vapour pressure**. Figure 20.1 shows the maximum pressure that can be exerted by water at each temperature.

In an open environment, the water vapour is generally not in equilibrium with any liquid water present. The air usually has less water vapour mixed with it than the maximum possible, so any liquid water in the environment will evaporate over time. An important measure of how much moisture is in the air is the **relative humidity**, RH. This is given by

$$\text{RH} = \frac{\text{Partial pressure of water vapour in the air}}{\text{Saturated water vapour pressure, same temperature}} \times 100\% \qquad (20.1)$$

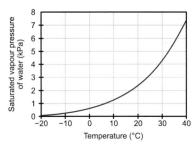

Figure 20.1 The saturation vapour pressure of water depends on temperature. Between 10 °C and 30 °C, the vapour pressure doubles approximately every 10 °C.

It is usually given as a percentage. At 100% relative humidity, the atmosphere has the maximum amount of water vapour possible at its current temperature. This is an important condition, because sweat will no longer evaporate from the skin. Recall from the previous chapter that the phase change from liquid to vapour requires a significant input of energy, which usually comes from the surrounding skin, cooling it. This is a very important mechanism for keeping body temperatures inside a suitable range. As the humidity gets higher, the evaporation of sweat slows down, and so even if the temperature is unchanged, changing the relative humidity can strongly affect how comfortable we feel in an environment.

20.3 Partial Pressure, Moisture Content

The partial pressure due to water vapour and the amount of moisture in the air are related, and we can apply the gas laws already encountered to derive a mathematical relationship. The **moisture content** is the mass of water vapour in the air per unit mass of (dry) air. As this is quite a small fraction, it is often given in units of grams of moisture per kilogram of air. The moisture content is also known as the **absolute humidity** (sometimes called the **humidity ratio**).

Suppose that a volume V contains n_w moles of water vapour with a partial pressure of P_w. The corresponding mass of water, $m_w = M_w n_w = 18 \times 10^{-3} \text{ kg mol}^{-1} \times n_w$ as the molar mass of water is 18 g mol^{-1}. We can obtain n_w from the ideal gas equation, $n_w = \frac{P_w V}{RT}$, so the mass of the water vapour is

$$m_w = 18 \times 10^{-3} \text{ kg mol}^{-1} \times n_w = 18 \times 10^{-3} \text{ kg mol}^{-1} \frac{P_w V}{RT} \qquad (20.2)$$

The mass of the dry air can be written similarly. Its partial pressure is $P_t - P_w$, the total pressure minus the partial pressure due to the water vapour, and its molar mass is 28.97 g mol^{-1}. (Air is mostly N$_2$, so the molar mass is close to the 28 g mol^{-1} of diatomic nitrogen molecules.) The air has mass

$$m_a = 28.97 \times 10^{-3} \text{ kg mol}^{-1} \times n_a = 28.97 \times 10^{-3} \text{ kg mol}^{-1} \frac{(P_t - P_w)V}{RT} \qquad (20.3)$$

Both gases share a common temperature, and occupy the same volume. Combining these two equations gives us the moisture content in terms of the partial and total pressures

$$\frac{m_w}{m_a} = \frac{18 P_w}{28.97(P_t - P_w)} = 0.621 \times \frac{P_w}{(P_t - P_w)} \qquad (20.4)$$

(Note that this is for the mass of water vapour in kg per kg of dry air. To get the moisture content in *grams* per kilogram of dry air, multiply by 1000.)

As the value of P_w is at most a few percent of the total pressure, the approximation for the relative humidity

$$\text{RH} \approx \frac{\text{moisture content}}{\text{saturated moisture content at that temperature}} \times 100\% \qquad (20.5)$$

is normally good enough to find the relative humidity from the moisture content.

20.4 Atmospheric Properties

There are a number of things we can measure that tell us the amount of water vapour mixed in with a sample of dry air. We can use these to calculate other quantities.

Dry-Bulb Temperature

The **dry-bulb temperature**, T_{db}, is the normal temperature that an ordinary thermometer reads. We will soon be introducing some other different kinds of temperature measurement, so the 'dry-bulb' label is nearly always used in psychrometrics (the study of gas–vapour mixtures) to reduce the possibility of getting them mixed up.

The Dew-Point Temperature

The **dew-point temperature**, T_{dew}, is the temperature to which a surface located in an air–water mixture needs to be cooled before condensation occurs on the surface. In other words, it is the temperature at which the saturated vapour pressure of water is the same as the partial pressure of water vapour in the atmosphere. Atmospheric samples with the same dew-point temperature have the same partial pressure of water vapour. If the total pressure is the same, these samples will all have the same moisture content, but can have different dry-bulb temperatures, and different relative humidities.

The dew point is measured by cooling a surface below the ambient temperature of the room and recording the temperature at which condensation first forms on a surface. In practice it is hard to measure accurately, and is easily under-estimated, because it requires looking for the temperature at which a very small amount of condensation has occurred. The dew-point temperature is always less than or equal to the dry-bulb temperature. When the dew-point and dry-bulb temperatures are the same, the relative humidity is 100%.

Wet-Bulb Temperature

The **wet-bulb temperature**, T_{wb}, is different to the dew-point temperature and is more easily measured. Wet-bulb temperature is the reading of a thermometer whose bulb is wrapped in cloth which is kept wet (see Figure 20.2). The wet-bulb temperature is always less than or equal to the dry-bulb temperature. It is also more than, or equal to, the dew-point temperature.

If the relative humidity is anything less than 100%, evaporation is possible, as the water vapour pressure is lower than the saturation water vapour pressure. The reason the wet bulb reads lower is that it is in a dynamic equilibrium with the surroundings. Because the thermometer is colder than its surroundings, it is constantly receiving heat from its surroundings. However, it is also constantly losing heat due to evaporation of water from the cloth that is wrapped around its base. The wet-bulb temperature is the equilibrium temperature such that the rate at which heat is lost by the thermometer due to evaporation is equal to the rate at which heat is transferred to the thermometer due to the temperature difference between it and its surroundings.

The evaporation rate is higher when the air is more able to receive extra water vapour, i.e. when the relative humidity is low. If the relative humidity is low, there will be a large difference between dry-bulb and wet-bulb temperatures (Figure 20.4). If the relative humidity is high, there will only be a small difference between dry-bulb and wet-bulb temperatures (Figure 20.3).

To give an accurate reading, the rate of air flow over the wet wick must be at least 3 m s^{-1}, and the wick must be clean and free of dirt and oils that affect the evaporation rate.

Humidity, Moisture Content and Partial Pressure

Relative humidity, moisture content and partial pressure due to water vapour are all difficult to measure directly, and are instead usually calculated from the temperatures

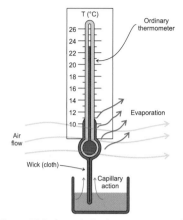

Figure 20.2 A wet-bulb thermometer usually reads a lower temperature than an ordinary dry-bulb thermometer due to evaporation.

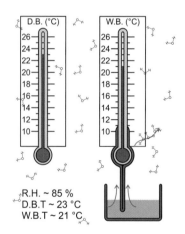

R.H. ~ 85 %
D.B.T ~ 23 °C
W.B.T ~ 21 °C

Figure 20.3 The wet-bulb reading is lower than the dry-bulb temperature when the relative humidity is less than 100%. The lower the humidity, the bigger the temperature difference.

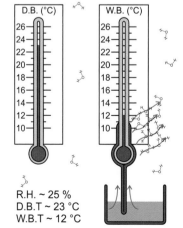

R.H. ~ 25 %
D.B.T ~ 23 °C
W.B.T ~ 12 °C

Figure 20.4 At lower humidity, the wet-bulb temperature is much lower than the dry-bulb temperature.

mentioned above. One way of determining these values is to use a **psychrometric chart**. This is the topic of the next section.

20.5   Psychrometry

Background

Psychrometry is the study of gas–vapour mixtures, usually mixtures of water vapour and air. The term vapour is used when a substance is in its gas phase at a temperature lower than its critical temperature. In this situation the vapour can be condensed into a liquid when the partial pressure is high enough. The critical temperature for water is 374 °C, so water vapour will always condense under typical atmospheric conditions if the partial pressure is increased sufficiently.

There are many possible quantities that we might want to specify for a given sample of water vapour and air: relative humidity, moisture content, wet-bulb temperature, dry-bulb temperature, dew point and energy content per unit mass, to name a few. Fortunately, these quantities are not independent, and so we need only measure a few to know the rest too. There is a simple postulate behind this, which is that to characterise the thermodynamic state of simple, single-component system requires determining the values of just two independent parameters. With a two-component mixture like we have here, that would require four, but because the temperatures of the two components are the same, we need only three. Total pressure is easily measured, so only two other measurements are required.

Plotting the relationships between various quantities simplifies the calculation, and this is the purpose of the psychrometric chart.

Psychrometric Charts

The **psychrometric chart** shows many of the quantities we mentioned earlier: moisture content (and/or partial pressure due to water vapour), relative humidity, wet-bulb temperature and dry-bulb temperature, typically, though others may be shown on more sophisticated charts. Each chart is valid for a particular total pressure, and all the ones shown in this book are for standard atmospheric pressure, 1.013×10^5 Pa. Every point on the chart below the 100% relative humidity line represents a possible state that a mixture of water vapour and air might be in. Each line on the chart corresponding to a stated property value (such as $T_{db} = 20$ °C) shows all the states that share that value. If we measure two independent properties of our mixture, we can find the one point on the chart which corresponds to the state of the mixture.

On the charts we will use for all our examples, the right-hand vertical axis shows the moisture content – how many grams of moisture per kilogram of dry air – so each horizontal line corresponds to a particular moisture content. The bottom horizontal axis shows the ordinary dry-bulb temperature, and so each vertical line is a particular temperature.

The curved lines show states that share a common relative humidity. For a given dry-bulb temperature, the relative humidity varies from 0% up to 100%, with the 100% value occurring at the moisture content that corresponds to the maximum moisture content of the air at that temperature. As the temperature rises, the amount of water vapour the air can hold rises, doubling every 10–12 °C, so the lines curve up.

The diagonal lines represent constant wet-bulb temperature. At 100% relative humidity, the wet-bulb temperature is the same as the dry-bulb temperature, as no evaporation occurs. The lines of constant wet-bulb temperature intersect with the 100% humidity line when the dry-bulb temperature has the same value. There is a minimum temperature that the wet-bulb thermometer can reach at a given room temperature. For example, at 24 °C, the minimum wet-bulb reading, which occurs when the relative humidity is zero, is approximately 7.5 °C.

The dew-point temperature is not labelled, but it is easy to determine from the other information. To measure the dew point, the mixture is cooled until 100% relative humidity (saturation point) is reached, without changing the total pressure or the

---

Closely related properties

On some charts the moisture content is supplemented or replaced by the vapour pressure. As shown by Eq. (20.4), they are closely related, and nearly proportional.

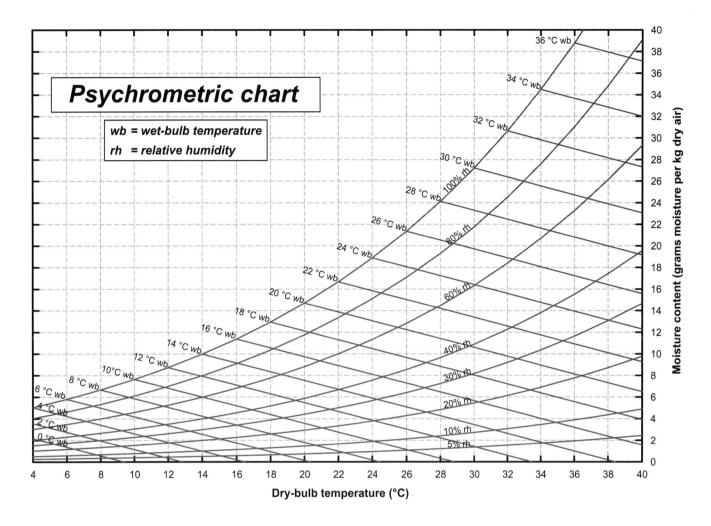

Figure 20.5 A psychrometric chart valid for water vapour and air mixtures at standard atmospheric pressure. [Created from data in *ASHRAE Fundamentals Handbook 2001*.]

moisture content. This process corresponds to moving the state of the mixture horizontally across the chart to lower temperature until the 100% RH curve is reached. The temperature of the point where this happens is the dew-point temperature. Hence measuring the dew point tells us which horizontal moisture content line indicates the state of our original (uncooled) mixture.

In summary, for mixtures of water vapour and dry air with a fixed total pressure:

- Two independent measurements define the state of a water-vapour–air mixture.

- Horizontal lines = lines of constant moisture content, dew-point temperature and partial pressure of water vapour.

- Vertical lines = constant dry-bulb temperature.

- Curved lines = constant relative humidity.

- Diagonal lines = constant wet-bulb temperature.

- If a mixture is heated or cooled without adding or removing water it moves horizontally across the chart.

- If water is added to or taken away from a mixture without changing its temperature, then its state moves vertically up or down the chart.

- If a mixture would have a humidity above 100% (which is equivalent to having a temperature lower than its dew-point temperature) then condensation will occur until the humidity is reduced to 100%. This is the process that produces clouds.

To use the psychrometric chart for some useful calculations, we often need to first find out the mass of dry air we are dealing with. This can be found from the volume, say the volume of the room in question, and the density of dry air, which is about $1.2 \, \text{kg m}^{-3}$ at 20 °C.

Example 20.1 *Moisture added to a room*

Problem: A room 3.5 m by 5 m by 2.4 m contains air and water vapour at a dry-bulb temperature of 20 °C with a dew-point temperature of 7 °C. The density of dry air is $1.2 \, \text{kg m}^{-3}$. Locate this point on a psychrometric chart, and determine the following:

(a) **The moisture content.**

(b) **The relative humidity.**

(c) **The total quantity of water vapour in the room.**

(d) **How much water, if any, would condense if a boiling kettle added 800 g of water vapour to the air without changing the temperature.**

Solution:

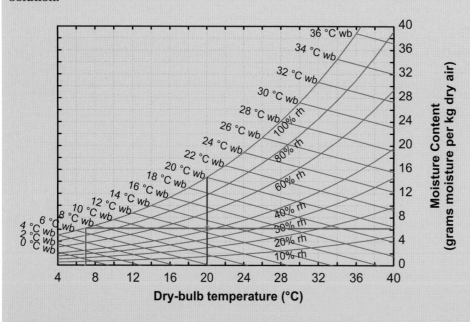

Figure 20.6 Use the dew-point temperature of 7 °C and dry-bulb temperature of 20 °C to determine the moisture content from the psychrometric chart.

(a) Find 7 °C on the horizontal axis and go up to the 100% humidity line. This gives us a moisture content of $6 \, \text{g kg}^{-1}$.

(b) The atmosphere in our room has the same moisture content as this, so moving across horizontally until we reach the correct air temperature of 20 °C, we can see that the relative humidity is 42%.

(c) Volume = $3.5 \, \text{m} \times 5 \, \text{m} \times 2.4 \, \text{m} = 42 \, \text{m}^3$
Mass of dry air = density × volume = $1.2 \, \text{kg m}^{-3} \times 42 \, \text{m}^3 = 50.4 \, \text{kg}$
Mass of water vapour = mass of dry air × moisture content = $50.4 \times 6 = 302.4 \, \text{g}$

(d) Total water vapour = $302.4 \, \text{g} + 800 \, \text{g} = 1102.4 \, \text{g}$
New moisture content = $1102.4 \, \text{g} / 50.4 \, \text{kg} = 22 \, \text{g kg}^{-1}$
Water will condense. There are several ways of seeing why:

- There is more water vapour ($22 \, \text{g kg}^{-1}$) in the room than the maximum ($14.5 \, \text{g kg}^{-1}$) possible at 20 °C.

www.wiley.com/go/biological_physics

- The dew point temperature for air with 22 g^{-1}kg of water vapour is 26 °C which is higher than the room temperature, so water vapour will condense.

- If we find the point on the chart with 22 g kg^{-1} of water vapour and a room temperature of 20 °C, it is above 100% relative humidity, so water will condense to reduce the relative humidity to 100%.

Water will condense until the relative humidity is 100%. At which point the moisture content is 14.5 g kg^{-1}. The difference between this and 22 g kg^{-1} tells us how much water has condensed per kilogram of air. So 22 g kg^{-1} – 14.5 = 7.5 g kg^{-1} of water has condensed. The total amount of water vapour that will condense is 7.5 g kg^{-1} × 50.4 kg = 378 g.

Example 20.2 *Humidity*

Problem: A bathroom with dimensions of 6 m × 3 m × 2.4 m is at a temperature of 28 °C and contains 1.24 kg of water vapour. (ρ_{air} = 1.2 kg m^{-3}.)

(a) What is the humidity in this bathroom?

Over time the bathroom cools down to 16 °C and some of the water condenses leaving the room at a relative humidity of 35%.

(b) What is the amount of water vapour in the bathroom now?

Solution: (a) we already know the dry bulb temperature in the bathroom is 28 °C, in order to find the humidity using the psychrometric chart we need to know either the wet-bulb temperature or the moisture content of the room. We have no information that will enable us to find the wet-bulb temperature, but we do know the volume of the room, the density of air and the total amount of water vapour in the room. By combining these three values we can calculate the moisture content

$$\text{M.C.} = \frac{m_{water}}{m_{air}}$$

The mass of water vapour is 1.24 kg or 1240 g. The mass of air in the room is

$$m_{air} = V_{room}\rho_{air} = (6\,\text{m} \times 3\,\text{m} \times 2.4\,\text{m}) \times 1.2\,\text{kg m}^{-3} = 51.8\,\text{kg}$$

so the moisture content (in g kg^{-1}) is

$$\text{M.C.} = \frac{1240\,\text{g water}}{51.8\,\text{kg air}} = 24\,\text{g kg}^{-1}$$

By finding the intersection of the 28 °C dry bulb line and the 24 g kg^{-1} moisture content line on the psychrometric chart we can find the relative humidity (and the wet-bulb temperature) from the psychrometric chart. As shown in Figure 20.7, the 28 °C dry-bulb line and the 24 g kg^{-1} moisture content line meet at the 100% relative humidity line. (b) We can solve this in a similar way to part (a). From the intersection of the 16 °Cline and the (estimated) relative humidity 35% line we can see (Figure 20.8) that the moisture content of the bathroom at this later time is 4 g kg^{-1}. Now we can work backwards to find the amount of moisture in the room.

$$\text{M.C.} = \frac{m_{water}}{m_{air}}$$
$$m_{water} = \text{M.C.} \times m_{air} = 4\,\text{g kg}^{-1} \times 51.8\,\text{kg} = 207\,\text{g}$$

Figures for the previous example:

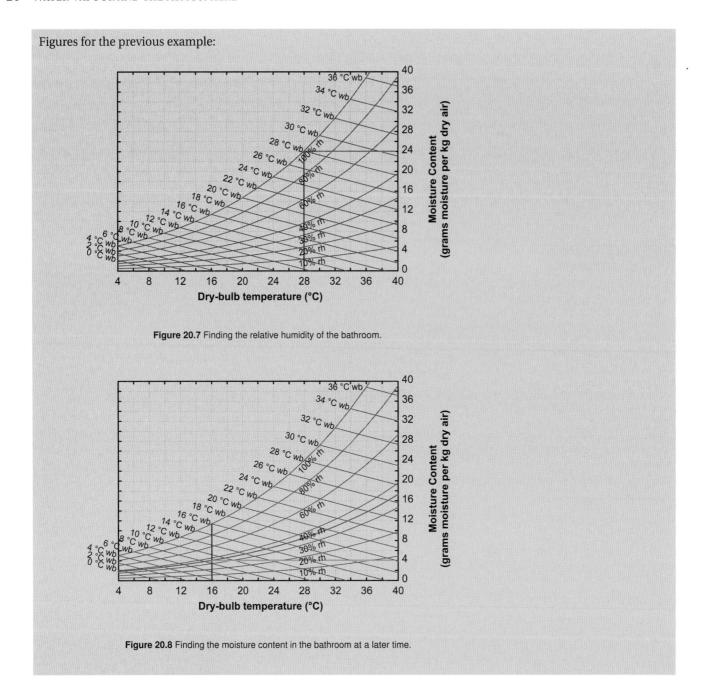

Figure 20.7 Finding the relative humidity of the bathroom.

Figure 20.8 Finding the moisture content in the bathroom at a later time.

Example 20.3 *Wet-Bulb Temperature*

Problem: The temperature in a lecture theatre filled with students is initially 22 °C and the relative humidity is initially 40%. Over the course of a 50 minute lecture, an air-conditioning system drops the temperature to 16 °C. A fault in the air-conditioning system means that the wet-bulb temperature has remained the same as when the lecture began.

(a) What is the wet-bulb temperature at the beginning of the lecture?

(b) What is the relative humidity at the end of the lecture?

Solution: (a) We know the dry-bulb temperature and the relative humidity at the beginning of the lecture and so we can use the psychrometric chart to find the wet-bulb temperature (and moisture content) at this time as shown in Figure 20.9.

www.wiley.com/go/biological_physics

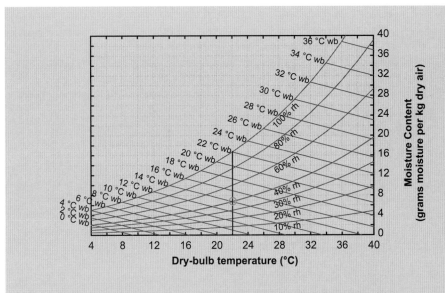

Figure 20.9 Wet-bulb temperature at the beginning of the lecture.

So the wet-bulb temperature at the beginning of the lecture was 14 °C.
(b)

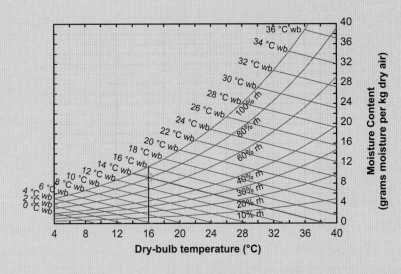

Figure 20.10 State of the air in the theatre at the end of the lecture.

The wet-bulb temperature remains constant over the course of the lecture, despite the reduction in dry-bulb temperature. We know that the difference between the wet-bulb and dry-bulb temperatures is indicative of the relative humidity and that the closer the two temperatures are, the higher the humidity. This indicates that the humidity in the lecture theatre will have risen. Figure 20.10 shows the state of the air in the lecture theatre after the lecture in which the relative humidity is 80%.

20.6 Applications

Medical Equipment: Humidification and Ventilators

As we breathe in air, it gets warmed to 37 °C and humidified to 100% humidity (this is 44 mg of water vapour per litre). If we breathe in air that is cooler than this, heat will be transferred from us to the air. If the air we breathe in has lower humidity than this, then we may lose water in the air we breathe out. In day-to-day life, we normally breathe in air that is cooler and less humid than the air we breathe out. This means that we lose water, which needs to be provided in our diet. It also takes some of our energy to heat this air. Patients requiring ventilatory support are often in a precarious position; they may be out of fluid balance and short on metabolic resources. Providing air at the same temperature and humidity as the air that will be breathed out reduces the heat and water loss by the patient. In designing breathing equipment for these patients, it is important to take these thermodynamic factors into consideration.

Combined Temperature Measures

The relative humidity is a key factor in determining how comfortable an environment is as well as the temperature, so a number of other measures have been developed to indicate when conditions are potentially hazardous. The *heat index* uses a formula based on temperature and relative humidity to try and indicate the air temperature that is perceived by the body, which differs from the true dry-bulb temperature.

Many organisations concerned with heat stress in industry and sports use a composite measure called the *wet bulb globe temperature* (WBGT) to estimate the combined effects of humidity, radiation and temperature. It uses a formula that is heavily weighted in favour of the wet-bulb temperature.

$$\text{WBGT} = 0.7\, T_{\text{wb}} + 0.2\, T_{\text{g}} + 0.1\, T_{\text{db}} \tag{20.6}$$

where T_{wb} is the wet-bulb temperature, T_{db} is the normal dry-bulb temperature and T_{g} is the temperature on a black globe thermometer, which measures solar (or other) radiation. In hot climates, a system of flags is used to indicate the WBGT at US military bases, and there are guidelines in place for what activities are considered appropriate for the situation. When the WBGT is over 32 °C (shown by a black flag), physical training and hard work are to be suspended for all personnel, excluding essential operational commitments.

20.7   Summary

Key Concepts

psychrometry The study of gas–vapour mixtures, usually mixtures of water vapour and air.

psychrometric chart A chart showing the relationships between various properties of dry-air/water-vapour mixtures. These often include moisture content, partial pressure due to water vapour, relative humidity, wet-bulb temperature and dry-bulb temperature, though others may be shown on more sophisticated charts. Each chart is valid for a particular total pressure.

relative humidity (RH) A measure of how much water vapour is present in a sample as a percentage of the maximum amount that could be present at that temperature.

dew-point temperature (T_{dew}) The temperature at which the partial pressure exerted by water vapour in the atmospheric sample in question is equal to the saturation vapour pressure of water. This is the temperature to which the atmosphere must be cooled for dew to begin to form.

dry-bulb temperature (T_{db}) The temperature measured by a standard thermometer.

wet-bulb temperature (T_{wb}) The temperature shown on a thermometer that has the bulb covered with a thin, wet cotton wick which has an air flow of at least 3 m s^{-1} past it. For humidities below 100%, evaporative cooling will cause the temperature reading to be lower than the dry-bulb temperature.

moisture content The mass of water vapour per unit mass of (dry) air, usually expressed in g kg^{-1}. It is also called the absolute humidity.

saturated vapour pressure The maximum water vapour pressure that can be sustained in a volume at a particular temperature.

Equations

$$RH = \frac{\text{partial pressure of water vapour}}{\text{saturated water vapour pressure at the same temperature}}$$

$$\text{moisture content} = \frac{\text{mass of water vapour}}{\text{mass of dry air}}$$

$$\frac{m_w}{m_a} = 0.621 \times \frac{P_w}{(P_t - P_w)}$$

20.8 Problems

20.1 You wish to measure the humidity in your flat. To do this you use two thermometers; one of which you wrap in a wet cloth and blow air over using a fan. The thermometer that is not wrapped in a wet cloth reads 18 °C, the one wrapped in a wet cloth reads 16 °C.

(a) What is the relative humidity in this room?

(b) You measure the temperature of the inner surface of one of the room's windows. At what window temperature would you expect water to condense on this window?

20.2 At a dry bulb temperature of 14 °C and at an atmospheric pressure of 101.3 kPa the moisture content of the air is 4 g (kg-dry air)$^{-1}$ ($\rho_{air} = 1.2 \, kg \, m^{-3}$ and $M_{H_2O} = 18 \, g \, mol^{-1}$).

(a) What is the relative humidity?

(b) What is the partial pressure of water vapour?

20.3 An infection control room in a paediatric ward has dimensions $3 \times 3 \times 4$ m. The dry bulb temperature is 20° and the wet bulb temperature is 18°. The temperature of the window surfaces drops to 15 °C overnight (the windows are not double glazed). What is the minimum volume of water that a dehumidifier must remove from this room such that condensation will not occur on the windows overnight?

20.4 At a dry bulb temperature of 30 °C, an atmospheric pressure of 101.3 kPa, and a humidity of 90% what is the partial pressure of water vapour in the atmosphere ($M_{H_2O} = 18 \, g \, mol^{-1}$)?

20.5 The air dry-bulb temperature in a room is 22 °C and the relative humidity is 60%. What is the minimum temperature of the inside surface of the glass in order to avoid the windows of the room getting fogged-up?

20.6 The bathroom mirror is all steamed up and the dry-bulb temperature is 28 °C. The mirror is slowly warmed from an initial cold temperature and it starts to clear when its temperature reaches 22 °C. What is the relative humidity in the bathroom?

20.7 The wet-bulb temperature in the room is initially 20 °C and the dry-bulb temperature is 24 °C. The surface temperature of the windows falls to 12 °C, while the room temperature remains constant. When moisture finally stops collecting on the windows, what is the wet-bulb temperature? (Assume that the dry-bulb temperature remains constant)

20.8 After taking a deep breath I exhale approximately 2.5 g of air. The intake air is at 20 °C at 40% relative humidity, and the air exhaled is at 100% relative humidity at a temperature of 34 °C. What is the net mass of water expelled during one breathing cycle?

20.9 The temperature of the air in a room is 13 °C and the partial pressure of water vapour is 1.10 kPa. The volume of the room is 60 m^3 and the molar mass of water is 18 g mol^{-1}. What is the mass of water vapour in the room?

20.10 The temperature of the air in a room is 14 °C and the relative humidity is 70%. The volume of the room is 60 m^3, the molar mass of water is 18 g mol^{-1}, and the density of air is 1.2 kg m^{-3}. What is the partial pressure of water vapour in the room?

20.11 On a cold evening, when the temperature of a particular room is 20 °C, condensation starts to form on the windows when the window surface temperature falls to 14 °C. Determine the relative humidity in the room.

20.12 The dry-bulb temperature of a sample of air is 30 °C and the relative humidity is 40%. Which of the following statements are correct? (Note: more than one statement may be correct.)

(a) The moisture content of the air is 27 g kg^{-1}.

(b) The dew-point temperature is 15 °C.

(c) If the moisture content is increased while keeping the wet-bulb temperature fixed, the relative humidity will reach 100% when the moisture content is 22 g kg^{-1}.

(d) The wet-bulb temperature is 20 °C.

(e) If the moisture content is increased keeping the dry-bulb temperature fixed at 30 °C, the maximum moisture content will be 27 g kg^{-1}.

20.13 Which of the following statements are correct? (Note: more than one statement may be correct.)

(a) An air condition with high relative humidity normally makes people feel uncomfortable because they are unable to lose heat easily through evaporation of perspiration.

(b) Except at 100% relative humidity, the dew-point temperature for air is greater than the wet-bulb temperature.

(c) When air flows over a wet surface it can cool the surface right down to the dew-point temperature, but no further.

(d) When the dew-point temperature is defined the moisture content of the air is also specified.

(e) When the wet-bulb temperature is defined the moisture content of the air is also specified.

20.14 On a certain day the dry-bulb temperature is 24 °C and the wet-bulb temperature is 20 °C. Which of the following statements is correct? (Note: more than one statement may be correct.)

(a) The dew-point temperature is 12 °C.

(b) The moisture content of the air is 6.5 g kg^{-1}.

(c) If the air is cooled to 6 °C at 100% relative humidity it will lose 9 g of moisture per kg of dry air.

(d) The relative humidity is 60%.

(e) At the same dry-bulb temperature the minimum wet-bulb temperature is approximately 8 °C.

HEAT TRANSFER

21

21.1 Introduction

Most biological systems need to exchange thermal energy with their surroundings in order to control their temperature. The main ways of transferring thermal energy from one place/object to another are conduction, convection, and radiation. Heat is energy transferred from hot materials to cooler ones, and so the driving force behind these methods is a temperature difference. The human body also transfers thermal energy to its surroundings by evaporation of perspiration, a phase change process. This is not driven by a temperature difference, but by a difference in the vapour pressure of water, so can be used to transfer heat to the environment even when the surroundings are hotter than the body.

Key Objectives

- To understand how conduction, convection and radiation transfer thermal energy.

- To understand the concepts of thermal conductivity and heat transfer coefficients.

- To be able to calculate heat-transfer rates under the following conditions: conduction through single and multiple layers, convection for air speeds up to $4\,\mathrm{m\,s^{-1}}$, radiation and a combination of all these.

21.2 Conduction

Heat Transfer by Conduction

Heat **conduction** is the transfer of thermal energy from an object at higher temperature to one at lower temperature by contact (Figure 21.1). The two objects can be in direct contact, or there may be a medium separating them. The material of the objects, or the medium connecting them, does not flow from one place to another, but the thermal energy is transferred through it, either by transfer of vibrational motion from molecule to molecule, or by conduction electrons wandering from atom to atom.

The rate of heat transfer through a material will depend on its microscopic structure. Metals, which are good conductors of electricity, are also good conductors of thermal energy as the electrons are able to move freely from atom to atom, carrying thermal energy. Materials which are poor electrical conductors (that is, good electrical insulators), such as plastics, wood and glass, also tend to be poor conductors of heat. The **thermal conductivity**, given the symbol k, is the property of the material that tells us how readily heat is conducted through it. The thickness of the material will also affect the rate of energy transfer, as will the cross-sectional area; more area means more molecules in contact, and more thickness means more molecules to pass the energy through. The temperature difference driving the transfer will also have an effect: a bigger difference means a higher transfer rate.

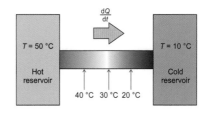

Figure 21.1 Heat transfer can take place by conduction when two bodies are in direct contact, or are connected by another medium. If the bodies are at different temperatures, there is a net flow of thermal energy at the rate $\frac{\Delta Q}{\Delta t}$.

Eq. (21.1) takes account of how these three factors influence the rate of heat transfer

$$\frac{\Delta Q}{\Delta t} = \frac{k A \Delta T}{d} = h_{\text{conduction}} A \Delta T \qquad (21.1)$$

where $\frac{\Delta Q}{\Delta t}$ is the rate of heat (Q) transfer by conduction, k is the thermal conductivity of the material, d is the thickness of the material, A is the cross-sectional area and ΔT is the temperature difference between the surfaces separated by thickness d. In the second part of Eq. (21.1) we have simplified the equation slightly, to put it in a form that will make it easier to combine with other types of heat transfer, using the **coefficient of conduction heat transfer**, $h_{\text{conduction}} = \frac{k}{d}$. When in doubt about which coefficient is you have been given, check the units: h has units of $\text{W m}^{-2}\,\text{K}^{-1}$, while k is measured in $\text{W m}^{-1}\,\text{K}^{-1}$.

Coefficients of Heat Transfer

The coefficient of heat transfer, h, has been introduced to make it easier to combine the effects of various heat-transfer mechanisms, and to deal with more than one material layer. The heat transfer coefficient tells us how rapidly heat is being transferred per square metre of surface area, when there is a 1 K temperature difference. To work out the rate of heat transfer we multiply the coefficient of heat transfer (h-value) by the surface area and by the appropriate temperature difference. For example, for a house, the h-value could be for a single insulation layer, or it might be a total h-value that takes into account several insulation layers as well heat loss by convection and radiation.

The h-value used in a calculation needs to match with the correct temperature difference. If, for example, the h-value you have is for conduction through a single layer, such as a thick woollen coat, then the corresponding temperature difference is the difference between outer surface of the coat and the inner surface. (Note that the temperature of the outer surface of the coat will normally be different from the surrounding air temperature.) If instead the h-value included conduction from a person's core through their tissue and clothing, plus convective and radiative heat losses from the outside of their clothes, then the temperature difference to use would be the person's core temperature minus the air temperature.

R-values

An alternate measure of the heat-transfer properties of a layer of material is the *R-value*, which is widely used in the building trades. The R-value is a measure of *thermal resistance*, and is the reciprocal of the *h*-value, i.e. $R = d/k$. The larger the R-value of the material, the better it is for insulation. Many countries use non-SI units for R-values, and some use a mixture, so some care is required.

Example 21.1 *Heat loss by conduction*

Problem: What is the rate of heat loss from a person with a core temperature of 37 °C and a skin temperature of 27 °C? Would you expect the surrounding air to be (a) hotter than 27 °C, (b) 27 °C or (c) colder than 27 °C?
The person's tissue is modelled as providing an insulating layer of thickness 1.0 cm, surface area 1.2 m^2 and conductivity $k = 0.2\,\text{W m}^{-1}\,\text{K}^{-1}$.

Solution:

$$h_{\text{conduction}} = \frac{k}{d} = \frac{0.2\,\text{W m}^{-1}\,\text{K}^{-1}}{0.01\,\text{m}} = 2\,\text{W m}^{-2}\,\text{K}^{-1}$$

$$\frac{\Delta Q}{\Delta t} = \frac{k A \Delta T}{d} = h A \Delta T = 20\,\text{W m}^{-2}\,\text{K}^{-1} \times 1.2\,\text{m}^2 \times (37 - 27)\,\text{K} = 240\,\text{W}$$

(c) They are losing heat from their core to their skin and their skin to their surroundings. You would expect their skin temperature to be intermediate between their core temperature and the temperature of the surroundings.

Conduction Through Multiple Layers

In many situations, heat is transferred from the body to the environment through multiple layers and we need to combine the effects of more insulation layers. For example, on a cold day heat is being conducted from a person's core through a layer of tissue to the skin surface, and then through a layer of clothing. When more layers are added, we would expect the rate of heat transfer to decrease. Therefore, we expect the h-value for multiple layers to be less than the h-value for any individual layer, so simply adding

h-values together is incorrect. The inverse of the h-value is a measure of thermal resistance, and we will see in the Electricity and DC Circuits topic that for resistors in series, the resistances add. Similarly, to get the total h-value for multiple layers we add the reciprocal h-values together:

$$\frac{1}{h_{total}} = \frac{1}{h_{layer\,1}} + \frac{1}{h_{layer\,2}} + \cdots \tag{21.2}$$

Example 21.2 *Heat loss by conduction through multiple layers*

Problem: What is the rate of heat loss from a person with a core temperature of 37 °C, a 1.0 cm thick tissue layer with conductivity of 0.2 W m^{-1} K^{-1}, and 5 mm thick clothing with conductivity of 0.04 W m$^{-1\cdot}$ K^{-1}? The outer surface of his clothing has a temperature of 10 °C. (We will assume a body surface area of 1.2 m^2.)

Solution:

$$h_{tissue} = \frac{k_{tissue}}{d_{tissue}} = \frac{0.2\ \mathrm{W\,m^{-1}\,K^{-1}}}{0.01\ \mathrm{m}} = 20\ \mathrm{W\,m^{-2}\,K^{-1}}$$

$$h_{clothing} = \frac{k_{clothing}}{d_{clothing}} = \frac{0.04\ \mathrm{W\,m^{-1}\,K^{-1}}}{0.005\ \mathrm{m}} = 8\ \mathrm{W\,m^{-2}\,K^{-1}}$$

therefore

$$h_{total} = \frac{1}{\frac{1}{h_{clothing}} + \frac{1}{h_{tissue}}} = 5.7\ \mathrm{W\,m^{-2}\,K^{-1}}$$

The rate of heat loss is therefore

$$\frac{\Delta Q}{\Delta t} = h A \Delta T = 5.7\ \mathrm{W\,m^{-2}\,K^{-1}} \times 1.2\ \mathrm{m^2} \times (37 - 10)\ \mathrm{K} = 185\ \mathrm{W}$$

If we add more clothing layers, what happens to the rate of heat loss? What happens to h_{total}? What happens to the skin temperature as we add more insulating layers of clothing?
The rate of heat loss decreases as there is more insulation. h_{total} gets smaller when we add more layers. The skin temperature gets higher. As the rate of heat transfer by conduction through the body tissue is lower than before, but h_{tissue} and A are the same, ΔT_{tissue} is less.

21.3 Convection

Heat transfer by fluid **convection** occurs as a result of the bulk motion of a fluid. For example, *natural* convective transfer occurs as a result of temperature-related density differences. A fluid that is heated becomes less dense, and hence more buoyant, rising upwards. (At least this will happen in an environment like the Earth's surface – this process will not happen somewhere like an orbiting space station.)

Forced convection happens when the fluid motion is caused by an external agent, such as a fan. In practice, the rate of heat transfer is measured for a particular circumstance and this is used to find the h-value for that convective process. For example, for air moving at speeds of less than 0.2 m s^{-1} over a surface like the human body, $h_{convection} = 3.1$ W m^{-2} K^{-1}. For higher air speeds v, in the range from 0.2–4.0 m s^{-1}, then $h_{convection} = 8.3 v^{0.6}$ W m^{-2} K^{-1}. These equations are examples of some useful approximations and apply for air; different values would be obtained if the fluid was, say, water.

As for conduction, the rate of heat transfer is

$$\frac{\Delta Q}{\Delta t} = h_{convection} A \Delta T \tag{21.3}$$

21.4 Radiation

All objects radiate energy in the form of **electromagnetic radiation**, with the rate and the frequency range being determined by temperature. Radiative heat transfer is especially important for the energy balance of the human body.

The Stefan–Boltzmann Law

The **Stefan–Boltzmann law** gives the rate at which an object *radiates* energy as electromagnetic radiation:

$$\frac{\Delta Q}{\Delta t}\bigg|_{\text{emitted}} = \varepsilon \sigma A T^4 \tag{21.4}$$

where ε is the surface emissivity, σ is the Stefan–Boltzmann constant, A is the surface area and T is the absolute temperature. The emissivity is between zero and one. (See the next section.) σ has the value $\sigma = 5.670 \times 10^{-8}$ W m^{-2} K^{-4}.

Eq. (21.4) is for the rate of *emission* of radiation. Any object that emits radiation will in fact be in an environment that is radiating energy back at it, which it will absorb. The *net* rate of heat transfer will be

$$\frac{\Delta Q}{\Delta t}\bigg|_{\text{net}} = \varepsilon \sigma A T^4_{\text{surface}} - \varepsilon \sigma A T^4_{\text{environ}} = \varepsilon \sigma A \left(T^4_{\text{surface}} - T^4_{\text{environ}}\right) \tag{21.5}$$

where the temperatures T_{surface} and T_{environ} are the surface and environment temperatures.

This is not in the form $\frac{\Delta Q}{\Delta t} = hA\Delta T$. However, we can re-write it in this form by using an approximation, and this will be useful in many situations when we wish to combine heat-transfer mechanisms. Suppose we write the temperature of the surface in terms of the temperature of the surroundings T and the difference, ΔT, so $T_{\text{surface}} = T + \Delta T$. Then the rate of heat loss by the surface is

$$\begin{aligned}
\frac{\Delta Q}{\Delta t}\bigg|_{\text{net}} &= \varepsilon \sigma A[(T + \Delta T)^4 - T^4] \\
&= \varepsilon \sigma A[T^4 + 4T^3\Delta T + \text{terms involving higher powers of } \Delta T - T^4] \\
&\approx 4\varepsilon \sigma A T^3 \Delta T
\end{aligned}$$

This assumes that ΔT is much smaller than T, so that higher powers of ΔT (e.g. ΔT^2) are small enough that we can ignore any terms involving them. We now have a useful form of Eq. (21.5):

$$\frac{\Delta Q}{\Delta t}\bigg|_{\text{net}} = h_{\text{radiation}} A \Delta T \tag{21.6}$$

where $h_{\text{radiation}} = 4\varepsilon \sigma T^3$, known as the **radiative surface heat-transfer coefficient**. The closer together the two temperatures are the more accurate this approximation is. A reasonable rule of thumb is that if the temperature difference is 10% or less of the absolute temperature of the surroundings then it is reasonable assumption. A good approximation for cases where we wish to calculate rates of heat loss from the body is to take $T = 290$ K and $\varepsilon = 0.9$, so that $h_r = 5.0$ W m^{-2} K^{-1}.

Emissivity Values

How well a surface emits radiation is normally determined by how well it absorbs it. The emissivity of a shiny, metallic (reflective) surface is very low (less than 0.1). Dull, black surfaces have the highest emissivities, between 0.9 and 1. The best emitters have high emissivity values. The theoretical perfect emitter/absorber has a emissivity of 1 and is known as a **blackbody**.

At the temperature of the human body and the environment, objects emit no visible light, only infrared. In this region of the electromagnetic spectrum, most surfaces have emissivities in the 0.9 to 1 range, with the exception of mirror-polished metallic surfaces. At infrared wavelengths, human skin has a high emissivity (about 0.97) which is why we can so easily feel the heat from a source like a radiant heater on our skin. Snow is an extreme example of the wavelength dependence of the emissivity. In the visible region its emissivity is very low, but in the infrared it is high, so to an infrared camera, snow looks black.

Example 21.3 *Heat loss by convection*

Problem: A person is running at 3 m s^{-1} on a calm 5celsius day. They have a surface area of 1.7 m^2. The outer surface of their clothing has a temperature of 9 °C. What is their rate of heat loss due to convection?

Solution: Temperature difference is between surface and surroundings, and is $\Delta T = 4$ K. For convection

$$h_{convection} = 8.3\, v^{0.6}\ \text{W m}^{-2}\ \text{K}^{-1} = 8.3 \times 3^{0.6}\ \text{W m}^{-2}\ \text{K}^{-1} = 16.0\ \text{W m}^{-2}\ \text{K}^{-1}$$

$$\frac{\Delta Q}{\Delta t} = h_{convection}\, A \Delta T = 16.0\ \text{W m}^{-2}\ \text{K}^{-1} \times 1.7\ \text{m}^2 \times 4\ \text{K} = 109\ \text{W}$$

Example 21.4 *Heat loss by radiation*

Problem: A person is running at 3 m s^{-1} on a calm 5 °C day. They have a surface area of 1.7 m^2. The outer surface of their clothing has a temperature of 9 °C. What is their rate of heat loss due to radiation?

Solution: The temperature difference between the person's clothing surface and surroundings is again 4 K. This time the heat loss is by radiation. Method 1: (You need to know both the surface and the surroundings temperature to use this method.) We expect the person's clothing to be effective at radiating in the infrared range, so we will use $\varepsilon = 1$

$$\frac{\Delta Q}{\Delta t} = \varepsilon A \sigma T^4_{surface} - \varepsilon A \sigma T^4_{surroundings}$$

$$\frac{\Delta Q}{\Delta t} = 1 \times 1.7\ \text{m}^2 \times 5.67 \times 10^{-8}\ \text{W m}^{-2}\ \text{K}^{-4} \times \left[(282\ \text{K})^4 - (278\ \text{K})^4\right] = 33.9\ \text{W}$$

Method 2: Using surroundings temperature to calculate $h_{radiation}$:

$$h_{radiation} = 4\varepsilon\sigma T^3_{surroundings} = 4.87\ \text{W m}^{-2}\ \text{K}^{-1}$$

$$\frac{\Delta Q}{\Delta t} = h_{radiation}\, A \Delta T = 4.87\ \text{W m}^{-2}\ \text{K}^{-1} \times 1.7\ \text{m}^2 \times 4\ \text{K} = 33.1\, W$$

Using the surroundings temperature to calculate $h_{radiation}$ underestimates heat loss slightly.

Colour and Temperature

The wavelength, and hence colour, of light emitted by a hot object is temperature dependent. For example, a piece of heated metal emits only infrared radiation at first, then as it gets hotter, it begins to glow red (as red is at the low-energy end of the spectrum), and if it is heated sufficiently, it changes colour to get closer to white. In fact, the temperature to which you would have to heat a perfectly-emitting *blackbody* source to get a particular colour of light is often used to describe the colour of light sources in the form of a colour temperature. A candle has a colour temperature of about 1900 K, a domestic tungsten filament bulb is about 2700 K, and full sunlight is 5500–6000 K, close to the surface temperature of the sun, but varying with the location of the sun. (Light on an overcast day is referred to as 'cooler' in hue, being bluer, but is actually spectrally similar to the radiation from a higher temperature source, at more like 7000 K.)

The wavelength at which the most emission occurs (λ_{max}) is related to the temperature by the **Wien displacement law**

$$\lambda_{max} = \frac{b}{T} \tag{21.7}$$

> **Radiation and frost formation**
> The effective temperature of the cloudless night sky is much lower than the local surroundings at ground level. This is because the atmosphere gets colder as you go up. Radiative heat loss is increased from surfaces facing the sky. This is why we get frost on clear nights, and why covering plants can protect them from frost.

where T is the absolute temperature and b is a constant, which has the value 2.9×10^{-3} m K. This puts the peak emission wavelength of the Sun (with effective surface temperature 5780 K) at 502 nm, which is in the green region of the visible spectrum where the human eye is most sensitive.

21.5   Combined Transfer Processes

Normally, conduction, convection and radiation all contribute to the heat transfer from an object.

Consider how weather affects you. On a cold, overcast day, you may have a high rate of heat loss due to radiation. On a windy day you may have a high rate of heat loss due to convection. A cold and windy day is when your rate of heat loss will be greatest, so we would expect the effects of convection and radiation to be additive, i.e. the h-value for convection and radiation from a surface will be greater than either of the individual h-values. Because the heat losses add cumulatively, the value for h_{surface}, the coefficient of thermal heat transfer for combined convection and radiation heat losses from a surface

$$h_{\text{surface}} = h_{\text{convection}} + h_{\text{radiation}} \qquad (21.8)$$

Take the more complex situation of heat transfer from a person's core to the surroundings. In this case we need to consider heat loss through many layers:

- Conduction

    - through their tissue
    - through their clothing.

- Convection from the surface of their clothing.

- Radiation from the surface of their clothing .

The temperature difference that we are likely to know here is $T_{\text{core}} - T_{\text{surroundings}}$, therefore the h-value that we need must take into account all the mechanisms listed. For conduction, $h_{\text{tissue}} = k_{\text{tissue}}/d_{\text{tissue}}$ and $h_{\text{clothing}} = k_{\text{clothing}}/d_{\text{clothing}}$. These values can be combined to give the total $h_{\text{conduction}}$:

$$\frac{1}{h_{\text{conduction}}} = \frac{1}{h_{\text{tissue}}} + \frac{1}{h_{\text{clothing}}} \qquad (21.9)$$

We can determine the surface coefficient for heat transfer by convection, $h_{\text{convection}}$, from the air speed and empirical data. $h_{\text{radiation}} = 4\varepsilon\sigma T^3$ (≈ 5.0 W m^{-2} s^{-1}), where T is the air temperature. We then combine the effects of convection and radiation.

$$h_{\text{surface}} = h_{\text{convection}} + h_{\text{radiation}} \qquad (21.10)$$

To combine the h-value for heat loss from the surface with the h-value for heat loss by conduction through insulating layers, we treat the surface h-value like we would another conducting layer

$$\frac{1}{h_{\text{total}}} = \frac{1}{h_{\text{tissue}}} + \frac{1}{h_{\text{clothing}}} + \frac{1}{h_{\text{surface}}} \qquad (21.11)$$

If we look at heat loss from a person's core to their surroundings, then the thermal energy is moving via conduction through their tissue, then by conduction through their clothes, then by a mixture of convection and radiation from the surface of their clothes to the surroundings. If we assume an equilibrium situation, so that the temperatures of surfaces and layers are constant, then heat must leave any layer at the same rate as it enters that layer. This means that the rate of heat transfer via conduction through the person's tissue is equal to the rate of heat transfer via conduction through their clothing, and is equal to the rate of heat transfer via convection and radiation from the clothing surface to the surroundings.

$$\frac{\Delta Q}{\Delta t} = h_{\text{tissue}} A (T_{\text{core}} - T_{\text{skin}})$$

$$= h_{\text{clothing}} A (T_{\text{skin}} - T_{\text{outer surface of clothing}})$$

$$= h_{\text{surface}} A (T_{\text{outer surface of clothing}} - T_{\text{surroundings}})$$

$$= h_{\text{total}} A (T_{\text{core}} - T_{\text{surroundings}})$$

If we know T_{core} and $T_{\text{surroundings}}$ we can use h_{total} to determine the rate of heat transfer. We can then determine the temperature at any intermediate surface.

Example 21.5 *Combined heat loss mechanisms*

Problem: At what rate does the runner in Examples 21.3 and 21.4 lose heat due to convection and radiation?

Solution: From Examples 21.4 and 21.3

$$h_{\text{radiation}} = 4.87 \text{ W m}^{-2} \text{ K}^{-1}$$

$$h_{\text{convection}} = 16.05 \text{ W m}^{-2} \text{ K}^{-1}$$

$$h_{\text{surface}} = h_{\text{radiation}} + h_{\text{convection}} = (4.87 + 16.0) = 20.9 \text{ W m}^{-2} \text{ K}^{-1}$$

$$\frac{\Delta Q}{\Delta t} = h A \Delta T = 20.9 \text{ W m}^{-2} \text{ K}^{-1} \times 1.7 \text{ m}^2 \times (9-5) \text{ K} = 142 W = 109 W + 33 W$$

How is this related to the rate at which they lose heat due to each of convection and radiation alone?
The rate at which heat is lost due to both convection and radiation = the rate at which heat is lost due to convection alone plus the rate at which heat is lost due to radiation alone.
In this example we have not considered the thickness and conductivity of the runner's clothes. Does this affect our ability to solve the problem?
No. We are told the temperature of the *outer* surface of their clothing and of the air so we are looking for the h-value for mechanisms of heat loss from the outer surface of the clothing to the surroundings.
What would change if the runner put on more clothes and why?
Heat transfer through the runner's clothing determines the temperature of the outer surface of their clothing. If they put on more clothes it would reduce their rate of heat loss from their core to the outer surface of their clothing. Heat must be transferred away from the outer surface of the clothing at the same rate as it arrives there, so this has also reduced the rate of heat loss from the outer surface of the clothing to the surrounding air. The temperature of the outer surface of their clothing would be closer to the surroundings when more clothes were put on.

Example 21.6 *Combined heat loss mechanisms*

Problem: A person's rate of heat transfer from core to surroundings is 100 W, their core temperature is 37 °C, and the external air temperature is 10 °C. What is the person's skin temperature? (Assume he has a 3 cm thick tissue layer with conductivity of 0.2 W m^{-1} K^{-1}, 5 mm thick clothing with conductivity of 0.04 W m^{-1} K^{-1}, and a surface area of 1.2 m^2.)

Solution: Rate of heat transfer from core to surroundings = Rate of heat transfer from core to skin = 100 W.
Rate of heat transfer from core to skin = $h_{\text{tissue}} A \Delta T$.

$$h_{\text{tissue}} = \frac{k}{d} = \frac{0.2 \text{ W m}^{-1} \text{ K}^{-1}}{0.03 \text{ m}} = 6.7 \text{ W m}^{-2} \text{ K}^{-1}$$

$$\Delta T = \frac{\frac{\Delta Q}{\Delta t}}{A h_{\text{tissue}}} = \frac{100 \text{ W}}{1.2 \text{ m}^2 \times 6.7 \text{ W m}^{-2} \text{ K}^{-1}} = T_{\text{core}} - T_{\text{skin}} = 12.5 \text{ K}$$

This gives us a skin temperature of

$$T_{skin} = T_{core} - \Delta T = 37\,°C - 12.5\,°C = 24.5\,°C$$

Example 21.7 *Combined heat loss mechanisms*

Problem: In Example 21.5 we were able to calculate the rate of heat loss from the runner without making any explicit calculations for conduction, but what if we did not know the runner's surface temperature, but we did know their core temperature?

In this example, our runner has a 1.0 cm thick tissue layer with conductivity of 0.2 W m^{-1} K^{-1}, and 2 mm thick clothing with conductivity of 0.04 W m^{-1} K^{-1}. Her surface area is 1.7 m^2, her core temperature is 37 °C, the surrounding air temperature is 5 °C, and she is running at 3 m s^{-1}.

Solution: From Example 21.21.5

$$h_{surface} = (4.87 + 16.0)\ W\,m^{-2}\,K^{-1} = 20.9\ W\,m^{-2}\,K^{-1}$$

Using the information we have about the persons tissue and clothes

$$h_{tissue} = \frac{k_{tissue}}{d_{tissue}} = \frac{0.2\ W\,m^{-1}\,K^{-1}}{0.01\ m} = 20\ W\,m^{-2}\,K^{-1}$$

$$h_{clothing} = \frac{k_{clothing}}{d_{clothing}} = \frac{0.04\ W\,m^{-1}\,K^{-1}}{0.002\ m} = 20\ W\,m^{-2}\,K^{-1}$$

$$\frac{1}{h_{conduction}} = \frac{1}{h_{tissue}} + \frac{1}{h_{clothing}}$$

$$= \frac{1}{20\ W\,m^{-2}\,K^{-1}} + \frac{1}{20\ W\,m^{-2}\,K^{-1}} = \frac{1}{10\ W\,m^{-2}\,K^{-1}}$$

therefore $h_{conduction} = 10\ W\,m^{-2}\,K^{-1}$

$$\frac{1}{h_{total}} = \frac{1}{h_{surface}} + \frac{1}{h_{conduction}}$$

$$= \frac{1}{20.9\ W\,m^{-2}\,K^{-1}} + \frac{1}{10\ W\,m^{-2}\,K^{-1}}$$

$$= 0.15\ W^{-1}\,m^2\,K$$

therefore $h_{total} = \frac{1}{0.15\ W^{-1}\,m^2\,K} = 6.7\ W\,m^{-2}\,K^{-1}$

$$\frac{\Delta Q}{\Delta t} = hA\Delta T = 6.7\ W\,m^{-2}\,K^{-1} \times 1.7\ m^2 \times (37 - 5)\ K = 360\ W$$

21.6   Summary

Key Concepts

heat transfer by conduction The transfer of thermal energy from a region of higher temperature to one of lower temperature through matter by direct contact.

convection The movement of molecules of a fluid.

heat transfer by convection The transfer of thermal energy assisted by the bulk movement of fluid molecules.

radiation Energy emitted by a body as electromagnetic waves that travel through a medium or through space.

heat transfer by radiation The transfer of thermal energy by the emission and absorption of electromagnetic radiation from the surfaces of objects.

thermal conductivity (k) The property of a material that relates to its ability to conduct heat. The thermal conductivity is measured in $W\,m^{-1}\,K^{-1}$.

coefficient of heat transfer (h) The proportionality constant relating the rate of heat transfer per unit area to the temperature difference. The heat transfer coefficient has units of $W\,m^{-2}\,K^{-1}$.

emissivity (ε) The ratio of the electromagnetic radiation emitted by an object to that emitted by a perfect emitter at the same temperature. The emissivity lies in the range from zero to one, is dependent on wavelength, and is dimensionless. The more reflective a material is in a particular wavelength range, the lower its emissivity.

Stefan–Boltzmann law The rate at which electromagnetic energy is radiated by an object per unit of surface area is proportional to the absolute temperature to the fourth power. See Eq. (21.4).

Stefan–Boltzmann constant (σ) The constant of proportionality in the Stefan–Boltzmann Law. $\sigma = 5.670 \times 10^{-8}\,W\,m^{-2}\,K^{-4}$.

Wien displacement law The wavelength at which the peak emission occurs for a blackbody is inversely proportional to its absolute temperature. See Eq. (21.7).

Wien's displacement constant (b) The constant of proportionality in Wien's law, which is equal to $2.898 \times 10^{-3}\,m\,K$.

Equations

$$\frac{\Delta Q}{\Delta t} = \frac{kA\Delta T}{d} = h_{\text{conduction}}\,A\Delta T$$

$$\frac{1}{h_{\text{total}}} = \frac{1}{h_{\text{layer 1}}} + \frac{1}{h_{\text{layer 2}}} + \dots$$

$$\frac{\Delta Q}{\Delta t} = h_{\text{convection}}\,A\Delta T$$

$$\left.\frac{\Delta Q}{\Delta t}\right|_{\text{emitted}} = \varepsilon\sigma A T^4$$

$$\left.\frac{\Delta Q}{\Delta t}\right|_{\text{net}} = h_{\text{radiation}}\,A\Delta T$$

$$h_{\text{radiation}} = 4\varepsilon\sigma T^4$$

$$\lambda_{\text{max}} = \frac{b}{T}$$

$$h_{\text{surface}} = h_{\text{convection}} + h_{\text{radiation}}$$

21.7 Problems

21.1 A penguin generates thermal energy through metabolic processes at a rate of 50 W, but loses 8 W due to respiration and moisture loss. Its surface area is 0.63 m² and the average thermal conductivity of its feather layer is 0.031 W m⁻¹ K⁻¹. The penguin maintains a skin temperature of 35 °C when the outer surface of its feather layer is −10 °C. What is the thickness of the penguin's layer of feathers?

21.2 The core temperature of an athlete is 37 °C and his surface area is 1.8 m². The thermal conductivity of his surface tissue, which is 6 mm thick on average, is 0.18 W m⁻¹ K⁻¹. If heat is transferred from his core to his skin at a rate of 270 W, what is the average temperature of his skin surface?

21.3 You have choice between wearing a dark woollen jersey (ϵ_{jersey} = 0.9, k_{wool} = 0.04 W m⁻¹ K⁻¹, and d_{jersey} = 5 mm) and a top made with a new high-tech material (ϵ_{htm} = 0.1, k_{htm} = 0.01 W m⁻¹ K⁻¹, and d_{htm} = 1 mm).

(a) Ignoring heat transfer due to radiative and convective processes, which top would you predict will keep you the warmest on a cold day?

(b) Which top would you predict will keep you warmest on a cold day if you take into account radiative and convective heat transfer (use $h_{convection}$ = 3.1 W m⁻² K⁻¹ and T_{air} = 2 °C)?

21.4 An Olympic swimmer is waiting to compete. Her average skin temperature is 33 °C and the average temperature of the pool hall is 23 °C. Her surface area is 1.6 m², her surface emissivity is 0.9 and where she stands the mean air speed is 1.5 m s⁻¹. Determine separately the rates at which she loses heat by radiative heat transfer and by convection.

21.5 A penguin is standing in a sheltered spot somewhere near the coast of Antarctica. The penguin is losing heat from it's core, which is at 39 °C, through a layer of fatty tissue and also through a layer of insulating feathers. The layer of fatty tissue is 2 cm thick, the layer of feathers is also 2 cm thick, the total rate at which the penguin is loosing heat is 70 W, and the total surface area of the penguin is 0.7 m² (k_{tissue} = 0.2 W m⁻¹ K⁻¹, $k_{feathers}$ = 0.035 W m⁻¹ K⁻¹).

(a) What is the rate of heat transfer across the layer of fatty tissue?

(b) What is the rate of heat transfer across the layer of feathers?

(c) What is the skin temperature of the penguin (in °C)?

(d) What is the temperature of the outside surface of the penguin (in °C)?

21.6 A person has a core body temperature of 37 °C, a tissue layer that is 5 mm thick with a thermal conductivity of 0.2 W m⁻¹ K⁻¹ and a surface area of 1.4 m². The average thickness of her clothes is 9 mm, with a thermal conductivity of 0.04 W m⁻¹ K⁻¹. The surface heat transfer coefficient of her clothes is 22 W m⁻² K⁻¹. Determine her rate of heat loss when the external temperature is 12 °C. What will be her average skin temperature under this condition?

21.7 The core temperature of a naked male, standing in air at 8 °C, is 37 °C. His surface area is 1.6 m². The thermal conductivity of his surface tissue, which is 1.2 cm thick on average, is 0.20 W m⁻¹ K⁻¹. If heat is transferred from his core to his skin at a rate of 195 W, what is his average skin temperature ?

21.8 You are working outside on a cold day (T_{air} = 2 °C). Your core body loses heat through a 2 cm thick layer of fatty tissue and a 1 cm thick layer of clothes. Your total surface area is 2 m² and the thermal conductivities of fatty tissue and your clothes are k_{tissue} = 0.2 W m⁻¹ K⁻¹ and $k_{clothes}$ = 0.01 W m⁻¹ K⁻¹. Your clothes are a light color (ϵ = 0.4) and it is a windy day (v_{air} = 3.6 m s⁻²).

(a) What is the total *surface* heat transfer coefficient?

(b) What is the combined heat transfer coefficient for heat *conduction* through your fatty tissue and clothes?

(c) What is the total heat transfer coefficient taking into account all methods of heat transfer?

(d) If the rate at which your core body is losing heat is 56.6 W, what is your core body temperature? Should you stay outside?

21.9 A 110 kg fisherman with a surface area of 2.2 m² falls from his boat into the cold southern ocean (T_{ocean} = 2 °C). The fisherman has a 4 cm thick layer of fatty tissue (k_{tissue} = 0.2 W m⁻¹ K⁻¹) and is wearing cold weather survival gear which is 1.5 cm thick and has a thermal conductivity of k_{sg} = 0.03 W m⁻¹ K⁻¹.

(a) If the fisherman's core body temperature is initially a healthy 37 °C estimate how long will it be before his body temperature drops to 34 °C (specific heat of the human body - c_{body} = 3.5 kJ kg⁻¹ K⁻¹)?

(b) Suppose the fisherman was wearing ordinary clothes ($k_{clothes}$ = 0.1 W m⁻¹ K⁻¹ and $d_{clothes}$ = 5 mm) instead of the survival gear. Estimate how long would it take for their temperature to drop to 34 °C in this case.

21.10 For each square metre of the sun's visible surface the rate of energy transferred to outer space is 63 MW. The emissivity of the sun's surface is approximately 1.0. Which of the following statements is correct?

(a) The temperature of the sun's surface is approximately 5766 K.

(b) The solar energy that reaches the earth from the sun involves radiative heat transfer primarily.

(c) Because the sun's surface temperature is relatively stable it must be receiving energy from space at a rate of 63 MW m⁻².

(d) The energy transferred from the sun is carried mainly by the solar wind, so the process is primarily convective.

(e) The rate at which energy leaves the surface if the sun cannot be calculated using the Stefan–Boltzmann law, because the sun's surface is white, not black.

THERMODYNAMICS AND THE BODY

22.1 Introduction

In this chapter we will apply the thermodynamics we have learned to help understand how the body regulates its temperature, and the energy balance of the human body.

Key Objectives

- To understand how the first law of thermodynamics applies to the body.

- To understand the mechanisms the body employs for thermoregulation.

- To be aware of the factors affecting human comfort levels in various environments.

22.2 The First Law

The **first law of thermodynamics** is a statement of conservation of energy. A system may be heated, cooled, have work done on it and do work on other objects. These may change the internal energy of the system. For a stationary system that can exchange energy, but not matter, with its surroundings, this change is equal to the energy transferred *to* the system as heat, less the energy transferred *from* the system to external agents on which the system does work.

The total internal energy of a system, which is given the symbol U, is a measure of the amount of energy in the system in the kinetic energy of molecules and the various forms of potential energy. Conservation of energy requires that

$$\Delta U = Q - W \tag{22.1}$$

where ΔU is the change in total internal energy of a system, Q is the net heat transferred *to* the system and W is the net work done *by* the system.

Eq. (22.1), which is an expression of the first law of thermodynamics, is a statement that energy can be transferred from one form to another, but it cannot be created or destroyed.

22.3 Energy and the Body

The concept of conservation of energy was first stated by a physician, Julius Robert von Mayer (1814–1878), studying the energy balance of the human body. First, the body loses or gains energy as heat from the environment, due either to the temperature difference between the body and the surrounding environment, or the latent heat associated with evaporation of perspiration. Second, when a person does physical work, some of their energy is transferred to the surrounding environment. Third, molecular energy is stored in the relatively weak bonds in food. We can obtain energy from food through a series of biochemical reactions which convert the potential energy of food molecules to other forms; this set of chemical reactions is called **metabolism**. Another way of looking at this is that in eating we add high-energy molecules to our bodies, and following metabolism, excrete lower energy molecules, with the difference in energy being transferred to our bodies.

These are three ways the body can gain or lose energy (ΔU):

1. Through heat transfer with its surroundings. The net heat transferred into the body is Q.

2. Through the body doing work on the surroundings. The net work done by the body is W.

3. Through gain of material by the system. The net gain of energy due to this type of process is E.

In this situation, the first law is

$$\Delta U = Q - W + E \qquad (22.2)$$

where E is the energy gained by the body as a result of ingestion and metabolism of food.

Metabolism, Hypothermia and Hyperthermia

The reason we can maintain a core body temperature that is usually higher than our surroundings is that we are constantly producing energy as a by-product of metabolism. Metabolic processes in the body convert stored energy into other forms of energy useful for biological function. This process is somewhat inefficient and much of the stored energy is not converted into useful forms, but is instead deposited in the body in the form heat. If this was not the case, we would quickly reach thermal equilibrium with our surroundings as we continued to lose heat via thermal transfer. Our metabolic rate is how many joules of energy per second that we are producing from metabolism. It is higher when we are exercising our muscles than when we are sitting still. If our metabolic rate is lower than our rate of heat loss to the environment then we will lose heat energy and our core temperature will decrease. If we stay in such a situation too long the result will be **hypothermia**, a potentially serious decrease in core body temperature below the normal 37 °C. (The prefix 'hypo' means low or below normal.) On the other hand, if our metabolic rate exceeds our rate of heat loss then we will gain energy, and our core temperature will increase. Remaining in an environment where this is the case for too long may result in **hyperthermia** and heat stroke. (The prefix 'hyper' is the opposite of hypo, and means over or excessive.)

We can calculate how much a person's core temperature will change from heat exchange and metabolism. First, we calculate the person's net rate of loss of energy from heat loss to the surroundings (due to conduction, convection, radiation as covered in the last chapter) and metabolism

net rate of energy loss = rate of heat loss − metabolic rate

total energy lost = net rate of energy loss × time

decrease in core temperature = $\dfrac{\text{total energy lost}}{(\text{mass of body}) \times (\text{specific heat capacity of the body})}$

Energy Value of Food

When food is fully oxidised, the amount of energy released per mole is the same whether that occurs rapidly in one step or slowly in many steps. The energy values of food are found by measuring the heat produced when food is oxidised by combustion. In cellular respiration, food is also oxidised to the same final products as in combustion, so the amount of energy released is the same, the difference being that in the body it is done in a controlled way. When listing the energy content of a particular food item, these values are adjusted to account for those parts of food, such as dietary fibre, which the body is unable to digest, and which are excreted unoxidised.

Energy values of major food components:

The second law

The second law of thermodynamics is beyond the scope of this book, but we will mention it here briefly. It states that for an isolated system that is not in equilibrium, the entropy will tend to increase over time. Entropy is a measure of the disorder of a system, and also a measure the unavailability of energy to do work.

Hyperthermia and fever

When the body has a raised temperature due to fever, it is because of the action of the hypothalamus. The body raises its temperature to better fight invading bacteria or viruses. In the case of hyperthermia, the rise in temperature is due to external factors and is not instigated by the body's control systems.

- Lipids (fats) 37 kJ g^{-1} (9 kcal g^{-1})

- Protein 17 kJ g^{-1} (4 kcal g^{-1})

- Carbohydrate 17 kJ g^{-1} (4 kcal g^{-1})

- Alcohol 29 kJ g^{-1} (7 kcal g^{-1})

The recommended daily intake (RDI) of energy for young adults is around 10 MJ for males and 8 MJ for females.

Efficiency

The **work efficiency**, η, of the body is defined as the ratio of the mechanical work done by the body (e.g. work done climbing up a mountain, running a marathon, or walking to university) to the energy used for mechanical work. The work done by the body is W, and the energy used will be the energy gained from food, plus any decrease in the internal stored energy, $E - \Delta U$. This is equivalent to the work output from the body plus the heat lost, which is $W - Q$, so

$$\eta = \frac{W}{E - \Delta U} = \frac{W}{W - Q} \qquad (22.3)$$

If the body uses up reserves to perform the work then $\Delta U < 0$ and the person reduces their stores of high-energy compounds in the short term and loses weight in the long term. If the body has more intake than required for the work to be done, this is stored long term as fat deposits. If the body neither uses stored reserves or adds to stored reserves over the time period of interest then $\Delta U = 0$ and in that case we have $\eta = \frac{W}{E}$.

The work efficiency of the body is typically around 2–10%, though it can be as high as 25% when engaged in sports. As the work efficiency is typically low, the heat loss from the body is normally about the same as the metabolic rate. For a person taking in enough food to maintain their stores, this suggests that the energy lost as heat will be about the same as the energy taken in, i.e., $E \approx -Q$.

Example 22.1 *Energy balance and hypothermia*

Problem: A 60 kg person has a core temperature of 37 °C. The onset of mild hypothermia occurs at a core body temperature of about 35 °C. The specific heat capacity of human tissue is 3500 J kg^{-1} K^{-1}. He is losing heat at a rate of 300 W to his surroundings, and will be exposed to this environment for two hours. Will he get hypothermia if he is sitting still? What about if he is walking or running?

Solution: Sitting still metabolic rate = 100 W; walking metabolic rate = 250 W; running metabolic rate = 600 W. Assume the work efficiency is negligible.

Sitting: Net rate of heat loss = 300 W – 100 W = 200 W

$\Delta T = \frac{200 \text{ W} \times 7200 \text{ s}}{60 \text{ kg} \times 3500 \text{ J kg}^{-1} \text{K}^{-1}} = 6.9 \text{ °C} \Rightarrow$ hypothermic.

Walking: Net rate of heat loss = 300 W – 250 W = 50 W

$\Delta T = \frac{100 \text{ W} \times 7200 \text{ s}}{60 \text{ kg} \times 3500 \text{ J kg}^{-1} \text{K}^{-1}} = 1.7 \text{ °C}$. They will be cold but above the hypothermia threshold.

Running: Net rate of heat loss = 300 W – 600 W = –200 W. They are producing heat more rapidly than they are losing it, so they will not become hypothermic. Thermoregulatory processes will normally intervene to prevent overheating.

Example 22.2 *Penguins*

Problem: How do male emperor penguins survive winter on the Antarctic ice? Males do not eat for four months while they incubate an egg in temperatures down to –60 °C. They are 100 km from open water. Determine their energy inputs and outputs. Can they survive?

Solution: To solve real world problems like this, we consider a model penguin with the following characteristics:

- To get from the sea to the breeding site and back to the sea takes the penguins five days walking in each direction with a metabolic requirement of 80 W.

- Over the 120 days the males lose 15 kg of body mass.

- The average air temperature is –20 °C and the penguin's body core temperature is 39 °C.

- The penguin's layer of feathers is 2 cm thick with $k = 0.03$ W m^{-1} K^{-1}.

- Penguins can be treated as cylinders 0.8 m high with 0.12 m radius.

- Metabolism produces 37 MJ per kg of lost body mass (fat).

Surface area of simplified model penguin = surface area of cylinder

$$A = 2\pi r h + 2\pi r^2 = 0.694 \text{ m}^2$$

$$\frac{\Delta Q}{\Delta t} = \frac{k}{d} A \Delta T = \frac{0.03}{0.02} \times (0.694 \times 59) \text{ W} = 61.4 \text{ W}$$

It will take about 10 days × 24 hours per day × 3600 seconds per hour × 80 J per second = 69 MJ of energy to make the five-day journey each way. The total energy available is 15 kg × 37 MJ kg^{-1} = 555 MJ. This leaves about 487 MJ for the 110 days they are incubating the eggs. At the rate of conduction loss we calculated, they will lose about $(110 \times 24 \times 3600 \times 61.4)$ J = 584 MJ. So, these model penguins are in energy deficit by 97 MJ. To avoid this problem, real emperor penguins huddle together in colonies to reduce their heat loss during the Antarctic winter.

22.4 Thermoregulation

Thermoregulation is the maintenance of a constant core body temperature in an organism. Humans and other warm-blooded organisms (homeotherms) thermoregulate. Organisms, such as reptiles, that do not are called thermoconformers, and their temperature changes according the environmental temperature.

Heat Sensors

The human body has temperature-sensitive receptors located in many places, such as the skin, hypothalamus, midbrain, spinal cord and abdominal cavity. The thermoreceptors sense either hot or cold, and most areas have both types. The sensors send information to the hypothalamus, which controls the body's response.

Vasoconstriction and Vasodilation

When the internal body temperature and/or the skin temperature gets too low, the blood vessels adjacent to the skin constrict, reducing the blood flow from the core to the surface. This response, called **vasoconstriction**, reduces the rate of heat loss from the skin surface by allowing the skin to get colder.

The minimum blood flow to the skin is about 3×10^{-3} kg s^{-1} for a male. If the temperature of the blood is reduced by 2 °C in transferring heat from the core to the skin, then the corresponding rate of heat loss is

$$4200 \text{ J kg}^{-1} \text{ K}^{-1} \times 3 \times 10^{-3} \text{ kg s}^{-1} \times 2 \text{ K} = 25 \text{ W}$$

To increase the rate of heat loss, when the body is too hot, then the blood supply to the skin can be increased by **vasodilation**. This increase in the blood flow to the skin, which reduces blood pressure, can be up to about 15 times the minimum blood flow, a maximum of 45×10^{-3} kg s^{-1} for a male. The corresponding rate of heat loss for a 2 °C temperature change between the core and the surface is about 378 W.

Piloerection and Shivering

Piloerection (which causes 'goosebumps') occurs when the fine body hairs stand on end in an attempt to reduce convective heat loss from the skin. The raising of the hairs is intended to create a stationary layer of air between the skin and the surroundings. As humans have little body hair and tend to be clothed, this doesn't normally play a major role in human thermoregulation.

Another physiological response to cold temperatures is **shivering**. Cold signals from the skin and spinal cord are transmitted to the part of the hypothalamus responsible for shivering, trigger the reflex. The body's muscles begin to make small movements in order to expend some metabolic energy and generate thermal energy. A person with a fever may experience shivering as a response to perceived cold, as the fever changes the temperature that the body's regulatory processes consider to be normal.

Perspiration

Perspiration, or sweating, is a very effective cooling mechanism, as the latent heat of evaporation of water is high. Heat loss by evaporation of liquid from the skin is the only way we can lose heat when the air temperature is above 37 °C. The human body has sweat glands in the skin which excrete fluid (mainly water with some dissolved salts). The rate of moisture loss by perspiration can reach up to 3 kg per hour, which is faster than the gut can absorb water (about 1 kg h^{-1}). This corresponds to a maximum rate of cooling of 2 kW, but this is obviously unsustainable.

A moisture loss rate of around 25 g h^{-1} for a male with skin temperature 30 °C is more typical, and this provides about 0.025 kg $\times$ 2430 kJ kg^{-1} $\times$ $\frac{1}{3600}$ s h^{-1} = 17 W. This is about 20–25% of the resting body heat loss, and the moisture loss is split 50/50 between losses through the respiratory tract and losses through the skin, normally without wetting the skin.

The effectiveness of sweating is dependent on the relative humidity of the local environment. In low humidity, the skin can remain quite dry even with significant sweating rates, but when the air is close to saturation with water vapour, the rate of evaporation is reduced. Thus, at high humidity, the area of the body that must be dampened by sweat for effective cooling is higher, and this is quite a lot less pleasant. When the surface area of the skin that is wet is more than a quarter of the total, most people experience discomfort.

Behavioural Responses

A highly effective way of regulating body temperature is for an organism to adjust its behaviour. Cold-blooded animals such as lizards bask in the sun or hide under rocks to control their temperature. Some animals, such as bats, hibernate through periods of low environmental temperature. Animals with high metabolic rates may spend parts of the day inactive to conserve energy.

In humans, adding or removing clothing layers, removing oneself from hot or cold environments, and changes of posture (such as folding arms) and exercise are ways in which we can affect the rate of heat exchange with the surroundings. A further behavioural response is the use of technology to regulate the environment with air-conditioning or heating systems.

Extreme Conditions: Wind-chill

The rate of heat loss due to convection is dependent on the air speed. The wind-chill temperature gives an estimate of the equivalent still-air temperature that would provide the same cooling rate on exposed 33 °C skin that the current conditions (air temperature and wind speed) provide. At higher wind speeds, the model for heat transfer

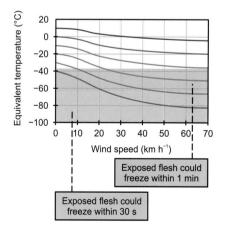

Figure 22.1 The effect of wind speed on the equivalent temperature. The wind-chill equivalent temperature provides an estimate of the still-air temperature that would provide the same cooling effect on exposed skin. [Created from data in *ASHRAE Fundmentals Handbook 2001*.]

by convection that we developed earlier no longer holds, and the wind-chill equivalent temperatures are the result of experiment. Figure 22.1 shows the effect of wind-chill on exposed skin.

22.5 Temperature and Health

Factors Affecting Comfort Level

A person feels comfortable when their temperature can be maintained in a narrow range, and they are not shivering or feeling sweaty. In high-humidity environments, low evaporation can lead to unpleasantly damp skin. At the other extreme, low humidity can cause unpleasant drying of the mucous membranes of the nose, and the eyes and throat. This is a problem that long-haul flight passengers are all familiar with. The relative humidity on an airplane can be a low as 1–2%. To humidify the air on planes would significantly increase costs.

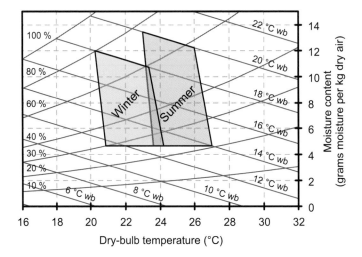

Figure 22.2 80% of people are thermally comfortable in the zones indicated. [Created from data in *ASHRAE Fundamentals Handbook 2001*.]

The effects of humidity on comfort levels can be slightly counter-intuitive. Even though a more humid environment may make it more difficult to lose heat by evaporation, in cold conditions it is more likely that the moisture that is lost through the skin will condense in and dampen the clothing. This can increase its thermal conductivity, making us feel cooler. Also, any significant wetting of the skin will increase discomfort, as will increased friction from clothing caused by the dampness.

Adverse Effects of Temperature

The elderly and infants are more susceptible to thermal stress, with the elderly at heightened risk for medical and often financial reasons. Many prescription drugs alter the body's ability to regulate temperature and increase the risks; these include sedatives, antidepressants, tranquilizers and cardiovascular drugs. Drugs can also increase the risk of heat stress. Medication for Parkinson's disease can decrease perspiration, and diuretics reduce the body's fluid reserves and decrease the blood flow to the skin.

The mortality rate for the elderly typically increases markedly during cold winters, particularly in temperate climates. In countries that do not have particularly cold winters, houses are often inadequately insulated, and populations may be under-prepared for the cold weather conditions. In these conditions, the leading cause of death amongst the elderly is not actually hypothermia from extreme cold, but stroke and heart attack. As vasoconstriction restricts the blood flow to the skin, this increases the amount of blood going to the central organs, putting them under stress. The body's response to this is to reduce the blood volume by excreting salt and water, which thickens the blood, increasing the risk of blood clotting. Older people's blood vessels often have rougher linings, so they are more at risk from clotting. In the UK, according to the Met Office, the winter death rate goes up about 1.4% for every degree drop in temperature below 18 °C.

This has significant public-health implications, as shown by two intervention studies by Professor Philippa Howden-Chapman of the Wellington School of Medicine and Health Sciences, University of Otago, which concluded that insulating houses led to a significantly warmer, drier indoor environment and resulted in improved health, less wheezing, fewer days off school and work, and fewer visits to general practitioners. It was also concluded that the use of non-polluting, effective home heating improved well-being and reduced symptoms of asthma and days off school.

Temperature can also affect health through air quality. Some of the contaminants that occur indoors are allergens such as dust, moulds and spores, and volatile organic compounds (e.g. acetone and toluene). High temperatures can increase the presence of volatile organic compounds in the air from sources such as carpets and furniture. On the other hand, low temperatures can increase relative humidity and condensation on surfaces, leading to the growth of moulds. There is also a risk of reducing air quality through decreased ventilation to maintain indoor air temperatures. Poor air quality can cause headaches, nausea, nasal congestion, fatigue and other symptoms. Individuals with asthma are more susceptible to adverse effects of poor air quality.

22.6 Summary

Key Concepts

work Energy transferred in a form that can perform mechanical work.

metabolism The series of biochemical reactions which convert stored energy in food molecules to other forms.

metabolic energy The amount of energy transferred to the body by oxidation of food.

first law of thermodynamics A statement that energy is conserved. Energy can be transferred from one form to another, but it can be neither destroyed nor created.

Equations

$$\Delta U = Q - W + E$$
$$\eta = \frac{W}{E - \Delta U} = \frac{W}{W - Q}$$

22.7 Problems

22.1 A cyclist does 2 MJ of mechanical work over the course of a day. His work efficiency is 20% and the metabolic energy of the food consumed during the day is 8 MJ.

(a) What is the net metabolic energy used during the day?

(b) How much heat does the cyclist transfer to his surroundings during the day?

(c) How much extra food will the cyclist need to consume at the end of the day in order not to loose weight?

22.2 A rugby player uses metabolic energy at a rate of 430 W while playing. At the same time he loses energy at an average rate of 210 W, due to work being done, convection, radiation and respiration. Assume that his core temperature does not change during an 80 minute game, and that the latent heat of vaporisation of perspiration is 2440 kJ kg^{-1}. How much moisture will he lose as a result of the evaporation of perspiration during the game?

22.3 During the course of a day a climber does 3.0 MJ of work with a mean work efficiency of 20%. How much heat must he lose during the day in order to avoid getting too hot?

22.4 On a long-distance polar trek an explorer has a nutritional intake of 15.6 MJ per day. She does mechanical work at an average rate of 100 W for 10 hours per day. Assuming she does not lose or gain weight, what is the average rate at which she loses heat to the environment (in W)?

22.5 During a tennis match the metabolic heat generation rate of a player is 450 W. She loses heat at a rate of 170 W by convection, radiation and respiration, how much moisture will she lose through evaporation of perspiration during a match lasting 3 hours? Assume that the core temperature of the tennis player does not change, and that the latent heat of vaporisation of water is 2440 kJ kg^{-1}.

22.6 After eating a 150 g pottle of yoghurt, Bob decides to go for a run to burn off the extra energy. The yoghurt provides 4.4 MJ kg^{-1}, and Bob is able to raise his metabolic rate to 220 W while running. How long does he have to run?

22.7 An athlete runs 5000 m in 30 mins, with a mechanical power output of 130 W. If she runs with a work efficiency of 15%, what is the net metabolic energy used?

22.8 The following statements are about how humans can transfer thermal energy to their surroundings when the air temperature is greater than the core body temperature, 37 °C. Which statements are correct? (Note: more than one statement may be correct)

(a) The body temperature is simply allowed to go up by a few degrees, which is OK for a few hours.

(b) Sweat glands release liquid that is hotter than the air onto the surface of the skin, so that the skin can then cool by convective transfer of heat to the cooler surrounding air.

(c) Water-like body perspiration evaporates into the surrounding air, provided it is unsaturated, producing a cooling effect at the skin surface

(d) It is impossible for the body to transfer thermal energy to the surroundings under these conditions if the relative humidity is 100%

(e) They cannot reject heat under these conditions, whatever the relative humidity, so they should just drink cold liquids to get cooler.

Activity	Metabolic Rate (W)
Standing relaxed	105 - 125
Walking 3 km h^{-1}	170 - 210
Walking 6 km h^{-1}	330 - 450
Running	500 - 800
Strenuous exercise	Exceeds 1000

Table 22.1 Metabolic rates associated with various levels of activity.

22.9 An endurance athletic event is conducted when the conditions are extremely stressful, the dry-bulb temperature being 45 °C. One of the participants in the event, a male, is able to absorb 1.1 litre of water per hour through the gut, although he can lose water through perspiration faster than this if necessary. The latent heat of vaporisation of water at the participant's skin temperature is 2440 kJ kg^{-1}.

(a) Determine the maximum rate at which he can reject heat by perspiration without becoming dehydrated.

(b) The athlete has a surface area of 1.8 m^2 and his surface heat transfer coefficient, including convection and radiation, is 22 W m^2 K^{-1}. During the endurance event his skin temperature is 37°C. Determine the rate of heat gain (or loss) due to convection and radiation alone, not including the effect of perspiration.

(c) The table above shows the metabolic rate of the athlete doing different activities. Assume this is the same as the rate of production of thermal energy in his body. Determine the maximum sustainable metabolic rate for the athlete during the endurance event without any temperature rise. What is the most vigorous activity that could be sustained under these conditions?

(d) Suppose the event is rescheduled for a cooler day. Again the athlete has sufficient water to perspire and evaporate 1.1 kg per hour and on this occasion he intends to run with a metabolic rate of 800 W. What is the maximum dry-bulb temperature at which he could do this sustainably, assuming his skin temperature is 37 °C?

IV

Electricity and DC Circuits

The two phenomena known to most people as electricity and magnetism are really both aspects of a single force – the electromagnetic force, one of the four fundamental forces mentioned previously in the book. The science of electromagnetism is the study of the interaction of particles which have an intrinsic property known as electric charge.

An understanding of how charges interact by electromagnetic forces is useful for the understanding of nearly every other branch of science. Light, as it will be explained in the Optics chapters, is an electromagnetic phenomenon. In Mechanics, we saw how the human body is a mechanical machine, and how the forces from muscle contractions allow us to move. The thoughts we have, the signals we send from the brain to the muscles and the changes that cause the muscle fibres to contract are all electrical.

In the next few chapters we will review the physics of electricity and DC (direct current) circuits, and present some examples of electric forces at work.

STATIC ELECTRICITY

23.1 Introduction

It has been known since ancient times that rubbing amber will allow it to attract other materials like straw, at least temporarily. This effect is due to the presence of stationary electric charges. The study of phenomena caused by stationary charges is known as **electrostatics**, and we will introduce the fundamental concepts of charge and the force that exists between charged particles in this chapter and those that follow.

Key Objectives

- To see that charge is a property of sub-atomic particles.

- To understand the types of charge.

- To understand the origins of the electric force.

23.2 Charge

The matter we are surrounded with is made up of molecules containing atoms, which are in turn made up of protons and neutrons (within the nucleus) and electrons (see Figure 23.1). An attractive force exists between the protons and the electrons which holds them together in the atom. There is some intrinsic property of these subatomic particles which causes them to be subject to this force, while the neutrons are not – we call this property **electric charge** and the force the **electrostatic force** or **electric force**. Atoms are also bound together into molecules by electric forces.

Electric charge comes in only two types; these types are labelled **positive** and **negative**. A proton has a positive charge, and an electron has a negative charge of exactly the same amount – they carry the same magnitude of charge. A particle that does not have an electric charge, such as a neutron, is said to be **neutral**. In an atom, there are equal numbers of protons and electrons (note that the atoms in Figure 23.1 have equal numbers of each), so while the atom contains charged particles, overall we say it is neutral since the positive and negative charges are balanced in magnitude and thus cancel each other out. In the case of an **ion**, which is formed when an atom loses or gains electrons, the number of negative charges is not equal to the number of positive charges; this means that the positive and negative charges do not cancel each other out and there is therefore a *net charge*.

The electric force between *like* charges, which may be both positive or both negative, is repulsive; it tends to push the charges apart. The force between *unlike* charges (i.e., a positive and a negative charge) is attractive, and it tends to draw the charges together. The force decreases in strength as the separation of the charges increases. We will consider the exact nature of this force in more detail in Section 24.2 on Coulomb's law.

The SI unit of charge is the **coulomb**, symbol C. This is defined as the quantity of charge that has passed in 1 s through the crosssection of an electrical conductor carrying one ampere of current. (Current is a measure of how much charge is moving through a fixed area per second, and will be covered in Chapter 27.) The charge on an

---

Naming the electron

The name 'electron' was proposed as the name for the fundamental unit of charge by Irish physicist G. J. Stoney in 1894. The name for electricity came from the greek word *elektron*, the Greek name for amber.

Positive and negative

The labelling of the charge carried by an electron as negative rather than positive was arbitrary and was thus determined by an agreed choice among scientists. The initial choice was made by Benjamin Franklin, who declared that rubbing silk on a glass rod left the glass positively charged. A consequence of this choice is that the electron is negatively charged and the proton is positively charged, even though Franklin made his choice well before the electron and proton were known to exist.

electron is -1.602×10^{-19} C (to 4 s.f.). A coulomb of negative charge would therefore require 6.24×10^{18} electrons – over 6 billion billion electrons.

There are two important experimental observations relating to charge, those of charge conservation and charge quantisation. In all observations ever carried out, the net charge of an isolated system has never been found to change. While charged particles such as electrons can be created and destroyed, other particles are created or destroyed at the same time to keep the overall charge of the system constant. This is known as **charge conservation**. In addition, there is a limit to the smallest amount of charge that can be measured on any free particle, and the net charge of any system is an integer multiple of this quantity; this is known as **quantisation of charge**, and the smallest amount of charge is the magnitude of the charge on an electron or proton, often called e, the elementary charge. (The charge on an electron is $-e$ and the charge on a proton is $+e$.)

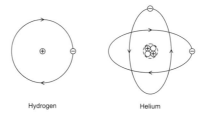

Hydrogen   Helium

Figure 23.1 A (very) simple representation of the sub-atomic constituents of the two simplest atoms, hydrogen and helium. Hydrogen has a nucleus which consists of a single positively charged proton and has a single electron 'orbiting' this proton. Helium has two protons and two neutrons in its nucleus, and two electrons 'orbiting' this nucleus. Each *atom* has no net charge, despite being made of charged parts.

Example 23.1 *Charge*

Problem: A 48 g ball of pure copper has a charge of +2 μC. Given that each copper atom has 29 protons and an atomic mass of 64 g mol^{-1}, what fraction of the electrons in the copper have been removed?

Solution: The magnitude of the charge on the copper ball represents the difference in the number of positive and negative charges it contains (protons and electrons). As the ball is positively charged, there must be *fewer* electrons than protons: 2 μC = magnitude of the charge on an electron × number of electrons removed, N_e.

$$N_e = \frac{2\mu C}{1.6 \times 10^{-19}} = 1.25 \times 10^{13}$$

Electron deficit, $N_e = 1.25 \times 10^{13}$.

The number of moles of copper atoms, n_{Cu} can be determined from the ball's mass and the molar mass of copper

$$n_{Cu} = \frac{m_{Cu}}{M_{Cu}} = \frac{48}{64} = 0.75 \,\text{moles}$$

so the number of copper atoms is

$$N_{Cu} = n_{Cu} N_A = 0.75 \times 6.0 \times 10^{23} = 4.5 \times 10^{23} \,\text{atoms}$$

The total number of electrons in the neutral copper ball are

$$N_n = 29 \times N_{Cu} = 29 \times 4.5 \times 10^{23} = 1.31 \times 10^{25} \,\text{electrons}$$

and the fraction of electrons removed, f_e is given by

$$f_e = \frac{N_e}{N_n} = \frac{1.25 \times 10^{13}}{1.31 \times 10^{25}} = 9.58 \times 10^{-13}$$

which is a very small fraction.

Example 23.2 *Moving charges across a cell membrane*

Problem: In order to convey signals a nerve cell transports ions across the cell membrane. If a particular cell transports 6.25×10^6 singly charged sodium ions (Na^+) across its cell membrane, what total charge has it transported?

Solution: Each Na^+ ion has a charge whose magnitude is equal to that of an electron, but of course positively charged ($Q_{Na^+} = +e = +1.6 \times 10^{-19}$ C). The total charge moved across the cell membrane is just the number of ions moved multiplied by the charge on each ion

$$Q = N \times Q_{Na^+} = 6.25 \times 10^6 \times \left(+1.6 \times 10^{-19} \, C\right) = 1 \times 10^{-12} \, C$$

23.3 Conductors and Insulators

Materials may be classified according to how easily charged particles move through them. A **conductor** is a material through which charge flows easily, and an **insulator** is one through which charge does not flow freely.

Metals are generally very good conductors. In a metal, the outermost electrons of each atom are only loosely bound to the nucleus, so they are able to hop from atom to atom, and are thus essentially free to move about inside the metal. The positively charged atomic nuclei remain fixed in place. The positive charge of the stationary nuclei is always cancelled by the negative electrons flowing past, so a metallic wire is electrically neutral even when a current flows through it.

Another type of conducting medium is a liquid containing ions in solution. This is rather different from conduction in a metal in that both the positively and negatively charged ions can move.

In insulators, there are no loosely bound electrons, and charge can't move around with the same ease. Common insulators are materials like glass, rubber and plastic. In Chapter 27 we will give a quantitative measure of conductivity, and will see that an insulator is simply a very poor conductor.

23.4 Charging of Objects

There are several methods for creating an excess positive or negative charge on an object. All of these methods involve moving charge from one object to another, and then separating the objects. Charge is conserved (as always), but one object is left more positive and the other more negative.

The simplest way to induce a charge on many common materials is to rub unlike materials together, as shown in Figure 23.2. As the electrons in the molecules on the surfaces are pushed close together, the surface molecules of one object attract electrons more strongly than they are held by the surface molecules of the other object, and the electrons transfer from one surface to the other. The direction of transfer depends largely on the property of atoms in a molecule known as *electronegativity*, which is a measure of how strongly they attract electrons. This leaves one material with an electron deficit (positively charged) and one with an excess of electrons (negatively charged). This is called **charging by friction**, or the **triboelectric effect**.

It is possible to create an ordered list of materials showing which are more likely to lose or gain electrons when rubbed against another material – this is called a *triboelectric series*. If two objects are rubbed together, the one made of the material closer to the positive end of the triboelectric series will be most likely be left more positive, and the other more negative.

An object may also be **charged by conduction**, as shown in Figure 23.3. If a conducting material is touched with a charged object, some charge will be transferred to the conductor. If the charged object is negative, some fraction of it's excess electrons move off it onto the neutral conductor as the repulsive force between the electrons

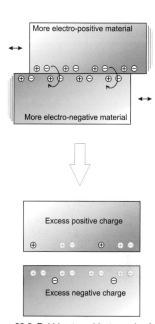

Figure 23.2 Rubbing two objects made of unlike materials together can result in charge transfer, leaving the object made of the most electropositive material with a net positive charge and the object made of the most electronegative material with a net negative charge. The total charge is conserved in this process; the two objects end up with equal and opposite charges.

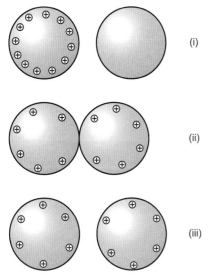

Figure 23.3 (i) A neutral conducting sphere (right) is brought close to an identical sphere with an excess positive charge (left). (ii) The two spheres are touched and the mobile charges redistribute over the surfaces of both, leaving both with an excess positive charge. (iii) The two spheres are separated, and each sphere has an excess positive charge of half the magnitude of the original excess charge on the left sphere.

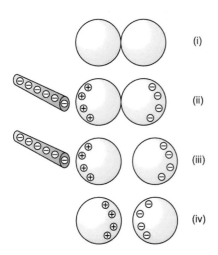

Figure 23.4 (i) Two identical uncharged conducting spheres are in contact with each other. (ii) A (negatively) charged object is brought close to the two spheres, but not so close that charge is transferred from the object to the spheres. This causes charges to separate within the two touching spheres. (iii) The two spheres are moved apart. As they are now no longer touching recombination of the separated charge is impossible. (iv) The charged object is removed and each of the conducting spheres finishes with an excess charge of equal magnitude but opposite sign.

causes them to move as far apart as they can. Similarly, if the charged object is positive, electrons flow off the conductor to fill the spaces left by the missing electrons and leave the conductor positive.

In most situations, the only charged particles moving about are negatively charged electrons, which is a consequence of the basic structure of atoms. In solid materials, the positive nuclei are not free to move about, but some electrons – the conducting electrons – don't have to stay bound to a particular nucleus, and can migrate through the material. However, if we remove an electron from a region which previously had an equal amount of positive and negative charge, this leaves it positive.

As electrons move about, the positive gap they leave behind behaves in the same way as a moving positive charge would: such a gap is usually referred as a 'hole'. Figure 23.5 shows how this works. As a negative charge moves right to fill a positive gap, leaving a similar positive gap behind it, this has the same effect as if the positive gap moved towards the left. When a conductor has a net positive charge, it is because there are holes that could be filled by electrons, but are not, leaving an excess positive charge. It is often easier to visualise the movement of these holes as being like positive charges moving about, which gives the same results.

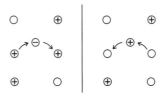

Figure 23.5 The movement of negative charge to the right has the appearance of positive charge moving to the left.

It is also possible to charge an object by a process known as **charging by induction**. In this process, two neutral conductors in contact are placed in the vicinity of a charged object, but not touching it. In the example shown in Figure 23.4, a negative rod is placed near the initially neutral conductors. Some of the negatively charged electrons move away from this negative rod, giving the conductor farthest away a negative charge, and leaving the nearer one positively charged. While the *total* amount of charge on the conducting spheres is unchanged, a *charge separation* has been *induced*. While the charges are still separated by the presence of the charged rod, the conductors can be physically separated, so they each retain a net charge, one sphere positive and the other negative.

Another way of using induction to create a permanent charge on a conductor is to 'earth' some part of the conductor while the charges are separated by the presence of a nearby charged object. The Earth acts like a giant source or sink of electrons. Just as the ocean contains so much water that a few drops more or less is unimportant, the ground can give up or accept quantities of charge without the change being noticeable. If a wire connected directly to the ground is touched to part of our conductor, that region of the conductor will lose excess electrons or gain extra ones to become neutral, which will affect the overall charge on the conductor once the wire is removed. In diagrams, a wire connected to the ground is represented by a line ending with the symbol ⏚.

23.5 Polarisation

It is possible to have a separation of charge within an object that, overall, has no net charge. This is called **polarisation**. Polarisation can be permanent, induced or instantaneous.

Permanent polarisation of a molecule occurs when the molecular structure is such that one side of the molecule is more positive than the other. Such molecules are called *polar* molecules. A common example is the water molecule, H_2O, shown in Figure 23.6. The end of the molecule where the oxygen atom is located is more negative than the ends with a hydrogen atom.

We already seen an example of induced charge separation in Figure 23.4. In this case, the charge in a conductor was separated by movement of the charges on a macroscopic scale. Induced polarisation can also occur by the separation of charges on a microscopic level. Any material, even an insulator, can become polarised by the distortion of the molecular electron clouds, so that on average one side of the molecule or atom is more positive, and the other more negative. The overall effect of these microscopic movements of charge is that the average position of all the positive charge and the average position of the negative charge no longer lie at the same place. In these situations, the polarisation persists for some time.

Figure 23.7 shows induced polarisation in both a conductor and an insulator in the presence of a charged rod. When the molecules in the material are already polar, then the material can become polarised in the presence of charged object by re-orientation of the molecules.

The changing nature of the instantaneous configuration of charges in an object may lead to it being polarised at some times, that is, an instantaneous polarisation may exist.

Because a charged object will cause a neutral object to become polarised, an attractive force can result. This is because, on average, the charges in the neutral object with same sign as the charged object are farther away than the charges with opposite sign. The force that charges exert on one another (which is covered in more detail in the next chapter) is dependent on distance, so the attraction dominates because the unlike charges are closer. This is easily demonstrated: a charged rod will pick up neutral paper, and a stream of water will be deflected towards a charged rod, regardless of which charge the rod carries.

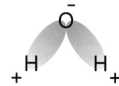

Figure 23.6 A simple representation of a very common permanently polarised molecule – the water molecule.

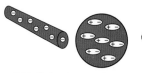

Figure 23.7 (i) Conductor: charges can move throughout the conductor and opposite charges accumulate on end of the conductor. (ii) Insulator: charges are still bound to a particular location in the insulator, but may spend more time on one side or other of a molecule.

Example 23.3 *Charging by induction*

Problem: A conducting, uncharged metal sphere (labelled A below) on an insulated base has an earth wire attached from the sphere to the ground. A second sphere, B, of the same shape, size and material, carrying a charge $+Q$, is brought close to, but not touching, the sphere A. Describe what happens. While keeping the sphere B close to sphere A, we remove the earth wire from the sphere A. If we now remove the sphere B, what is the charge on sphere A? How is this charge (if any) distributed?

Solution:

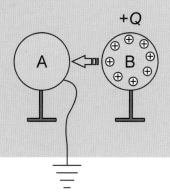

When sphere B is brought close to sphere A, electrons in sphere A are attracted to the side of sphere A close to sphere B, leaving a positive charge on the opposite side of sphere A.

Electrons are attracted from the ground via the earth wire onto sphere A to counter this positive charge on the side away from sphere B. When the earth wire is removed, there is no longer a path for electrons to move to or away from sphere A.

When we remove sphere B from the vicinity of sphere A the electrons on sphere A spread uniformly around the outside of sphere A. Sphere A has a charge of $-Q$. This is known as charging by induction.

23.6 Summary

Key Concepts

charge (Q or q) Electric charge is a fundamental property of matter, and comes in two types, known as positive and negative, which are referred to as the *sign* of the charge. The SI unit of charge is the coulomb (symbol C).

elementary charge (e) The smallest (non-zero) charge magnitude that can be carried by any observable elementary particle. Electrons have a charge of $-e$ and protons have a charge of $+e$. It has the value of 1.602×10^{-19} C.

electron A fundamental subatomic particle that carries a *negative* electric charge of $-e = -1.602 \times 10^{-19}$ C, and has mass 9.109×10^{-31} kg.

atom Usually defined as the smallest entity that retains the chemical properties of an element. An atom consists of a nucleus and electrons, with the number of electrons equalling the number of protons in the nucleus.

ion An atom that has gained or lost electrons and consequently carries a net positive or negative electric charge.

triboelectric effect A form of electrification by contact, which is the charging that happens when two materials are rubbed together. Some charges are transferred from one object to the other, leaving one object more positive than before and the other more negative.

conductor A material that will readily permit the flow of electric charges.

insulator A material which does not readily permit the flow of electric charges.

polarisation In electrostatics, the partial or complete separation of positive and negative electric charge in a system.

Equations

$$\text{charge on an electron} = -e = -1.602 \times 10^{-19} \text{ C}$$
$$\text{charge on a proton} = +e = +1.602 \times 10^{-19} \text{ C}$$

23.7 Problems

23.1 An uncharged metal sphere, A, is on an insulated base. A second sphere, B, of the same shape, size and material carrying a charge $+Q$ is brought *into contact* with sphere A.

(a) Describe what happens to charges on spheres A and B as they are brought into contact.

(b) If we now remove sphere B and place it far away, what is the charge on sphere A?

(c) How is this charge (if any) distributed?

23.2 An uncharged metal sphere, A, is on an insulated base. A second sphere, B, of the same shape, size and material carrying a charge $+Q$ is brought *close to, but not touching*, sphere A.

(a) Describe what happens to the charges on spheres A and B as they are brought close together *but not touching*.

(b) If we now remove sphere B, what is the charge on sphere A?

(c) How is this charge (if any) distributed?

23.3 A positively charged metal sphere, sphere A, is held close to *but not touching* an identical uncharged sphere, sphere B. Sphere A is now removed. After sphere A has been removed Sphere B is touched to an initially uncharged sphere, sphere C. What is the sign of the charge (if any) on sphere C after it has been touched to sphere B?

23.4 A physicist traps an ionized atom in a magnetic trap. She performs an experiment and finds that the atom has a charge of $+3.2 \times 10^{-19}$ C. If the atom has 12 protons and 12 neutrons, now many electrons must it have at the time it was trapped by the physicist?

23.5 A small sheet of aluminium foil measuring 2×2 cm is charged by rubbing it on some plastic material. The charge on the small sheet of aluminium foil is then measured and found to be $+Q$. An *uncharged* sheet of gold foil measuring 4×4 cm is brought close to, but not touching, the sheet of aluminium.

(a) What is the total charge on the sheet of aluminium?

(b) What is the total charge on the sheet of gold?

The gold foil and aluminium foil are now allowed to touch before being separated again.

(c) What is the total charge on the sheet of aluminium now?

(d) What is the total charge on the sheet of gold now?

(Note: both aluminium and gold are good conductors)

23.6 The following pairs of materials were rubbed together and the sign and approximate magnitude of the charge on each material noted. Use this information to rank these materials from *least electronegative* (most likely to lose electrons) to most electronegative. In other words construct a small triboelectric series.

Paper and synthetic rubber: paper, small +ve charge; rubber, small -ve charge.

Paper and polypropylene material: paper, medium +ve charge; polypropylene, medium -ve charge.

Rabbit fur and synthetic rubber: fur, medium +ve charge; rubber, medium -ve charge.

Rabbit fur and polypropylene material: fur, large +ve charge; polypropylene, large -ve charge.

23.7 To answer these questions use the small triboelectric series constructed in Problem 23.6.

(a) What would the sign of the charge on synthetic rubber be if it were rubbed against polypropylene material?

(b) What would the sign of the charge on paper be if it were rubbed against rabbit fur?

ELECTRIC FORCE AND ELECTRIC FIELD

24.1 Introduction

We have already stated that there is a force between charged objects; this electric force decreases with increasing separation. In this chapter, we will take a closer look at this force, which is described by Coulomb's law, and present an equation that tells us precisely how it changes with charge separation.

Also in this chapter we will introduce the concept of the electric field. A field is a physical quantity that is associated with each point in space (and time, though we will be looking at only static fields in this text). Charges are able to influence each other at a distance, and one way of describing this process is to picture a group of charges generating a field, which then creates a force on other charges placed in the field. This is a useful concept in many situations, as the field description allows the inclusion of a time delay to the interaction between particles. We will take a look at both the size and strength of the field created by simple configurations of charge, and look at the force that an electric field exerts on a charged object.

Key Objectives

- To be able to use Coulomb's law to calculate the size of the electric force.

- To understand the type of electric fields that exist around simple charge configurations, such as isolated positive and negative charges.

- To be able to calculate the size and direction of the force a field exerts on charges placed within it.

24.2 Coulomb's Law

A force exists between charged particles. It has been experimentally determined that the *magnitude* of this force depends on the magnitude of each charge and how far apart they are (see Figure 24.1) according to the following formula, known as **Coulomb's law**

$$|F_{1 \text{ on } 2}| = |F_{2 \text{ on } 1}| = k\frac{|q_1 q_2|}{r^2} \tag{24.1}$$

where $F_{1 \text{ on } 2}$ is the force on charge 2 exerted by charge 1, which is the same in magnitude as $F_{2 \text{ on } 1}$, the force exerted on charge 1 by charge 2. The magnitudes of the two charges are q_1 and q_2, r is the distance separating the charges, and k is an experimentally determined constant of proportionality.

The constant k equals $\frac{1}{4\pi\varepsilon_0} \approx 9 \times 10^9$ N m^2 C^{-2}. (The quantity ε_0 is a constant known as the permittivity of free space, which we will encounter again when we look at capacitors. $\varepsilon_0 = 8.854 \times 10^{-12}$ F m^{-1}).

The direction of the force is along the line joining the two charges. The force is attractive if the two charges have opposite signs and repulsive if they have the same sign. (Figure 24.1 shows the direction of these forces for like charges. If the charges had

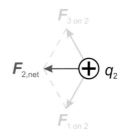

$$|\boldsymbol{F}| = k \frac{|q_1 q_2|}{r^2}$$

Figure 24.1 The Coulomb force between two 'like' charges (i.e., both positive, or both negative).

opposite signs, the forces would be attractive.) This force is exerted on both charges, that is, the first charge exerts a force on the second charge, and the second charge exerts a force on the first. These two forces are equal in size, but point in opposite directions – recall Newton's third law from Mechanics.

24.3   Superposition of Electric Forces

The **principle of superposition** applies here as it does whenever we combine forces

> **Key concept:**
> The net force on an object that is interacting with more than one other object is the *vector* sum of the forces from all the interactions.

For a charged particle, the *net* electric force on the particle is the *vector* sum of all the electric forces due to all the other charges present:

$$\boldsymbol{F}_{\text{on 1, net}} = \boldsymbol{F}_{2 \text{ on } 1} + \boldsymbol{F}_{3 \text{ on } 1} + \boldsymbol{F}_{4 \text{ on } 1} \dots \tag{24.2}$$

where $\boldsymbol{F}_{2 \text{ on } 1}$ is the force that charge 2 exerts on charge 1 and so on.

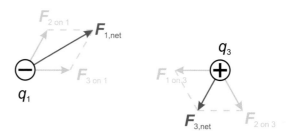

Figure 24.2 A system of three charges, q_1, q_2 and q_3. All three charges have the same magnitude, but not the same sign. The net force on each charge is due to the superposition of the two individual Coulomb forces applied to that charge by each other charge

This is illustrated in Figure 24.2. For example, the total force on charge 1 (the negative charge) is the sum of the forces exerted on it by both of the positive charges. Charge 3 has the opposite sign to 1, so it pulls charge 1 to the right. Charge 2 pulls charge 1 directly towards itself also. In this particular case, the two forces on 1 are the same size, as the charges are all equal in magnitude and separation. The resulting force, however, is *not* simply twice the size of $\boldsymbol{F}_{2 \text{ on } 1}$; it is a vector sum, so the magnitude is in fact less than the simple numeric sum. The direction of the resultant force must also be found from a vector addition.

Example 24.1 *Coulomb's law*

Problem: A proton and an electron are separated by the Böhr radius (the radius of a hydrogen atom). What electric force does the proton exert on the electron?

Solution: The charge on a proton is $+e = 1.6 \times 10^{-19}$ C, the charge on an electron is $-e = 1.6 \times 10^{-19}$ C and the Böhr radius is 5.29×10^{-11} m. To determine the electrical force that the proton exerts on the electron we use Coulomb's law

$$F = \frac{k|q_1 q_2|}{r^2}$$

$$F_{electron} = \frac{9 \times 10^9 \times 1.6 \times 10^{-19} \times 1.6 \times 10^{-19}}{(5.29 \times 10^{-11})^2} = 8.23 \times 10^{-8} \text{ N}$$

The proton exerts an 8.23×10^{-8} N attractive force on the electron.

Example 24.2 *Coulomb's law*

Problem: A 2 C and 3 C charge are separated by 8 cm. If we double the distance between the charges, what happens to the force on the 2 C charge? What about the force on the 3 C charge?

Solution: Because the magnitude of the Coulomb force follows an inverse square law, when we double the charge separation, the force between the charges decreases by a factor of two squared (= four). The force on the 2 C decreases by a factor of four. Newton's third law tells us that the forces on each charge are equal in magnitude, but opposite in direction so if the force on the 2 C charge has been reduced by a factor four, so too has the force on the 3 C charge.

Example 24.3 *Coulomb's law*

Problem: A 5 C charge is positioned at $x = 0$ cm, a –2 C charge is positioned at $x = 3$ cm, and a 7 C charge is positioned at $x = 8$ cm as shown in Figure 24.3. What is the net force on the –2 C charge due to the other charges?

Solution:
We need to calculate the net force acting on the –2 C charge. In order to do this we first calculate the forces on the –2 C charge due to each of the other charges, we then add up these individual forces taking their directions into account.

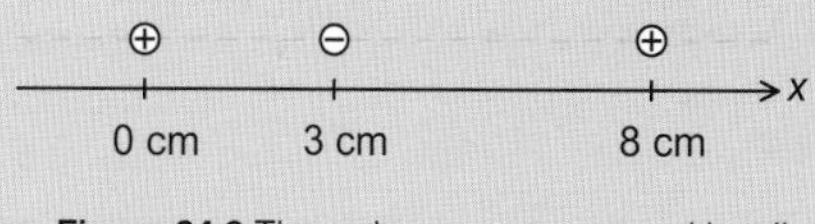

From Coulomb's law we have that the electrical force on the –2 C charge due to the 5 C charge is:

Figure 24.3 Three charges are arranged in a line.

$$F = \frac{k|Q_{0\,cm}Q_{3\,cm}|}{r^2} = \frac{9 \times 10^9 \times 5 \times 2}{0.03^2} \text{ N} = 1.00 \times 10^{14} \text{ N}$$

This force is to the left.
From Coulomb's law we have that the electrical force on the –2 C charge due to the 7 C charge is

$$F = \frac{k|Q_{8\,cm}Q_{3\,cm}|}{r^2} = \frac{9 \times 10^9 \times 7 \times 2}{0.05^2} \text{ N} = 5.04 \times 10^{13} \text{ N}$$

This force is to the right.
The net force acting on the –2 C charge is the vector sum of these two forces, 4.96×10^{13} N towards the left.

Example 24.4 *Coulomb's law*

Problem: Three charges are positioned as shown in Figure 24.4. A +3 C charge at $x = 2$ cm, a test charge q_{test} at $x = 0$ cm, and an unknown charge Q_u at $x = 8$ cm. If the net force on the test charge, q_{test} is zero, what is the magnitude of the unknown charge Q_u?

Solution:
There are two forces acting on the test charge. In order for the net force on the test charge to be zero, these two forces must be equal in magnitude, but opposite in direction. In order for the magnitude of the two forces to be equal we need

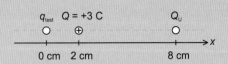

Figure 24.4 Two charges arranged such that a test charge placed at $x = 0$ cm will experience no net electrostatic force.

$$F_{3\,C} = \frac{9 \times 10^9 \times 3 \times q_{\text{test}}}{0.02^2} = F_{Q_u} = \frac{9 \times 10^9 \times Q_u \times q_{\text{test}}}{0.08^2}$$

Cancelling out common factors we have

$$\frac{3}{0.02^2} = \frac{Q_u}{0.08^2}$$

so

$$Q_u = \frac{3 \times 0.08^2}{0.02^2} = 48\ \text{C}$$

The unknown charge has a magnitude of 48 C.

In order to determine the sign of the charge we need to consider the direction of the forces acting on the test charge. If the test charge is positive, the 3 C charge will exert a repulsive force towards the left on the test charge. The force exerted by Q_u must therefore be to the right which in this case requires Q_u to be negatively charged. This reasoning still works if we assume the test charge is negative. In order for the net force on a test charge located at $x = 0$ to be zero, Q_u must be −48 C.

24.4 Inverse Square Laws

Gravitational analogy

The analogy between gravity and the electric force can be useful. Another situation where it helps is in the case of a uniform electric field, which has similarities with the near-uniform gravitational field near Earth's surface (provided the change in height is small). We will resort to using it on occasion, but the usefulness of the analogy is limited given that the existence of charges of different signs, + and −, make it a more complex situation.

The form of the mathematical expression of Coulomb's law bears a close resemblance to the mathematical form of Newton's law of gravitation. Both forces depend on the inverse of the square of the distance between the two objects, and both depend on the product of some property that the objects possess. This means that the electric force scales in a similar way to the gravitational force: if the distance is doubled, then the magnitude of the force drops to one quarter of the original value; if one charge is doubled, the force doubles; and so on.

There are important differences, however; primarily, that there is only one type of mass, whereas there are two kinds of electrical charge. Also, for the typical sizes of separation and charge encountered on a human scale on the surface of the Earth, the electrostatic interaction between two objects is much, much stronger than the gravitational attraction between them, as the masses are quite small. (We are not including interactions *with the Earth* when we make this statement.)

There are a number of physical quantities that obey an inverse square law: electric field strength from a point source falls with distance squared; the intensity of electromagnetic radiation (light) from a point source falls with distance squared; and, as already mentioned, gravitational attraction to a mass falls with distance squared. The common factor in these situations is that something, be it light rays or field lines, is spread out in three-dimensional space from a point. Whatever is spreading out is doing so over the surface of a sphere, and so the surface area over which it is spread depends on the radial distance *squared* from the source, because the surface area of a sphere is also proportional to r^2.

24.5 The Electric Field

Coulomb's law is very useful for calculating the force on charged particles, but it is not always the most convenient approach. Sometimes what we would really like to do is to describe the overall effect of a large collection of charges, where the summation of superposed forces is inconvenient to calculate. We are also typically interested only in the force *on* a charge, and not the force exerted *by* it. This is where the field concept becomes very handy.

The **electric field** is a way of describing how a configuration of charges affects the surrounding space. It is defined in terms of the magnitude and direction of the electric force per unit of charge that a charged object would be subjected to at a given point in space. In the SI system, it has units of N C^{-1}, so it is a measure of how much force a +1 C charge would experience. The force on a charge is a vector quantity, and so the electric field is also a vector quantity, with its direction being that in which a *positive* charge would experience a force.

$$E = \frac{F}{q} \tag{24.3}$$

It is important to keep in mind the distinction between the 'test charge' that is often referred to, and the charges creating the field (the source charges). We can imagine putting a one coulomb test charge at some place in order to ascertain how it would be affected by a field. However one coulomb of charge is a very large amount in most contexts and would likely alter the configuration of charges we were interested in investigating, as well as polarise any nearby material, and thus make it more difficult to obtain useful information. The electric field describes the modification of the properties of the space by some collection of charges, and this exists whether this test charge is there or not. The field strength in N C^{-1} tells us how much force +1 C would be subjected to, but the equation above (Eq. (24.3)) can be used to find out how much force *any* amount of charge would be subjected to. If +1 C would have a force of 1 N on it, then +0.01 C would have 0.01 N, and so on. A 'test charge' then is a nominal (and possibly imaginary) charge placed at a point in space that is small enough not to cause any significant change in the system of charges which are being investigated, but which allows us to investigate the electric field at a point in space.

The **electric field due to a point charge** is easy to derive from Coulomb's law. If the point charge creating the field is Q, and we imagine placing a test charge q a distance r from the first charge, the magnitude of the force on the test charge is

$$F = k\frac{Qq}{r^2} \tag{24.4}$$

The magnitude of the electric field is

$$E = \frac{F}{q}$$

and for a point charge Q we therefore have

$$E = k\frac{Q}{r^2} \tag{24.5}$$

The direction of the field is the same as the direction that the force would be on a positive charge – radially outward from the source charge if it is positive, towards the source charge if it is negative.

24.6 Electric Field Diagrams

There are a few common methods of trying to represent electric fields on a flat page. This is not a simple thing to do, because we are trying to represent a quantity that has both a size and a direction at every point in space. One way of illustrating the field is to draw a **vector field**. Representing the field at a single point in space is no problem – we can draw a simple vector showing the direction (which way it points) and the relative size of the field (how long the vector is).

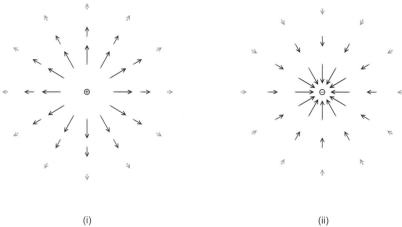

(i)   (ii)

Figure 24.5 The electric field at various points in space around (i) an isolated positive charge and (ii) an isolated negative charge. Each electric field vector represents the direction an magnitude of the electric field at the base of the vector.

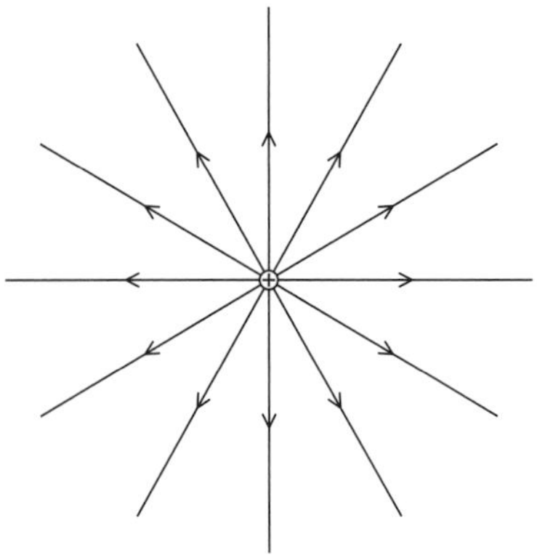

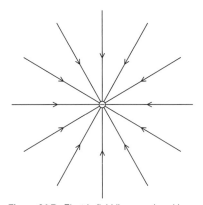

Figure 24.6 Electric field lines produced by an isolated positive point charge.

Figure 24.7 Electric field lines produced by an isolated negative point charge.

E

Figure 24.8 Uniform electric field lines produced by parallel plates carrying uniform opposite charges. (Note that the non-uniform field formed at the fringes of the plate is not shown here.)

Doing this for every single point on a two-dimensional page is not possible; there would be ink everywhere! When vectors are used to indicate electric field, the best we can do is to use a small number of vectors to indicate what is going on at only a few places, and leave it to the viewer to interpolate what is going on everywhere else. It should be borne in mind, though, that the vector represents the field *at a single place* – the base of the vector – and that no vector drawn in is *not* the same as no field at that place. Figure 24.5 shows a vector field representation of the electric field created by single point charges.

Another method of showing the field at different points in space is to draw **field lines**. The line is intended to represent the direction of the coulomb force on a positive charge. The lines are drawn with arrows so that the direction along the line is known, but these field lines are *not* vectors; they have no length to indicate how large the electric field vector is. Instead the strength of the field on a diagram of this type is indicated by how close together the lines are: the closer together the field lines are, the stronger the field is.

Electric field lines are drawn *starting on positive charges and ending on negative charges*. An electric field line is drawn starting or ending in free space only if the source charges are outside the region being depicted, not because the lines don't end on charges. Electric field lines cannot cross, as this would indicate that the direction of the net force on a charge could be in two directions at once, which is not the case. The arrows which are drawn to represent electric field vectors can cross, as long as their bases are not on the same spot. These vectors represent the size and direction of the electric field at their base and *not* at any other point. If the *arrows* cross, it is only as a consequence of the scale used to draw the vector, and has no special significance. Figures 24.6 and 24.7 show the field line representation of the field for the same point charges as before.

Some other charge configurations we will be interested in at times are the uniform electric field in which the electric field has the same magnitude and direction at all points, and the dipole field which is created between pairs of positive and negative charges. A nearly **uniform electric field** can be generated by having parallel plates carrying a uniform charge distribution (as in Figure 24.8), with positive charge on one plate and negative on the other. The field between the plates points from the positive plate to the negative, and provided the distance between the plates is small in comparison with the plate area, nearly the same magnitude and direction everywhere between the plates. (In problems involving uniform fields, it can help to sketch in where such plates creating the field might be.) The field is significantly non-uniform at the plate edges – the field lines curve in this fringe area.

An **electric dipole field** can be created by equal magnitude positive and negative charges separated by some distance. The field lines point from the positive to the negative as shown in Figure 24.9.

24.7 Superposition of Electric Fields

The principle of superposition holds for the electric field just as it does for the electric force. The net electric field at any point in space can be found by a vector sum of the fields at that point due to all the charges present. Figure 24.10 demonstrates how this superposition principle can be used to find the dipole field shown in Figure 24.9. At each point in space, the resultant electric field vector (dark blue) is the sum of the electric field vector that would be created by the lone positive charge (shown in green) and that we would get from just the negative charge (shown in red).

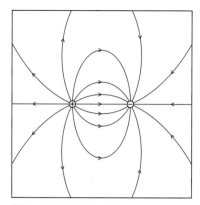

Figure 24.9 Electric field lines produced by equal magnitude positive and negative charges (electric dipole field). The Coulomb force on a positive charge placed on a electric field line always points along the line.

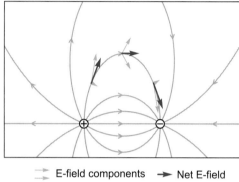

⇉ E-field components   ➡ Net E-field

Figure 24.10 Electric field vectors are shown for several points on an electric field line between two charges. Notice that the electric field vector is parallel to the electric field line (at the vector's base). Also shown are the electric field components due to each individual charge.

24.8 Summary

Key Concepts

Coulomb's law The law that describes the force between two point charges. The force is proportional to the product of the magnitudes of the charges and inversely proportional to the square of the distance between them.

field A numerical quantity associated with each point in space. A field can be scalar (as in the case of a temperature field) or vector (like a velocity or electric field).

electric field (E) The vector field produced by electric charge. The electric field vector is defined as the Coulomb force per unit charge that a 'test charge' would experience if placed at that point in space, and its magnitude is the same as the magnitude of the Coulomb force that would be exerted on a +1 C 'test charge'. The direction of the electric field vector is in the same direction as the Coulomb force on a *positive* 'test charge'. The electric field is measured in units of $N\,C^{-1}$ (or equivalently in volts-per-meter, $V\,m^{-1}$: see next chapter).

Equations

$$F_{1\,on\,2} = F_{2\,on\,1} = k\frac{|q_1 q_2|}{r^2}$$

$$E = \frac{F}{q}$$

$$E = k\frac{Q}{r^2}$$

24.9 Problems

24.1 A $+3 \times 10^{-6}$ coulomb charge is placed 5 centimetres due west from a $+2 \times 10^{-6}$ coulomb charge.

(a) What is the force the $+2\,\mu C$ charge exerts on the $+3\,\mu C$ charge?

(b) What is the force the $+3\,\mu C$ charge exerts on the $+2\,\mu C$ charge?

24.2 Two charges, one +4 C and the other +2 C, are separated by some distance R_0. If we increase the distance between the charges by a factor of 5, what happens to the magnitude of the force on the +2 C charge?

24.3 Two charges, Q_A and Q_B, are separated by a distance x. If we double the distance between the charges and triple the magnitude of charge A, what happens to the magnitude of the force that charge A exerts on charge B? What happens to the magnitude of the force that charge B exerts on charge A?

24.4 Three charges, Q_A, Q_B, and Q_C are positioned as shown in Figure 24.11

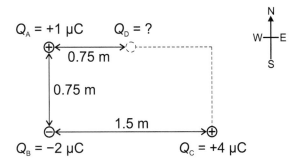

Figure 24.11 Three charges are arranged in 2 dimensions. What fourth charge will ensure that the net electrostatic force on charge B is zero?

(a) Determine the net force acting on charge Q_B due to the other two charges.

(b) A fourth charge, Q_D, is placed at position D shown in Figure 24.11 such that the net electrostatic force on charge Q_B is zero. What is the sign and magnitude of Q_D?

24.5 A mad scientist invents a device that is able to teleport every electron in their body to the center of the Earth (6.37×10^6 m below). We can make the assumption that the human body is mostly water with a bit of carbon and as such there will be about 3.2×10^{26} electrons per kilogram of body. If the mad scientist, who weighs 65 kg is so unwise as to actually use this device on himself what will the magnitude of the attraction between his body (stripped of all electrons) and all the electrons newly deposited at the center of the Earth? (How does this compare with the gravitational attraction between the mad scientist and the Earth?)

24.6 A small charged particle of mass 9×10^{-6} kg and charge of magnitude -3×10^{-6} C is placed in a chamber in which there is a uniform electric field. If the charge accelerates due north at a rate of $250\,\mathrm{m\,s^{-2}}$ what is the magnitude and direction of the electric field inside the chamber? (ignore gravitational forces)

24.7 Answer the following:

(a) What is the magnitude and direction of the electric field 10 m away from a +0.1 mC charge?

(b) What is the magnitude and direction of the electrostatic force on a +1.5 mC charge placed at this point (10 m from the charge in (a))?

(c) What is the magnitude and direction of the electrostatic force on a −3.5 mC charge placed at this point (10 m from the charge in (a))?

24.8 What is the magnitude and direction of the electric field at each of the three points, A, B, and C shown in Figure 24.12?

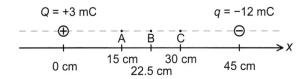

Figure 24.12 Two charges of opposite signs are placed 45 cm apart.

24.9 What is the magnitude of each of the charges q and Q in Figure 24.13?

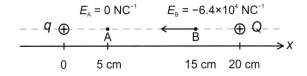

Figure 24.13 Two charges are placed 20 cm apart. The magnitude of the electric field at point A is zero, while the electric field at point B is non-zero.

24.10 An electric dipole shown in Figure 24.14.

(a) Calculate the electric field strength at points A, B, C, and D.

(b) What electrostatic force will a $+0.2\,\mu C$ charge experience when placed at each of the four points in part (a)?

(c) What electrostatic force will a $-0.1\,\mu C$ charge experience when placed at each of the four points in part (a)?

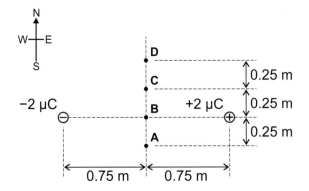

Figure 24.14 Find the electric force on charges placed at the points A to D.

ELECTRICAL POTENTIAL AND ENERGY

25

25.1 Introduction

In the previous chapter, we introduced the electric or Coulomb force and developed the idea of the electric field. In the chapters on Mechanics, we saw that the idea of potential energy may be associated with a conservative force. In this chapter, we will apply the idea of potential energy to the electric force. We will define a new concept, the electrical potential, and show how to use this to solve problems using energy methods. As in mechanics, using the idea of potential energy makes it easier to solve some problems.

Key Objectives

- To understand electrical potential and electrical potential energy.

- To be able to apply the idea of electrical potential to calculate work done by electric fields.

- To understand representation of electrical potential by equipotential lines and the relationship of electrical potential to electric field lines.

25.2 Electrical Potential Energy

The electric force has a great deal in common with the force of gravity. They are both conservative, action-at-a-distance forces for which we can use a field description, and associate a potential energy.

Imagine for a moment a universe containing nothing but a single, positive, point charge. If another positive charge was to be placed anywhere in this universe, we can see from Coulomb's law that it would have a force pushing it towards or away from the original charge. This force would cause it to accelerate, which would be an increase in the kinetic energy of the new particle. From this information we can infer that the charge gained potential energy when it was put in the electric field created by the first charge. This **electrical potential energy** is very similar to the more familiar gravitational potential energy. The electrical potential energy is often represented by the symbol U.

The potential energy of any system, gravitational, electrical or otherwise, can be found by calculating the amount of work needed to put that system together (see Figures 25.1 and 25.3). We can use this idea to calculate the potential energy of a system of charges. Imagine we take each charge from infinity (where the force on it from the other charges is zero) and bring it in to its final position. By multiplying the force applied to the charge by the distance through which it is moved, we calculate the total work done on the charge. We could then repeat this procedure for all of the charges that make up the system we are analysing. However, the strength of the force varies with distance; this means that the electric force that we have to move the charges against changes considerably while we are moving our charge into position. We are still able to do the calculation, but we need to use calculus, which is outside the scope of this book.

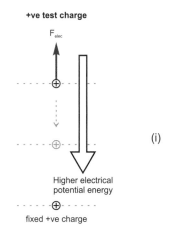

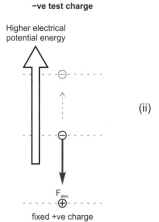

Figure 25.1 (i) Moving two 'like' charges closer together increases the electrical potential energy of each charge. (ii) Moving two 'unlike' charges further apart increases the electrical potential energy of each charge

Introduction to Biological Physics for the Health and Life Sciences   Franklin, Muir, Scott, Wilcocks and Yates
©2010 John Wiley & Sons, Ltd

As the calculation of the potential energy can be difficult, we will restrict ourselves mainly to the simple case where the force is not changing, i.e., the electric field is uniform. This is much like we did in the case of gravity (see Figures 25.2 and 25.3), where we restricted ourselves to cases dealing with objects near the Earth's surface, where the force changes slowly over the distances we are usually interested in, due to the enormous size of the Earth compared to the relatively tiny distances we are moving things on the surface.

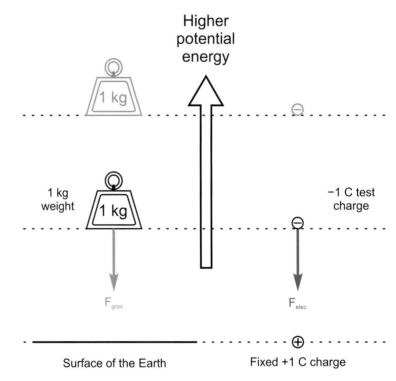

Figure 25.2 A useful analogy can be drawn between the concept of gravitational potential energy and electrical potential energy. Like most analogies it is imperfect, however. When considering electrical potential energy, we have both positive and negative charges which means both attractive and repulsive forces are possible, unlike gravitational potential energy, when all forces are attractive.

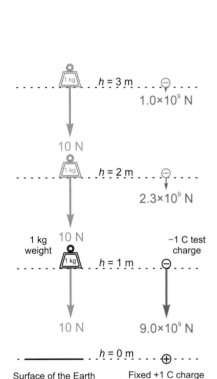

Figure 25.3 The gravitational force acting upon an object can be considered constant when one object is much more massive than the other and only is moved over a short distance. In most cases such a simplification cannot be made for the electrostatic force between two charges, so we will restrict ourselves mostly to the simple case of a uniform electric field.

For a simple gravitational example, such as a billiard ball on a table top, we would calculate how much work was done against the gravitational attraction of the Earth to raise the billiard ball to its position on the table. We calculate the work done on the billiard ball by multiplying the force applied to the ball by the distance through which it moves. We know that the path we take to put the ball where we want it doesn't matter because gravity is a conservative force. The magnitude of the force we would need to use to overcome the force of gravity is just $F = mg$, so the work done to raise the ball to a height h is just $W = mgh$, and we recognise this as the increase in the potential energy of the ball as it is moved to its final position.

In the rest of this chapter we will develop an understanding of the relationship between *electrical* work, potential energy and potential using the example of a constant electric field.

25.3 Electrical Potential

The concept of the electrical potential is one of the more difficult in introductory-level physics for those who haven't encountered it before. We will discuss the electrical potential and why it is useful, starting with a gravitational analogy.

The gravitational potential energy of a mass near the Earth's surface increases as its height above the surface increases. It makes some sense to think of greater height being equivalent to greater potential energy. The difficulty with this view is that a 10 kg mass only 10 m above sea level has more potential energy than a 1 kg mass which is 90 m above sea level. The gravitational potential energy depends on the mass of the object as well as its position. It would be useful sometimes to have a quantity that is related to the potential energy that is dependent only upon position relative to a charge or charges.

The electrical potential is such a measure. The electrical potential at a point in space does not depend on the charge of any object placed at that point, but is a measure of the electrical potential energy that a charge would have if it were placed at that point. In this it is very similar to the concept of the electric field which is a measure *force* on a charge if it were placed at a particular point in space. The electrical potential has a value at each point in space whether or not it is occupied; just like the electric field is present all through a region of space. The potential is a property associated with a point in space caused by the presence of an electric field (which is due in turn to some configuration of charges). The electrical potential is a *scalar field* – it has a value at each place, but is directionless.

We define the **electrical potential**, symbol V, by

$$V = \frac{U}{Q} \tag{25.1}$$

where U is the electrical potential energy of charge Q. The units of the electrical potential are **volts**, where $1\,\text{V} \equiv 1\,\text{J}\,\text{C}^{-1}$.

The electrical potential is a useful tool in many situations as it is independent of the charge. (Note that we mean the 'test' charge here, not the source charges creating the field.) Two different charges placed in the same position would have different potential energies, but they would have identical electrical potentials.

> **Conservative fields**
>
> The electric field, like the gravitational field, is a conservative field (see Figure 25.13). This means that the work done to move a charge from one place to another depends only on the start and finish points, not the path taken. We will not attempt to prove this in this book; however, it is an important fact about the electric field since it is this fact that allows us to use the concept of potential energy.

25.4 Electrical Potential and Work

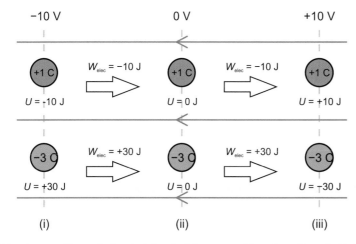

Figure 25.4 Two charges, with magnitudes of +1 C and −3 C, are moved through a uniform electric field. Shown is the amount of work done by the *electric field* when moving the charges, and the electrical potential energy *at those points in space* where the electrical potential is : (i) −10 V, (ii) 0 V and (iii) +10 V.

When an object falls freely in the Earth's gravitational field, work is done on the object by that gravitational field. This work is the amount of gravitational potential energy that is converted into kinetic energy. In the same way, when a charge moves due to the influence of an electric field, work is done on the charge by the electric field. The work done on the charge *by the electric field* is equal to the reduction in electrical potential energy

$$\Delta U = -W_{\text{elec}} \tag{25.2}$$

For example, if the charge moves to a position where its electrical potential energy is lower, the change in electrical potential is negative, and the charge has positive work done on it by the field. Figure 25.4 shows a constant electric field, and the work done moving different charges from place to place. For example, moving a +1 C charge through a potential difference of +10 V, the work done by the electric field is −10 J. This amount is only the work done by the electric field – we have made no statements about any other forces acting at the same time. In fact, some external force would be needed to achieve the movement of the positive charge in the direction shown.

In order take a better look at this and similar situations, we will derive some useful expressions for the electrical work, the change in electrical potential energy and the difference in electrical potential between points in a constant electric field pointing in the x direction. Consider the work done by the electric field on a charge which is moved from one point to another in the constant electric field. This is just

$$W_{\text{elect}} = \text{force} \times \text{distance} = F \times \Delta x \qquad (25.3)$$

In this equation, Δx is the distance that the charge is moved *in the direction of the force*, and F is the force exerted by the electric field on the charge. This force is given by $F = qE$, so our equation for the work done by the electric field becomes

$$W_{\text{elect}} = qE\Delta x \qquad (25.4)$$

where Δx is the direction moved in the direction of the electric field.

Now that we have an expression for the work done by the electric field, we can derive an expression for the change in the electrical potential energy of the charge as it moves from one position to another

$$\Delta U = -qE\Delta x \qquad (25.5)$$

and the potential difference between the two points is (from $V = U/q$)

$$\Delta V = -E\Delta x \qquad (25.6)$$

Note that it is common to use almost any common distance symbol for distance moved, so d, Δx or l may all be seen in other textbooks in equations similar to Eq. (25.6). Rearranging this

$$E = -\frac{\Delta V}{\Delta x} \qquad (25.7)$$

This equation gives an alternate way of writing the units for electric field strength; we can write the units as N C^{-1} or the equivalent V m^{-1}. This provides us with another way of thinking of electric field – the strength of the field tells us how rapidly the electrical potential changes with distance. At right angles to the electric field direction, therefore, the electrical potential is unchanging.

Example 25.1 *Potential difference*

Problem: What is the electrical potential difference between point A and point B as shown in Figure 25.5?

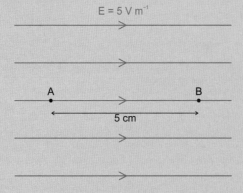

Figure 25.5 Two points in a region of uniform electric field.

Solution: The electrical potential difference, $\Delta V = -E\Delta l = -5\,\text{V m}^{-1} \times 0.05\,\text{m} = -0.25\,\text{V}$. Note that the negative sign indicates that a positive charge would have lower electrical potential energy at point B than point A.

Problem: A muscle cell in the heart typically has a potential difference of 90 mV between the inside and outside of the cell membrane. In order to function, the cell must 'pump' ions from one side of this membrane to the other.

(a) How much energy is required to pump a single Na^+ ion from the outside of the cell membrane to the inside if the outside of the cell is negatively charged with respect to the inside?

(b) A typical chocolate bar will release around 1000 kJ of energy into the body once metabolised. How many ions can be transported across a heart cell membrane with this amount of energy?

Solution: (a) If the outside of the cell membrane is more negatively charged than the inside, the electric field in the vicinity of the cell membrane will be pointing from the inside of the cell, out. Moving a positively charged into the cell thus requires it be moved against the direction of the electric field and so work will need to be done on the cell

$$\Delta U = q\Delta V = 1.6 \times 10^{-19}\,\text{C} \times 90 \times 10^{-3}\,\text{V} = 1.44 \times 10^{-20}\,\text{J}$$

(b) 1000 kJ is enough energy to transport $\frac{1000 \times 10^3\,\text{J}}{1.44 \times 10^{-20}\,\text{J per Na}^+\text{ ion}} = 6.9 \times 10^{25}\,\text{Na}^+$ ions

25.5 Equipotential and Field Lines

In addition to showing the electric field on a diagram, we can also indicate the electrical potential at various points in space. The electrical potential is a *scalar* quantity, so rather than representing it with vectors or field lines, we can draw lines that show places where the electrical potential is the same. These are called **equipotential lines** (see Figures 25.6, 25.7 and 25.8).

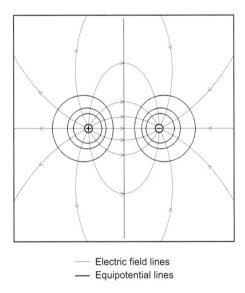

— Electric field lines
— Equipotential lines

Figure 25.6 A selection of electric field lines and equipotential lines for two unlike charges forming an electric dipole.

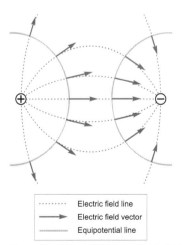

.... Electric field line
——▶ Electric field vector
——— Equipotential line

Figure 25.7 Electric field lines, electric field vectors, and equipotentials around an electric dipole. The electric field vector is perpendicular to the equipotential line that it lies upon. Note that the electric field strength is not the same at all points on a single equipotential line.

Because any change in position that has a component in the direction of the electric field will result in a change in potential ($\Delta V = E\Delta x$), equipotential lines are *always* *perpendicular* to electric field lines (see Figure 25.7). It is common to draw in equipotentials at regular voltage spacings or to label them with the voltage. The electric field points in the direction of *decreasing* electrical potential.

It may help to think of the equipotential lines as being like contour lines on a topographical map, and the electric field direction as the direction that points downhill. The contour lines are closest together where the hills are steepest; the equipotential lines are closest where the field is strongest, i.e., more change in volts per metre.

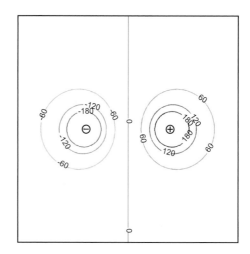

Figure 25.8 Equipotentials lines around an electric dipole. Each equipotential is labeled with the potential in volts. The magnitude of the charges is |2|μC and they are separated by a distance of 0.5 m.

25.6 Electrical and External Forces

Now let's look put all this together by looking at some possible scenarios.

Positive Charge, No External Force

Take a look at Figure 25.9. A positive charge, initially at rest at A, moves from A to B (in the direction of the field), and there are no other external forces acting. Here the work done on the charge by the field is positive; the charge moves in the same direction as the force exerted by the field. If it helps, remember that we can create a uniform field with parallel metal plates carrying positive and negative charge. To create this field, the positive plate would have to be on the right, as field lines point from positive charge to negative charge. The force must therefore be to the left on our charge.

If positive work is done by the field, the change in the electrical potential energy of the charge is negative. It loses electrical potential energy and will gain kinetic energy. The electric field will accelerate the charge, and the charge will reach point B with kinetic energy equal to the lost electrical potential energy. Since the change in electrical potential energy is negative, the potential difference between point A and B, $V_B - V_A$, will also be negative – point B is at a lower potential than point A. We could work this out from the electric field direction, too.

Positive Charge, External Force

Consider the charge shown in Figure 25.10, which is just like the previous figure, only the final velocity is different. A positive charge, initially at rest at A, moves from A to B (in the direction of the field), and finishes at rest at B. Unless there is another force acting in this situation, this is not possible! A force must be applied to stop the particle accelerating, or to slow it and stop it at B. In order to understand this scenario, we will now consider how to include non-electrical external forces into our energy equation.

The electrical force is conservative and we can write an equation for energy conservation in the following form:

$$(KE)_f + U_f = (KE)_i + U_i + W_{applied} \tag{25.8}$$

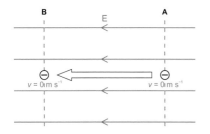

Figure 25.9 A positive charge, initially at rest at A, moves from A to B, and there are no external forces acting. The electric field will accelerate the charge and the charge will reach point B with kinetic energy equal to the lost electrical potential energy.

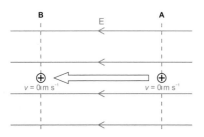

Figure 25.10 A positive charge, initially at rest at A, moves from A to B, and finishes at rest at B. The work done by the non-electrical external force is negative and is equal in magnitude to the positive work done by the electric field on the charge.

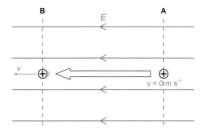

Figure 25.11 A negative charge at rest moves from A to B, and finishes at rest at B. The work done by the non-electrical external force is positive and is equal in magnitude to the negative work done by the electric field on the charge.

The work done by some non-electrical force is labelled $W_{applied}$. The work done by the electric field is built-in to this equation as the change in electrical potential energy, $W_{elect} = -\Delta U = -(U_f - U_i)$.

In the proposed situation, the electrical potential energy is initially greater than it is after the move from A to B. This means that the change in electrical potential energy is negative, and the work done by the field is positive. The initial and final kinetic energies are zero, so the change in kinetic energy is also zero and so this can't be used to balance the equation. The only way that the energy equation can be balanced is if the work done by the non-electrical external force is negative and is equal in size to the positive work done on the charge.

To see that this makes sense, consider that if you had to apply a force to the charge such that it could get from A to B without a change in kinetic energy, that force would have to be exactly the same size as the electric force, so that the net force was zero. (Remember, no net force, no acceleration, no change in kinetic energy.) Work done equals force times distance, and here the external force is in the opposite direction to the motion, hence negative work is done by it.

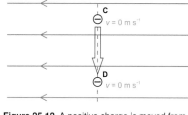

Figure 25.12 A positive charge is moved from C to D along an equipotential line. There is no work done by the electric field.

Negative Charge, External Force

This scenario is illustrated in Figure 25.11. A negative charge at rest at A moves from A to B (in the direction of the field), and finishes at rest at B. This time, we can see once again that an external force must be acting. Picture the charge configuration that would create our uniform field, and you can see that this would be like our negative charge moving towards other negative charges, which will not happen without another force being applied.

The external force is in the opposite direction to the force from the electric field, but this time it is in the same direction as the motion, so the work done on the charge by the external force is positive. The work done on the charge by the electric field must be negative, as we have no change in kinetic energy –from Eq. (25.8), if $W_{applied}$ is positive, $\Delta U = U_f - U_i$ is positive, and W_{elec} is negative.

We can check to see that this really makes sense by looking at it from another angle. Moving the charge against the direction of the electrical force is like lifting a rock against the direction of the gravitational force. The final potential energy must indeed be higher, and the external force is adding energy to the rock/charge, so it does do positive work.

Also, the electrical potentials at both points, A and B, are completely independent of the charges we put there. The electric field still points from A to B, so V_A is still higher than V_B. From our stated relationship between V and U in Eq. (25.1),

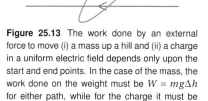

$$\Delta U = U_f - U_i = Q \times (V_B - V_A) \tag{25.9}$$

We know $V_B - V_A$ is negative, and the charge Q is negative also, so ΔU is positive.

Figure 25.13 The work done by an external force to move (i) a mass up a hill and (ii) a charge in a uniform electric field depends only upon the start and end points. In the case of the mass, the work done on the weight must be $W = mg\Delta h$ for either path, while for the charge it must be $W = -q\Delta V$.

Charge Moving Perpendicular to the Field Direction

Figure 25.12 shows a positive charge moving from C to D, two points at the same potential, with no change in velocity. In this case, there is no change in potential energy, and hence no work done by the electric field.

> **Example 25.3 *Electron energy***
>
> **Problem: An electron travelling at 7 260 000 m s^{-1} is fired into a chamber like that shown in Figure 25.14 in which there is a uniform electric field. The furthest the electron makes it into the chamber is 10 cm. What is the magnitude of the electric field, *E*?**

Solution:

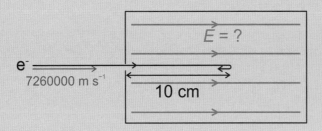

Figure 25.14 An electron is stopped and turned around by a uniform electric field.

The electron starts off with some kinetic energy which is converted into electrical potential energy by the electrostatic force. When the kinetic energy is completely gone, the electron will change direction and start moving back towards the entry point. (Note that the electron was always *accelerating* back towards this point.)
Given this we can say that

$$\mathrm{KE_i} = \Delta U = q\Delta V = qE\Delta x$$

$$E = \frac{\mathrm{KE_i}}{q\Delta x} = \frac{\frac{1}{2}mv^2}{q\Delta x}$$

$$= \frac{\frac{1}{2} \times 9.11 \times 10^{-31}\,\mathrm{kg} \times \left(7\,260\,000\,\mathrm{m\,s^{-1}}\right)^2}{1.6 \times 10^{-19}\,\mathrm{C} \times 0.1\,\mathrm{m}} = 1500\,\mathrm{V\,m^{-1}}$$

Example 25.4 *Potential difference and energy*

Problem: An electron, initially at rest, is accelerated by a 10 kV accelerating potential (i.e., placed in a electric field such that the electron accelerates from an original position to a second position at a potential 10 kV higher that the original).

  (a) **What is the change in electrical potential energy of the electron?**

  (b) **What is the change in kinetic energy of the electron?**

  (c) **What is the final speed of the electron?**

  (d) **If the 10 kV potential is generated by a uniform electric field over a 0.2 m distance, what is the electric field strength?**

  (e) **What is the force on the electron due to the electric field?**

Solution: Change in electrical potential energy = charge × change in electrical potential = $-1.6 \times 10^{-19} \times 10 \times 10^3$ J $= -1.6 \times 10^{-15}$ J.

This makes sense as the positive work done on the electron by the electric field corresponds to the decrease in its electrical potential energy.

As no other forces act on the electron, the change in kinetic energy of the electron is equal to the work done on the electron by the electric force. In other words, the change in the kinetic energy of the electron is equal to the negative of the change in the electrical potential energy of the electron. The electron's electrical potential energy has been converted into the electron's kinetic energy. We can determine the electron's final speed from

$$\frac{1}{2}mv^2 = 1.6 \times 10^{-15}\,\mathrm{J} \tag{25.10}$$

$$v = \sqrt{\frac{2 \times 1.6 \times 10^{-15}}{9.1 \times 10^{-31}}} = 5.9 \times 10^7\,\mathrm{m\,s^{-1}} \tag{25.11}$$

The electron's final speed is $5.9 \times 10^7\,\mathrm{m\,s^{-1}}$.

25.7 The Heart and ECG

In this section we will give a very brief introduction to the electrical system of the heart and discuss how this system produces the signals seen in an electrocardiogram (ECG).

When the heart is beating normally, an electrical signal is generated at the sino-atrial (SA) node of the heart which travels through the heart muscle (myocardium). The electrical signal causes the contraction of the heart muscle cells in sequence, and is ultimately responsible for the synchronised beating of the heart (see Figure 25.15). For efficient pumping of the blood without backflow, the muscle fibres in the atria contract first, and after a delay, the ventricles contract. The delay is introduced in the electrical signal by the atrioventricular (AV) node. The 'pacemaking' signal from the SA node occurs at a rate of 60–100 beats per minute. This is known as *normal sinus rhythm*. When the impulse occurs at a lower rate, this is called *sinus bradycardia*, and when it happens more rapidly, this is known as *sinus tachycardia*.

A typical heart muscle cell (Figure 25.16) is about 100 μm long, and 15 μm wide. The outer membrane around the cell is 8–10 nm thick, and in its resting state, a potential difference of around 90 mV exists between the inside and outside of the cell membrane, with the outside of the cell being positive. This is called the 'resting potential', and a cell with this potential difference is called 'polarised'. When the transport of ions from one side of the membrane to the other changes the sign of this potential difference, the cell membrane becomes 'depolarised' and the cell contracts.

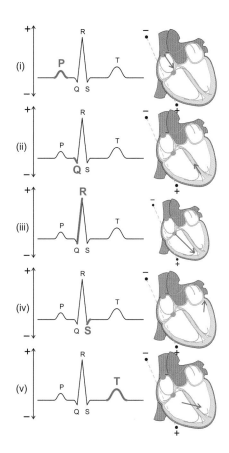

Figure 25.15 The heart vector at various times during the cardiac cycle.

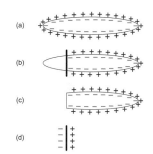

Figure 25.16 (a) Polarised cell. (b) The cell depolarises from one end. (c) The charge distribution in (b) can be thought of as the sum of (c) and (d).

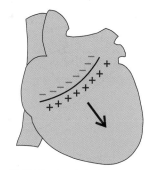

Figure 25.17 The depolarisation wave sweeps down the heart.

Before the wave of electrical activity (Figure 25.17) that causes atrial contraction, all the heart muscle cells are polarised. When the nearby nerve cells send a signal, this causes the muscle cell membrane to become permeable to the charges that are sitting on the surfaces. The charge distribution that results as this depolarisation sweeps along the cell (see Figure 25.16 (b)) can be thought of as the sum of a similar charge distribution to the original one, but truncated (c), and an extra bit that is like a flat membrane with a reversed charge distribution (d). The net result is that the charge distribution looks like a line of separated positive and negative charges that sweep down the heart as in Figure 25.17.

The travelling electrical signal can be modelled as an electric dipole (a separation of positive and negative charge) that changes strength and size with time. This dipole creates an electrical potential throughout the body cavity, and causes a pattern of equipotential lines on the surface of the body. By using a series of electrodes placed on the skin, the potential differences can be measured, and the evolution of the dipole moment (the 'heart vector') can be measured.

The electrical potential differences that are seen on the surface of the skin are of the order of a few millivolts. Voltages of this size are easily detected. The measurement of these voltages by the placing of electrodes on the skin is called an **electrocardiogram**

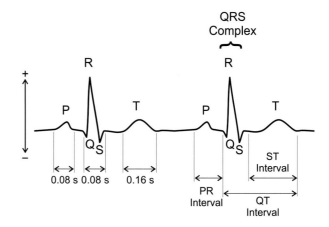

Figure 25.18 A ECG trace showing a complete cycle with the P, Q, R, S and T peaks.

or **ECG**, and is the primary tool used for detecting abnormal heart rhythms. The more common ECG machines in use measure the potential at the end of the arms and left leg, and in six locations across the chest, mostly on the left side. These signals can be plotted separately, but are usually combined in a single trace, as shown below on a 'normal' ECG. The first bump (the P wave) is caused by the depolarisation of the atria. The sharper feature in the middle (the QRS complex) shows the depolarisation of the ventricles (and masks the repolarisation of the atria). The last bump is the T wave, caused by the repolarisation of the ventricles (see Figures 25.19 and 25.20).

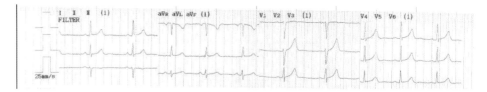

Figure 25.19 The individual voltage traces from a six-lead ECG machine.

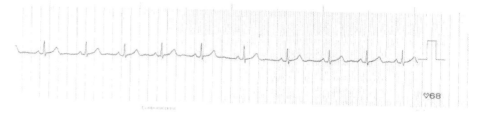

Figure 25.20 The combined trace from a normal ECG.

25.8 Summary

Key Concepts

electrical potential (V) The potential energy per unit charge at each point in space. The electrical potential is a scalar field. The SI unit of electrical potential is the volt, symbol V. One volt is equivalent to one joule per coulomb.

electrical potential difference (ΔV) The difference in electrical potential between two points. The potential difference is measured in volts (V). The electrical potential energy is often just referred to as the 'voltage'.

electrical potential energy (U) The potential energy stored in a system of charges. The electrical potential energy may be thought of as the energy required to bring the charges to positions they occupy. The electrical potential energy is also the maximum amount of

www.wiley.com/go/biological_physics

work that a system of charges may do if unconstrained. It is usually given the symbol U and is measured in joules (J).

equipotential lines Lines of equal electrical potential. Equipotential lines are perpendicular to the direction of the electric field.

Equations

$$\text{KE}_f + U_f = \text{KE}_i + U_i + W_{\text{applied}}$$

$$\Delta U = -W_{\text{elec}}$$

$$\Delta V = \frac{\Delta U}{q}$$

$$E = -\frac{\Delta V}{\Delta x}$$

25.9 Problems

25.1 A 20000 N C^{-1} uniform electric field does +5000 J of work on a +0.20 C charged object.

(a) Did the charged object move in the direction of the electric field or against it?

(b) How far did the object move?

(c) What was the change in electrical potential through which the object moved?

(d) If the object was initially at at point with an electrical potential of −2000 V, what was the electrical potential its end point?

25.2 A proton is moved at a constant velocity from a position at which the electrical potential is 100 V to one at which the electrical potential is −50 V.

(a) How much work was done on the proton by the electric field?

(b) How much work was done on the proton by the external force?

25.3 In a region of space there is a uniform 6000 N C^{-1} electric field like that shown in Figure 25.21.

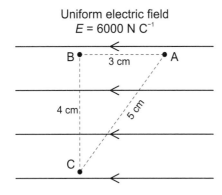

Figure 25.21 Three points in a region of uniform electric field.

(a) What is the potential difference between points A and B? Which point is at the lower electrical potential?

(b) What is the potential difference between points A and C? Which point is at the lower electrical potential?

(c) What is the potential difference between points B and C? Which point is at the lower electrical potential?

25.4 A charge of +0.1 μC is placed at point A in Figure 25.21.

(a) How much work is done on the charge by the electric field when moving the charge from point A to point B?

(b) How much work is done on the charge by the electric field when moving the charge from point B to point A?

(c) How much work is done on the charge by the electric field when moving the charge from point B to point C?

(d) How much work is done on the charge by the electric field when moving the charge from point C to point B?

25.5 What is the potential difference between points A and D in Figure 24.14?

25.6 Use Figure 25.22 to answer the following questions.

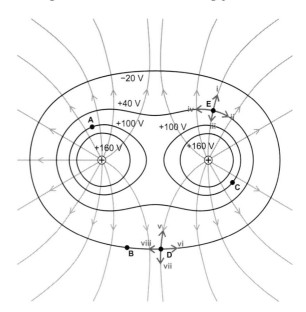

Figure 25.22 The electric field and some equipotentials in the region around two positive charges.

(a) What is the potential difference between points A and B?

(b) How much work does the electric field do on a −0.5 C charge that is moved from A to C?

(c) If a −5 C charge is released from point D which path would it take?

(d) If a +6 C charge is released from point E which path would it take?

25.7 A charge $Q_0 = -0.5\,\mu$C is placed in a region of space far from any other charges and is fixed so that it cannot move. Some equipotential lines around this charge are shown in Figure 25.23. A small object with a mass of $m_{obj} = 5$ mg and a charge $Q_{obj} = -0.8$ nC is placed at point A ($V_A = -800$ V) and released. The repulsive force between the two charges causes Q_{obj} to accelerate along the path shown towards point B. What is the velocity of the object when it reaches point B ($V_B = +100$ V)?

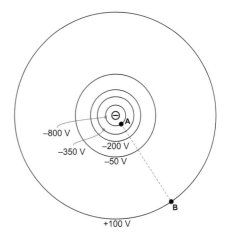

Figure 25.23 Equipotentials around an isolated point charge.

25.8 At rest, the potential inside a nerve cell is lower than that of the extracellular fluid. The membrane potential, the potential difference between the inside and outside of the cell membrane, is 70 mV.

(a) What is the change in electrical potential energy of the sodium ion when moving from inside the cell to outside the cell?

(b) How much work must be done on a sodium ion (Na^+) to move it from inside the cell to outside the cell?

25.9 An electron at an initial electrical potential of $0\ V$ is fired towards a second electron which is held *fixed* in space. The moving electron was fired at an initial speed of $1 \times 10^4\ m\,s^{-1}$. When the moving electron is $5.11 \times 10^{-6}\ m$ from the fixed electron it's speed has been reduced to $1 \times 10^3\ m\,s^{-1}$. What is the electrical potential $5.11 \times 10^{-6}\ m$ at the point $5.11 \times 10^{-6}\ m$ away from an isolated electron?

CAPACITANCE

26.1 Introduction

Picture charged plates carrying the same magnitude charge but with opposite signs separated by some distance. There is energy stored in this charge configuration, and there exists an electric field and a potential difference between the plates. We can define a quantity associated with such a charge configuration that is the ratio of the magnitude of the charge on each plate to the potential difference between the plates. We call this ratio the capacitance. With this definition, the smaller the potential difference created for a given magnitude of charge, the larger the capacitance of the pair of plates. In this way, the capacitance is a measure of the capacity of the circuit element to store charge. Circuit elements may be designed specifically to store charge, and these are called capacitors. In this chapter we will develop these ideas and show how to use them to calculate the capacitance of capacitors and the energy stored in them.

Key Objectives

- To understand the nature of electrical capacitance.

- To be able to describe the relationship between the stored charge, potential difference and capacitance of a capacitor.

- To be able to calculate the energy stored in a capacitor.

- To be able to understand how capacitors in circuits can be combined in series and parallel.

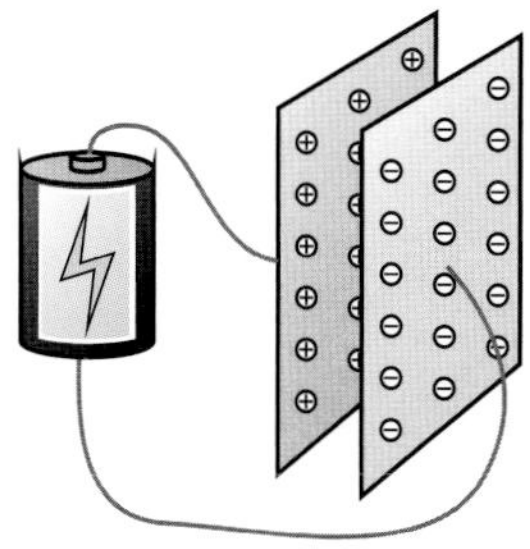

26.2 The Capacitor

A **capacitor** is a device which stores electrical potential energy in the form of a separation of some charge. The simplest devices have two metal plates, separated so that charge cannot flow between them. The conductors can be any shape, but are still referred to as plates. When a battery is connected to the plates, as shown in Figure 26.1, electrons are removed from one plate leaving it positive, and added to the other plate leaving it negative. A capacitor in such a state is said to be 'charged'. It is easy to see that connecting the charged plates together will allow the charges to move from one plate to the other until the plates are neutral again, so a capacitor is storing energy by separating charges.

The region of space between the plates has an electric field present (see Figure 26.2), and hence there is a potential difference between the plates. The derivation of the exact form of the electric field between the plates is outside the scope of the book. We therefore will state the result without derivation. For the case of plates separated by a vacuum

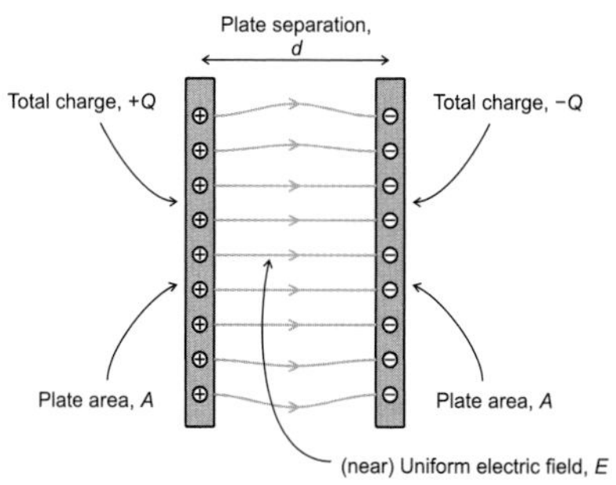

$$E = \frac{1}{\varepsilon_0} \frac{Q}{A} \qquad (26.1)$$

where E is the electric field strength, A is the area of each plate, Q is the magnitude of the charge on *each* plate and ε_0 is a constant known as the *permittivity of free space*

Figure 26.1 A simple capacitor.

Figure 26.2 A capacitor is formed by two conducting plates are separated by a distance d, and having equal magnitude $|Q|$, but opposite sign charge on each plate.

$(\varepsilon_0 = 8.854 \times 10^{-12} \text{ F m}^{-1})$. The field between two charged parallel plates is uniform; this is one way to generate the uniform fields discussed in the chapter on electrical potential energy and work.

Because the field is uniform, finding the potential difference between the plates is straightforward and was discussed in the previous chapter. If the plates are separated by distance d, the potential difference, V, is

$$V = Ed \tag{26.2}$$

so that

$$E = \frac{V}{d} \tag{26.3}$$

(Note that ΔV would probably be more correct here, but it is customary to use V for the potential difference in capacitor equations.)

Equating this expression (Eq. (26.3)) with the expression for the electric field between the plates of a capacitor (Eq. (26.1)), gives

$$\frac{1}{\varepsilon_0}\frac{Q}{A} = \frac{V}{d} \tag{26.4}$$

We can then rearrange this expression to give

$$\frac{Q}{V} = \varepsilon_0\frac{A}{d} \tag{26.5}$$

For a particular arrangement of plates, the values of A and d don't change. Thus there is some fixed quantity that we can associate with a particular arrangement of plates, and this relates the amount of charge that is stored on the plates to the potential difference between them. This is the **capacitance**, C, which we define

$$C = \frac{|Q|}{|V|} \tag{26.6}$$

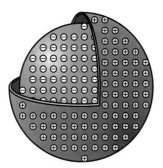

Figure 26.3 Two concentric conducting spherical shells can act as a capacitor. Provided the plate separation is much smaller than the radius of the sphere, we can even treat it as an approximate parallel-plate capacitor.

The SI unit of capacitance is the **farad**, symbol F. Most capacitors found in circuits have only small fractions of a farad capacitance, and so typically have values in the range from picofarads (pF, meaning 10^{-12} F) to microfarads (μF, meaning 10^{-6} F). For a **parallel-plate capacitor**

$$C = \varepsilon\frac{A}{d} = \varepsilon_r\varepsilon_0\frac{A}{d} \tag{26.7}$$

where A is the area of each plate, d is the separation distance and ε is permittivity of the material between the plates (ε_0 for a vacuum). The ratio of the permittivity of the material (ε) to the permittivity of free space (ε_0) is known as the **relative permittivity**, ε_r

$$\varepsilon_r = \frac{\varepsilon}{\varepsilon_0} \tag{26.8}$$

As mentioned earlier, a capacitor need not be constructed from parallel plates (see Figure 26.3). However, regardless of the shape of the capacitor, the capacitance is always Q/V.

In a biological cell, the intracellular fluid and the extracellular fluid are both conducting electrolytes, separated by the cell membrane which maintains across itself a potential difference of around 100 mV. The cell membrane is not a perfect insulator, but is still a much poorer conductor than the other media, so the cell membrane acts like a leaky capacitor.

Example 26.1 *Electric field between plates of a capacitor*

Problem: A simple parallel plate capacitor with a plate area of 0.5 m² is constructed and charged such that it holds +8 μC of charge on one plate and –8 μC on the other. The gap between the plates is filled with air.

(a) What is the magnitude of the electric field between the plates of this capacitor?

(b) What is the potential difference between the plates of this capacitor if they are 0.05 m apart?

(c) What is the capacitance of this capacitor if the plates are 0.05 m apart?

Solution: (a) We can find the electric field by using Equation (26.1). The charge on a capacitor is defined as the magnitude of the charge on each plate, so $Q = 8\,\mu\text{C}$. In this case with air filling the gap between the plates, $\varepsilon_r = 1$.

$$E = \frac{1}{\varepsilon_0}\frac{Q}{A} = \frac{1}{8.854 \times 10^{-12}\,\text{F}\,\text{m}^{-1}}\frac{8 \times 10^{-6}\,\text{C}}{0.5\,\text{m}^2}$$
$$= 1.8 \times 10^6\,\text{V}\,\text{m}^{-1}$$

(b) With $d = 0.05$ m, the potential difference between the plates will be

$$\Delta V = E\Delta x = 1.8 \times 10^6\,\text{V}\,\text{m}^{-1} \times 0.05\,\text{m} = 90 \times 10^3\,\text{V}$$

(c) The capacitance can be found using the Equation 26.1

$$C = \varepsilon_r\varepsilon_0\frac{A}{d} = 1 \times 8.854 \times 10^{-12}\,\text{F}\,\text{m}^{-1}\frac{0.5\,\text{m}^2}{0.05\,\text{m}} = 8.9 \times 10^{-11}\,\text{F}$$

As we have calculated the potential difference between the plates we could also have used $C = \frac{Q}{V}$ to answer this question.

26.3 Energy Stored in a Capacitor

In order to charge a capacitor, work needs to be done to separate the charges. Let's examine the charging process in more detail.

To begin with, the plates have no charge, and there is no potential difference. At a later time, the plates carry charge Q' and the potential difference $V' = Q'/C$. Now that the plates have charge, it is going to be more difficult than before to force more charge onto each plate, as the charge already there is repelling the new charge (see Figure 26.4). Recall that the work done moving a charge Q across a potential difference of V is

$$|W| = |\Delta U| = |VQ| \tag{26.9}$$

So to add charge ΔQ to the charge Q_1 will take work input

$$\Delta W = V'\Delta Q = \frac{Q'}{C}\Delta Q \tag{26.10}$$

provided that ΔQ is sufficiently small that V' is essentially unchanging while the new charge is added to the plates. The total work done to get a final charge of Q onto each plate will be the sum of all the little bits of work required to add on each little bundle of charge, which is the area under the plot of the voltage versus charge

$$W_{\text{total}} = \sum_{i=1}^{n}\Delta W_i = \frac{1}{2}\frac{Q^2}{C} \tag{26.11}$$

We can see the origin of this result graphically in Figure 26.5. Because the capacitance is constant, the graph of Q versus V is a straight line, starting at the origin. The increment of work to transfer ΔQ at a particular voltage is equal to the area of the little rectangle shown with area $W = V\Delta Q$. The total work is the area of all the rectangles,

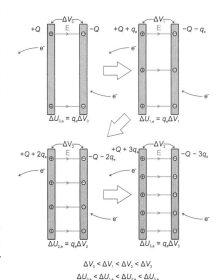

Figure 26.4 When charging a capacitor, each additional electron requires slightly more work to add or remove than the previous electron. Each time an electron is added or removed the charge on each plate increases, which increases the potential difference between the plates, which increases the work required to move the next electron.

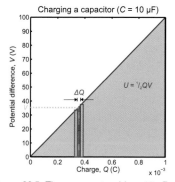

Figure 26.5 The energy stored in a 10 μF capacitor can be found from the area under the QV plot. As the capacitance is a constant, this plot is a straight line and the area is given by $U = \frac{1}{2}QV$.

i. e. the area under the line. This is a triangle with base length Q and height $V = Q/C$. Using area equals half base times height, we get the same result as above.

The energy stored in the capacitor is the same as the work done in charging it. Because it is a form of potential energy, U is the symbol typically used to represent the energy stored in a capacitor. Using $Q = CV$,

$$U = \frac{1}{2}\frac{Q^2}{C} = \frac{1}{2}QV = \frac{1}{2}CV^2 \qquad (26.12)$$

26.4 Capacitors in Series and Parallel

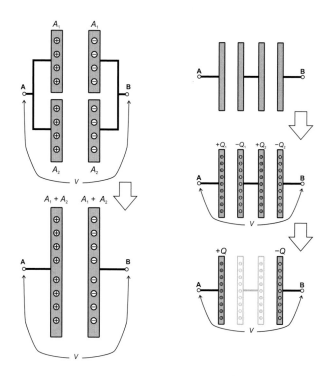

Figure 26.6 (Left) Adding capacitors together in parallel is like using single larger capacitor, increasing the total capacitance. (Right) Adding capacitors together in series reduces the total capacitance.

Just like with resistors, which we will come to in the next chapter, we often need to evaluate the overall capacitance of a number of capacitors grouped together. There are two basic ways this can be done, which we call **series** and **parallel**. Two capacitors are in series if they are placed consecutively in a circuit, so that any current flowing in the circuit will pass through each capacitor in turn. Two capacitors are in parallel if they are placed so that current must split and pass through either one capacitor or the other before rejoining and that the potential difference across each capacitor is the same. In diagrams, a capacitor is usually represented by two parallel lines, as shown in the summary in the next chapter (Figure 27.3), though sometimes one of the lines is drawn curved.

In the parallel case, the wire connecting all the positive plates together ensures that they are at the same potential, and the same is true of the negatively charged plates. In effect, it is like adding all the plates together to make one big pair of plates with the combined area of all the individual capacitors (see Figure 26.6). Because the capacitance is directly proportional to the plate area, it follows that the total capacitance is

$$C_{\text{total, parallel}} = C_1 + C_2 + C_3 + \dots \qquad (26.13)$$

When capacitors are wired in series, things are different. Imagine that the wire that connects the two middle plates in the diagram is made shorter until the plates are touching. In effect, we just have a piece of uncharged metal in the middle of two charged plates with a separation equal to the sum of the separations of the original two capacitors. As capacitance is inversely proportional to separation distance d, adding capacitors in series decreases the capacitance and the result is that

$$\frac{1}{C_{\text{total, series}}} = \frac{1}{C_1} + \frac{1}{C_2} + \frac{1}{C_3} + \dots \qquad (26.14)$$

To see this for the case of two capacitors in series, take a look at Figure 26.6. The diagram shows two capacitors. Before any voltage difference is applied between points A and B, the two middle plates and the wire connecting them are neutral. When a potential difference is created between A and B, the outer plates gather charge, and so the charge migrates between the two inner plates also. This means that whatever charge is left on the positive plate of the right-hand capacitor must be equal to the amount of negative charge that has moved to the negative plate of the left-hand one. In other words, $Q_1 = Q_2 = Q$. Clearly

$$C_1 = \frac{Q_1}{V_1} = \frac{Q}{V_1} \quad \text{and} \quad C_2 = \frac{Q_2}{V_1} = \frac{Q}{V_2}$$

To find the effective capacitance between A and B, we need the potential difference and the amount of charge separated. The charge on the outer plates is Q and the potential difference between A and B is $V_1 + V_2$. So

$$C_{\text{total}} = \frac{Q}{V_1 + V_2} \qquad (26.15)$$

Using $V_1 = Q/C_1$ etc.

$$C_{\text{total}} = \frac{Q}{Q/C_1 + Q/C_2}$$
$$= \frac{1}{1/C_1 + 1/C_2}$$
$$\frac{1}{C_{\text{total}}} = \frac{1}{C_1} + \frac{1}{C_2}$$

Example 26.2 *Energy in capacitors*

Problem: A capacitor of capacitance 50 mF is to be charged to a potential of 12 V.

(a) How much energy does this capacitor store when fully charged?

(b) How many electrons need to be moved from one plate to the other to charge this capacitor to a potential of 12 V?

(c) How much work is required to move the 10th electron from one plate to the other?

(d) How much work is required to move the 10^{18}th electron from one plate to the other?

Solution: (a) The energy stored in a capacitor is given by Equation (26.12)

$$U = \frac{1}{2}CV^2 = \frac{1}{2} \times 50 \times 10^{-3}\,\text{F} \times (12\,\text{V})^2 = 3.6\,\text{J}$$

(b) We can find the number of electrons that need to be moved by first finding the charge on each plate of the capacitor

$$Q = CV = 50 \times 10^{-3} \, \text{F} \times 12 \, \text{V} = 0.6 \, \text{C}$$

As each electron carries a charge of 1.6×10^{-19} C, 0.6 C corresponds to $\frac{0.6 \, \text{C}}{1.6 \times 10^{-19} \, \text{C per e}} = 3.75 \times 10^{18}$ electrons.

(c) We can use Equation (26.12) to find the amount of work done as the 10th electron is moved from one plate to the other. The charge already on the capacitor is $Q_0 = 9 \times 1.6 \times 10^{-19} \, \text{C} = 1.44 \times 10^{-18} \, \text{C}$

$$W = \frac{Q_0}{C} \Delta Q = \frac{1.44 \times 10^{-18} \, \text{C}}{50 \times 10^{-3} \, \text{F}} \times 1.6 \times 10^{-19} \, \text{C} = 4.61 \times 10^{-36} \, \text{J}$$

A very small amount of energy indeed!

(d) For the 10^{18}th electron, $Q_0 = 1 \times 10^{18} \times 1.6 \times 10^{-19} \, \text{C} = 0.16 \, \text{C}$

$$W = \frac{Q_0}{C} \Delta Q = \frac{0.16 \, \text{C}}{50 \times 10^{-3} \, \text{F}} \times 1.6 \times 10^{-19} \, \text{C} = 5.1 \times 10^{-19} \, \text{J}$$

This is still a very small amount of energy but is $10^{17} \times$ larger than the case in (c).

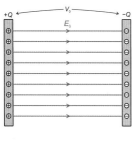

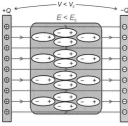

Figure 26.7 *(Top)* A capacitor without a dielectric. *(Bottom)* Inserting a dielectric into this capacitor reduces the electric field between the plates thereby increasing the capacitance.

26.5 The Dielectric in a Capacitor

When Michael Faraday (1791–1867), after whom the farad is named, was investigating capacitance, he realised that filling the space between the capacitor's plates with some material increased the value of the capacitance. A material that is put between the conductors for this purpose is called a **dielectric**. The factor by which the capacitance is increased relative to the value for a vacuum is known as the **dielectric constant**. Because the capacitance is higher, this means that for a particular potential difference, the presence of a dielectric increases the amount of charge that the plates carry. The dielectric also affects the electric field between the plates. The presence of the applied field causes the dielectric to become polarised (see Section 23.5). The extra electric field created by the polarised medium is in the opposite direction to the applied field (see Figure 26.7), and so the net field is smaller.

In a practical sense, the dielectric also helps by stopping the plates from touching, reducing the separation that can be used.

When choosing an appropriate dielectric, in addition to the dielectric constant, a major consideration is the *dielectric strength*. This is a measure of the maximum applied field the material can withstand before it breaks down and becomes conducting, measured in V m^{-1}. In recent years, good dielectric materials have been developed, leading to 'super capacitors'. Their ability to store electrical energy has many applications in the electricity generation sector.

Example 26.3 *Dielectric materials in capacitors*

Problem: A capacitor is constructed using an air gap of 1 mm between parallel plates.

(a) **What is the capacitance per square metre for such a capacitor?**

(b) **If the capacitor had an area of 0.75 m^2, what would it's capacitance be?**

The space between the plates of our capacitor is now filled with water ($\varepsilon_r = 80$).

(c) **What is the capacitance per square metre now?**

(d) **If the capacitor had an area of 0.02 m^2, what would it's capacitance be?**

Solution: (a) We can see from Equation (26.1) that the capacitance of a capacitor is directly proportional to the area of its plates. Because of this we can say that if a capacitor of $x\,m^2$ has a capacitance of C_x, then an otherwise identical capacitor of $2x\,m^2$ will have a capacitance of $2C_x$ and so forth. For the capacitor in this question, the capacitance per square metre is

$$C_{per\,m^2} = \varepsilon_r \varepsilon_0 \frac{1\,m^2}{d} = 1 \times 8.854 \times 10^{-12}\,F\,m^{-1} \frac{1\,m^2}{0.001\,m} = 8.9 \times 10^{-9}\,F$$

(b) So if the capacitor had an area of $0.75\,m^2$, it would have a capacitance of $0.75\,m^2 \times 8.9 \times 10^{-9}\,F\,m^{-2} = 6.6 \times 10^{-9}\,F$.

(c) Inserting a dielectric will increase the capacitance per square metre of the capacitor:

$$C = \varepsilon_r \varepsilon_0 \frac{A}{d} = \varepsilon_r C_0 = 80 \times 8.9 \times 10^{-9}\,F\,m^{-2} = 712 \times 10^{-9}\,F\,m^{-2}$$

(d) $C = 0.02\,m^2 \times 712 \times 10^{-9}\,F\,m^{-2} = 14 \times 10^{-9}\,F$

Example 26.4 *Breakdown potential*

Problem:

(a) **A large capacitor used to store energy in a physics lab is made of two concentric spheres spaced 1 mm apart. The radius of the capacitor is 0.75 m. If the space between the plates of the capacitor is filled with air ($\varepsilon_r = 1$) how much energy does this capacitor store when charged to the maximum 3000 V?**

(b) **A dielectric which has a relative permittivity of $\varepsilon_r = 1.2$ but a reduced 'breakdown potential' of $2.5 \times 10^6\,V\,m^{-1}$ is inserted into the space between the plates of the capacitor. How much energy is stored in the capacitor now?**

Solution: (a) The maximum potential difference that can be maintained between the plates of the capacitor depends upon the 'breakdown potential' (which is expressed in volts per meter, $V\,m^{-1}$) of the dielectric between them. Too large a potential difference between the plates can ionise the dielectric between them and result in a spark or electrical arc leaping from one plate to the other. This spark or arc will transfer charge from one plate to the other. This maximum potential difference is due to the breakdown potential of air. Air is capable of supporting a potential difference of $3 \times 10^6\,V$ *per meter* and so its 'breakdown potential' is $3 \times 10^6\,V\,m^{-1}$.

The energy stored is given by Eq. (26.12)

$$U = \frac{1}{2}CV^2$$

We know that $V = 3000\,V$, but we do not yet know the capacitance. We can find this by calculating the area of each plate (which will be approximately the same as $r \gg d$) and using Eq. 26.7.

$$A_{sphere} = 4\pi r^2 = 4 \times \pi \times (0.75\,m)^2 = 7.07\,m^2$$

$$C = \varepsilon_r \varepsilon_0 \frac{A}{d} = 1 \times 8.854 \times 10^{-12}\,F\,m^{-1} \frac{7.07\,m^2}{0.001\,m} = 6.26 \times 10^{-8}\,F$$

The energy stored is thus

$$U = \frac{1}{2}CV^2 = \frac{1}{2} \times 6.26 \times 10^{-8}\,F \times (3000\,V)^2 = 0.28\,J$$

(b) With the dielectric inserted, the capacitance of the capacitor changes to $1.2 \times 6.26 \times 10^{-8}$ F $= 7.51 \times 10^{-8}$ F. This is not the only effect of the dielectric, however, as it also has reduced breakdown potential of 2.5×10^6 V m^{-1}. With the 1 mm gap between the plates this corresponds to a maximum potential of 1×10^{-3} m $\times 2.5 \times 10^6$ V m$^{-1} = 2500$ V. The energy stored is now

$$U = \frac{1}{2}CV^2 = \frac{1}{2} \times 7.51 \times 10^{-8} \text{ F} \times (2500 \text{ V})^2 = 0.23 \text{ J}$$

26.6 Summary

Key Concepts

capacitor A device which stores energy by separating charge. One kind is the parallel plate capacitor which stores charge $+Q$ and $-Q$ on two parallel metal plates separated by air or a dielectric material.

capacitance (C) A measure of the amount of charge on each plate of a capacitor for a given electrical potential across that device. The SI unit of capacitance is the farad, symbol F.

dielectric An insulating material used in a capacitor, usually one which does not break down at large voltages. The dielectric material becomes polarised, causing the capacitor to have a larger charge accumulation for a given potential difference.

permittivity A quantity that describes how a material changes an electric field. The higher the permittivity of a material, the more the electric field within it is reduced. The permittivity is usually given the symbol ε. The relative permittivity, ε_r, gives the permittivity as a fraction of the permittivity of free space, $\varepsilon_0 \approx 8.854 \times 10^{-12}$ F m^{-1}.

dielectric constant Another term for the relative permittivity.

Equations

$$C = \frac{|Q|}{|V|}$$

$$C = \varepsilon \frac{A}{d} = \varepsilon_r \varepsilon_0 \frac{A}{d}$$

$$U = \frac{CV^2}{2} = \frac{Q^2}{2C} = \frac{QV}{2}$$

26.7 Problems

26.1 A 9 V battery is connected to a capacitor which subsequently has a *magnitude* of 0.5 µC of charge on each plate, with the charge on each plate having an opposite sign. What is the capacitance of this capacitor?

26.2 You have some metal shelving with two shelves, each of which measures 0.5 m × 0.2 m and the shelves are 0.3 m apart. (assume the shelves are electrically insulated from each other and any other objects)

(a) What is the capacitance of your shelves?

(b) How much charge would they hold if you connected each terminal of the shelves to a 1.25 V battery?

26.3 You wish to construct a capacitor which has a capacitance of 1 F. You intend to construct your capacitor using two parallel sheets of tinfoil held 1 mm apart by placing plastic wrap between them (the plastic wrap fills the space between the tinfoil). Both tinfoil and plastic wrap come in rolls which measure 30 cm by 10 m. The relative permittivity of the plastic wrap is $\epsilon_r = 2.9$.

(a) How many rolls of tinfoil and plastic wrap would you need? Is your plan feasible?

(b) How much charge could you store on this capacitor if you connect it to a standard 9 V battery?

26.4 A large industrial capacitor with an air gap between the plates stores 140 kJ of energy at a potential difference of 1200 V. A dielectric is inserted into the space between the plates of this capacitor and it can now store 11200 kJ of energy at 600 V. What is the relative permittivity of the dielectric inserted between the plates of the capacitor?

26.5 A particular capacitor stores 1 J of energy when charged to a potential difference of 12 V.

(a) What is the capacitance of this capacitor?

(b) What is the charge stored on this capacitor?

26.6 You measure the electric field between the plates of a 5 nC capacitor to be 2000 N C^{-1}. If the charge on this capacitor is 2 µC, how far apart are the plates?

26.7 You place books in the shelving described in Problem 26.2. The relative permittivity of paper is 2.4. How much charge does each of your bookshelves hold when connected to the same battery as before?

26.8 You have a sheet of paper which measures 15 cm × 20 cm and a piece of nylon sheeting which has the same dimensions. You charge these objects by rubbing them together and when you separate them and hold them parallel to each other and 1 cm apart you find that there is an electric field of magnitude 1500 V m^{-1} between them (you can assume that $\epsilon_r = 1$ for air).

(a) What is the potential difference between the paper and nylon sheets?

(b) What is the capacitance of these sheets in this position?

(c) What is the magnitude of the charge on each sheet?

(d) How much electrical energy is stored in the sheets?

The two sheets are now moved further apart to a separation of 5 cm.

(e) What is the capacitance of the sheets now?

(f) What is the potential difference between the sheets now?

(g) What is the magnitude of the electric field between the sheets now?

(h) How much electrical energy is stored in the sheets now?

26.9 Nerve cells maintain a charge separation across their cell membrane. The cell membrane of a particular cell is 10 nm thick and the cell can be modeled as a cylinder with a diameter of 12 µm and a length of 80 µm. If the potential difference across the cell membrane is 90 mV, what is the charge stored on the cell? (you can assume that $\epsilon_r = 1$ for the cell membrane)

26.10 How much energy is stored in the form of charge separation in the cell in Problem 26.9?

DIRECT CURRENTS AND DC CIRCUITS

27

27.1 Introduction

Electrical forces may be used to transfer energy and information between two points. The analysis of these processes using Coulomb's law is extremely complex. The theory of electrical circuits has been developed specifically to simplify these calculations. In this chapter we will introduce the basics of circuit theory.

Key Objectives

- To understand the nature of electric circuits.

- To understand the nature of the electrical current.

- To understand the idea of electrical resistance.

- To be able to use Ohm's law to calculate currents through and potential differences across a resistor.

- To be able to apply Kirchhoff's laws to analyse circuits.

- To be able to combine resistors in series and parallel into a single equivalent resistor.

27.2 Electric Current

Most of the common uses of electricity involve charges moving through conductors. The flow of charge is known as an **electric current**, symbol I. The electric current is a measure of the rate at which charge moves across a given crosssectional area

$$I = \frac{\Delta Q}{\Delta t} \tag{27.1}$$

where ΔQ is the net charge crossing the area in time Δt.

The SI unit of current is the **ampere**, symbol A, which corresponds to a flow of one coulomb of charge per second. The ampere is one of the base units of the SI system, so the size of coulomb is actually defined in terms of the ampere. One ampere is defined as the steady current that, when flowing in straight parallel wires of infinite length and negligible cross-section, separated by a distance of 1 m in free space, produces a force between the wires of 2×10^{-7} N per metre of length.

The direction of the current is defined to be the direction that positive charges would flow (see Figure 27.1). Often, the only charges that are actually moving are the negative electrons, so the direction of the so-called 'conventional current' is opposite to the way the electrons are moving.

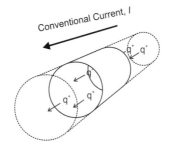

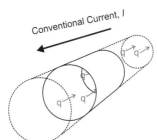

Figure 27.1 Current is a flow of charge, either positive charges (top), negative charges (bottom), or a combination of both. The 'conventional current' is in the same direction as a flow of positive charge, or in the opposite direction to a flow of negative charge.

27.3   Current Flow and Drift Velocity

Consider a small segment of metal wire, with length l and cross-sectional area A. The volume of the segment is $V = Al$. If the density of conduction electrons is n, then in the volume V there are nV electrons. Note here that most often when we talk about density, we mean the mass density or mass per unit volume, but in this case we mean the **number** density, the number of electrons per unit volume.

The total amount of mobile charge in the segment is equal to the number of charge carriers times the charge on each charge carrier, so

$$\Delta Q = qnV \tag{27.2}$$

Suppose each charged particle travels at velocity v. The time it would take for all the charge in this volume to cross the area at the end of the segment is the length of time it would take to travel a distance l at v.

$$\Delta t = \frac{l}{v} \tag{27.3}$$

and the current is therefore

$$I = \frac{\Delta Q}{\Delta t} = \frac{|q|nV}{l/v} = |q|nAv \tag{27.4}$$

This expression shows how the current is related to the number density of charge carriers, and how fast they travel on average (see Figure 27.2). To give some idea of scale, for a 1 mm-radius copper wire carrying 10 A, the average velocity of the electrons in the wire is 2.4×10^{-4} m s^{-1}. This is *very* slow. The slow speed of the charge carrier movement is emphasised by referring to it as the **drift velocity**.

Given that this velocity is so low, why does a light bulb turn on as soon as we flick the switch? If the electrons that made the light glow had to travel all the way from the power station, it would take a rather long time for the bulb to begin to glow. This is not what happens. Nearly as soon as the light switch is closed, a potential difference is created across the light bulb; at close to the speed of light, an electric field is produced throughout the circuit. All the electrons in the bulb begin to move almost instantaneously at once under the influence of this electric field and cause the light bulb to glow. Whether the source of power is a local battery or a power station kilometres away, what is supplied by such a source is not electrons – the power supply gives the electrons inside the light the energy to move. Electric power companies sell energy, not electrons.

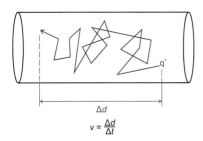

$$v = \frac{\Delta d}{\Delta t}$$

Figure 27.2 The average (or drift) velocity of the moving charges that constitute a current is much lower than the instantaneous velocity of these charges. This is due to the many collisions with atoms/molecules that occur as the charge moves along its path.

Example 27.1 ***Current and drift velocity***

Problem: A cylinder of fluid contains 4×10^{24} sodium ions, which are the only charge carriers, i.e., the only charged particles which are free to move. The dimensions of the cylinder are 20 cm long, with a circular crosssection of radius 5 mm. If a 5 A current flows along the length of the cylinder, what is the drift velocity of the sodium ions?

Solution:

$$I = nqvA$$

n is the charge carrier density, which is the number of charge carriers per unit volume

$$n = \frac{N}{V}$$

The volume of the cylinder is given by

$$V = l \times \pi r^2 = 0.2 \, \text{m} \times \pi \times (5 \times 10^{-3} \, \text{m})^2 = 1.57 \times 10^{-5} \, \text{m}^3$$

The volume of the cylinder is 1.57×10^{-5} m^3, so the charge carrier density is:

$$n = \frac{4 \times 10^{24}}{1.57 \times 10^{-5}} = 2.55 \times 10^{29} \, \text{m}^{-3}$$

Rearranging gives:

$$v = \frac{I}{nqA} = \frac{5}{2.55 \times 10^{29} \times 1.6 \times 10^{-19} \times \pi \times (5 \times 10^{-3})^2} = 1.56 \times 10^{-6} \, \text{m s}^{-1}$$

The drift velocity of the sodium ions is 1.56×10^{-6} m s^{-1}.

Example 27.2 *Current and charge carriers*

Problem: A copper wire of radius 2 mm carries a 10 A current. There is one conduction electron per copper atom. The density of copper is 8.94×10^3 kg m^{-3}. The molar mass of copper is 64 g mol^{-1}. What is the drift velocity of the conduction electrons in the copper wire?

Solution:

$$v = \frac{I}{nqA}$$

To determine n, we need to consider the charge carriers in copper. The charge carriers are electrons. In most metals there is one electron per atom which is free to move, so the charge carrier density is the same as the number of copper atoms per cubic metre. We can determine the number of copper atoms per cubic metre from the density and the molar mass of copper.

The number of moles of copper per cubic metre is equal to the mass of copper per cubic metre divided by the mass of a mole of copper.

$$\text{moles of copper per cubic metre} = \frac{\rho}{M} = \frac{8.94 \times 10^3}{0.064} = 140 \times 10^3$$

The number of copper atoms per cubic metre is equal to the number of moles of copper per cubic metre times Avogadro's number

$$140 \times 10^3 \times 6.02 \times 10^{23} = 8.43 \times 10^{28} \, \text{copper atoms per cubic metre}$$

The charge carrier density, n is the same as the number of copper atoms per cubic metre = 8.43×10^{28}. The drift velocity is then

$$v = \frac{10}{8.43 \times 10^{28} \times 1.6 \times 10^{-19} \times \pi \times (2 \times 10^{-3})^2} \, \text{m s}^{-1} = 5.9 \times 10^{-5} \, \text{m s}^{-1}$$

The drift velocity of the conduction electrons in the copper wire is 5.9×10^{-5} m s^{-1}.

27.4 Direct Versus Alternating Current

We are not going to cover the details of alternating current (AC) in this book, but for the sake of interest, it is worth pointing out a few things. The electricity that is available from the wall plugs in your home is AC, even though you have probably noticed that a lot of appliances you own run on direct current (DC), and you need a special power converter to plug them into the wall. In the early days of domestic electricity, there was fierce debate about whether the supply should be AC or DC, but AC has a very distinct advantage over DC – the use of transformers. A transformer is a device which can change the voltage of an AC power supply (with accompanying change in current). This makes it straightforward to transmit the power at high voltage (with lower current and hence less heating and energy loss) and to then transform the voltage down to something less hazardous at the user's end. An exception is the cable that transmits power from the South Island hydro lakes to New Zealand's North Island, which uses DC for economic reasons.

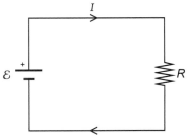

Figure 27.3 A (very) simple circuit containing just one battery and one resistor. Current is defined as flowing from the positive terminal of the battery to the negative terminal and (assuming zero resistance in the wires) the potential difference across the resistor is the same as the potential difference across the battery ($\mathcal{E}$). (Note: the 'actual' current of course is made up of electrons moving from the negative terminal to the positive terminal.)

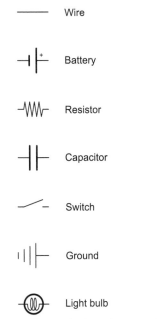

Figure 27.4 The usual schematic representation for common circuit elements.

27.5 Circuits and Circuit Diagrams

Generally, when we talk about an **electric circuit**, we are talking about some closed path through which charge may flow (see Figure 27.3). This path is composed of some combination of conductors and components such as resistors, capacitors, or batteries (see Figure 27.4). In this chapter we will represent circuits using circuit diagrams. These are diagrams which show the circuit elements being used and the order in which they are wired together to allow current flow.

We will begin our discussion of circuit diagrams by looking at simple combinations of power sources and resistors. In the next chapter we will expand this discussion to include capacitors.

27.6 Power Sources

Imagine for a moment a capacitor consisting of two parallel plates carrying charges $+Q$ and $-Q$. In the region of space between the plates there is an electric field, and a potential difference exists between the plates. If the plates were connected by a conducting wire, the charges would move from one plate to the other. This is a circuit, of sorts. However, in this case, the charges would eventually stop moving as the electric field got smaller and smaller. In most electric circuits, a device exists to maintain a potential difference between two places, so the flow of charges can be continuous. This is the role of a **battery** in a circuit. The battery in a circuit diagram is an idealised battery that maintains a specified potential difference between two points. Note that a particular battery may supply different amounts of *current* or *electrical power* depending on the circuit it is in, but it will maintain a certain *fixed potential difference* between the two points specified.

We have already seen that charged particles can move about under the influence of an electric field. The field does work on the charges, so some external energy source is needed to maintain this field. A non-electrical energy source that provides such energy is known as a source of **emf** or **electromotive force**. This emf is defined as the work done per unit charge by non-electrical forces. It is usually given the symbol $\mathcal{E}$. It is measured in volts, so it is not actually a force, but the historical name has stuck. For a particular device, the net emf is the energy gained per unit charge (U/Q) when a charge Q passes through that device and gains an energy U. The source of the energy may be electrochemical reactions (in the common battery), radiant energy (in the case of a solar cell) or thermal energy (in a thermocouple).

The purpose of a power source like a battery is to provide the energy to move charges. It produces a potential difference between two points in a circuit which causes the charges to move. This is why batteries are labelled with a voltage, not a current. A battery is *not* a source of charge; it does not store up electrons and then release them like water from a dam! It is more like a pump that takes in water as its input and raises the water up to greater height so that it can flow around a set of pipes back to the input.

A real battery of the type a consumer would buy at a supermarket has a fixed amount of energy that it may supply over its lifetime (or per charge if it is rechargeable). If you look at a battery (particularly a rechargeable one) you may see a rating in *milliamp hours* (mAh), which is a measure of the total charge (as current times time gives charge) and thus energy they can supply over their lifetime. If the current drawn from the battery is high, the length of time this can be kept up is reduced.

In circuit diagrams, a voltage source such as a battery is represented by two parallel lines of different length. The longer line is the positive terminal, that is, the terminal that is at the higher potential. The other is the negative terminal.

27.7 Resistance and Ohm's Law

It has been observed that for many materials there is a linear relationship between the potential difference across the material and the current flow through an object made of that material. That is, $V \propto I$. This relationship is known as **Ohm's law**. To turn this

www.wiley.com/go/biological_physics

relationship into a useful equation, we associate a property known as resistance with the object. This **electrical resistance**, R, is defined as

$$R = \frac{\text{potential difference between the ends of the object}}{\text{current flowing through the object}} = \frac{V}{I} \qquad (27.5)$$

This resistance is the opposition to the flow of electrical current through an object, which causes electrical energy to be converted to heat; an object which does this is called, not surprisingly, a **resistor**. Resistance is measured in **ohms**, symbol Ω.

Not all circuit elements obey Ohm's law (see Figure 27.5), and those for which I is not proportional to V are known as non-ohmic. One example is the **diode**, which is circuit element that has a resistance that depends on the direction of current flow, being very high in one direction and very low in the other. The graphite you would find in a pencil also shows non-ohmic behaviour.

27.8 Resistors and Resistivity

A resistor is a circuit device that has resistance. The symbol used in circuit diagrams is a zig-zag line as shown in Figure 27.6.

When a potential difference is established between the ends of a resistor, current flows through it, with the relationship between them being given by Ohm's law

$$V = IR \qquad (27.6)$$

For simple resistors, the value of the resistance R depends on the material the resistor is made from, and the shape into which it is made. It should be readily apparent that if two identical resistors are placed end on end (i.e., in series), to maintain the original current flow would require that the same potential difference be applied across each resistor, and hence the total potential difference would be twice that through each individual resistor. So the total resistance of this arrangement is twice the resistance of each resistor, suggesting that the resistance is proportional to length.

Now imagine two identical resistors side by side (i.e., in parallel), with the same potential difference across each. In this situation, the same current will flow through each resistor, just as it would if the other wasn't there. This means that the total current flow is twice that which flows through each single resistor. For the same potential difference, twice the current flows, so the total resistance is in effect halved. This indicates that doubling the cross-sectional area of a resistor halves the resistance, i.e., resistance is inversely proportional to cross-sectional area.

The resistance can therefore be expressed in terms of material and shape properties by

$$R = \rho \frac{l}{A} \qquad (27.7)$$

where A is the cross-sectional area, l is the length of the resistor and ρ is called the **resistivity** which is a property of the material. Resistivity is measured in Ω m. Good conductors have low resistivity (about 10^{-8} Ω m) and good insulators have very high values (more like 10^{14} or 10^{15} Ω m). Table 27.1 gives some values of the electrical resistivity for various metals.

Most materials have a resistivity that has some dependence on temperature. This is the basis of the thermistor for measuring temperature that was mentioned in Section 17.3.

When examining circuit behaviour, we often need to determine the net effect of a certain configuration of resistors. We'd like to know what single resistor could replace the ones that are there such that nothing else would change, that is, the voltages and currents in the rest of the circuit would be unaffected. As for capacitors, there are only two ways that a pair of resistors may be wired together – in series and in parallel. From the discussion above, it can be seen that putting resistors in series will increase the resistance, as it is much like increasing the length of a resistor. Putting resistors in parallel will decrease the overall resistance, as it is like increasing the area through which current can flow.

> **Conductance**
>
> Resistance (R) a measure of the opposition to electrical current. The reciprocal property is conductance (G), the ease with which electric charges can flow. $G = 1/R$.

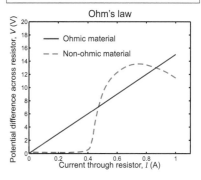

Figure 27.5 The linear relationship between current and potential difference for many materials is called Ohm's law. The plot shown is for a resistor with a resistance of 15 Ω. Also shown is the relationship between I and V for a material that is non-ohmic. A non-ohmic material is any material which does not obey Ohm's law.

Resistivity in $\mu\Omega$ cm at 20 °C	
Metal	Resistivity
Aluminium	2.824
Brass	7
Copper	1.771
Gold	2.44
Iron (99.98% pure)	10
Lead	22
Silver	1.59
Zinc	5.8

Table 27.1 Resistivity of selected metals

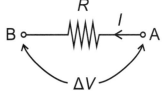

Figure 27.6 There is a potential difference of ΔV across a resistor of resistance R resulting in a flow of current of magnitude I through the resistor. (Note: The current is flowing from point A to point B, this implies that the electrical potential at point A is higher than that at point B.)

To examine the effect of a combination of resistors in detail, it will help to know the basic principles of DC circuits, known as **Kirchhoff's laws**. We will return to resistor combinations in series and parallel in Section 27.11.

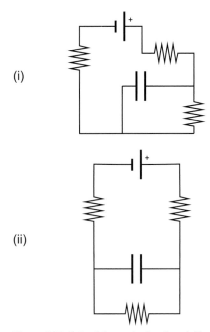

(i)

(ii)

Figure 27.7 Both of these circuits, (i) and (ii), are identical. When making the assumption that wires have zero resistance you may redraw a circuit however you wish provided you maintain the relative positions of circuit elements (series, parallel etc.) which maintains the electrical potential at each point.

27.9 Wires

In a circuit, the elements such as batteries, resistors, light bulbs, etc. are connected with **wires**. These may be plastic-coated pieces of copper, or printed silver tracks on a circuit board, but the purpose is to connect together two points in a circuit with as little resistance to current flow as possible, so low-resistivity materials are used. Half of the world's copper is used in the electrical industry, and something like 6000 million US dollars worth of gold is used each year in electronics. Of the 45% of the world's silver production that is used by industry, a large amount is used for electrical contacts.

In an ideal situation, the wires connecting circuit elements would have no resistance. In this case we can see that Ohm's law suggests that no resistance implies no potential difference between the wires ends. In other words, all the parts of a circuit that are connected by only wire are *at the same potential*. Because adding or removing some wire from a circuit diagram doesn't have any significant effect, as long as the correct arrangement of series and parallel elements is maintained, it can be useful to redraw a circuit to clarify things.

An example of how re-drawing a circuit can make it easier to interpret is shown in Figure 27.7. The diagrams show circuits that are drawn differently, but have identical behaviour in terms of currents and voltages through the resistors. Some ways of drawing diagrams can make it more clear for the purpose of analysis, so it is important to understand how these circuits are essentially the same circuit. A circuit diagram may be redrawn however you like, so long as the circuit elements are still connected together in the same way. Notice that in this example, the left side of the capacitor is connected to the left side of the lower resistor and the and the bottom of the top-left resistor both times. If the capacitor was to be connected by a wire directly to the battery, say, the circuit would not be the same.

In the real world, wires do have a small amount of resistance, and so there is some potential drop along them, and some energy lost in the form of heat.

Example 27.3 *Current and charge*

Problem:

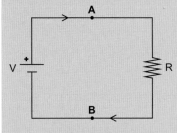

Figure 27.8 A simple circuit.

The current at point A in the circuit shown in Figure 27.8 is 3 A.

(a) In five minutes, how much charge passes point A?

(b) If the charge carriers are electrons, how many electrons pass point A in 5 min?

(c) What is the current at point B?

Solution: Current is the amount of charge passing a point in a given time, in other words

$$Q = I \times \Delta t = 3\,A \times 5\,min \times 60\,s\,min^{-1} = 900\,C$$

In 5 min (300 s), 900 C of charge pass point A. The number of electrons passing point A in 5 min is the number of coulombs of charge passing point A in 5 min divided by the charge on one electron.

$$N = \frac{Q}{q_e} = \frac{900 \text{ C}}{1.6 \times 10^{-19} \text{ C}} = 5.6 \times 10^{21}$$

The current at B is the same as the current at A, as each charge carrier that passes A must also pass point B. This is the case whenever two points are in series.

27.10 Kirchhoff's Laws

Kirchhoff's Law of Voltages

This rule is essentially the law of energy conservation applied to a circuit. It can be stated:

> **Key concept:**
> The sum of the directed potential differences around any closed loop is zero.

By directed potential differences, we mean that a direction around the closed loop must be chosen, and all the potential differences must be evaluated with respect to the direction chosen.

Resistance rule: Moving through a resistor in the direction of the current, the change in potential is $-IR$. In the opposite direction to the current the change in potential is $+IR$.

EMF rule: Moving through a source of emf, the change in potential is $+\mathcal{E}$ if going from the negative terminal to the positive, or $-\mathcal{E}$ if going positive to negative.

> **Internal resistance**
>
> The voltage difference between the battery terminals and the emf may not be the same. When current is flowing, a voltage drop will occur across any *internal resistance* in the battery. If we treat the battery as ideal, we can use $\mathcal{E}$ or V for the battery voltage.

Kirchhoff's Law of Currents

This rule is the law of charge conservation applied to a circuit. It says:

> **Key concept:**
> In any electrical circuit where no build up of charge is occurring, the sum of electric currents flowing into a point equals the sum of the electric currents flowing away.

The application of these rules can be seen in the end-of-chapter worked examples. We will now show how they can be applied to simple resistor configurations to derive rules for resistance in series and parallel.

Kirchhoff's law of voltages   Kirchhoff's law of currents

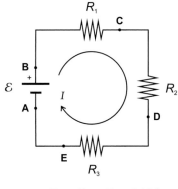

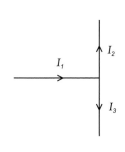

$$\mathcal{E} - IR_1 - IR_2 - IR_3 = 0 \text{ Volts} \qquad I_1 = I_2 + I_3$$

Figure 27.9 Kirchhoff's laws. Potential changes around a closed loop sum to zero, and the current into a junction equals the current out.

27.11   Resistors in Series and Parallel

Resistors in Series

Consider a simple **series** circuit with resistors R_1, R_2 and R_3 and an emf source, $\mathscr{E}$ (Figure 27.10). The current flowing around this loop is I. An equivalent circuit exists that has a single resistor R_S, but has the same emf and current flowing – this is the effective resistance that we are trying to find. For our series circuit, we can apply Kirchhoff's law of voltages

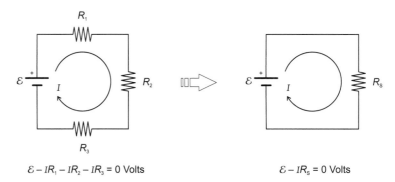

$$\mathscr{E} - IR_1 - IR_2 - IR_3 = 0 \text{ Volts} \qquad\qquad \mathscr{E} - IR_s = 0 \text{ Volts}$$

Figure 27.10 The three resistors R_1, R_1 and R_1 are in series. They can be replaced by a single resistor of resistance R_S such that the same current is drawn from the battery. The rule used to add resistors in series is $R_S = R_1 + R_2 + R_3$.

$$\mathscr{E} - IR_1 - IR_2 - IR_3 = 0$$
$$\mathscr{E} = I\left(R_1 + R_2 + R_3\right)$$

For the equivalent circuit with one resistor,

$$\mathscr{E} = IR_S$$

so by comparison

$$R_S = R_1 + R_2 + R_3$$

and in general, for any number of series resistors we have

$$R_S = R_1 + R_2 + R_3 + \ldots \tag{27.8}$$

Resistors in Parallel

Let's look now at three resistors in **parallel** (Figure 27.11). Applying Kirchhoff's current law to the junction point

$$I = I_1 + I_2 + I_3$$

The three resistors share a common potential difference, $V = \mathscr{E}$. Applying Ohm's law

$$I_1 = \frac{\mathscr{E}}{R_1} \text{ and } I_2 = \frac{\mathscr{E}}{R_2} \text{ and } I_3 = \frac{\mathscr{E}}{R_3}$$
$$I = \frac{\mathscr{E}}{R_1} + \frac{\mathscr{E}}{R_2} + \frac{\mathscr{E}}{R_3} = \mathscr{E}\left(\frac{1}{R_1} + \frac{1}{R_2} + \frac{1}{R_3}\right)$$

If the three resistors were replaced with a single resistor, R_P, Ohm's law gives us

$$I = \frac{\mathscr{E}}{R_P}$$

so

$$\frac{1}{R_{\mathrm{P}}} = \frac{1}{R_1} + \frac{1}{R_2} + \frac{1}{R_3}$$

and generalising to any number of resistors in parallel

$$\frac{1}{R_{\mathrm{P}}} = \frac{1}{R_1} + \frac{1}{R_2} + \frac{1}{R_3} + \cdots \qquad (27.9)$$

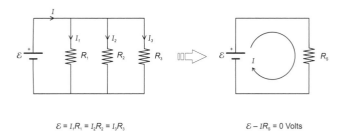

$\mathcal{E} = I_1 R_1 = I_2 R_2 = I_3 R_3$ $\mathcal{E} - I R_s = 0$ Volts

Figure 27.11 The three resistors R_1, R_2 and R_3 are in parallel. They can be replaced by a single resistor of resistance R_{P} such that the same current is drawn from the battery. The rule used to add resistors in parallel is $\frac{1}{R_{\mathrm{P}}} = \frac{1}{R_1} + \frac{1}{R_2} + \frac{1}{R_3}$.

Example 27.4 *Circuits*

Problem: In the circuit shown is Figure 27.12, an 18 V battery is connected in series with three resistors: a 3 Ω resistor, a 6 Ω resistor and a 9 Ω resistor. Calculate the following quantities:

(a) **The total resistance of the circuit.**

(b) **The current drawn from the battery.**

(c) **The current through the 3 Ω resistor, the 6 Ω resistor and the 9 Ω resistor.**

(d) **The voltage across the 3 Ω resistor, the 6 Ω resistor and the 9 Ω resistor.**

(e) **How is the voltage across the 3 Ω resistor related to the voltage across the 6 Ω resistor?**

Solution:

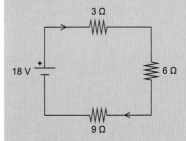

Figure 27.12 A simple circuit

(a) For resistors in series, the total resistance is the sum of the resistances in the circuit

$$R_{\mathrm{total}} = R_1 + R_2 + R_3 = 3 + 6 + 9 = 18\,\Omega$$

(b) The current drawn from the battery is given by Ohm's law as the battery voltage divided by the total resistance

$$I_{\mathrm{battery}} = \frac{V}{R_{\mathrm{total}}} = \frac{18}{18} = 1\,\mathrm{A}$$

(c) Resistors in series have the same current through them so the current through all the resistors is the same as that through the battery, or 1.0 A.

(d) The voltage across a given resistor is determined from Ohm's law, $V = IR$. The voltage across the 3 Ω resistor is $1\,\mathrm{A} \times 3\,\Omega = 3\,\mathrm{V}$. The voltage across the 6 Ω resistor is $1\,\mathrm{A} \times 6\,\Omega = 6\,\mathrm{V}$. The voltage across the 9 Ω resistor is $1\,\mathrm{A} \times 9\,\Omega = 9\,\mathrm{V}$. The sum of the voltages across the three batteries is 18 V.

(e) In series the same current flows through each resistor, so the voltage is shared between the resistors based on their size. The 6 Ω resistor has twice the resistance of the 6 Ω resistor so the voltage across it is twice the voltage across the 3 Ω resistor. Similarly the voltage across the 9 Ω resistor is three times the voltage across the 3 Ω resistor. The sum of the voltage drops across each of the resistors is the same as the voltage rise across the battery.

27.12 Power Dissipation

In an electrical circuit, energy is transferred from some source, such as a battery or generator, and is used by other devices such as resistors (the 'load'). Load devices change electrical energy into other forms like heat, light and movement. In most situations however, it is the *rate* at which the energy is moved about that is of most interest, more than the total amount. The rate at which energy is produced or consumed in circuits is called the **power** (symbol P).

$$\text{Power} = \frac{\Delta E}{\Delta t} = \frac{\text{Work done on charges}}{\text{Time}} = \frac{\text{Energy dissipated by load}}{\text{Time}}$$

Power is measured in units of watts, symbol W, where $1\,\text{W} = 1\,\text{J\,s}^{-1}$.

There is a straightforward relationship between power lost or gained, current and potential difference. The current flowing between two points in a circuit is a measure of how much charge passes a given point each second, that is

$$I = \frac{Q}{t}$$

The difference in electrical potential energy between these points is

$$\Delta U = QV$$

where Q is the charge that moved from one point to the next, and V is the potential difference between the two points. With the exception of a simple circuit, this is generally not the same as the potential difference across the battery. So, if power is the change in energy per unit time

$$P = \frac{QV}{t} = VI$$

The power dissipated by a resistor is

$$P = VI = \frac{V^2}{R} = I^2 R \qquad (27.10)$$

Here we have used Ohm's law to find several alternate ways of expressing the relationship.

Example 27.5 *Power*

Problem: In the circuit shown in Figure 27.12:

(a) **What is the power dissipated in the 3 Ω resistor?**

(b) **What is the power dissipated in the 6 Ω resistor?**

(c) **What is the power dissipated in the 9 Ω resistor?**

(d) **What is the power supplied by the battery?**

Solution: The power dissipated in a resistor is given by

$$P = VI$$

(a) The power dissipated in the 3 Ω resistor is: $P = 3\,\text{V} \times 1.0\,\text{A} = 3\,\text{W}$

(b) The power dissipated in the 6 Ω resistor is: $P = 6\,\text{V} \times 1.0\,\text{A} = 6\,\text{W}$

(b) The power dissipated in the 9 Ω resistor is: $P = 9\,\text{V} \times 1.0\,\text{A} = 9\,\text{W}$

The power supplied by the battery is the sum of the power dissipated (or stored if the circuit contains capacitors) in each of the circuit elements = 3 W + 6 W + 9 W = 18 W.

Alternatively, the power supplied by the battery is equal to the battery voltage times the current drawn from the battery = 18 V × 1.0 A = 18 W.

27.13 Alternate Energy Units

Normally, energy is measured in joules, but for convenience, some other units are in use when talking about electricity. A commonly used unit found on bills for domestic power supply is the *kilowatt-hour*. This is the amount of energy used by a 1 kW load in one hour:

$$1\,\text{kW h} = 1000\,\text{J s}^{-1} \times 3600\,\text{s} = 3\,600\,000\,\text{J}$$

Clearly, if you want to know how much energy a 100 W light bulb uses in four hours, the kilowatt-hour is a unit of energy which greatly simplifies things.

27.14 Electric Shock Hazards

Electric current passing through the human body can produce a variety of effects. Low currents may be detected as merely a mild tingling sensation, whereas higher currents can cause fatalities. The most important factors in determining the effect are the amount of current passing through the body, the path taken and the duration of the current flow. If the shock is produced by an alternating current (AC) source (like a wall socket), then the frequency of the current can be important as well.

The amount of current passing through the body is determined (through Ohm's law) by the voltage of the source and the resistance of the body. This resistance can vary greatly – dry skin is a good insulator whereas skin wet with sweat is a much better conductor. The resistance of the skin surface can vary from 10^3 Ω to 10^5 Ω or more. It is important to understand that a voltage source which may be safe under some conditions may be deadly under others. Table 27.2 summarises the effects of different currents, and lists the potentials which will produce these currents for the lower and upper possible values of the resistance of human skin.

When an electric current passes through any material with resistance to the flow of current, energy is dissipated in the form of heat. If current passes through the body, this can cause severe burns, which may appear on the skin, or may be internal.

Often a more important effect of a significant current is the disruption of the functioning of the central nervous system. Nerve cells in the body transmit electrical signals throughout the body to regulate a wide variety of body systems and functions. An electric current passing through the body can override these neural signals and cause involuntary muscle contractions. In the muscles, this involuntary contraction is called *tetanus*. Involuntary muscular contraction is particularly dangerous when the conductor supplying the current to the person is held in the hand. Since the muscles which close the hand into a fist are much stronger than the ones which open the hand, involuntary muscular contraction can make a person grip the current source harder, mak-

> **Bird on a wire**
>
> A small bird can sit on a high voltage power line without harm because both its feet are at the same potential. As long as there is no path to ground through the bird, everything is fine. Things are not so rosy for large birds like eagles – if they touch two wires the results are often fatal.

Electric current	Physiological effect	Voltage required	
		Resistance 10 000 Ω	Resistance 1 000 Ω
1 mA	Threshold of feeling	10 V	1 V
5 mA	Maximum harmless current	50 V	5 V
10 - 20 mA	Start of sustained muscular contraction	100-200 V	10-20 V
50 mA	Ventricular interference	500 V	50 V
100-300 mA	Ventricular fibrillation, possibly fatal	1000-3000 V	100-300 V
6 A	Sustained ventricular contraction followed by normal beat rhythm. Operating parameters for a defibrillator	60 000 V	6 000 V

Table 27.2 Physiological effects of electric shock

ing it very difficult to break the connection. The loss of control over motor function can persist for some time after the current has ceased – this is how stun guns or tasers work.

Even worse is the effect that the current can have if the path through the body is through the heart muscles. In the case of sufficient direct current, the heart can be forced into a sustained contraction which will cease blood flow around the body. Even if the shock current is not strong enough to cause sustained contraction, the effect on the nerve cells around the heart can still send the heart into a state of *ventricular fibrillation*. The heart flutters rather than beating properly and blood flow becomes ineffective. The risk of fibrillation is higher with alternating currents than direct currents. The defibrillating equipment used in hospitals utilises direct current. This is meant to temporarily cease fibrillation and give the heart a chance to resume a normal rhythm.

It is worth noting that most electrocutions (electric shocks that cause death) are due to ventricular fibrillation and that this condition is caused by quite a narrow range of current flows, around 50–200 mA. Higher currents, i.e., more than 300 mA generally cause burns and heat damage rather than electrocution.

27.15 Electricity in Cells

Cell Membranes

The functioning of cells, the building blocks of life, relies on the establishment and maintenance of potential differences across the membranes that surround the interior of the cell. In addition, signals sent along nerve cells are electrical; while the signals between cells are chemical. In a clinical setting, human life is often defined by the presence of electrical activity in the cells of the brain. We will describe here some of the basic functioning of the cell membrane, mostly focussing on the mechanisms of most importance to nerve cells.

The **cell membrane** itself consists of a phospholipid bilayer, that is a two-molecule thick membrane formed from a phosphate group and a hydrocarbon chain, i.e., a lipid. The lipid ends of the molecules are hydrophobic and the phosphate groups hydrophilic, and so a membrane is formed from two layers of phospholipid molecules, where the hydrocarbon ends of each molecule are in the middle of the layer, and the phosphate ends of each molecule are on the outside. Pure phosolipid bilayers are good insulators, having no free ions in the membrane, and the conductance per unit area is around $1 \times 10^{-13} \Omega^{-1} m^{-2}$. Real cell membranes have much higher conductances, having pores and ion channels that allow the transport of charge across the membrane in complex ways.

Because the cell membrane can maintain a separation of positive and negative charges, it has a capacitance. This capacitance is on the order of $1 \times 10^{-2}\,F\,m^{-2}$. This capacitance is not much influenced by the various biological processes occurring in

Signal speed

When researchers began trying to measure the speed of electrical signals through the body, they found that they did not move anywhere close to that they achieved down copper wire. Signals in the body travelled at a positively sedate pace – more like the speed of a pretty fast car – compared to the million times quicker that they could move through copper. The speed of transmission down nerve depends on its size, with larger meaning faster, so to find nerves large enough to study properly, pioneering physiologists needed to find a creature that was both long and with super-fast reflexes (and thus with nerve cells big enough to study). They found what the needed in the squid. Alan Lloyd Hodgkin and Andrew Huxley earned a Nobel Prize for their pioneering work on squid nerve cells.

a cell, and is close to what would be found for a simple lipid layer of the appropriate thickness.

When a nerve cell is inactive, an electrical potential gradient exists across the membrane, and this is called the **resting membrane potential**. When a nerve cell becomes active, the polarity of the charge across the membrane changes, and this is the **action potential**. The resting membrane potential is on the order of 70–90 mV, and the outside of the cell is positive with respect to the interior.

The cell is essentially a small bag of saline solution in a pool of saline solution, where the solutions inside (intracellular) and outside (extracellular) differ slightly in their chemical constituents, specifically in the relative abundance of various ions. In its normal resting state, the extracellular fluid is high in sodium (Na^+) and low in potassium (K^+), while the intracellular fluid is high in potassium, low in sodium. The exchange of substances between the cell interior and the environment is regulated by the cell membrane. If the cell membrane were made permeable to sodium, diffusion due to a concentration difference would provide a means to push sodium into the cell. The electrostatic potential across the membrane would also push the positive sodium ions into the cell. In the case of potassium, the concentration gradient would drive it out of the cell, while the electrostatic forces would drive it in.

The establishment of this separation of electrical charge and corresponding electrical potential difference requires energy, and it is created and maintained by the action of a pump – sodium and potassium ions are actively pumped across the membrane, with sodium pumped out and potassium in.

The cell membrane has many channels formed by proteins, known as ion channels, that can selectively allow the passage of ions through the cell membrane. In the resting state, these channels are closed and the membrane is impermeable to the flow of ions. Even though the chemical concentration gradient and electrical potential gradient would drive sodium into the cell, sodium cannot cross the membrane.

The action potential is a 1 ms long increase in the cell membrane's permeability to sodium. The sudden increase in permeability of the cell allows positively-charged sodium ions to flood into the cell, making the interior of the cell briefly more positive. This is followed by an increase in the cell membrane's permeability to potassium, and these ions subsequently move out of the cell. This cycle of changing potential is shown in Figure 27.13.

Circuit Models of the Cell and The Cell Membrane

The cell membrane itself can be modelled as a combination of driving potentials, resistors and a capacitor. The current is a combination of the movement of sodium, potassium and chlorine ions, and membrane potential V_m is the potential difference this creates between the interior and exterior of the cell. A commonly used model is shown in Figure 27.14.

In discussion the properties of cell membranes, conductance (G) is more often used that resistance. Conductance is the reciprocal of resistance, so Ohm's law becomes

$$I = GV \tag{27.11}$$

The force on an ion will depend on the equilibrium potential for the ion (E_{ion}) and the membrane potential and so

$$I_{ion} = G_{ion} \times (V_m - E_{ion}) \tag{27.12}$$

The electrical properties of the axon of a nerve fibre can also be modelled by a resistor-capacitor network, as shown in Figure 27.15. We will state without proof that the velocity at which a signal can move along such a cell depends on the relationship between the resistance across the membrane and the axial resistance along the nerve fibre, as well as the capacitance.

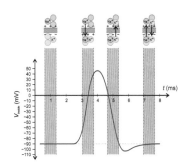

Figure 27.13 As the cell membrane becomes permeable to sodium allowing charge to flow across the membrane, the potential changes. This is known as the action potential.

Puffer fish

One of the most deadly toxins encountered in the animal kingdom – tetrodotoxin, produced by the Japanese puffer fish – causes the sodium pumps in the cell membrane to stop working. Without a means to create a potential difference between the interior and exterior of the cell, nerve cells are unable to perform their vital function, and death by suffocation results.

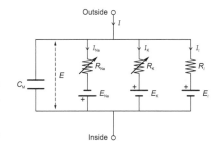

Figure 27.14 A Hodgkin-Huxley-style circuit model of the cell membrane. The arrows through the resistors indicate that they have variable resistances/conductances.

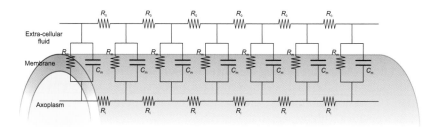

Figure 27.15 The transmission of a signal along a nerve axon depends on the relationship between axial resistance (along) and transmembrane resistance. It can be modelled like a network of resistors and capacitors.

Higher resistance across the membrane and lower capacitance are associated with higher speed. To allow fast signal conduction, either a reduced resistance along the axon or an increased membrane resistance is desirable. Lower axial resistance can be achieved with larger diameter axons, like those in squid. In complex organisms, rather than increased size, increased trans-membrane resistance and decreased capacitance increase the speed of signal transmission. Layers of insulating myelin play a crucial role in altering the resistance and capacitance in human nerve cells.

27.16 Summary

Key Concepts

electromotive force or emf ($\mathscr{E}$) The work done per unit charge by non-electrical forces. It is given the symbol $\mathscr{E}$ and is measured in volts. The source of the energy can be electrochemical reactions (as in the lead–acid battery), magnetic, thermal or radiant energy (as in a solar cell).

earth/ground (With reference to electricity) In electrical circuits, voltages are typically measured relative to a point that is considered to have zero potential, known as the ground or earth. This is often a direct physical connection to the Earth.

circuit Generally, a closed path through which current can flow, is composed of some combination of conductors and other components such as resistors, capacitors or batteries.

circuit element A single component of an electrical circuit, such as a resistor or a capacitor.

direct current (DC) Electric current that flows in one direction only.

alternating current (AC) Electric current that reverses direction periodically, usually many times a second. (Alternating currents are outside the scope of this book.)

charge carrier A particle carrying an electric charge which is free to move in response to an electric field, such as an electron or ion.

drift velocity The *average* velocity of charge carriers moving through a conductor under the influence of an electric field.

electrical resistance (R) The opposition to the flow of electric current through a material. Electrical resistance causes electrical energy to be converted to other forms such as thermal energy. Resistance is measured in units of ohms (symbol Ω).

Ohm's law The relationship between direct current, electrical resistance and applied voltage across a circuit element. The flow of direct current through a circuit element is proportional to the applied voltage. The constant of proportionality is called the resistance.

resistivity (ρ) A tendency of a material to oppose the flow of electrical current. The resistivity has the symbol ρ and is measured in Ω m.

Kirchhoff's law of currents In any electrical circuit where no build up of charge is occurring, the sum of electric currents flowing into a point equals the sum of the electric currents flowing away. This is a consequence of charge conservation.

Kirchhoff's law of voltages The sum of the directed potential differences around any closed loop is zero. This is a consequence of the conservation of energy.

electrical power (P) The rate at which energy is transferred, dissipated or absorbed by a circuit element.

resistors in series Two or more resistors are in series if electrical current goes through them sequentially.

resistors in parallel Two resistors are in parallel if the circuit branches splitting the current such that each resistor has the same potential difference across it and the circuit subsequently rejoins so the current recombines also.

Equations

$$R = \rho \frac{l}{A}$$

$$V = IR$$

$$I = |q| n A v$$

$$P = VI = I^2 R = \frac{V^2}{R}$$

$$R_S = R_1 + R_2 + R_3 + \ldots$$

$$\frac{1}{R_P} = \frac{1}{R_1} + \frac{1}{R_2} + \frac{1}{R_3} + \ldots$$

27.17 Problems

27.1 Sodium ions (Na^+) are flowing through a cylindrical ion channel which has a diameter of $0.85\,\mu m$ and is $5\,\mu m$ long. There is a potential difference of $225\,mV$ between the ends of the channel. The sodium ions have a drift velocity through the channel of $0.015\,m\,s^{-1}$ and in a period of $1\,ms$ a total of 15×10^6 ions exit the channel.

- (a) What total charge exits the channel in a time period of $1\,ms$?
- (b) What is the current in the ion channel?
- (c) What is the 'resistance' of the channel to the flow of sodium ions?
- (d) What is the number density, n, of ions in the channel?
- (e) How many sodium ions are in the channel at any one time?

27.2 Given that the potential is $+18\,V$ at the point shown, what is the electrical potential at the points A, B, C, and D in the circuit shown in Figure 27.16?

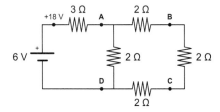

Figure 27.16 A circuit with 5 resistors and a single battery.

27.3 A simple circuit is constructed in which a $1\,\Omega$ resistor is connected across a $1\,V$ battery and so draws a $1\,A$ current. What is the current drawn from the battery if:

- (a) an extra $1\,\Omega$ resistor is connected in series with the existing one?
- (b) four extra $1\,\Omega$ resistors are connected in series with the existing one?

27.4 A simple circuit is constructed in which a $1\,\Omega$ resistor is connected across a $1\,V$ battery and so draws a $1\,A$ current. What is the current drawn from the battery if:

- (a) an extra $1\,\Omega$ resistor is connected in parallel with the existing one?
- (b) four extra $1\,\Omega$ resistors are connected parallel with the existing one?

27.5 Circuit A shown in Figure 27.17 consists of a single light bulb ($R = 288\,\Omega$) and a $12\,V$ battery.

- (a) What current is drawn from the battery?
- (b) What is the power dissipated in the light bulb?

An identical light bulb is now added to Circuit A in series with the first (see Circuit B in Figure 27.17).

- (c) What is the current drawn from the battery in this case?
- (d) What is the current passing through each light bulb?
- (e) What is the potential difference across each light bulb?
- (f) What is the power dissipated in each light bulb?
- (g) What is the total power supplied by the battery?

An identical light bulb is added to Circuit A in parallel with the first (see Circuit C in Figure 27.17).

- (h) What is the current drawn from the battery in this case?
- (i) What is the current passing through each light bulb?
- (j) What is the potential difference across each light bulb?
- (k) What is the power dissipated in each light bulb?
- (l) What is the total power supplied by the battery?

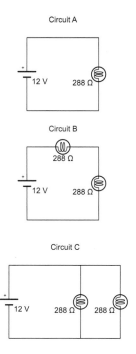

Figure 27.17 Three circuits showing different arrangements of light bulbs.

27.6 Five resistors are connected to a $24\,V$ resistor as shown in Figure 27.18.

- (a) What is the total resistance of the circuit?
- (b) How much current flows from the battery?
- (c) How much power is supplied by the battery?
- (d) What is the potential difference across each resistor?
- (e) What is the current through each resistor?
- (f) What is the power dissipated in each resistor?

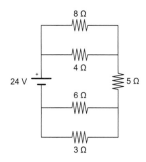

Figure 27.18 A circuit with 5 resistors in both series and parallel.

TIME BEHAVIOUR OF RC CIRCUITS

28

28.1 Introduction

An electrical circuit containing resistors and capacitors displays characteristic charging and discharging behaviour. This behaviour is seen in such biological systems as the changing potential across a cell membrane during neural activation and the discharging of the energy stored in a defibrillator through the resistance of the human body. In this chapter we will introduce the concepts and techniques needed to analyse this time-dependent behaviour.

Key Objectives

- To understand the time dependence of the charging and discharging of a capacitor.

- To be able to calculate the time constant of an RC circuit.

- To be able to calculate the time dependence of the charge, voltage and current in an RC circuit.

28.2 The RC Circuit

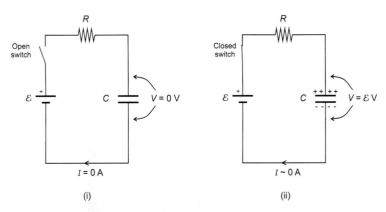

Figure 28.1 (i) A simple RC circuit before the switch has been closed. There is no current flowing, and the capacitor is uncharged (so there is zero potential difference between its plates). (ii) The switch is closed and after some time the capacitor is fully charged such that the potential difference across it is the same as the battery voltage. At this point in time there is also (approximately) zero current flowing.

The simple resistor circuits we've looked at so far have been ones with no time variation. The only values of current and voltage we were concerned with were the steady-state values. In many cases, such as biological cells, there is important variation of these values with time. This is often due to the presence of capacitance in these systems. Circuits that contain combinations of resistors and capacitors are known as **RC circuits**.

Consider a simple example with a resistor and a capacitor in series with a battery as shown in Figure 28.1. Suppose that the capacitor is initially completely uncharged. When the switch in the circuit is closed so that the capacitor and resistor are connected to the battery, charge builds up on the plates of the capacitor until the potential difference across it reaches the maximum value possible, $\mathscr{E}$, and the capacitor is fully charged. This charging doesn't happen instantly; we will find that it takes a greater or lesser time depending on both the capacitance of the capacitor and the resistance of the resistor. Similarly, when the capacitor is already fully charged and is connected in series with just a resistor, the charge flows off the capacitor over a time period that depends on the size of R and C.

28.3 Discharging RC Circuit

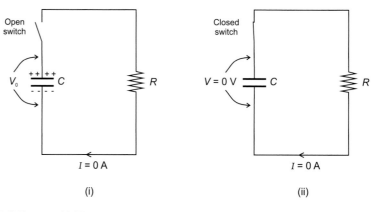

Figure 28.2 (i) The potential difference across the charged capacitor will cause a current to flow around the circuit. The current is proportional to the potential difference across the resistor (which is the same as that across the capacitor at all times). As current flows it reduces the potential difference across this resistor, which reduces the current and so on. (ii) After some time all of the charge on the capacitor is gone (there is zero potential difference across the capacitor) and there is no longer a current flowing around the circuit.

Consider the circuit shown in Figure 28.2, with the switch open with a capacitor that has been fully charged and then disconnected from the battery. What happens to the charge which is accumulated on the capacitor's plates? There is no net charge to this circuit; there are equal and opposite amounts of charge on the capacitor plates. When the switch closes, a current will flow around the circuit to neutralise the charge on the capacitor plates. That is, electrons will flow from the negative plate, through the resistor, to neutralise the positive charge on the other plate. Figure 28.3 shows how the charge on the capacitor, the potential difference across the capacitor and the current through the circuit change over time.

The speed with which this happens will be limited by the size of the current that can flow in the circuit. The size of the current that flows can be determined using Kirchhoff's laws, and will depend on the resistance of the resistor and the potential difference across this resistor.

The potential difference across the resistor will depend on the electric field due to the charge built up on the capacitor plates. This will change as the capacitor discharges, which means that the current will change over time. The greatest current will flow when the capacitor is fully charged (and the potential difference is highest). This current will decrease as the capacitor discharges.

We can use Kirchhoff's laws to determine the time dependence of the discharging current in this circuit. To begin with, we add up the potential drops around the circuit with the switch closed. The voltage loop rule tells us that the sum of these potential drops will be zero

$$\Delta V_{\text{capacitor}} + \Delta V_{\text{resistor}} = 0 \tag{28.1}$$

At any given instant, the current I flowing through the resistor is going to be

$$I = \frac{\Delta V_{\text{resistor}}}{R} \tag{28.2}$$

Discharging a capacitor

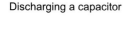

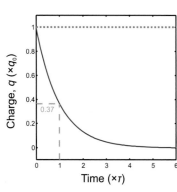

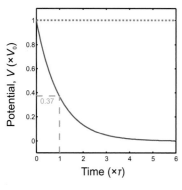

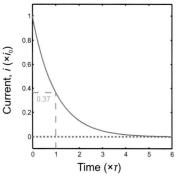

Figure 28.3 The charge on the capacitor, the electrical potential difference across the capacitor and the current in the RC circuit as a function of time for a discharging capacitor.

The potential difference across the capacitor is related to the capacitance C and the charge q by

$$q = C\Delta V_{capacitor} \qquad (28.3)$$

Putting this into Eq. (28.1) gives

$$\frac{q}{C} - IR = 0$$

The current I at any instant in time is related to the change in the amount of charge leaving the capacitor. Over a time interval Δt, the charge changes by Δq, and

$$I = \frac{\Delta q}{\Delta t}$$

Putting this all together

$$\frac{\Delta q}{\Delta t} = \frac{q}{RC} \qquad (28.4)$$

This equation fits a well-known form. In any situation where the rate of change of a value is proportional to the value, the quantity in question is changing exponentially with time. The result is an equation for the time-dependent charge, $q(t)$

$$q(t) = q_0 e^{-t/\tau} \qquad (28.5)$$

where the characteristic timescale is determined by

$$\tau = RC \qquad (28.6)$$

After the characteristic time, τ, has passed, we can see that

$$e^{-t/\tau} = e^{-1} \approx 0.37 \qquad (28.7)$$

and after 2τ

$$e^{-2t/\tau} = e^{-2} \approx (0.37)^2$$

After any time interval equal to τ, the amount of charge on the capacitor plates decreases to about 37% of its previous value. The current and voltage follow a similar relationship. Because the voltage across the capacitor is $V = q/C$, we have

$$V(t) = V_0 e^{-t/\tau} \qquad (28.8)$$

In this circuit, the voltage across the capacitor is the same as the voltage across the resistor (by Kirchhoff's voltage law), so using $V = IR$ it is also true that

$$I(t) = I_0 e^{-t/\tau} \qquad (28.9)$$

To summarise, for a discharging capacitor, the voltage, current and charge are all exponentially decaying with time, and the voltage across the resistor in series with the capacitor is also decaying.

28.4 Charging RC Circuit

For charging a capacitor, the situation is similar, but with some notable differences. Suppose now that the capacitor in the circuit is completely discharged, so there is no charge on the capacitor plates at all. At some time we call $t = 0$, the switch in the circuit is closed so that the resistor and capacitor are connected to the battery. The voltage loop law still applies and tells us that the sum of the potential changes around the circuit sum to zero, but now there is a new potential change to include – the increase in potential as we go through the battery. PLots of the charge on the capacitor, the voltage across the capacitor and the current in the circuit are shown in Figure 28.4.

The sum now becomes

$$\Delta V_{battery} + \Delta V_{resistor} + \Delta V_{capacitor} = 0$$

Time constant of the cell membrane

The capacitance per unit area of a typical cell membrane is $1~\mu F~cm^{-2}$. The membrane resistance is more varied, in the range 1–$100~k\Omega~cm^{-2}$. This puts the time constant for the cell membrane on the order of 1–100 ms.

Charging a capacitor

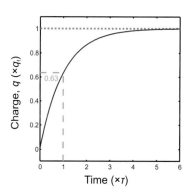

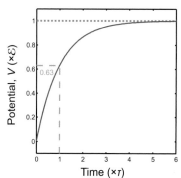

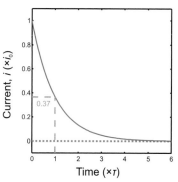

Figure 28.4 The charge on the capacitor, the electrical potential difference across the capacitor, and the current in the RC circuit as a function of time for a charging capacitor.

$$\mathcal{E} - IR - \frac{q}{C} = 0 \qquad (28.10)$$

Proceeding in a similar fashion to the discharging RC circuit, the change in the charge on the capacitor plates is determined by the current flowing *onto* the capacitor plates

$$\mathcal{E}C - \frac{\Delta q}{\Delta t}RC - q = 0$$

$$\Rightarrow \frac{\Delta q}{\Delta t} = \frac{1}{RC}\left(\mathcal{E}C - q\right) \qquad (28.11)$$

Applying a bit of calculus gives the result

$$q(t) = q_{\mathrm{f}}\left(1 - e^{-t/\tau}\right) \qquad (28.12)$$

where again $\tau = RC$ and the final amount of charge is $q_f = C\mathcal{E}$. In other words, with time, the charge on the plates gets closer and closer to a final value. It now takes a constant time, $\tau = RC$, to get to 63% of the final value, since

$$1 - (e)^{-1} = 1 - 0.37 = 0.63$$

The voltage across the capacitor follows a similar pattern. The current flowing around the circuit, however, must drop off to zero again when the capacitor is charged. As this happens, the voltage across the resistor gets less and less. As before

$$I(t) = I_0 e^{-t/\tau} \qquad (28.13)$$

Example 28.1 *Charging a capacitor (initial state)*

Problem: A capacitor of capacitance C and a resistor of resistance R are in series with a battery of voltage $\mathcal{E}$ and a switch. The capacitor is initially completely discharged and the switch is open. Describe each of the following quantities at this point:

(a) Current through the circuit.

(b) Voltage across the resistor.

(c) Charge on the capacitor.

(d) Voltage across the capacitor.

(e) Energy stored in the capacitor.

Solution: While the switch is open, no current can flow through the circuit and so (by Ohm's law) there can be no potential difference across the resistor. This indicates that the answers to parts (a) and (b) are 0 A and 0 V respectively.

The question also states that the capacitor is uncharged, i.e., the answer to part (c) must be 0 C. Given that there is no (excess) charge stored on the capacitor, and hence no charge separation across the plates, there can be no potential difference between them and so part (d) is 0 V.

If there are no excess charges on either plate of the capacitor and there is no difference in potential between the plates there can be no work done by this capacitor, and so the answer to part (e) must be 0 J.

Example 28.2 *Charging a capacitor (initial stages)*

Problem: The switch in the circuit from Example 28.1 is now closed. Describe each of the following quantities the instant after the switch has been closed:

(a) **Current through the resistor**

(b) **Voltage across the resistor.**

(c) **Charge on the capacitor.**

(d) **Voltage across the capacitor.**

(e) **Energy stored in the capacitor.**

Solution: Immediately after the switch has closed current is free to flow through the circuit onto one plate of the capacitor and off the opposite plate. As it does so the capacitor is charged.

Initially there is no charge on the capacitor and the instant after the switch has been closed there is still no charge on the capacitor ((c) 0 C).

Given this the potential difference across the capacitor and the energy stored in the capacitor must also be zero just after the switch has been thrown ((d) 0 V and (e) 0 J).The potential difference across the resistor must be the same as the potential difference across the battery ((b) $\mathscr{E}$ V) and so the current through the resistor is $I = \frac{\mathscr{E}}{R}$ ((a)).

Example 28.3 *Charging a capacitor (final state)*

Problem: It has now been some time since the switch in the circuit from Example 28.1 has been closed. Describe the same quantities as the previous examples a long time after the switch has been closed.

Solution: After the switch was closed current started flowing through the circuit and charging the capacitor. As the capacitor charged, the potential difference across the capacitor increased, thus reducing the potential difference across the resistor. This has the effect of reducing the current over time, and so reducing the rate of increase of the charge on the capacitor (see Section 28.4).

After a long time the potential difference across the capacitor approaches that of the battery ((d) ε V), which means that the potential difference across, and hence current through, the resistor must approach zero ((a) 0 A (b) 0 V).

The charge on the capacitor ((c)) is given by $Q = CV = C\varepsilon$ and the energy stored ((e)) by $U = \frac{1}{2}CV^2 = \frac{1}{2}C\varepsilon^2$.

Example 28.4 *Charging and discharging a capacitor*

Problem: A 3 µF capacitor and a 4 kΩ resistor are in series with a 12 V battery. The capacitor is initially uncharged.

(a) **What is the time constant of this circuit?**

(b) **What is the voltage across the capacitor after one time constant?**

(c) **What is the voltage across the capacitor after two time constants?**

(d) **What is the voltage across the capacitor after four time constants?**

Solution: The time constant, τ is given by

$$\tau = RC = 3 \times 10^{-6} \times 4 \times 10^3 \text{ s} = 12 \times 10^{-3} \text{ s}$$

The voltage across the capacitor after one time constant is 0.632 of the maximum voltage. The maximum voltage across the capacitor occurs when there is no current flowing, hence no voltage drop across the resistor, and is equal to the battery voltage, 12 V.

After one time constant the voltage across the capacitor = $(1 - 0.368) \times 12\,\text{V} = 0.632 \times 12\,\text{V} = 7.59\,\text{V}$.

After two time constants the voltage across the capacitor is $(1 - 0.368^2) \times 12\,\text{V} = 10.38\,\text{V}$.

After four time constants, the voltage across the capacitor is $(1 - 0.368^4) \times 12\,\text{V} = 11.78\,\text{V}$.

Example 28.5 *Capacitance*

Problem: A circuit contains an unknown capacitor, initially uncharged and in series with a 3 kΩ resistor and a 12 V battery. If after 16 ms, the current flowing in the circuit is 0.541 mA, what is the capacitance of the capacitor?

Solution: The initial current is $\frac{12\,\text{V}}{3 \times 10^3} = 4 \times 10^{-3}\,\text{A} = 4\,\text{mA}$.

After one time constant the current will have dropped to $4\,\text{mA} \times 0.368 = 1.472\,\text{mA}$.

After two time constants the current will have dropped to $4\,\text{mA} \times 0.368^2 = 0.541\,\text{mA}$.

This is the same as the current after 16 ms so we now know that the time constant of the circuit is 8 ms

$$C = \frac{\tau}{R} = \frac{8 \times 10^{-3}}{3 \times 10^3} = 2.67 \times 10^{-6}\,\text{F}$$

28.5 Summary

Key Concepts

RC circuit A circuit containing a combination of resistors (R) and capacitors (C).

RC time constant (τ) The characteristic time of an RC circuit. In a time equal to one time constant, an initially uncharged capacitor will charge to 63% of its maximum charge and voltage and the current flowing through the capacitor will drop to 37% of its initial value.

Equations

$$I(t) = I_0 e^{-t/\tau}$$
$$q(t) = q_0 e^{-t/\tau}$$
$$q(t) = q_\text{f}\left(1 - e^{-t/\tau}\right)$$
$$\tau = RC$$

28.6 Problems

28.1 An 18 μF capacitor has been charged to 100 V. A 15 kΩ resistor and a 5 kΩ resistor, are connected in series with the capacitor.

(a) What is the time constant of this circuit?

(b) Approximately how long (in terms of τ) will it take for the charge stored on the resistor to drop to 0.1% of its original charge?

Once the capacitor is fully discharged the 5 kΩ resistor is removed and replaced with a 12 V battery.

(c) What is the time constant of the circuit now?

(d) Approximately how long will it take for the charge stored on the resistor to rise to 95% of its maximum charge (in terms of τ)?

28.2 The circuit shown in Figure 28.5 is constructed using a 6 V battery, a 3 kΩ resistor, and a 6 kΩ resistor. Initially both switches are open and the capacitor is uncharged.

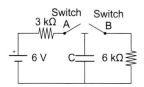

Figure 28.5 An RC circuit

Switch A is closed and the capacitor begins to charge.

(a) After $\frac{1}{2}$ second the potential difference across the capacitor is 3.78 V. What is the capacitance of the capacitor?

After several minutes switch A is opened.

(b) What is the approximate potential difference across the capacitor?

Switch B is now closed and the capacitor begins to discharge.

(c) What is the characteristic time for discharging this capacitor?

(d) What will the current through the 6 kΩ resistor be after 2 seconds?

(e) What will the charge on the capacitor be after 3 seconds?

28.3 A defibrillator can be modelled as a capacitor which discharges through the patient, inducing an electrical current in the chest. The resistance of the path which the electrical current takes through the chest of a typical adult is 50 kΩ. A particular defibrillator has a capacitance of 16.67 nF and is designed to be charged to an electrical potential of 900 V before discharging. The defibrillator includes a 'ballast' resistor of 40 kΩ which is connected in series with the patient.

(a) What is the maximum current I_{max} that will pass through the typical adult patient's chest?

(b) How long will it take the current passing through the typical adult patient's chest to drop below 1.37 mA?

You wish to redesign this defibrillator so it can be used on a child. The resistance of the typical child's chest is around 40 kΩ and the maximum current that be allowed through the chest is 8 mA. The current must drop to 0.15 mA in 4 ms. If the defibrillator is charged to a maximum potential of 900 V as before, then

(c) What is the required resistance that must be connected in series to the child patient's chest?

(d) What is the required capacitance of the defibrillator?

28.4 A 10 μF capacitor has been charged to a potential difference of 12 V. A 100 kΩ resistor is connected in series with the capacitor.

(a) What are the potential difference across the resistor, current through the resistor, and charge on the capacitor just after the circuit has been closed?

(b) What are the potential difference across the resistor, current through the resistor, and charge on the capacitor 1 second after the circuit has been closed?

(c) What are the potential difference across the resistor, current through the resistor, and charge on the capacitor 5 minutes after the circuit has been closed?

28.5

(a) How much energy does the defibrillator in Problem 28.3 deposit in the patient (before it has been redesigned)?

(b) What fraction of this total energy does this defibrillator deposit in the patient before the current drops below 1.37 mA?

28.6 The circuit shown in Figure 28.6 is used to charge a capacitor from a potential of 0 V to a potential of 95 V in a period of 1 s. Furthermore I_1, the current through resistor R_1, and I_2, the current through resistor R_2 satisfy the relation $I_2 = 3I_1$ at all times. What is the capacitance of the capacitor?

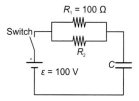

Figure 28.6 An RC circuit featuring a pair of resistors in parallel.

V

Optics

Introduction

Though we experience the world through many senses, perhaps the most evocative is vision. Vision is our long-range sense; we are able to perceive the world at far greater distances with sight than with any of our other senses. Vision is also the primary sense through which we measure the world. Optics is the study of light and its interaction with matter. It is light that allows vision, and an understanding of light and its behaviour is essential to an understanding of vision.

In this topic we will gain an understanding of what light is, how optical components such as lenses and mirrors interact with light, and how the human eye makes use of the properties of light to allow us to see. We will also study the eye as an optical instrument and come to an understanding of both the normal functioning of the eye and what happens when this function is disrupted.

The topic begins with a discussion of light as an electromagnetic wave. This first section will focus on the fundamental characteristics of light like wavelength, frequency and the speed of light. We will also discuss the scattering and absorption of light by matter and how these processes allow us to see the world around us. An understanding of the scattering and reflection of light will also allow us to understand how the world comes to be coloured. We will introduce the law of reflection, Snell's law of refraction and the phenomenon of dispersion. These last properties of light will allow us to understand the behaviour of mirrors and lenses, the subject of the next section.

The next section covers geometric optics. Geometric optics is a method for analysing optical systems that takes advantage of the fact that light propagates in straight lines under a wide range of conditions. Geometric optics will provide us with the tools we need to understand the way in which microscopes magnify objects so that they may be observed and measured by the human eye. Geometric optics also allows us to explain the functioning of the human eye.

The physics of human vision is the subject of the next chapter. Here we will come to an understanding of the normal function of the eye and some of the ways that defects in the eye may be corrected. We will discuss in particular how the eye's lens system allows the focussing of light onto the retina, and some of the limitations that this system places on eyesight.

In the final section of this topic, wave optics, we will investigate those situations where the straight-line approach to light propagation is insufficient. In these situations light must be treated as a wave and wave phenomena like interference and diffraction become important. An understanding of the wave properties of light is important for the understanding of the limitations both of optical instruments such as the microscope and the telescope, and of the human eye. These limitations are due to the limitations on the resolving power of optical systems due to diffraction by apertures.

THE NATURE OF LIGHT

29.1 Introduction

The study of light has been central to the development of physics for centuries. It was through a thought experiment involving the propagation of light that Albert Einstein first came to an understanding of special relativity. It was the analysis of the photoelectric effect, in which light ejects electrons from the surface of metals, that led Einstein to suggest the quantisation of energy and to postulate the existence of the photon. Light is the locus of much of the strangeness of modern physics. This can most clearly be seen in the dual nature of light; it is a wave and at the same time it is a particle. This is true of all matter, but it is particularly clear in experiments involving light. As such, optical experiments have always been at the forefront of the investigation of quantum mechanics. Light is also key to the strangeness in special relativity. Because the speed of light does not vary with the velocity of observers, some arcane behaviour can result from relativistic situations.

In this chapter we will discuss the nature of light. It will be identified as a wave, and the propagation of light will be seen as an example of wave propagation. However, we will emphasis that light is a wave unlike the other waves of our common experience. It is an electromagnetic wave, and some of the oddness of light will be shown to be a consequence of this.

Key Objectives

- To understand the basic nature of light.

- To understand the place of visible light in the electromagnetic spectrum.

- To understand how the change of light speed in different materials leads to refraction and dispersion.

29.2 Electromagnetic Waves

In the everyday world, there seem to be just two ways in which energy can be transmitted from place to place. Either an object physically moves from one place to another (transmission by particle) or energy can be sent as a disturbance through some medium (transmission by wave). Once the questions of how we see and what light really is began to be addressed by scientists, these two alternate models seemed to be the only sensible competing frameworks, and over the last few centuries the results of experiments were variously used to support one or other of the two views. The modern view is that *both* the wave and particle views are necessary for a full description of the behaviour of light, and this has resulted in the **photon** model of light, where we regard it as a stream of particle-like units which have wave properties.

For the purposes of understanding most common optical phenomena, however, a wave model is sufficient. Light is an **electromagnetic wave**, that is, light is a self-propagating combination of oscillating electric and magnetic fields. It can be shown that a changing electric field causes a changing magnetic field, and vice-versa. A 'waving' electric field causes a similarly waving magnetic field at right angles to it, and these

---

Radiation

The terms 'electromagnetic waves' and 'electromagnetic radiation' are used more or less interchangeably. Although the word radiation has acquired a somewhat negative connotation, the term electromagnetic radiation can be applied to harmless forms, such as radio waves or visible light, as well as gamma radiation (which falls in the dangerous 'ionising radiation' category).

(sinusoidally) oscillating fields travel through space at a fixed speed, known as c, the speed of light.

The existence of electromagnetic waves was predicted by a set of equations developed by James Clerk Maxwell in 1861 and extended in 1865. The equations were largely the work of previous scientists (notably Ampère and Faraday), but Maxwell's new derivation and additions enabled him to predict the speed of these waves and to show that light was an electromagnetic wave.

The Constant Speed of Light

In other types of wave motion, some medium is required in which the waves are able to propagate. To understand exactly what the wave medium does that allows the wave to propagate, take as an example the propagation of water waves. In a water wave, the up-and-down oscillations at one point in the wave cause neighbouring parts of the wave to oscillate up and down as well. The connection between these spatially separate parts of the wave is provided by the wave medium, in this case water. As one set of molecules moves up or down, the bonds between adjacent molecules drag the neighbouring molecules along as well. Clearly, the medium is required for this type of wave to propagate.

Light, on the other hand, is self-propagating, and this is due to the fact that it is an electromagnetic wave. The oscillation of the electric field in this wave causes the oscillations of the magnetic field, even when travelling through empty space. In turn, the oscillations of the magnetic field cause the oscillations of the electric field. It is this relationship between the magnetic and electric fields which connect spatially separate parts of the wave. Thus light is unlike other wave phenomena in that it is not a disturbance of a medium, like waves travelling through water, or sound waves passing through air.

This self-supporting process is only possible when the electromagnetic wave propagates at a certain speed, $299\,792\,458$ m s^{-1} (in a vacuum). If the electric wave propagated at any other speed, it would not produce a magnetic wave which would allow the combination of the electric and magnetic fields to be self-supporting. Similarly, if the magnetic wave propagated at any other speed, it would not produce an electric wave which allowed the combination to be self-supporting. This means that the speed of light is an extremely strange quantity; the measured speed of light does not change when the velocity of the emitter, or the device we use to measure it, changes.

For this reason, the speed of light (or any electromagnetic wave) is of fundamental importance in physics. It is the same for all observers, regardless of their direction or speed. As an example, consider two cars travelling along a straight stretch of road. Suppose they are travelling toward each other, and they are both travelling at a speed of 50 km h^{-1}. The driver of each car would see the other car approaching them at 100 km h^{-1} (i.e., about 28 m s^{-1}). Now suppose that each driver turns on their headlights. It may seem reasonable to presume that each driver would measure the speed of light from the headlights of the oncoming car as being $(299\,792\,458 + 28)$ m s^{-1}. This does not happen. In an enormous number of experimental tests it has been found that the speed of light is unchanged by changes to either the velocity of the observer or the velocity of the light source, and remains exactly $299\,792\,458$ m s^{-1}. This strange fact of nature is described by the **special theory of relativity**, which, while fascinating, is outside the scope of this book.

Because of its importance and invariance, the speed of light is defined as *exactly* $c = 2.99792458 \times 10^8$ m s^{-1}, and the metre is defined in terms of this speed. In many cases, approximating the speed of light as 3×10^8 m s^{-1} is sufficiently accurate.

Wavelength and Frequency

The rate of the oscillations is the **frequency** of a wave, f. As for any wave, the frequency is related to the **wavelength**, λ, by $v = f\lambda$, where v is the wave speed. For light in a vacuum, this gives us

$$c = f\lambda \tag{29.1}$$

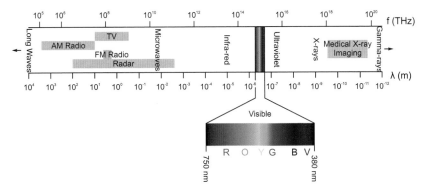

Figure 29.1 The electromagnetic spectrum.

where λ and f are the vacuum wavelength and frequency, respectively.

The Electromagnetic Spectrum

Electromagnetic waves can have a large range of wavelengths, and what humans regard as **visible light** occupies only a small piece of the **electromagnetic spectrum**, lying between about 380 nm and 750 nm. The spectrum itself stretches over a vast range of wavelengths/frequencies. Different frequency ranges in the spectrum are usually generated by different physical processes, and interact differently with matter, so are given different labels (such as X-rays, or radio waves), but they are all still electromagnetic waves. The names given to the different spectral regions are shown in Figure 29.1.

29.3 Reflection

When light hits the surface of almost any material, some of that light 'bounces back' off the surface. This is called **reflection** and is a very familiar phenomenon to the sighted – it is how we are able to see objects. For highly polished metallic surfaces, the amount of light that bounces off is very high. At the other extreme, an object that appears black under white light does so because it absorbs nearly all the light, reflecting very little back.

When light falls on a smooth surface from a particular direction, the reflected light also travels in a particular direction away from the surface. The reflected light leaves the surface at the same angle that the incident light falls on it. This is called the **law of reflection**.

> **Key concept:**
> The **law of reflection**: the angle of incidence is equal to the angle of reflection.

The angles are always measured from the **normal** to the surface. The normal is a line drawn perpendicular to the surface at the exact point where the light ray meets the surface. The normal, the **incident ray** and the **reflected ray** are all *coplanar*, that is, in three dimensions, they all lie in a single plane, as shown in Figure 29.2.

Reflection can be described as **specular** or **diffuse** (see Figure 29.3). Specular reflection is what happens when light hits a very flat, reflective surface, such as a mirror. All light coming from a single direction is reflected in a single direction. This occurs when the surface is smooth on a scale comparable with about quarter of the wavelength of the light. When the surface is rougher, as in the case of white paper such as that on which this text is printed, the light is reflected in a wide range of directions – this is called diffuse reflection. The law of reflection still applies in both cases, but when the surface is rough, the normals to different areas of the surface point in many directions, and so the reflections are fairly randomly oriented. It is this scattered light that allows us to see most objects.

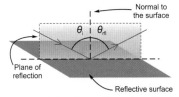

Figure 29.2 Light reflects off a surface. The incident angle (as measured from the normal to the surface in the plane of reflection) is the same as the angle of reflection.

> **Rays**
> In maths, a ray is a line that starts at a point and travels off in one direction to infinity. In optics, we use lines called rays to show the path and direction of travel of a light wave.

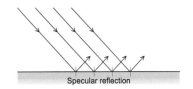

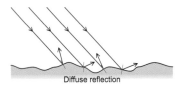

Figure 29.3 Specular reflection is when parallel light rays remain parallel after reflection. Smooth surfaces reflect in this manner. Diffuse reflection results is when parallel light rays are reflected at different, usually random, angles. Rough surfaces reflect in this manner.

29.4 Refraction

Refraction is the change in direction of a light ray at the interface between two media, which occurs when there is a change in *wave speed*. The frequency of the wave (how many cycles every second) stays the same, but the distance between successive peaks (i.e., the wavelength) changes. The wavelength decreases if the speed reduces and increases if it gets faster.

If the ray meets the interface along the normal, there is no change in direction. At any other angle, the ray is either bent, or does not travel into the next medium at all – this last case is called **total internal reflection**. In more complex cases where the change in speed is continuous rather than sudden, there is also a continuous change in ray direction. Materials that are uniform and have the same wave propagation speed everywhere are called *isotropic* media, and we will mostly restrict our discussions to these.

Instead of listing the speed of light in a particular material, data sheets will usually give the **refractive index**, n, of the material instead. This is the ratio of the speed of light in a vacuum, which is very close to the speed of light in air, to the speed of light in the material.

$$n = \frac{c}{v} \tag{29.2}$$

where c is the speed of light in a vacuum and v is the speed of the light wave in the material. Table 29.1 contains the refractive index of a selection of materials.

Material	n-value
Air (0 °C, 1 atm)	1.000293
Liquids at 20 °C:	
Benzene	1.501
Ethanol	1.361
Water	1.333
Solids at 20 °C:	
Diamond	2.419
Glass, crown	1.52
Glass, flint	1.66
Quartz, fused	1.458
Zircon	1.923

Table 29.1 Refractive index values of some selected solids, liquids and air. [Reprinted with permission from *College Physics*, Paul Peter Urone, Copyright (1990) Brookes/Cole.]

Example 29.1 *Refractive index*

Problem: When yellow light of wavelength 580 nm passes from air ($n = 1.00$) into water ($n = 1.33$), how do each of the following change (i.e., do they increase, decrease or not change), and what are their values in water?:

- **Speed of light.**

- **Wavelength of light.**

- **Frequency of light.**

Solution: The speed of light decreases, the wavelength of the light decreases, and the frequency is unchanged

$$v_{\text{water}} = \frac{v_{\text{air}} n_{\text{air}}}{n_{\text{water}}} = 2.25 \times 10^8 \text{ m s}^{-1}$$

$$\lambda_{\text{water}} = \frac{\lambda_{\text{air}} n_{\text{air}}}{n_{\text{water}}} = 435 \text{ nm}$$

The frequency is unchanged

$$f = \frac{v_{\text{air}}}{\lambda_{\text{air}}} = \frac{v_{\text{water}}}{\lambda_{\text{water}}} = 5.17 \times 10^{14} \text{ Hz}$$

Snell's Law

Snell's law relates the angle of incidence and **angle of refraction** for wave propagation at the boundary between isotropic media (see Figure 29.4). The ratio of the sines of the angles is the same as the ratio the wave speed in the two media, and is inversely related to the refractive indices. (See Section 32.3 for a mathematical explanation of why this is so.)

$$\frac{\sin\theta_1}{\sin\theta_2} = \frac{v_1}{v_2} = \frac{n_2}{n_1} \tag{29.3}$$

where θ_1 and θ_2 are the angles the ray makes with the normal to the boundary in the two media, v_1 and v_2 are the wave speeds, and n_1 and n_2 are the refractive indices. This is more often seen in a simplified form

$$n_1 \sin\theta_1 = n_2 \sin\theta_2 \qquad (29.4)$$

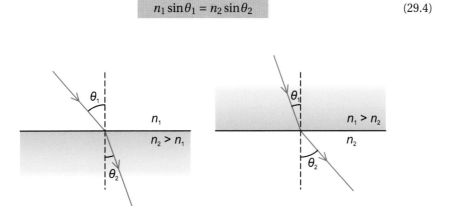

Figure 29.4 (Left) When light passes from a region of lower refractive index (such as a vacuum or air) to a region of higher refractive index (such as water or glass) so that $n_1 < n_2$, the light is bent towards the normal to the surface and $\theta_1 > \theta_2$. (Right) When light passes from a region of higher refractive index to one of lower refractive index so that $n_1 > n_2$, the light is bent away from the normal, and $\theta_1 < \theta_2$.

Example 29.2 *Snell's law*

Problem: Light is incident on a water surface at an angle of 30° to the normal. What is the angle of the light to the normal in the water? The refractive index of air is 1.00. The refractive index of water is 1.33.

Solution: We use Snell's law to solve this problem

$$n_{air} \sin\theta_{air} = n_{water} \sin\theta_{water}$$

So

$$\theta_{water} = \sin^{-1}\left(\frac{n_{air} \sin\theta_{air}}{n_{water}}\right) = \sin^{-1}\left(\frac{1.00 \times \sin 30°}{1.33}\right) = 22°$$

Light travels through the water at an angle of 22° to the normal.

Total Internal Reflection

Total internal reflection is the complete reflection of an incident light ray at a boundary, with no transmission. The phenomenon of total internal reflection occurs only for waves incident on a boundary with a medium where the refractive index is reduced. If we label the media in the order the wave encounters them as 1 and 2, then if $n_1 > n_2$, we can specify a **critical angle**, θ_c, given by

$$\sin\theta_c = n_2/n_1 \qquad (29.5)$$

If the angle of incidence is *larger* than this critical angle, then none of the wave is transmitted through the boundary, and only reflection occurs (see Figure 29.5). In the case of a water–air boundary, the critical angle is 48.6°.

Total internal reflection is utilised in many optical devices. Optical fibres use total internal reflection to confine light to a narrow glass rod, allowing the light to be transmitted very long distances. The main source of loss is absorption in the glass, rather than loss through light leaving the fibre. Total internal reflection can hamper efforts to see into parts of the eye, as seen in Figure 29.6.

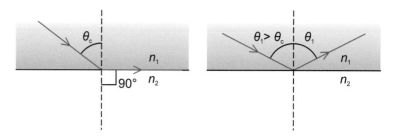

Figure 29.5 (Left) When passing from a region of higher refractive index to one of lower refractive index there is some critical incident angle θ_c for which the refracted angle $\theta_2 = 90°$. (Right) When the incident angle θ_1 is greater than θ_c, there are no (real) solutions to Snell's law (Eq. 29.4) and all of the incident light is reflected.

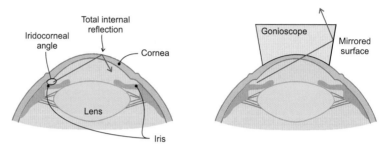

Figure 29.6 To diagnose and classify glaucoma, the ophthalmologist needs to view the angle between the iris and the cornea. Normally this angle cannot be directly viewed, as light originating in this part of the eye is totally internally reflected. In order to view the area, a prism called a gonioscope or a special contact lens is placed against the eye.

Example 29.3 *Critical angle and total internal reflection*

Problem: What is the critical angle for total internal reflection when light of wavelength 550 nm travels through plastic with a refractive index of 1.2 to air with a refractive index of 1.0?

Solution:

$$n_{air} \sin 90° = n_{plastic} \sin \theta_c \tag{29.6}$$

$$\theta_c = \sin^{-1} \frac{n_{air}}{n_{plastic}} = \sin^{-1} \frac{1}{1.2} = 56° \tag{29.7}$$

The critical angle is 56°.

Example 29.4 *Critical angle and total internal reflection*

Problem: Red light of wavelength 700 nm is travelling through water and is incident at an angle of 75° on a water–air interface. Describe what happens to this light.

Solution: To determine the critical angle we have

$$n_{air} \sin 90° = n_{water} \sin \theta_c$$

$$\theta_c = \sin^{-1} \frac{n_{air}}{n_{water}} = \sin^{-1} \frac{1}{1.33} = 49°$$

It gets totally internally reflected because it is incident at an angle greater than the critical angle.

29.5 Dispersion

Although when we defined the refractive index there was no mention of any dependence on wavelength, in real materials the refractive index is not a constant for all colours and wavelengths of light. Figure 29.7 shows the variation of refractive index for a specific kind of glass – each type of glass has its own characteristics.

For a material to be transparent, visible wavelengths of electromagnetic radiation must pass through without being absorbed. This transparency in the visible wavelengths is not very common for solids, which is why we can see most objects in our environment easily enough. Materials transparent to visible light will generally absorb in the wavelength bands on either side, in the infrared and ultraviolet. The refractive index of a material varies in a characteristic way around these wavelength bands where absorption occurs, being higher on the shorter wavelength side and lower on the long wavelength side. This increases the refractive index at the blue end of the spectrum, and the red end of the visible spectrum has a lower refractive index. When the wave speed is dependent on frequency, this is known as **dispersion**. Materials with this property are called *dispersive media*.

> **Glass**
>
> Glass is not a substance with a specific chemical composition, but rather a type of material with an amorphous, non-crystalline structure. The simplest recipe combines silica (SiO_2), soda (Na_2CO_3) and lime ($CaCO_3$). Other chemicals, such as boric oxide and lead oxide may be added, and these change the colour and refractive index. Hence glass does not have a single refractive index, but in the case of common, inexpensive glasses, a refractive index of about 1.5 is likely. Some glasses may have refractive indices of 2 or more.

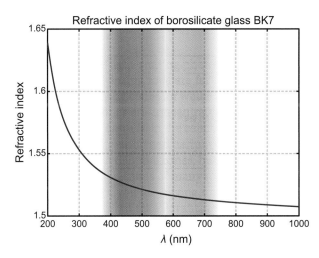

Figure 29.7 The refractive index variation with wavelength a type of glass used in lenses.

Example 29.5 *Dispersion*

Problem: When white light passes through a prism, blue light is bent more than red light. What does this tell us about how the refractive index of glass depends on the wavelength of light?

Solution: A larger change in angle indicates a larger difference in refractive index between the glass and air at the blue end of the spectrum. As the refractive index of glass is higher than that of air, this means that the refractive index of glass is higher for blue light than red light.

Examples of Dispersion

Chromatic aberration

Figure 29.8 shows how dispersion spreads out the different wavelengths passing through a prism. Light passing through lenses can be similarly affected, causing the focal length of lenses to differ for different wavelengths. This can make the images formed on film or a screen fuzzy. If the lens–screen distance is correct for the image to be sharply in focus for blue, parts of the image with red colour will tend to be out of focus.

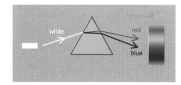

Figure 29.8 Dispersion in a prism separates white light into the visible spectrum. Blue light is refracted more than red light.

Optical fibres

Dispersion is a problem in optical communications. The varying speeds of the different wavelengths causes temporal spreading of a light pulse travelling through a transparent material such as an optical fibre. No light pulse is ever fully monochromatic (i.e., single coloured and therefore single frequency), so some wavelengths travel down a fibre

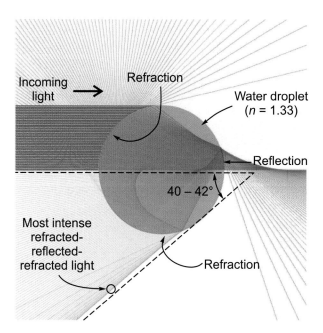

Figure 29.9 A rainbow is formed by light that is first refracted at the air–water boundary, then reflected at the opposite surface, and refracted a second time. Many of the rays entering near the top of the droplet end up coming out close together and nearly parallel after being refracted at the air–water boundary, then reflected at the opposite surface, and refracted a second time at the water–air boundary. It is this 'bunching' of rays in combination with dispersion that is responsible for the bright arc of reflected light we call a rainbow.

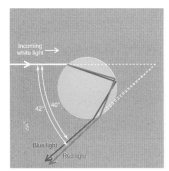

Figure 29.10 White light is separated into its component colours when it leaves the water droplet. Only the extreme ends of the spectrum, red and blue light, are shown here. Blue light is refracted more than red light and so ends up 'above' the red light exiting the droplet.

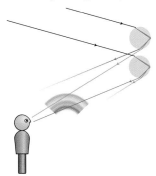

Figure 29.11 At first glance, Figure. 29.10 makes it look as though rainbows should be blue at the top, which is not what happens. The red light from those droplets higher up and further away hits the eye at a higher angle than the blue light from lower and closer droplets. This gives the familiar appearance of a rainbow.

faster than others. This causes the sharp on–off pulses required for digital communications to become spread out and information to be lost. Due to dispersion and loss of signal strength, it was previously necessary to put repeaters on optical fibres to collect and re-transmit the signals. Some of the materials now being used are able to transmit light pulses with minimal dispersion, and can even amplify the signal periodically, reducing the need for repeaters.

Gemstones

Diamond is valued for several reasons: it has a high refractive index (~2.4), it is the hardest substance, and it has high dispersion. It is the high dispersion that gives diamonds their 'fire', the flashes of colour that make them so pretty, so high dispersion is valued in gemstones.

Rainbows

For the primary rainbow, the light from the Sun enters the water droplets (and is refracted), a portion is reflected off the back surface, and then it exits again through the front surface (being refracted a second time). If the paths of a great many rays of light incident upon a spherical water droplet are plotted as in Figure 29.9, it can be seen that there are many approximately parallel and co-incident rays exiting the droplet at an angle of around 40–42° to the incident angle. It is this phenomenon, in combination with dispersion, that forms a rainbow.

Upon each refraction, the different wavelengths are bent varying amounts, with blue changing path more than red. As a result, the exact angle at which the refracted–reflected–refracted light is most intense is different for different wavelengths of light (it is 40° for blue light and 42° for red light). Figure 29.10 shows how the refraction and reflection off a curved surface in a water droplet cause white light to be split into its component colours. Figure 29.11 shows how this forms a rainbow.

Lightning

Some interesting dispersive effects can be observed in the radio waves that are generated by lightning strikes. Because a lightning strike is rapid, the radio waves are all broadcast at the same time, and so sound like a loud 'click' when played through a radio receiver. Some of the radio-wave energy escapes into space, and is guided by the Earth's magnetic field back to another part of the Earth's surface. The lower-frequency radio waves arrive a long time after the higher-frequency ones; what started out as a radio-wave burst lasting less than a thousandth of a second is dispersed into a signal which lasts one or two seconds. When played through a receiver and converted to sound waves, these sound like someone whistling a sliding scale from high to low frequencies, so they are called 'whistlers'.

29.6 Summary

Key Concepts

light Electromagnetic radiation in and around the wavelength range visible to humans, which is from around 380 nm to 750 nm. Even though they are not visible, the term 'light' is often also applied to the ultraviolet and infrared parts of the electromagnetic spectrum.

electromagnetic waves As a time-varying electric field generates a magnetic field and vice–versa, these oscillating fields together form an electromagnetic wave. All electromagnetic waves travel at the same speed in a vacuum, c, whatever their wavelength and frequency.

electromagnetic spectrum The range of possible frequencies of electromagnetic waves.

photon A discrete packet of electromagnetic radiation.

speed of light (c) All electromagnetic radiation travels at the same speed in a vacuum. This speed is denoted by the symbol c, and equals 299 792 458 m s^{-1} exactly.

wavelength (λ) The distance between two consecutive points on a wave that are in phase, measured in the direction of propagation.

frequency (f) The number of repetitions of a complete waveform per unit time. Measured in cycles per second, or hertz (symbol Hz).

reflection A change in the direction of light when it hits and is turned back from a surface or boundary. The angle of incidence is equal to the angle of reflection.

refraction The change in propagation direction of a wave due to a change in the wave speed as the wave passes from one transparent medium into another.

specular reflection When radiation is reflected from a surface that is flat compared with the wavelength of the radiation, and light coming in from a single direction is reflected in a single direction.

diffuse reflection When a surface is rough compared to the wavelength of the radiation, light coming from a single direction is scattered in many directions.

total internal reflection When the angle of incidence at a boundary from one medium (refractive index n_1) to a medium with a lower refractive index, n_2, is greater than a critical angle θ_c given by $\sin\theta_c = n_2/n_1$, none of the wave is transmitted through the boundary, and only reflection occurs.

dispersion Spreading. In optics, dispersion refers to the spreading of light due to different wavelengths travelling at different speeds, and having different refractive indices. This can cause angular separation of light by wavelength under refraction, and temporal spreading of a light pulse travelling through a transparent material.

monochromatic 'Single coloured'. Light which is monochromatic contains electromagnetic radiation with only a single wavelength. In the real world, no light is truly monochromatic, and a small spread of wavelengths is always present.

Equations

$$c = f\lambda$$

$$n = \frac{c}{v}$$

$$\theta_{\text{incidence}} = \theta_{\text{reflection}}$$

$$n_1 \sin\theta_1 = n_2 \sin\theta_2$$

$$\sin\theta_c = \frac{n_2}{n_1}$$

29.7 Problems

29.1 A diagnostic device uses a bright red laser light to illuminate structures just under the surface of the skin. Light from the laser passes first through the air, and then the skin, to scatter off the subcutaneous structures that are to be imaged. The scattered light passes back through the skin and into an optical device which forms an image of the scattered light on a CCD array. The laser light used has a wavelength of 633.0 nm in a vacuum. The refractive indices of air, the glass used in the imaging optics, and skin are 1.008, 1.700, and 1.381 respectively (use $c = 2.998 \times 10^8$ m s^{-1} for this question).

(a) What is the frequency of the red light when it passes through each material?

(b) What is the wavelength of the red light as it passes through each material?

(c) How fast does the red light travel through each material?

29.2 It is not possible to make images of, and therefore see, arbitrarily small objects using visible light. The minimum size of an object that can be 'seen' by light using conventional optics is roughly equal to a few times the wavelength of the light used. If a bacterium that is 1.2 μm across can *just* be seen using a particular optical system when the bacteria is floating in a watery solution ($n_{solution} = 1.35$), what will be the minimum size of bacterium that this optical system could 'see' in air ($n_{air} = 1.0$)?

29.3 Light strikes a mirror as shown in Figure 29.12. This mirror has another mirror placed at right angles to it. Such an arrangement of mirrors is known as a corner reflector. At what angle does the light get reflected back (i.e., what angle is the outgoing light at when it crosses the dotted line)?

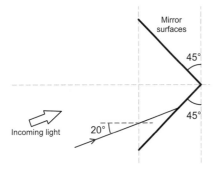

Figure 29.12 Two mirrors are placed a right angles to one and other. This arrangement of mirrors reflects light in a particular fashion, making them useful for a range of purposes.

29.4 The glass half-cylinder prism shown in Figure 29.13 is used to measure the critical angle for light of various wavelengths. For red light the critical angle measured was 36.78°. For blue light the critical angle was 36.28° (The refractive index of air is $n = 1.0$).

(a) What is the refractive index of the glass for red light?

(b) What is the refractive index of the glass for blue light?

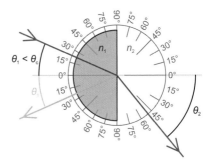

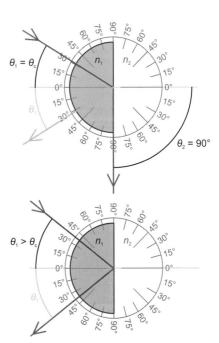

Figure 29.13 A glass half-cylinder prism is constructed from a section of a glass half cylinder. A beam of light aimed towards the center of the apparatus will not be refracted at the first air-light interface as the incident angle will be 0° and $\sin 0° = 0$.

29.5 A beam of white light passes through a 1.5 cm thick pane of glass at an angle of 45° as shown in Figure 29.15. The refractive index of the glass for light of wavelength 470 nm (deep blue) is 1.66 while the refractive index of the glass for light of wavelength 630 nm (bright red) is 1.60.

(a) What is the spacing, S, between the red and blue components of a narrow beam after they have passed through the pane of glass?

(b) Use your answer in (a) to explain why we do we not ordinarily see the effects of dispersion when looking through flat panes of glass.

(c) How thick would the pane of glass need to be for the separation of the red and blue rays to be 1 cm?

29.6 A beam of light of wavelength 550 nm strikes a water droplet as show in Figure 29.14. What are the angles θ_A and θ_B at which the reflected and refracted beams travel?

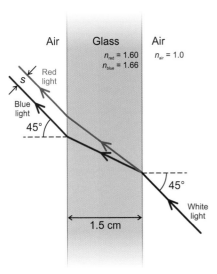

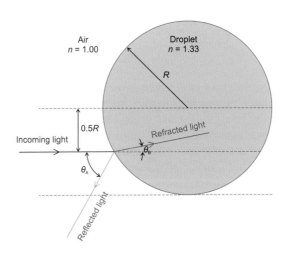

Figure 29.14 A beam of light hits a spherical water droplet.

Figure 29.15 A beam of white light passes through a glass pane. Dispersion causes the red and blue components of the light to be bent at different angles. After passing through the pane of glass the red and blue components are slightly offset.

29.9 By what angle (θ_{cornea}) is the beam of light shown in Figure 29.16 deviated as it passes from air to the cornea if the incident angle is $\theta_i = 23.6°$? The refractive index of air is $n_{air} = 1.00$, the refractive index of the cornea is $n_{cornea} = 1.38$. Ignore further deviation of light as it passes from the cornea into the aqueous humour, etc.

29.7 A fish in a pond looks up and sees the light from a street lamp at an angle of 35° to the vertical. If the street light is 5.5 m tall and the fish is 30 cm below the surface of the pond and 3 m from its edge, how far from the *edge of the pond* is the street lamp? ($n_{air} = 1$, and $n_{water} = 1.33$)

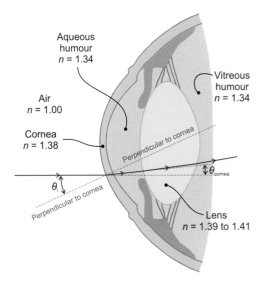

Figure 29.16 Most of the bending of light in the eye is done at the air-cornea interface. The lens is responsible for only a small amount of the bending, but of course is adjustable.

29.8 Two divers jump out of their boat and swim straight down to a depth of 10 m. The water surface becomes calm again very quickly after the divers jump in. Once the divers reach their final depth they begin to swim in opposite directions at the same rate while periodically stopping to shine a flashlight back at the surface of the water where they had jumped in. After the divers have swum far enough apart they begin to notice a strong reflection from the other diver's flashlight that was not present before. How far apart are divers when this starts to happen? ($n_{air} = 1.0$, and $n_{water} = 1.33$)

29.10 The ability of your eyes to focus is impaired when you attempt to look around underwater (if you are not wearing a pair of swimming goggles). Recalculate your answer for Problem 29.9 for the case in which the eye is submerged in water ($n_{water} = 1.33$).

www.wiley.com/go/biological_physics

GEOMETRIC OPTICS

30.1 Introduction

Geometric optics is one of the two branches of classical optics, and in it a ray treatment of light is used to predict the path of light waves through an optical system. When light interacts with objects larger than a few times the wavelength, its path is straight enough for a ray approximation. Geometric optics fails to explain some optical behaviour, such as polarisation and diffraction, for a which a *wave optics* treatment is needed, and this will be covered in Chapter 32.

In order for the words in this document to be seen by the eye, light must travel from each point on the page to the eye. This remains true if the page is tilted, so light is travelling out in all directions from each point on the page, and it is doing so in *straight lines*. Each of these straight-line paths through space is a *ray*.

Most of the basic behaviour of light was covered in the previous chapter: light can travel unimpeded through space; it can travel through some medium other than free space, such as air or glass, changing speed and direction at any interfaces; it can bounce off a surface; it can also be absorbed. At the boundary between two materials, a combination of these things happens. For example, when sunlight shines on a window, about 4% of the visible light bounces back off each surface, most of the visible light travels through the glass but has its path bent, and large proportions of infrared and ultraviolet light are absorbed.

It this chapter we are most concerned with applying the rules of reflection and refraction to systems of mirrors and lenses to see how these can be used to manipulate light.

Key Objectives

- To understand how to draw ray diagrams for mirrors and lenses.

- To be able to use the thin lens equation to calculate the type and position of an image.

- To be able to calculate the magnification of an image.

30.2 Ray Diagrams

Ray diagrams are a useful tool to figure out what kind of image will be formed, and roughly where. They show an object (simplified) at the correct location, the kind of mirror or lens being used, and the path that rays of light take on leaving the object *from one or two representative locations on the object*. The place where rays from one point of the object cross after leaving the mirror or lens is the location of the *image point* that corresponds to the object point chosen. If the rays never cross (as for a diverging mirror), the image location is the place where the rays appear to have a common origin.

It is common to use an arrow with its base on the optical axis as a representation of the object. The advantage of this is that all the rays originating from a point at the base and travelling along the optical axis will be reflected back along this line, so the image must also have its base on this line. All that remains is to locate the image point

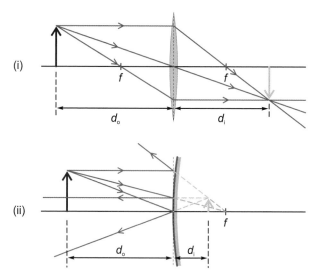

Figure 30.1 (i) A typical ray diagram showing the object (heavy arrow on left), a converging lens of focal length f, three principal rays refracted through the lens, and an image (light arrow on right), which in this case is a real image. The lens is assumed to be thin in comparison to all other distances (d_o, d_i and f) and the ray diagram is drawn so that the rays refract in one plane (the dotted line) rather than at the drawn surface of the lens. (ii) An equivalent ray diagram for a diverging mirror. In this case the mirror *reflects* light and forms a virtual image.

that corresponds to the top of the arrow. This can be done by drawing in three appropriate rays and seeing where they meet. Using an arrow makes it easy to see at a glance whether the image is upright or inverted. Figure 30.1 shows some sample ray diagrams, illustrating the various distances

There are a number of terms (specific to spherical mirrors and lenses) that we will need to use in this chapter:

Optical axis The line connecting the centres of curvature of the lens surfaces, or the centre of curvature to the middle of the mirror.

Centre of curvature The point that forms the centre of the sphere that the mirror or lens surface lies on.

Radius of curvature (R) The distance from the lens or mirror surface to the centre of curvature.

Focal point The point(s) on the optical axis halfway between the centre of curvature and the mirror or lens. Rays coming into a concave mirror parallel to the optical axis are reflected towards the focal point. This is often indicated on the ray diagram with an F.

Focal length (f) The distance from the lens or mirror to the focal point.

Object distance (d_o) The distance from the lens or mirror to the object.

Image distance (d_i) The distance from the lens or mirror to the image . (Sign conventions for when these distances are considered to be negative will be covered later.)

Principal rays The rays with easy to predict paths that are usually drawn in a ray diagram: parallel to the optical axis, through the focal point and through the centre of the mirror or lens.

Figure 30.2 A plane mirror forms an unmagnified virtual image the same distance behind the mirror as the object is in front of it.

30.3 Plane Mirrors

A **plane mirror** is a flat, specularly-reflecting surface. Plane mirrors form upright, virtual images of objects in front of them, with the image formed the same distance behind the mirror as the object is in front. By **image**, what we mean is an artifact that

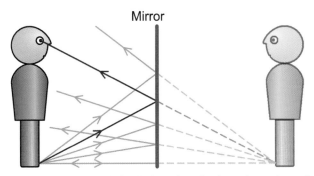

Figure 30.3 Light rays scattering off a person's foot hit the surface of a plane mirror and are reflected off it. Some of these rays will hit the persons eye and contribute to the image of the person's foot formed in the mirror.

resembles a real object. To an observer looking into the mirror, it appears that there is an object resembling them that seems to be located behind the mirror's surface (see Figure 30.2). This is what we term a **virtual image**, because the light rays that are travelling from the object only appear to have come from the image location, but never truly passed through that location. The light didn't pass through the mirror and come back, so it could never have been where the virtual person appears to be (see Figure 30.3).

Example 30.1 *Plane mirror*

Problem: How tall must a mirror be , in order for a person to be able to see their own feet and the top of their head within it?

Solution: About half their height.

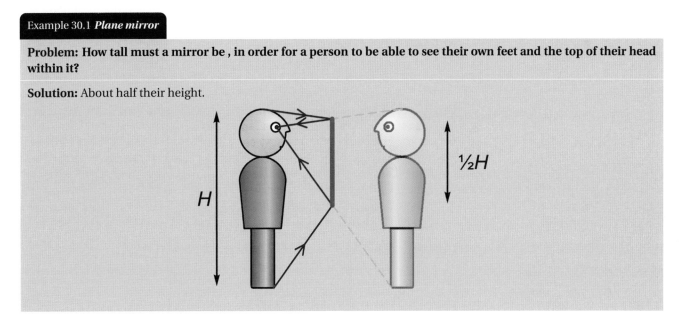

30.4 Spherical Mirrors

Concave and Convex Mirrors

A **spherical mirror** is one in which the reflecting surface forms part of the surface of a sphere. In other words, it has a crosssection that forms part of a circle, and the radius of that circle is called the **radius of curvature** of the mirror. The point that would be the centre of the sphere is called the **centre of curvature**. Spherical mirrors can be either **concave** or **convex**. A concave mirror reflects light rays parallel to the optical axis in the direction of the focal point; a convex mirror reflects light rays parallel to the optical axis away from the focal point (see Figures 30.4 and 30.5). These types of mirrors are also known as **converging** (for the concave) and **diverging** (for the convex).

Image Formation By a Concave Mirror

There are some basic rules for drawing ray diagrams to show what kind of image is formed and where:

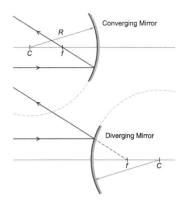

Figure 30.4 Converging and diverging mirrors with spherical surfaces of radius R showing the centre of curvature C and focal point f. A ray parallel to the optical axis is incident on, and reflected off, each of the mirrors.

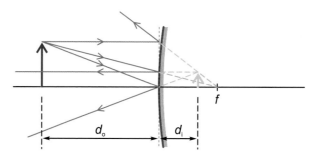

Figure 30.7 A virtual image formed by a diverging mirror. Notice that the rays leaving the mirror are diverging as if coming from a point 'behind' the mirror. This image cannot be seen on a screen because at no point do the rays actually pass through that position.

- Rays coming in parallel to the optical axis go out though the focal point (Figure 30.8 (i)).

- Rays coming from the centre of curvature are reflected back along their path.

- Rays hitting the centre of the mirror act as though they are hitting a plane mirror (as in Figure 30.8 (ii)).

- Rays coming in through (or as though they passed through) the focal point are reflected parallel to the optical axis (Figure 30.8 (iii)).

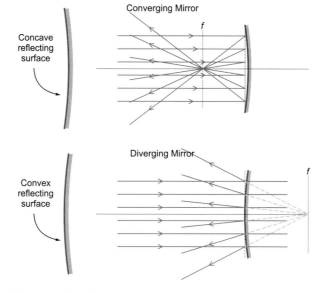

Figure 30.5 Converging (concave) and diverging (convex) mirrors with spherical surfaces. Parallel light rays incident on a converging mirror *converge* to a single point after being reflection. Parallel light rays incident on a diverging mirror appear to *diverge* from a point after reflection.

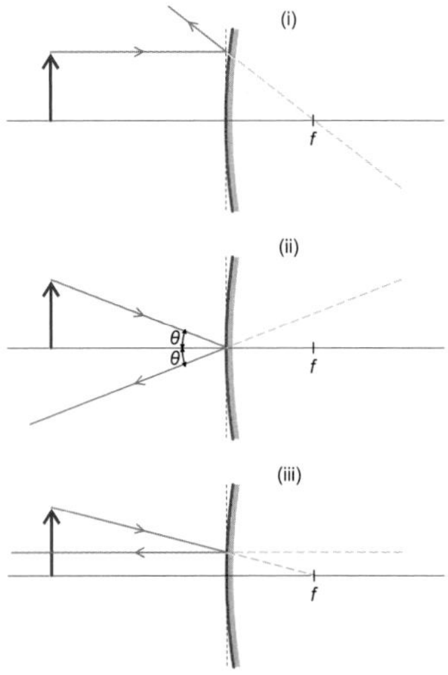

Figure 30.6 The three principal rays for drawing a ray diagram, shown here for a convex mirror.

Image Formation By a Convex Mirror

The same basic rules that apply to concave mirrors apply also to convex mirrors. A distinction is that the focal point of a convex mirror is on the opposite side of the mirror surface to the object (see Figure 30.7). Figure 30.6 shows the three rays used to construct a ray diagram for a convex mirror.

Light from an object that is reflected from a convex mirror is always diverging and so this kind of mirror will always form a virtual image.

Types of Image—Real and Virtual

There are two kinds of image that can be formed, which are illustrated in Figure 30.9 and Figure 30.10. These are called **real** and **virtual**. A real image is one through which light rays actually pass, and which could therefore be seen on a screen placed at that position. An example of a real image is the image formed on the film in a camera, or on the sensors inside a digital camera – the light really reaches the film, and a representation of some real object is formed. A virtual image is one which can't be seen on a screen, as the light rays never pass through the location where the image sits. An example is the image seen in a plane mirror. When you look in the mirror, you see something that resembles yourself on the other side of the mirror surface, but the light never went through that location.

The Mirror Equation

The mathematical relationship between the location of a point on the object and the corresponding point on the image is

$$\frac{1}{d_o} + \frac{1}{d_i} = \frac{1}{f}$$

(30.1)

This mirror equation has exactly the same form as the thin lens equation which we will meet soon.

Sign Convention for Mirrors

- All figures are drawn with light initially travelling from left to right, so the object is to the left of the mirror.

- The distance from the object to the mirror is positive.

- The distance from the mirror to a real image is positive. (Image is located to the left of the mirror.)

- The distance from a virtual image to the mirror is negative. (Image is located to the right of the mirror.)

- For a concave (converging) mirror, f is positive.

- For a convex mirror (diverging) mirror, f is negative.

The image distance is positive if it is in the same direction as the outgoing light, and negative if in the other direction (see Figure 30.11).

Example 30.2 *Mirrors*

Problem: Determine whether each of the following statements are true or false:

(a) **A concave mirror causes light to diverge.**

(b) **A convex mirror always forms virtual images.**

(c) **A concave mirror has a positive focal length.**

(d) **A concave mirror can form both real and virtual images depending on the position of the object.**

Solution:

(a) False, a concave mirror causes light to converge.

(b) True, a convex mirror causes light to diverge so always forms a virtual image.

(c) True, a concave mirror has a positive focal length, as by convention a mirror or lens which causes light to converge has a positive focal length.

(d) True, a concave mirror forms a real image when the object distance is greater than the focal length and forms a virtual image when the object distance is smaller than the focal length (see Figures 30.9 and 30.10).

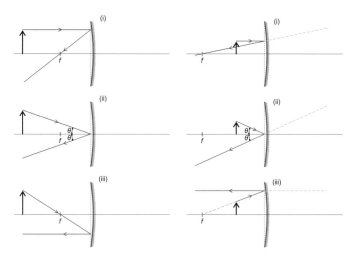

Figure 30.8 Three rays used to construct ray diagrams for a converging mirror. The object is placed outside the focal plane of the mirror in the example on the left and inside it on the example on the right. The rules are the same in each case however.

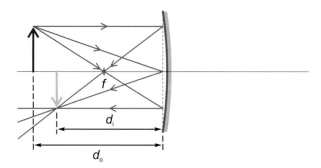

Figure 30.9 A real image formed by a converging mirror. Notice that the rays leaving the mirror are converging towards a point. The image could be viewed on a screen at that point.

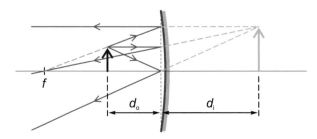

Figure 30.10 A virtual image formed by a converging mirror. Notice that the rays leaving the mirror are diverging as if coming from a point 'behind' the mirror. This image cannot be seen on a screen because at no point do the rays actually pass through that position.

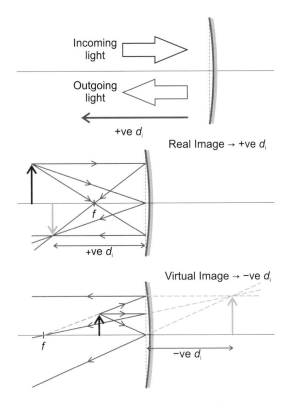

Figure 30.11 The sign convention for use with image formation by mirrors. While a converging mirror is shown here, the same sign conventions for d_i and d_o apply for diverging mirrors, with f negative.

30.5 Magnification

The image formed by a converging or diverging mirror will, in general, be of a different size to the object. The single factor which sets the relative sizes of the object and the image is the placement of the object with reference to the focal plane of the mirror. Figure 30.12 shows the images formed by a converging mirror when an object is placed at three different positions, $d_o > 2f$, $d_o = 2f$ and $f < d_o < 2f$. All three images are real images, and so are inverted, and their relative size varies.

The magnification, M, of an image tells us the relative size of the image with respect to the object. In other words, the magnification is equal to the ratio of the image and object heights. A negative number indicates that the image is inverted with respect to the object. M is expressed as

$$M = \frac{\text{image height}}{\text{object height}} = \frac{h_i}{h_o}$$

Close inspection of Figure 30.12 will show that the ratio of the image and object distances is the same as the ratio of the image and object heights, but when both distances are positive, the image is inverted, so we need to include a negative sign

$$M = -\frac{\text{image distance}}{\text{object distance}} = -\frac{d_i}{d_o}$$

We can express the magnification in terms of either the object and image heights, or the object and image distances

$$M = \frac{h_i}{h_o} = -\frac{d_i}{d_o} \tag{30.2}$$

A magnification of $M = 2.5$ means that the image is upright, 2.5 times larger than the object, and 2.5 times farther away from the mirror than the object. A magnification of $M = -0.333$ means that the image is inverted, only $\frac{1}{3}$ as tall as the object, and one third the distance to mirror that the object is. The sign tells us whether the image is real

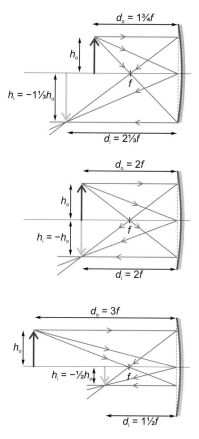

Figure 30.12 The same object is placed at three different positions with respect to a converging mirror. When $2f > d_o > f$, the object is enlarged ($|M| > 1$), whereas when $d_o > 2f$, the object is reduced ($|M| < 1$). A special case exists when $d_o = 2f$. In this case $M = -1$.

or virtual. A negative sign indicates that the image is inverted, and so is a *real* image, while a positive sign indicates the image is upright, which must be a *virtual* image.

Using the magnification equation, Eq. (30.2), and the mirror equation, Eq. (30.1), we can develop a general expression for the magnification of an image given d_o and f. The mirror equation may be rewritten as

$$\frac{1}{d_i} = \frac{1}{f} - \frac{1}{d_o}$$

and from Eq. (30.2)

$$\frac{1}{M} = -\frac{d_o}{d_i}$$
$$= -d_o \frac{1}{d_i}$$
$$= -d_o \left(\frac{1}{f} - \frac{1}{d_o} \right)$$
$$= -\frac{d_o}{f} + 1$$

This gives us a useful result:

$$M = \frac{1}{1 - \dfrac{d_o}{f}} \qquad (30.3)$$

30.6 Lenses

Lenses use refraction to bend light rays. Lenses designed to work with visible wavelengths of light are typically made with types of glass or plastic – materials that have

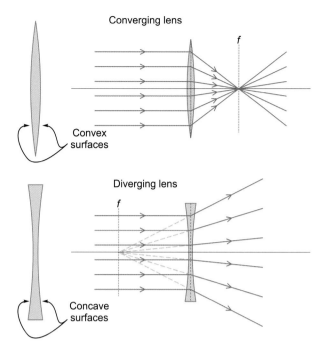

Figure 30.13 Parallel light passing through a converging lens converges to a point on the opposite focal plane of the lens. Parallel light passing through a diverging lens appears to diverge from a point on the focal plane of the lens that is on the side the light entered.

a higher refractive index than air. As shown in Figure 30.13, a lens that causes parallel rays to become convergent is called a **converging lens**, and one that causes them to be divergent is called a **diverging lens**. Rays that are convergent will eventually cross, whereas rays that diverge will never cross as they continue to propagate through space.

The simplest lenses have both sides with the same kind of curvature. A lens with both sides curving so that the middle is thicker than the edges is called a **convex lens**. Convex lenses are converging. A lens with the sides curving so that the edges are thicker than the middle is called a **concave lens**, and is diverging. (This is easy to remember – a concave lens has surfaces shaped like caves.)

Image Formation By Lenses

Lenses can form images in a similar manner to mirrors. Of course, a lens forms an image by refracting light instead of reflecting it.

The set of rules for drawing rays is very similar to that used for mirrors and is as follows:

- Rays coming in parallel to the optical axis are refracted such that they exit heading towards (or away from) the focal point.

- Rays hitting the centre of the lens act as though they are passing through a flat piece of glass.

- Rays coming in through (or as though they passed through) the focal point are refracted so that they exit parallel to the optical axis.

Image Formation by a Converging Lens

A converging lens has *convex* surfaces which bulge outwards. A converging lens bends light towards the optical axis and so will focus parallel light to a point on the focal plane (the plane perpendicular to the optical axis and which intersects the optical axis at the focal point) as shown in Figure 30.13. The distance between the focal plane and the centre of the lens is the focal length of the lens.

Because light passes through the lens there is a focal plane on either side of the lens, unlike mirrors for which there is only one focal plane.

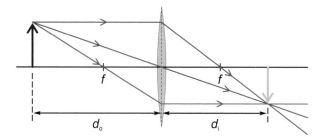

Figure 30.14 A real image formed by a converging lens. Notice that the rays leaving the lens are converging towards a point. If a screen were placed here, an image of the object could be seen on it.

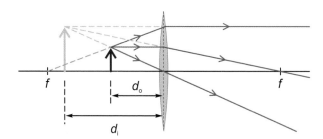

Figure 30.15 A virtual image formed by a converging lens. Notice that the rays leaving the lens are diverging. Because of this, the image formed cannot be seen on a screen at the image position, as could be done with the real image in Figure 30.14.

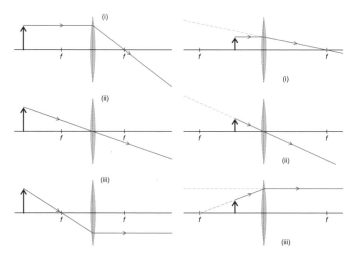

Figure 30.16 Three principal rays used to construct ray diagrams for a converging lens. The object is placed outside the focal plane of the lens in the example on the left and inside it in the example on the right. The rules are the same in each case, however.

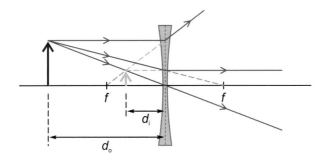

Figure 30.17 The image formed by a diverging lens. A diverging lens always produces a virtual image when a 'real' object is used. Figure 30.19 shows the principal rays used to construct this ray diagram.

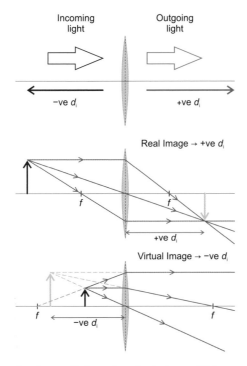

Figure 30.18 The sign convention for use with image formation by lenses. While a converging lens is shown here, the same sign conventions for d_i and d_o apply for diverging lenses, with f negative.

Figure 30.14 shows the formation of a real image by a converging lens and Figure 30.15 shows the formation of a virtual image. Just as for a converging mirror, placing the object so that $d_o > f$ will result in a real image whereas placing the object so that $d_o < f$ will result in a virtual image.

Figure 30.16 shows the three principal rays used to construct a ray diagram when the object is placed such that either $d_o < f$ or $d_o > f$.

Image Formation by a Diverging Lens

A diverging lens has *concave* surfaces. A diverging lens bends light away from the optical axis and will focus parallel light to a point on the focal plane as shown in Figure 30.13. A diverging lens will create a virtual image for all (real) object positions. Figure 30.17 shows a ray diagram for a diverging lens and Figure 30.19 shows the principal rays used to construct such a ray diagram. There are a few subtle differences in the correct use of the principal rays for a diverging lens that relate to the fact that a diverging lens has a negative focal length.

Sign Convention for Lenses

Figure 30.18 shows the sign convention used for lenses. The convention is the same for both converging and diverging lenses.

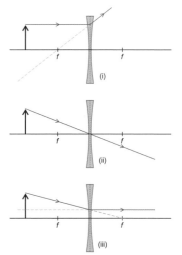

Figure 30.19 The principal rays used when constructing a ray diagram for a diverging lens.

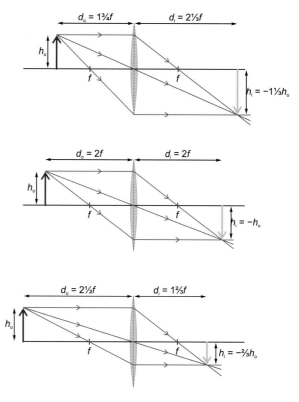

Figure 30.20 The same object is placed at three different positions with respect to a converging lens. When $2f > d_o > f$ the object is enlarged ($|M| > 1$) whereas when $d_o > 2f$ the object is reduced ($|M| < 1$). A special case exists when $d_o = 2f$. In this case $M = -1$.

- All figures are drawn with light initially travelling from left to right, so the object is to the left of the lens.

- The distance from the object to a lens is positive.

- The distance from a lens to a real image is positive. (Image is located to the right of the lens.)

- The distance from a lens to a virtual image is negative. (Image is located to the left of the lens.)

- For a convex (converging) lens, f is positive.

- For a concave (diverging) lens, f is negative.

Magnification and Lens Power

The image produced by a lens is magnified in the same way as the image produced by a mirror. The arguments and equations given in Section 30.5 hold for lenses as well. That is, the magnification, M, of the image produced by a lens is given by

$$M = \frac{h_i}{h_o} = -\frac{d_i}{d_o} \tag{30.4}$$

and

$$M = \frac{1}{1 - \frac{d_o}{f}} \tag{30.5}$$

Figure 30.20 shows how the placement of the object can affect the size of the image produced by a converging lens. Notice that when $d_o = 2f$ then $M = -1$ and that moving the object closer to the lens results in a more magnified image, just as for a converging mirror.

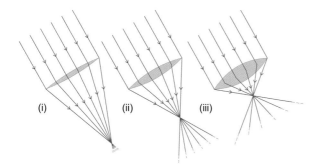

Figure 30.21 Three lenses of different optical power are shown. The most powerful lens (iii) bends the light to a greater degree than the weakest lens (i).

The *power* of a lens is a measure of how 'strong' it is, or more formally how much it bends the light passing through it. In Figure 30.21, three lenses are shown focussing parallel light to a point. It makes intuitive sense to call lens (i) the weakest lens and lens (iii) the strongest lens. Lens (i) has a long focal length and bends each ray less than lens (iii), which has a much shorter focal length.

Given that a 'long' focal length means that the lens is 'weak', then there is an inverse relationship between optical power and focal length. The equation expressing the relationship between optical power, P, and focal length, f, is

$$P = \frac{1}{f} \tag{30.6}$$

The units of optical power as defined above are dioptres (D). When using Eq. (30.6) the focal length should always be expressed in the correct S.I. unit of metres. A lens with a power of $P = 20$ D would have a focal length of $f = \frac{1}{20\text{D}} = 0.05$ m, and a lens with a focal length of $f = -0.25$ m (a diverging lens) would have an optical power of $P = \frac{1}{-0.25\text{m}} = -4$ D. Just as a negative focal length indicates a diverging lens/mirror, so does a negative optical power.

Example 30.3 *Image formation*

Problem: Determine whether each of the following statements is true or false for images formed by a single lens or mirror:

  (a) **Real images can be seen on a screen at the image position because different light rays from a point on the object meet at the image point.**

  (b) **All virtual images are upright.**

  (c) **All real images are inverted.**

  (d) **Mirrors (of all types) can only form virtual images.**

Solution:

  (a) True, to form an image on a screen light rays from a single point on the object must converge to a single point at the screen location. This is the case for a real image.

  (b) True, virtual images are always upright.

  (c) True, real images are always inverted.

  (d) False, concave mirrors can form real images when the object distance is greater than the focal length.

> **Example 30.4 *Lenses***
>
> **Problem: How could you experimentally determine the approximate focal length of a converging lens? Would your method work for a diverging lens? How could you adapt your method to work for a diverging lens?**
>
> **Solution:** The approximate focal length can be found by measuring how far from the lens (parallel) light from a distant object is focussed. You could, for example, focus a landscape on the wall).
>
> This does not work for a diverging lens, as this method is reliant on forming a *real* image of the distant object. To determine the focal length of a diverging lens one could combine it with a stronger converging lens (higher dioptre, shorter focal length) and determine the focal length of the combination lens (which will be converging). The power of the diverging lens can be inferred from this and the power of the converging lens.

30.7 Summary

Key Concepts

geometric optics One of the two branches of classical optics, where a ray treatment of light is used to predict the path of light waves through an optical system. Geometric optics fails to explain some optical behaviour such as polarisation and diffraction, for a which a *wave optics* treatment is needed.

image (from the latin *imago*, likeness) An artifact (usually two-dimensional) that resembles a real object.

lens A transparent object with axial symmetry which refracts light, converging or diverging light rays.

converging lens A lens which causes light rays passing through it to converge. Any lens that is thicker in the centre than the edges will act as a converging lens.

diverging lens A lens which causes light rays passing through it to diverge. Any lens that is thinner in the centre than the edges will act as a diverging lens.

concave lens A lens in which both surfaces are curved and are depressed into the lens. More correctly called biconcave. A lens that has one flat surface and one concave surface is called plano-concave. Both sorts of concave lens are diverging.

convex lens A lens in which both sides are curved and bulge out. More correctly called biconvex. A lens that has one flat surface and one convex surface is called plano-convex. Both sorts of convex lens are converging.

focal point Rays of light entering the lens parallel to the optical axis will be converged to a point, or diverge from a point, known as the focal point.

focal length (f) The distance along the optical axis from the centre of the lens or mirror to the focal point.

optical axis The line that passes through the centres of curvature of a lens or mirror. Light passing along this axis is not refracted.

optical power (P) The amount that a lens converges or diverges light. It is the inverse of the focal length, $\frac{1}{f}$, and is measured *diopters*, which are inverse metres. Specifying the power rather than the focal length is useful because when thin lenses are placed close together, their powers approximately add.

real image An image in which the light rays actually pass through the image. A real image can be seen on a screen placed at the image position.

virtual image An image in which the light rays originating at a point on the object never reconverge at another point, but appear to have had a common origin at the image location.

thin lens equation This equation shows the relationship between object distance, image distance and focal length for thin lenses.

magnification A measure of the size of an image relative to the original object. The magnification is negative if the image is inverted, and has magnitude less than 1 if the image is smaller than the object.

mirror A surface with good specular reflection, so that an image can be formed. The surface may be flat (a plane mirror) or curved.

radius of curvature (spherical mirror or lens) The distance from the surface to the centre of the sphere that the curved surface forms a part of.

www.wiley.com/go/biological_physics

Equations

$$\frac{1}{d_o} + \frac{1}{d_i} = \frac{1}{f}$$

$$P = \frac{1}{f}$$

$$M = \frac{h_i}{h_o} = -\frac{d_i}{d_o}$$

$$M = \frac{1}{1 - \frac{d_o}{f}}$$

30.8 Problems

30.1 Is it possible for a converging lens to form a virtual image? If so, under what conditions is the image virtual? If not, why not?

30.2 Is it possible for a diverging lens to form a real image of a physical object? If so, under what conditions is the image real? If not, why not?

30.3 An object is placed 0.25 m away from a lens. The lens forms an image that is 0.167 m away from the lens, upright, and on the same side of the lens as the object.
  (a) What is the focal length of the lens?
  (b) What kind of lens is used?

30.4 You wish to produce inverted real images of an object with the given magnifications using a converging mirror. How far from the mirror must you place the object in each case (express your answer in terms of the focal length of the mirror)?
  (a) $M = -0.5$
  (b) $M = -1$
  (c) $M = -2$
  (d) $M = -4$

30.5 You wish to produce upright virtual images of an object with the given magnifications using a diverging lens. Where must you place the object in each case (express your answer in terms of the focal length of the lens)?
  (a) $M = 0.1$
  (b) $M = 0.25$
  (c) $M = 0.5$
  (d) $M = 0.75$

30.6 What is the largest magnification attainable when imaging a real physical object using a diverging mirror, and how far from the mirror must the object be placed to attain this magnification?

30.7 You are standing 5 m from the edge of a very large, 150 m diameter hemispherical building which is coated in a reflective material. You are carrying a small laser pointer which you hold 1.5 m above the ground.
  (a) You point the laser pointer at the building and the reflected beam travels straight back at the pointer. At what angle below the horizontal are you holding the pointer?

  (b) You point the laser pointer at the building and the reflected beam is traveling parallel to the ground. At What angle below the horizontal are you holding the pointer?

30.8 Light from a distant source enters a 0.5 dioptre lens parallel to the optical axis.
  (a) How far from the first lens must a second, 1.2 D lens be placed such that the light leaving the second lens is also parallel to the optical axis?
  (b) How far from the first lens must a second, -1.5 D lens be placed such that the light leaving the second lens is also parallel to the optical axis?
  (c) A second 1.2 D lens is placed 1.2 m behind the first. Is the light leaving this lens, converging, diverging, or parallel to the optical axis?

30.9 A converging lens with a focal length of 30 cm is used to create an image of a 2 mm long ant.
  (a) If the lens is placed so that the image of the ant is 8 mm long, upright, and viewed by looking through the lens, how far away from the ant was the lens placed?
  (b) If the lens is placed so that the image of the ant is 8 mm long, inverted, and viewed on a screen held some unspecified distance on the other side of the lens to the ant, how far away from the ant was the lens placed?

30.10 When you look at the back of a spoon you see an upright image of yourself. This is because the reflective curved surface of the metal acts as a diverging mirror. This image does tend to be distorted because spoons seldom have the spherical or parabolic curvature required for an undistorted image. Ignore these distortions when answering the following questions.

  (a) If the image of your head is 3 cm tall, your head is 22 cm tall, and you are holding the spoon 16 cm away from your head, what is the focal length of the back of the spoon?

  (b) When you flip the spoon around it now acts like a converging mirror and you see an inverted image. Assuming that the curvature of the inside of the spoon is the same as the curvature of the outside of the spoon how large is the image of your head?

The Eye and Vision

31

31.1 Introduction

Anyone reading this text on the page knows already how important our vision is to us; it is the most important of our senses, when it is intact. In this chapter we will cover the basic structure of the human eye, and apply what we have learned about converging lenses to show how images are formed on the retina. We will then take a look at how refractive defects can be corrected with different types of lenses. As humans are one species that has colour vision, we will also see how it is that we distinguish between different colours, and how we can define colours – a challenge, as colour perception varies from individual to individual.

The human eye is very good at what it does. Our eyes are able to not only detect the presence or absence of light, but are able to detect light that varies in intensity a great deal. We can see in bright sunlight, or spot a lone candle flame kilometres away. We can see movement, as vision is a continuous process, unlike exposing the film in a camera. We can see shapes and colours, due to the number and type of detector cells we have at the back of our eyes. Using our binocular vision, and a sense of the state of focus of our eyes, we can estimate distance. We also have the brain power to interpret the images we see, and compare them to ones we have seen previously.

Our eyes, as good as they are at their job, are not the best around. There are examples of better eyesight in other animals, particularly birds. There are also many other types of eye out there, and nearly every imaginable example of a way in which an image could be formed is used by some species. We will give a few examples of other eye types and focussing methods to show how these differ from the human eye.

Key Objectives

- Learn the parts of the eye and their functions.

- Understand image formation by the human eye.

- Understand the causes of vision defects and how they may be corrected.

31.2 The Parts of the Eye

Figure 31.1 shows a crosssection through the human eye, with the main parts labelled. The key parts for image formation are the **lens** and **cornea**, which together act like a thin lens to focus incoming light onto the **retina**. The clear cornea is a specialised part of the outer eyeball, and is made mainly of collagen. This collagen is arranged in crisscrossing sheets, which makes it tough. If removed, sliced open and immersed in water, the cornea will swell to many times its normal size, and will become slightly opaque. There are membranes on the inner and outer surfaces that prevent this happening in our eyes. The rest of the outer surface of the eye (the fibrous, white part) is the **sclera**. Its thickness varies from 0.3 mm to 1 mm, and it is to the sclera that the outside muscles attach.

Immediately behind the cornea is a chamber (the **anterior chamber**) filled with a watery salt solution, called the **aqueous humour**. This liquid is being constantly replaced, and is important for supplying nourishment to the cornea and the lens, neither

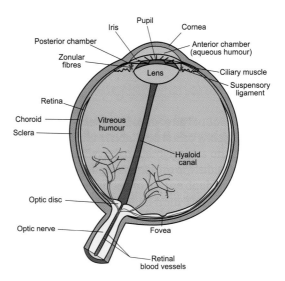

Figure 31.1 The human eye. [Public domain picture courtesy of Wikipedia.].

of which are equipped with a blood supply. The aqueous humour contains many of the substances found in the blood, including glucose and dissolved oxygen. The aqueous humour also plays a role in regulating the pressure in the eye. It normally drains from the anterior chamber through a channel in the angle between the cornea and the iris. If this becomes blocked, the pressure may increase, and glaucoma can result.

The **lens** (see Figure 31.2) is a collection of transparent cells suspended in place by **suspensory ligaments** connected to the **ciliary muscle**. When the eye is relaxed, the lens is flattened slightly by the tension in the ligaments (caused by pressure from the vitreous humour). The ciliary muscle allows focussing of the eye. When it contracts, it reduces the tension in the fibres, making the lens more spherical.

Slightly in front of the lens is the **iris**, which connects to the sclera and ciliary body at its outer edge, and is made up of a pigmented, fibrous part known as the stroma, and muscles which constrict and dilate the **pupil**, the gap in the iris through which light passes. The pupil appears bigger than it really is, due to magnification by the cornea. The back of the iris is strongly pigmented and is nearly black. The colour as seen from the front depends on the amount of the pigment melanin present. There is no blue pigment in the eyes of people with blue eyes, but this colour results from selective absorption and scattering in the blood vessels (from haemoglobin in the blood and collagen in the vessel walls, for example). Other eye colours are the result of the deposition of melanin in the front layer of the iris. This pigmentation is not always present at birth in Caucasian babies, so they may be born with blue eyes that later change colour. The pattern of pigmentation in the iris is fixed by one year of age. (This suggests that the alternative medicine technique of diagnosing illness from the pattern of the iris (iridology or iridodiagnosis) may have limited practical use.)

The expansion and contraction of the pupil is involuntary in humans, and occurs in response to changing light levels. The diameter of the pupil is typically around 3–4 mm, but varies from about 1.5 mm to 8 mm. There is some cross-over in the sensory pathways from both eyes, so light entering one eye causes both pupils to change. The narrowing of the pupil not only stops too much light from entering the eye, but also changes the depth of field, the range of object distances over which the image is still acceptably well in focus. In animals which need to be able to see well in the daytime as well as hunt at night, the pupils may be slit-shaped instead of round to block more light.

Filling the bulk of the centre of a eye is a gelatinous, transparent material called the **vitreous humour**. It is mostly water, but contains some salts, sugars, collagen and hyaluronic acid. Its refractive index is 1.336, which is close to that of water. Before birth, an artery (the **hyaloid artery**) supplies blood to the developing lens. This usually disappears, leaving a clear zone through the vitreous called the hyaloid canal. Sometimes, though, clumps of cells from these blood vessels are left behind, which we can

---

The pupil

The name 'pupil' comes from the latin *pupilla*, meaning 'little doll', which refers to the tiny reflection of oneself that can be seen in another person's eye.

---

www.wiley.com/go/biological_physics

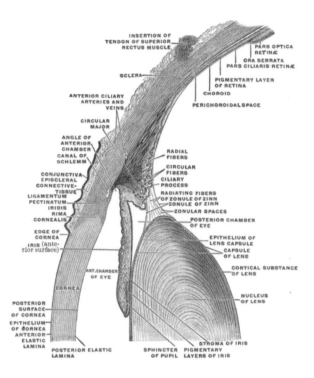

Figure 31.2 Schematic of the front of the human eye. [Reprinted from www.bartleby.com. Copyright 2010.]

see occasionally in our eyes as 'floaters'.

The back surface of the eye is covered by a tissue-thin membrane called the **retina**. This contains the light-sensitive cells that allow us to see. The retina has a mixture of two types of these light-sensitive cells, one type that responds slowly to most of the visible wavelengths (the **rods**), and another that responds faster, but only selectively to regions of the visible spectrum (the **cones**). The **macula** is a small (about 1.5 mm across) spot on the retina that appears yellow, and contains the **fovea**, a small pit with the highest concentration of cone cells. This is the area responsible for central vision. The fovea makes up only 1% of the area of the retina, but half the visual cortex of the brain is devoted to processing its signals.

Between the retina and the sclera lies a layer that supplies blood, known as the **choroid**. It is darkly pigmented by melanin in humans.

The majority of the refractive power of the eye is provided by the cornea. For a relaxed eye, it is responsible for about 2/3 of the power. The refractive index of the stroma, the thickest part of the cornea, is about 1.376. (This similarity between the refractive index of water and of the cornea is why we see so poorly under water. The lens alone is not powerful enough to do all the focussing.)

> **Eyeshine**
>
> In some animals, such as cats, the pigmentation of the choroid is absent in places, and there is a reflective layer that improves night vision, called the *tapetum lucidum*. This is what makes some eyes so strongly reflecting at night – an effect called eyeshine. In the case of cats, their eyes appear green; the colour varies in other animals. In humans, a similar effect may be seen in the case of very bright illumination ('red-eye' from camera flashes) and in the case of abnormalities like cataracts (which can appear like white eyeshine).

31.3 Emmetropia (Normal Vision)

By treating the cornea and lens of an eye as a single idealised converging thin lens, and the curved retina as a planar screen, we can produce a model of the eye sufficiently simple to be understood using only those aspects of optics covered so far, yet sufficiently accurate to be useful.

While the human visual system is considerably more complicated than a single idealised thin lens, such an approximation allows a good explanation of the general geometric properties of vision and an easy understanding of some aspects of physical visual defects/errors such myopia and hypermetropia.

By modelling the eye as an idealised thin lens the *image distance* becomes fixed. The eye *must* form an image on the retina or else the object cannot be viewed clearly unaided. From the content of the preceding chapters, it could be taken that the eye would only view objects clearly (form an image on the retina) if the object was placed a certain distance away from the eye. We know from everyday experience that this is not true. An aspect of the eye that is quite different to the simple lenses covered so far is that the optical power of the eye is variable. This is what allows the eye to clearly view objects over a range of different distances. This variation in the optical power (focal length) of the eye is achieved by distortion of the eye lens by a ring muscle called the ciliary muscle and is called **accommodation**.

When the ciliary muscles are relaxed, the tension in the suspensory ligaments (see Figure 31.2) that hold the lens in place keeps the lens more flattened. As the ciliary muscle tightens, the ligaments relax, increasing the curvature of the lens, and its power.

As shown in Figure 31.3, a more curved lens is suited to viewing nearby objects as it has a higher optical power (shorter focal length) and thus bends incoming light more. A lens under tension (which, remember, happens when the ciliary muscle is relaxed) has reduced curvature, and hence lower optical power (a longer focal length) and bends incoming light less to form a clear image of objects further away.

The eye does not have an infinite range of accommodation, but instead has both minimum and maximum optical powers (maximum and minimum focal lengths respectively). This imposes limits on the ability of the eye to form a clear image of an object at any arbitrary position. The terms *near point* and *far point* describe these limits.

Near Point The closest point an object can be placed at such that the eye can form a clear image of it on the retina.

Far Point The furthest point at which an object can be placed such that the eye can form a clear image of it on the retina.

When viewing an object at the near point, the ciliary muscle is tensed, and so the ligaments place the minimum tension on the lens, and the eye is at its maximum attainable optical power (it is at its most curved). The light scattered from objects closer than this point is not focussed on the retina, but instead at a point behind the retina. This results in a blurred image. The near point will vary from person to person and will likely be significantly shorter for younger readers than for older readers, but for the purposes of this text we will base calculations on a 'normal' near point of 25 cm.

For people with normal vision, the far point lies at infinity. In other words, for a normal eye, the fibres stretching the lens are capable of creating sufficient flattening of the lens to focus parallel light, like that from a distant star, onto the retina. In cases where the far point is closer than infinity, the *minimum* optical power of the eye is *too great* and the image of an object beyond the far point will be formed in front of the retina, resulting in a blurred image.

While the size of the human eye varies from person to person, we can model a 'normal' human eye as having a lens-to-retina distance of around 20 mm. Using the thin lens equation we can then calculate the 'normal' maximum and minimum power of the eye. The accommodation range is about 4 D.

$$P_{\text{max}} = \frac{1}{\text{near-point distance}} + \frac{1}{\text{lens–retina distance}}$$
$$= \frac{1}{0.25 \text{ m}} + \frac{1}{0.020 \text{ m}}$$
$$= 54 \text{ D}$$

$$P_{\text{min}} = \frac{1}{\text{far-point distance}} + \frac{1}{\text{lens–retina distance}}$$
$$= \frac{1}{\infty \text{ m}} + \frac{1}{0.020 \text{ m}}$$
$$= 50 \text{ D}$$

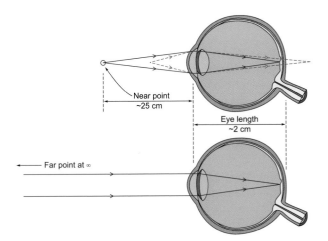

Figure 31.3 In a person with normal vision, the optical power of the eye can change over a range such that the person will be able to clearly view objects from as close as 25 cm all the way to infinity. An object is clearly viewed when an in-focus image is formed on the retina. When attempting to view an object that is closer than the near point, the cornea and lens have insufficient optical power to produce an image on the retina and an image would instead be formed behind the retina (if it were not there). A normal relaxed eye will have a minimum optical power sufficient to focus parallel light (such as that originating from an object at infinity) onto the retina.

31.4 Myopia

Myopia is a condition that causes objects that are *far away* to appear blurred, so is often referred to as nearsightedness. It occurs when collimated light rays entering the eye (such as rays from a distant object) are focussed to a point in front of the retina (see Figure 31.4). We can make a distinction between the causes:

Axial myopia The eye is too long.

Refractive myopia The problem is caused by a refractive error. This could be due to excessive curvature of one of the refractive parts of the eye, most often the cornea. A change in the index of refraction in the refractive media, such as caused by cataracts, can also cause myopia.

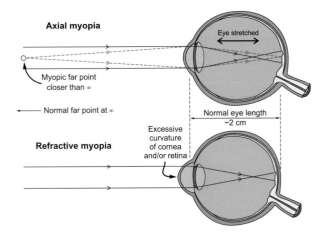

Figure 31.4 Myopia is a result of light being bent too strongly, even at the minimum optical power given by a relaxed eye, to form an image on the retina. The image is instead formed in front of the retina. The far point of a myopic person's eye is less than infinity, and an object further away than the far point cannot be viewed clearly.

Both genetic and environmental factors are believed to contribute to myopia, caused by an incorrect pairing of axial length and optical power. Some causes of myopia are degenerative, such as posterior staphyloma, a bulging of the sclera at the rear of the eyeball. Some pharmaceuticals can also induce myopia.

The degree of myopia is characterised by the power of the lens required to correct the defect. Low myopia is -3 D or less. High myopia is more than -6 D. Myopia is corrected with a *diverging* lens.

Example 31.1 *Myopia*

Problem: A man with refractive myopia cannot see objects clearly if they are further away than 10 m. In order to be able to drive safely and obtain his driving licence he needs to be able to clearly see objects up to 100 m away. What is the minimum optical power of the corrective lenses he needs? (Solve this problem by finding the combined power of his eye and the corrective lenses, and the actual minimum power of the man's eye. You may assume that the distance between the corrective lenses and the man's eyes is unimportant.)

Solution: We are told to use the combined power of the corrective lenses and eye (P_{combo}) and the minimum power of the eye alone (P_{eye}) to solve this problem.

This is a good approach, as when lenses are combined by placing them close together we can approximate the combined optical power of the system of lenses as the sum of the individual optical powers of each lens. In this case

$$P_{combo} = P_{eye} + P_{glasses} \tag{31.1}$$

Refractive myopia has been specified as the cause of the man's vision problems. We can therefore assume that his eye is of normal size and the lens–retina distance is 20 mm.

The man's far point is at 10 m, which means that, even when his eye is at its minimum optical power, it is bending incoming light too strongly. An image of any object that is 10 m or further away will form in front of the retina.

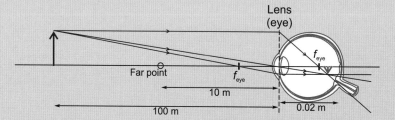

Without surgery, the minimum optical power of the man's eye is fixed. This means that for an object further away than his far point, a corrective lens must create a virtual image at, or closer than, his far point.

We can calculate the minimum power of the man's eye using the thin-lens equation. We know that the image distance is fixed by the size of the eye, $d_i = 0.02$ m. We also know that when $P_{eye} = P_{min}$ (or alternatively $f_{eye} = f_{max}$), $d_o = 10$ m. Using the thin lens equation we have

$$\begin{aligned}
P_{eye} &= \frac{1}{f_{eye,max}} = \frac{1}{d_o} + \frac{1}{d_i} \\
&= \frac{1}{10\,\text{m}} + \frac{1}{0.02\,\text{m}} \\
&= 50.10\,\text{D}
\end{aligned}$$

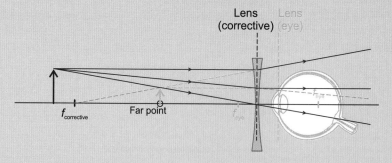

When using glasses, the eye 'sees' the virtual image created by the corrective lenses, and not the object. The image produced by the corrective lenses becomes the 'object' for the eye. This makes sense, as the eye knows nothing about the history of any rays that hit it, only that they hit the eye at a certain position and angle. So, any image created by the corrective lens at or nearer than the far point can be treated as if it were an object.

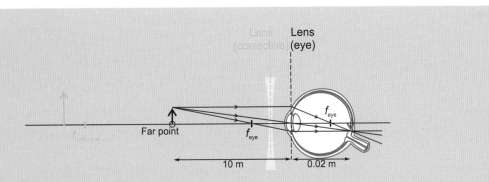

The combination of the corrective lenses and the man's own eye is able to form a clear image of an object at infinity on the retina, which is 0.02 m from the eye's lens. The minimum combined optical power must be

$$P_{combo} = \frac{1}{f_{combo,max}} = \frac{1}{d_o} + \frac{1}{d_i}$$
$$= \frac{1}{100 \text{ m}} + \frac{1}{0.02 \text{ m}}$$
$$= 50.01 \text{ D}$$

We can now return to Eq. (31.1) to calculate the required optical power of the corrective lenses:

$$P_{combo} = P_{eye} + P_{glasses}$$
$$P_{glasses} = P_{combo} - P_{eye}$$
$$= 50.01 \text{ D} - 50.10 \text{ D}$$
$$= -0.09 \text{ D}$$

The required power of the corrective lenses is -0.09 D. The negative sign indicates that the lenses should be *diverging* lenses and a power of -0.09 D equates to a focal length of $\frac{1}{-0.09 \text{ D}} = -11$ m.

The fact a diverging lens is required makes sense, as the man's eye is too powerful: it bends incoming light too much. A diverging lens will increase the divergence of incoming light, offsetting the too-large optical power of the eye itself.

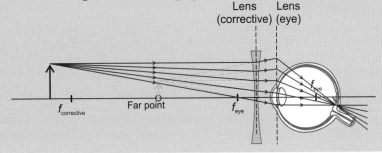

31.5 Hypermetropia (or Hyperopia)

Hypermetropia is a condition that causes objects that are *near* to become blurry in appearance. It is often called farsightedness, although this term is too general, as other conditions may cause poor near vision. It occurs when light rays from an object that would be clearly visible and focussed to a normal eye cannot be focussed strongly enough to form a clear image on the retina. This can be because the lens and cornea are unable to bend the light enough (refractive hypermetropia), or because the eye is too short (axial hypermetropia).

The degree of hypermetropia is characterised by the power of the lens needed to correct it. Mild or low hypermetropia is up to $+3$ D, moderate from $+3$ to $+10$ D and high hypermetropia is more than $+10$ D. A *converging* lens is needed to correct for hypermetropia.

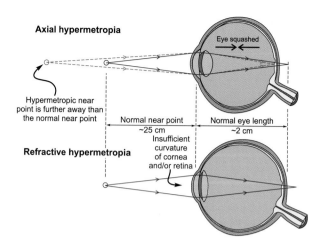

Figure 31.5 A hypermetropic eye cannot bend light enough to form an image of a nearby object on the retina. An image is instead formed behind the retina. The near point of a person with hypermetropia is further away than the typical 25 cm.

31.6 Presbyopia

The term **presbyopia** refers to the changes to the focussing ability of the eye that occur with age, and is merely the result of the same aging process that gives us wrinkles and grey hair. It is not due to 'wearing out' the eyes with use. As we age, the lens loses elasticity, and eventually will not curve enough during accommodation to allow near objects to be seen clearly. This is why some elderly people may need to hold the newspaper further away to see it. For those who started with myopia, it may be possible to see near objects without trouble even with this diminished flexibility of the lens.

Example 31.2 *Presbyopia/hypermetropia*

Problem: As Beth has aged, she has noticed that in order to read the newspaper in the morning she must hold it further and further away from herself. She has recently bought a cheap set of reading glasses, as in order to read the paper without them, she must hold the newspaper at arm's length. With her new glasses on, she can hold the paper as close as 30 cm from her eyes and still read it clearly. If Beth's arms are 55 cm long and she wears her glasses such that they are 2.5 cm in front of her eyes, what is the optical power of her new reading glasses?

Solution: The maximum optical power of Beth's eyes is now no longer large enough to form clear images of nearby objects. Like many people, Beth has developed presbyopia as she has aged. If Beth can just read the newspaper when she holds it in her outstretched arms, we know that her near point is now 0.55 m. When the newspaper is closer than this point the image formed by Beth's unaided eyes would be behind the retina.

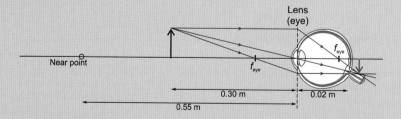

When she wears her new glasses, Beth's eye is able to form an image of the newspaper on her retina when it is as close as 0.3 m. Beth's eye has not changed, but the corrective lenses are producing a *virtual* image at Beth's near point which is acting as a new 'object' for her eye. The image produced by the glasses must be virtual as a real image would be inverted.

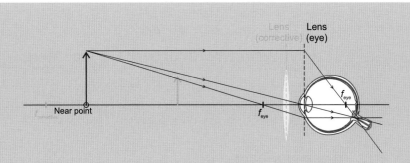

We can calculate the optical power of her new glasses by using the known object and image distances as shown in the diagram below. Beth holds the newspaper 0.3 m away from her eye to read it using her new glasses. Given that her glasses sit 0.025 m in front of her eye, the object distance must be $d_o = 0.275$ m. Similarly the image distance must be $d_i = -0.525$ m. The negative image distance is quite important, as this indicates the corrective lenses are producing the required virtual image.

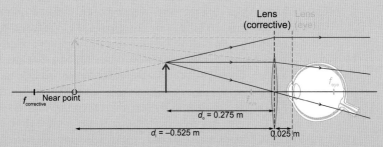

The thin-lens equation will enable us to calculate the optical power of Beth's new reading glasses:

$$P_{glasses} = \frac{1}{f_{glasses}} = \frac{1}{d_o} + \frac{1}{d_i}$$
$$= \frac{1}{0.275 \text{ m}} + \frac{1}{-0.525 \text{ m}}$$
$$= 1.73 \text{ D}$$

The optical power of the corrective lenses in Beth's new glasses is 1.73 D. The positive sign indicates that the lenses are *converging* lenses and a power of 1.73 D equates to a focal length of $\frac{1}{1.73 \text{ D}} = 0.578$ m.

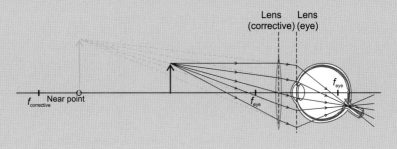

31.7 Astigmatism

Astigmatism occurs when the curvature of a focussing optical element differs along different axes. Imagine that instead of the surface of the cornea or lens being like part of a sphere, it is more ellipsoidal. This gives the eye different focal lengths in different planes. It is usually hereditary, but may be caused by injury, surgery or keratoconus, a disease that causes a weakening of the corneal tissue.

Mild astigmatism may not be a problem at all. In more noticeable cases, it can be corrected with glasses and contact lenses. In the case of contact lenses, the orientation of the lens matters, and these lenses are designed to maintain their position by having

Astigmatism in aquatic mammals

It has been reported that many species of seal have severe corneal astigmatism. This is not a problem under water, as the cornea and the water have similar refractive indices. In the air the orientation of the slit pupil may diminish the effects, but seals could well suffer from myopia in low light conditions. Otters have normal vision in the air, and they compensate under water by being able accommodate further than humans are able to.

Birds and eyesight

It has long been recognised that birds have better eyesight than us. Says Gordon Walls, in his book *The Vertebrate Eye and its Adaptive Radiation*, 'In this respect, man acknowledged even the small birds to be his superior, centuries ago – it was the habit of the medieval falconer to carry a caged shrike on his saddle to keep track of the falcon. As long as the shrike acted fearful and excited, the hawker knew that his proud tiercel was in sight – though not to *him*!'

a thickened bottom that tends to stay in the 6 o'clock position.

31.8 Alternative Structure & Placement

Throughout the animal kingdom, there are examples of light-sensing organs that differ quite a bit from our own. The most interesting differences have to do with how image formation is managed, and the placement of the eyes.

Focussing Ability

The more sophisticated visual systems incorporate some method of focussing so images can be formed of objects at differing distances. There are two main ways of achieving this: changing the power of the focussing part of the eye, or changing the distance to the light-sensing part.

As humans, our eyes can change their focal length by changing the shape of the lens. This is a technique we have in common with other mammals, birds and reptiles. Some invertebrates, and the lamprey (an eel-like creature), can alter the curvature of the cornea. Birds and lizards are able to squeeze their lenses to achieve much higher curvature than we can, and so are able to see objects that are very close.

In most fish, the lens-to-retina distance can be varied by moving the lens. Some fish use muscles to pull the lens backward to focus on more distant objects, and some push the lens forward to see nearer. Frogs and snakes also use this method. These creatures have lenses that are near spherical, rather than biconvex like ours.

Having a tilted retina is another option. This would mean that altering the direction of the eye would change where on the retina the image is and hence the image distance, and for quite some time it was reported that horses used this method, though that idea has now been discarded. The eyes of some rays (the fish kind) seem to have tilted retina, though.

Another way of having vision that works for a range of distances is to have detector cells that are long, so that light is focussed somewhere onto the cells for objects at a range of distances. This requires no active focussing, and is used by some deep-sea fishes and geckoes. The fruit bat has a unique variation on this, with conical deformations of the visual layer, which put retinal cells at a range of depths, so the image is formed sharply on the rods at a some level on these little mountains.

Eye Placement and Field of Vision

In humans, the eyes are both forward-looking and placed at the front of the head. This serves us well, as it allows for **stereoscopic vision**. Our two eyes see slightly different views of the world, and this allows us judge distance quite well. There is a trade-off for this; we can only see in front of us. Our field of vision covers only the 180° or so that lies in front of the direction we face.

Many other animals have a vastly improved field of vision. Most fish and birds have their eyes placed to the side of their heads, which means that they can see most of the way behind and even above them. In the case of the hare, the field of vision from the two eyes is nearly circular; your chances of sneaking up on a hare are not good!

In many species, the eye placement is an indicator of whether the species is the hunter or the hunted. Eyes placed in front for overlapping fields and improved depth perception indicates predator (owls, cats, etc.), while eyes to the side indicates prey (mouse, pigeon, and so on). There are some cases where we see both. The swallow has eyes placed so that the inner halves of the visual fields overlap, but it can also see a long way to the side. Each of the swallow's eyes has two maculae (areas on the retina with a high density of cells for acute vision), one set for looking straight forward, and one set giving a clear side view. Some birds even have their maculae spread out in bands across the retina. If we humans had this, it would possibly be like being able to clearly see all the books on a shelf at once, rather than having to concentrate on one at a time. Horses also have a macular streak. The ability to keep a better eye on the horizon seems to be a good survival trait.

An interesting, but not well-known, fact about human vision is that the half of the visual field interpreted by each eye is sent to the other side of the brain. Because the image formed by the cornea and lens is inverted, the left-hand side of what we see before us is imaged onto the right side of the retina in each eye. This is sent to the right side of the brain *from both eyes*. The images from the nasal side of what we see with each eye is sent to the other side of the brain. In the case of brain injury, this can result in extra difficulty reading that is culturally dependent. If, like native English readers, you are used to reading left-to-right, the inability to see the text that is coming next to the right will slow your reading speed a lot, whereas the inability to see to the left will not matter so much. The side of the brain that is injured can produce different degrees of difficulty based on which direction you learned to read.

31.9 Colour Vision

Detector Types

The human eye has two types of light-sensing photoreceptors in the retina, known as **rods** and **cones**. The intensity-sensing rods are far more numerous (on the order of 100 million) than the colour-sensing cones for which estimates vary, but the number is something like 5 million. The cones are concentrated in the area of the retina known as the macula, and in the fovea centralis, in the centre of the macula, there are no rods.

Rods are responsible for night vision and peripheral vision. They are about a thousand times more sensitive to light than the cones, but take longer to adapt to changing light conditions. Because of the presence of many rods in the regions of the retina responsible for peripheral vision, if you are having trouble seeing at night, it is a good idea to look off to the side a little, to take advantage of the increase in light sensitivity. Interestingly, rods respond to the yellow-green area of the spectrum strongly, and moderately to the blue end, but not the red. This is the reason for the increasing brightness of green grass and trees as twilight approaches. It is also useful in situations where good dark vision is required – red lights can be used without affecting the eyes adaptation to low light levels, whereas white light would spoil it for up to half an hour. The pigment in the rods in vertebrates (called rhodopsin) which allows them to respond to light is sometimes referred to as *visual purple* for this reason – it doesn't absorb well in the red, reflecting it instead and thus appearing purplish. Vitamin A is required to make this pigment, and so one of the first symptoms of a deficiency in this vitamin is night-blindness.

The cones show some peculiarities in distribution and sensitivity. The long-wavelength cones (designated L, and peaking in the yellow) and medium-wavelength cones (designated M, and peaking in the green) make up by far the largest numbers (about 95% in total). The short-wavelength (S) cones are found scattered slightly further out than the other types. These blue-sensitive cones are much more sensitive than the others, but not enough to compensate for the reduced number. It is believed that some compensation happens in the brain. This difference in the number of S-cones is partly responsible for blue objects being less distinct than other colours, but the difference in focal length for the lens–cornea system at different wavelengths also has an effect. Graphic arts experts suggest caution when putting red and blue in close proximity, particularly for text. The difference in accommodation in the eye needed to focus on the different colours is tiring for the viewer, as in Figure 31.6.

DNA coding for some of the proteins involved in vision are found on the X-chromosome, which is the reason for the sex-linked nature of many colour-vision abnormalities. The proteins that are needed for red and green detection are found only on the X-chromosome, so only one copy exists in males. If this copy contains a defect, red-green colour blindness is the result. This occurs in something like 5% of the male population.

Colour Science

The colour-receptive cones in the retina come in three varieties, which have peak responses in different areas of the visible spectrum. When light of all wavelengths enters

Vitamins and vision

An important vitamin for eyesight is vitamin A. The retinal group that is a key chemical part of the light-sensitive molecules in our rods and cones is produced from vitamin A, and night-blindness is an early symptom of an insufficient supply. A good source of vitamin A is liver, though is it wise to avoid the livers of animals adapted for polar environments, as they contain too much. Less than 100 g of polar bear liver is enough to be fatal in humans.

Another vitamin deficiency that is linked to eyesight problems is lack of vitamin B_1 (thiamin). A lack of thiamin causes problems with neural functioning, and so can interfere with proper transmission of the signals along the optic nerve.

Riboflavin (vitamin B_2) deficiency can cause the small blood vessels that normally stop at the sclera–cornea junction to begin to grow onto the cornea, and can cause some opaque spots to form in the cornea. Riboflavin is also needed to prevent cataracts.

Figure 31.6 The eye has to change accommodation to focus properly at the two ends of the visible spectrum, which can make some colour combinations jarring to look at when in close proximity.

Camouflage

In the Second World War, objects were camouflaged by painting them the same shade of green as grass. However, the spectrum of wavelengths scattered from the paint and the grass were not the same, even if they appeared that way to the naked eye. By using filters to block some wavelengths, the differences could be seen.

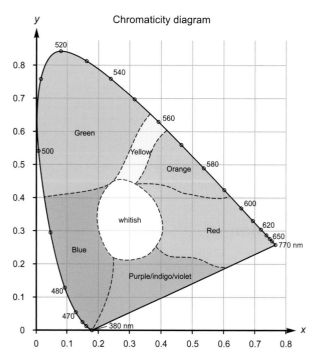

Figure 31.7 An approximation of the 1931 CIE (Commission Internationale de l'Èclairage) chromaticity diagram. The colours that we see from single wavelengths are around the top edge. Mixtures of two wavelengths appear to lie on the line between these points. There are many different mixtures of wavelengths in varying intensities that will have the same appearance to the eye.

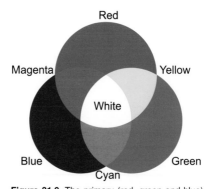

Figure 31.9 The primary (red, green and blue) and secondary (magenta, cyan and yellow) colours. Adding two primary colours of light together will produce the secondary colours, adding all three primary colours together will produce white. Adding all three secondary colours together will also produce white.

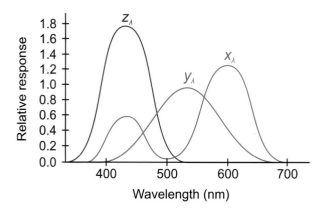

Figure 31.8 Functions used to model the response of the human eye to colour.

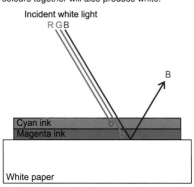

Figure 31.10 A combination of cyan and magenta ink appears blue.

the eye, all the receptors are stimulated, and the brain interprets this as close to white. If only some wavelengths are present, the receptors are stimulated different amounts and this is interpreted as a particular colour.

For example, most of the emission for orange sodium street lamps is at a *single wavelength* (589 nm) which stimulates the red and green receptors, and most people interpret this as orange. By contrast, an orange sweater looks orange in white light because the pigments in the fibres absorb some light, reflecting a *mixture of wavelengths* that stimulates the red and green cones. A chromaticity diagram, like the one sketched in Figure 31.7, is a way of showing how any mixture of wavelengths of visible light will appear to the eye. A nearly monochromatic (that is single wavelength) source, like a laser, will lie on the upper border of the shaded area. Any mixture of two wavelengths will lie on the line between the two wavelengths, and will lie closer to the more intense of the pair.

Figure 31.8 shows the functions used for modelling the response of the typical eye to the visible spectrum. By multiplying these functions by the actual light spectrum from a given object, three values can be assigned to the colour. These three numbers are known as the *tristimulus values*. These values are weighted so that they add up to one, which is how we are able to produce a two-dimensional plot – the third value can be found from the two that are plotted.

Because the eye responds primarily to red, green and blue light, these are known as the **primary colours**. Almost any colours that the brain can differentiate between can be produced with mixtures of monochromatic red, green and blue light. This is why if you look very closely at your TV screen you will see that it is made up of little areas of red, green and blue.

Three *secondary colours* are defined also, these being cyan, magenta and yellow. These secondary colours can be thought of a mixtures of pairs of the primary colours of light, or the absence of one of the colours from white light.

The usefulness of these secondary colours becomes readily apparent in the printing process, and indeed the three secondary colours will be very familiar to anyone with a colour printer. If yellow ink is printed onto a page, it appears yellow under white light because red and green light are reflected by the pigment and blue light is absorbed, hence you can think of yellow as red and green mixed, or the absence of blue. Similarly, magenta ink looks that way because green is absorbed by it, and blue and red are reflected. Cyan ink absorbs red light, reflecting green and blue. This means that having a layer of magenta ink and then cyan ink will absorb green and red light from a white light mixture, leaving blue. In the same way, cyan and yellow mixed look green, and yellow and magenta will look red. This ability of pairs of secondary colour pigments to appear as any of the primary colours means they can produce almost any visible colour when mixed in the right proportions.

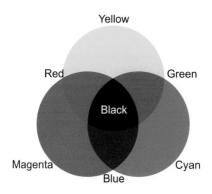

Figure 31.11 Subtracting two secondary colours from white light will produce the primary colours, subtracting all three secondary colours will produce black. Subtracting all three primary colours will also produce black.

31.10   Summary

Key Concepts

myopia A vision defect which causes collimated light to be focussed before the retina even with the accommodation muscles relaxed, making distant objects appear blurred. It can be the result of a refractive problem, or because the eye is too long. Sometimes called 'short-sightedness'. Myopia can be compensated for with a diverging lens.

hypermetropia (or hyperopia) A vision defect that causes light rays to be focussed behind the retina when the accommodation muscles are in a relaxed state, either due to a refractive error or the eye being too short. Distant objects can be brought into focus by increasing the lens power by accommodation, so it is sometimes called 'farsightedness'. Hypermetropia can be compensated for with a converging lens.

presbyopia Age related loss of flexibility of the lens, causing blurred near vision.

emmetropia Normal vision.

astigmatism A visual defect caused by an irregularly shaped cornea or lens, causing a difference in curvature along different axes.

31.11 Problems

For the purposes of answering these questions you can assume that a normal human eye has a minimum optical power of 50 D, a maximum optical power of 54 D, and that the normal distance between the retina and the lens is 2 cm. **31.1** A person with axial hypermetropia has a lens-retina distance of 1.9 cm and the maximum optical power of their eye is the same as that for a normal person.

(a) What is the near point of this person?

(b) What is the range of accommodation this person needs to see objects from their near point all the way up to their far point (which is the same as for a normal eye)?

(c) What is the optical power of the contact lenses used to treat this person and give them a normal near point of 25 cm?

31.2 A person with a normal lens-to-retina distance wears contact lenses with an optical power of 1.2 D in order to be able to clearly see objects 25 cm in front of them.

(a) What kind of vision defect does this person have?

(b) What is this person's near point (without the contact lenses)?

31.3 A person has refractive myopia with a far point of only 5 m. They are to be prescribed a set of glasses that will enable them to see distant objects clearly and the person's glasses will typically sit 2 cm in front of their eyes.

(a) What is the minimum optical power of this persons eyes?

(b) What is the optical power of the glasses required?

31.4 A person has refractive hypermetropia with a near point at 3 m. They are to be prescribed a set of glasses that will enable them to have normal close-in vision and the person's glasses will typically sit 2 cm in front of their eyes.

(a) What is the maximum optical power of this person's eyes?

(b) What is the optical power of the glasses required?

31.5 A person who had normal vision when they were younger now has age related presbyopia. They can still see distant objects clearly but have a reduced range of accommodation of just 1.0 D.

(a) What is this person's near point?

(b) What is the power of the contact lenses needed to correct this presbyopia (so that person has a normal near point of 25 cm)?

31.6 When driving you need to be able to clearly see road signs and traffic some distance ahead of you as well as the dashboard in your car. Assume that you have a normal lens-to-retina distance.

(a) What minimum optical power of the eye is needed to clearly see a road sign 200 m ahead?

(b) What maximum optical power is needed to clearly see the dashboard 40 cm away?

(c) What range of accommodation is needed?

(d) If you can accommodate at a maximum rate of 1.1 dioptres per second, how long does it take your eyes to adjust when looking up at the road from the dashboard?

(e) If you are traveling on the open road at 100 km h^{-1}, how far do you travel in the time it takes your eyes to accommodate between the dashboard and the road?

31.7 Many automated industrial engineering plants use high definition cameras on the production line in order to monitor the quality of products on the assembly line. A particular plant manufactures small machined products which have a maximum depth of 5 cm. If the cameras used have a single lens which is around 3 cm from the CCD array on which the image is projected (and captured), and are placed such that the lens is 8 cm above the conveyor belt on which the circuit boards rest. What is the necessary range of accommodation of these cameras if they are required to clearly image details over the whole range of depth of the circuit boards? How does this compare to the accommodation range of the human eye?

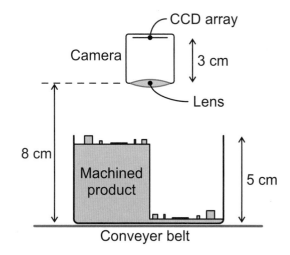

Figure 31.12 A machined product passing underneath an automated camera. The camera needs to be able to focus on all parts of the object (although not necessarily at the same time).

31.8 A person with refractive myopia can see objects as close as 25 cm clearly, and objects as far away as 3 m clearly.

(a) What is the maximum optical power of this person's eye?

(b) What is the minimum optical power of this person's eye?

(c) What is the range of accommodation of this person's eye?

The person gets a set of contact lenses for their eyes to correct their far vision. When they are wearing the contact lenses they can see objects in the distance clearly.

(d) What is the optical power of the contact lenses?

(e) What is the person's new near point?

(f) What is the range of accommodation of this person when wearing their contact lenses?

WAVE OPTICS

32

32.1 Introduction

In some situations, it is not possible to explain the behaviour of light without using a proper wave description. When light interacts with structures on a similar scale to its wavelength, its wave nature can no longer be ignored. In this chapter we will look at some diffraction and interference effects, and explain how the diffraction of waves limits the resolution with which we detect location.

Key Objectives

- To understand how Huygens' principle can be applied to explain wave phenomena.

- To understand wave diffraction.

- To understand the pattern formed by diffraction of monochromatic light through a single slit.

- To understand how diffraction limits resolution.

32.2 Superposition and Interference

Before we go on to look at wave effects in more detail, a quick reminder about superposition and interference of waves.

When more than one wave propagates through a medium, the waves pass through one another unchanged, and the resulting disturbance is the sum of the individual displacements. This is called the principle of **superposition**.

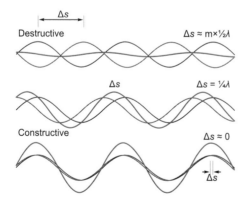

Figure 32.1 The superposition of two waves. When the waves are nearly out of phase they add together destructively (the sum of the two waves ≈ 0 at all points). When the waves are nearly in phase they add together constructively (the sum of the two waves is close to its maximum at all points).

At each moment in time, the displacement of a particular point in a medium is the sum of the displacements caused by of all the waves passing that point (see Figure 32.1). The resulting patterns of **constructive** and **destructive** addition are called **interference**.

32.3 Huygens' Principle

In geometric optics, we treated light as something that travelled in straight lines (rays) from point to point. An analysis method named after Christiaan Huygens (1629–1695) allows us to follow the propagation of not just a simple ray, but an entire **wave front**. Figure 32.2 shows a glimpse of how such wave fronts propagate out from a single point in simple circular case, rather like the ripples spreading out in a pond when a stone is dropped in. We can use the idea of overlapping circular wavelets like these to understand the way much more complex wave fronts will move through space.

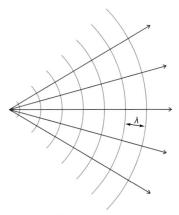

Figure 32.2 Circular waves propagating out from a source point. The curved lines represent the crests of the wave. A wave front is all the points of a wave that are in phase, such as these crests. The rays show the direction of propagation which is everywhere perpendicular to the wave fronts.

> **Key concept:**
> **Huygens' principle**: Every point of a primary wave front serves as the source of spherical secondary wavelets such that the primary wave front at some later time is the envelope of these wavelets. Moreover, the wavelets advance with the speed and the frequency equal to that of the primary wave at each point in space.

By treating *every* point on the wave front as a source of secondary circular wavelets, and calculating the mathematical sum of these secondary wavelets, the position of the wave front at a later time can be determined. Figure 32.3 shows some representative points on the wave front, and where the secondary wave fronts from those points will have reached at a later time. In most places, the wavelets will show some degree of destructive interference, except at the lines, which are where the new wave front will be at particular times. This is how we expect a plane wavefront to move forward in an isotropic medium.

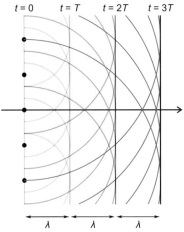

Figure 32.3 The application of Huygens' principle to the propagation of a plane wave in an isotropic medium. After one time period, $T = 1/f$, each wavelet has advanced radially by a distance equal to the wavelength of the wave, λ. Each wavelet will interfere constructively only along the straight line representing the translated wave front.

Refraction Revisited: Proof of Snell's Law

We can use this idea of secondary wavelets to help us in the proof of Snell's Law. Figure 32.4 shows a few selected wavelets propagating in the second medium, which has a slower wave speed. From this, we can build up a picture of where the single wavefront will be at various times. The direction of propagation is perpendicular to the wavefront, so the light bends at the interface.

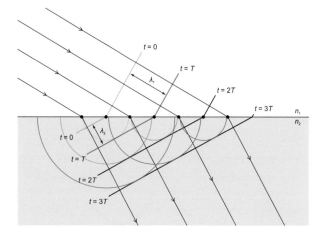

Figure 32.4 As a wave front advances at an angle along a boundary between two media of different refractive index, Huygens' wavelets are produced which propagate through the second medium at a different speed (slower in this case as $n_2 > n_1$). The wave front of the wave in the new medium can be reconstructed from these wavelets.

Figure 32.5 shows just the wavefronts, with the distances travelled in the same time by different parts of the wave front. The angles of incidence (i) and refraction (r) are defined as the angles the light rays make with the normal. By looking at the shaded triangles constructed, we can see that the angles i and r are the same as the interior angles of the triangles that have been similarly labelled. The sines of the two angles are

$$\sin i = \frac{x}{h} \longrightarrow h = \frac{x}{\sin i}$$
$$\sin r = \frac{y}{h} \longrightarrow h = \frac{y}{\sin r} \tag{32.1}$$

If t is the time taken for the wave to travel the distance labelled x, and the wave travels at speed v_1 in the first medium, then $x = v_1 t$. Similarly, $y = v_2 t$, as y is the distance the wave travels in the second medium *in the same time interval*. By the definition of the refractive indices,

$$v_1 = \frac{c}{n_1} \text{ and } v_2 = \frac{c}{n_2} \tag{32.2}$$

so

$$x = \frac{ct}{n_1} \text{ and } y = \frac{ct}{n_2} \tag{32.3}$$

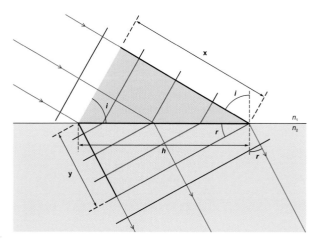

Figure 32.5 Geometric construction that allows the derivation of Snell's law, as given in Eq. (29.4)

Putting Eq. (32.1) together and substituting in Eq. (32.3)

$$\frac{x}{\sin i} = \frac{y}{\sin r} \tag{32.4}$$
$$\frac{ct}{n_1 \sin i} = \frac{ct}{n_2 \sin r}$$

This results in Snell's Law

$$n_1 \sin i = n_2 \sin r \tag{32.5}$$

which was presented previously in the ray optics section of the text.

32.4 Diffraction

When waves pass by an obstacle or through a gap, there is some bending into the shadowed region, as can be seen in Figure 32.6. This is known as **diffraction**. The other properties of the wave (speed, wavelength, frequency) are not changed.

The amount of bending is greatest when the gap through which the wave passes is around the same size as the wavelength. An example of this is the way we can hear around corners, but not see around them. The sound waves audible to humans have wavelengths on the order of centimetres or metres, and will diffract around corners, with the effect being stronger for bass (lower) frequencies. Light waves, with wavelengths less than a micron, will not bend anywhere near enough. (Sound waves also echo off walls in a more coherent manner as the surfaces are very flat compared with the wavelength)

The technique of using rays travelling in straight lines, which was sufficient for explaining image formation by mirrors and lenses, cannot be used to explain diffraction.

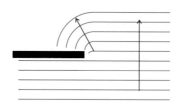

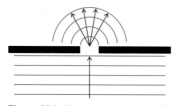

Figure 32.6 Huygens' wavefronts 'diffracting' around corners. Subtraction of all the wavelets blocked by the obstruction results in a curved wavefront when the remaining wavelets are added together.

Diffraction confusion

Don't confuse diffraction with dispersion. Dispersion in glass is caused by different wave frequencies having different speeds. Diffraction bends different frequencies different amounts, but there is no change in speed. Having said this, some definitions of dispersion include any effect that separates the different frequencies, and include diffraction as a cause of dispersion. In this book, when we refer to dispersion, we mean separation caused by a frequency-dependent velocity.

Using Huygens' principle to draw the secondary wavelets can give an explanation for this phenomenon, though.

32.5 Young's Double-Slit Experiment

Thomas Young (1773–1829) made significant contributions to various areas of optics (such as the investigation of colour vision) and is also widely regarded as being the first person to decipher some of the Egyptian hieroglyphics on the Rosetta Stone. There is one particular demonstration of the wave nature of visible light that was so significant that his name is still nearly always added to it – the double slit experiment.

Figure 32.7 shows a version of the sketch from Thomas Young's 1803 paper on the experiment. The two points labelled A and B are *coherent* point sources of waves. By coherent, we mean that the sources are in phase – they emit waves that go up and down at exactly the same time. The stripes and white spaces represent the crests and troughs of waves. Any line (such as the centre line from between A and B to between D and E) where the crests always meet crests and the troughs always meet troughs will have the most variation in wave height. The points labelled C, D, E and F are places where the opposite happens, and crests always meet troughs. At these points complete destructive interference always occurs, and there is no resulting displacement. The points C, D, E and F are what we call **nodes**. The points halfway between them where the displacement is the largest are called **antinodes**.

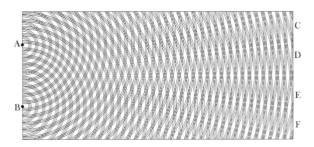

Figure 32.7 A copy of the sketch from Young's original work.

This effect is easily seen with water waves, but to observe these areas of constructive and destructive interference on a screen with light is a little trickier. The light must be monochromatic – a single wavelength – otherwise the places that are nodes for one wavelength still won't be dark because they are not nodes for others.

Monochromatic light can be produced with spectral lamps, which produce light at only a limited number of frequencies, and a suitable filter. Nowadays it is easy to get monochromatic light from lasers. The two sources must also be the right spacing apart. For visible light wavelengths the separation between sources needs to be on the order of a few microns. The best way to do this in practice is to use a single light source and to let this light pass through two very thin slits. If the slits are thin enough, they will act rather like point sources of spherical wave fronts, as in Figure 32.7. This will also ensure that the two 'sources' are in phase.

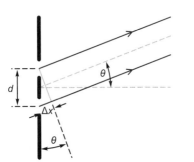

Figure 32.9 Close up of the slit, showing the angles.

Figure 32.8 and the magnified section in Figure 32.9 will allow us to use some geometry to find the criteria for constructive and destructive interference. We have squeezed Figure 32.8 up quite a lot to show the labelling of the distances better. The equations we are developing are for a much bigger L than that shown, which is why the two figures show different ray directions.

For a small angle, θ, if we use the dimensionless radian to measure the angle, then

$$\sin\theta \approx \tan\theta \approx \theta \text{ (in radians)} \tag{32.6}$$

From Figure 32.8

$$\tan\theta = \frac{y}{L} \tag{32.7}$$

 www.wiley.com/go/biological_physics

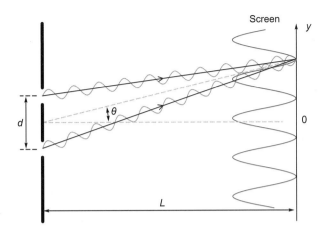

Figure 32.8 Light from each of the slits leaves *in phase* with the light from the other and hits the screen. At some points the phase difference caused by the different paths taken causes destructive interference, at others it causes constructive interference.

and from Figure 32.9

$$\sin\theta = \frac{\Delta x}{d} \tag{32.8}$$

The distance Δx is called the **path difference** – it is the difference in distance travelled for light from the top slit and bottom slit to reach the same place on the distant screen. When the path difference Δx is a whole number of wavelengths, that is $\Delta x = m\lambda$ where m is an integer, the interference will be constructive and there will be an intensity maximum on the distant screen. To find the places where there is destructive interference, the path difference needs to be such that a wave crest meets a wave trough, i.e., an integer-plus-a-half number of wavelengths. Assuming the angle is small enough for the small angle approximation (Eq. (32.6)) to apply, we have the following criteria

Constructive:

$$\Delta x = m\lambda = d\sin\theta = \frac{dy}{L}, \; m \in (\ldots, -1, 0, 1, 2 \ldots) \tag{32.9}$$

Destructive:

$$\Delta x = \left(m + \frac{1}{2}\right)\lambda = d\sin\theta = \frac{dy}{L}, \; m \in (\ldots, -1, 0, 1, 2 \ldots) \tag{32.10}$$

We can summarise the necessary information in one equation

$$\boxed{m\lambda = d\sin\theta} \tag{32.11}$$

as long as we remember what value of m corresponds to each feature of interest that we might want to locate, such as $m = 2$ for the second intensity peak off the central axis.

32.6 Single-Slit Diffraction

Of much more interest for practical purposes is the pattern of light and dark created by just *one* gap. This is of crucial importance for real-world applications. There is a relationship between the aperture (gap) size that we use to collect light, and how much information gets lost due to the wave fronts being bent.

Just as interference between waves originating from each slit forms alternating bands of light and dark, an interference pattern is observed when light passes through just one slit. This pattern takes the form shown in Figure 32.10.

The single-slit diffraction pattern is formed by the mutual interference of light that passes through the slit. We can gain a qualitative understanding of this effect by considering many rays passing through the slit at different positions and different angles. Figure 32.11 shows how the bright central peak, the first dark band and the first bright fringe are formed.

> **Fresnel and Fraunhofer**
>
> Here we are concerned only with what is called the 'far-field' or Fraunhofer diffraction pattern. Close to the slits, the pattern is quite different, and is known as the 'near-field' or Fresnel diffraction pattern.

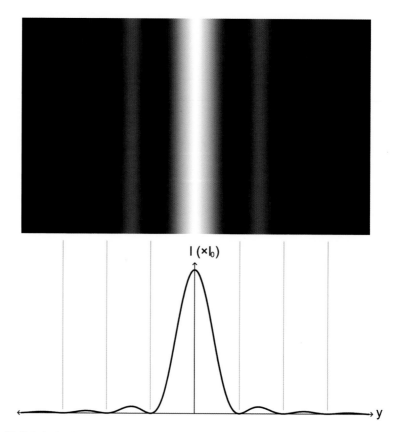

Figure 32.10 A single-slit diffraction pattern is characterised by a large central peak flanked by a series of increasingly weak bands. At the top of the figure is a simulated diffraction pattern for a slit of width 50μm being illuminated by light of wavelength 633 nm. Below this the intensity profile showing the variation in the intensity of the pattern.

Provided that the distance to the screen on which the diffraction pattern is being viewed is much larger than the slit width, we can say that the light hitting any point on the screen is the sum of all rays exiting the slit at some angle, θ.

The bright central peak is created by the constructive interference of all the rays exiting the slit straight ahead. Each ray is exactly in phase with the others and so the central peak is very bright.

The first minima either side in the diffraction pattern can be understood in terms of destructive interference between pairs of rays, each with a path length difference of $\frac{\lambda}{2}$. For every ray in the 'top' half of the slit, a ray may be found in the bottom half that has a path length difference of $\frac{\lambda}{2}$. These pairs of rays destructively interfere and the result is a minimum in the diffraction pattern. In this case the two 'extreme' rays passing through each edge of the slit will have a path length difference of λ and

$$\sin\theta = \frac{\lambda}{D}$$

In contrast to the double slit case, $\lambda = D\sin\theta$ results in *destructive interference*.

As the angle increases slightly, the number of pairs of rays which destructively interfere decreases, and the number of pairs of rays which constructively interfere increases. The first bright fringe is formed at an angle such that the path-length difference of the two extreme rays is close to $\frac{3\lambda}{2}$ and

$$\sin\theta = \frac{3\lambda}{2D}$$

Each ray in the 'top' third of the slit will *destructively* interfere with a ray from the middle third of the slit. This bright fringe is not as bright as the central peak, partially because not all rays interfere constructively.

As the angle continues to increase, other regions of destructive interference and partial constructive interference are found which cause the second, third, fourth, etc. bright/dark fringes.

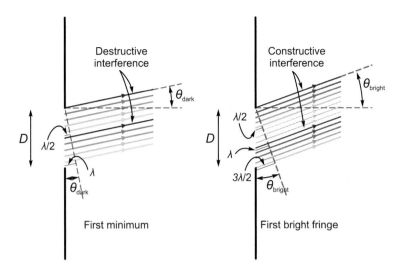

Figure 32.11 The first dark band (θ_1 in Figure 32.12) in the single-slit diffraction pattern is formed because of destructive interference between pairs of rays exiting the slit at an angle such that each ray destructively interferes with another that has a path-length difference of $\frac{\lambda}{2}$. The first bright side band (between θ_1 and θ_2 in Figure 32.12) is formed because of destructive interference between pairs of rays from the top third and middle third that have a path difference of $\frac{\lambda}{2}$. Some rays do not destructively interfere, and this bright band is less intense than the central peak.

Generalising this, the condition for *destructive* interference for a single slit is

$$D\sin\theta_m = m\lambda, \quad m = \pm 1, \pm 2, \pm 3, ... (m \text{ not zero}) \tag{32.12}$$

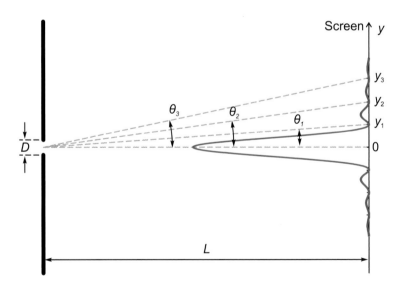

Figure 32.12 Destructive interference is seen at the angles $\theta_1, \theta_2, \theta_3$, etc. These angles correspond to the integers $m = 1, 2, 3, ..$ in Eq. (32.12)

In the real world, the two slits that are used to produce an interference pattern each have a finite width, and would individually produce a single-slit diffraction pattern like Figure 32.10. This means that the real pattern produced is not an infinite series of light and dark bands on a screen, but is a combination of the light and dark bands that would be expected from Eq. (32.11), with an intensity pattern that is shaped like a single-slit pattern with nodes at the positions predicted by Eq. (32.12).

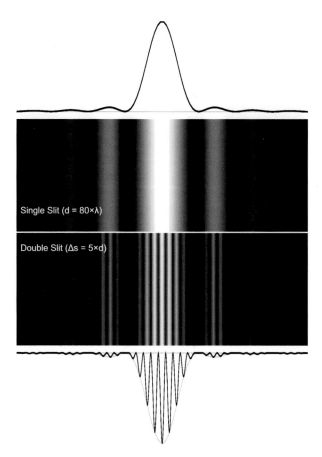

Figure 32.13 A comparison of the single and 'real' double slit diffraction patterns. When we have a pair of slits, each slit individually produces a pattern like that in the top of the figure. When two such patterns are overlapping, there is interference between the two patterns and a pattern like our theoretical double slit diffraction pattern is produced. The spacing of the thin bands is dictated by the double-slit interference pattern, but these peaks lie within the profile of a single-slit diffraction pattern, so some have lowered intensity.

Example 32.1 *Single-slit diffraction and double-slit interference*

Problem: Two slits, each 90 μm wide, are separated by a distance of 360 μm. The diffraction pattern produced when 750 nm light passes through the slits consists of a series of narrow peaks whose peak intensities vary inside an 'envelope' whose shape is the same as the single-slit diffraction pattern for each slit (see Figure 32.14).
In this particular case, as in Figure 32.14, the first minimum of the single-slit 'envelope' happens to coincide with the a peak in the double-slit pattern. Because of this there is a missing fringe. What is the order number m of the missing double-slit interference pattern peak produced through the double slits specified in the question? (Note: it is not the same as that shown in Figure 32.14.)

Solution: For light of wavelength λ, the *minimum* in the single slit envelope are given by

$$D \sin \theta_n = n\lambda$$

where D is the width of each slit, n is the order number of the minimum ($n = 0, 1, 2, ...$) and θ_n is the angle to the n-th minimum. The *maxima* in a double slit pattern are given by

$$d \sin \phi_m = m\lambda$$

where d is the spacing between the slits, m is the order number of the maximum ($m = 0, 1, 2, ...$), and ϕ_m is the angle to the mth maximum.

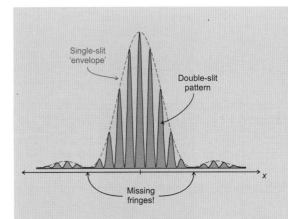

Figure 32.14 A double-slit pattern consists of a series of narrow peaks, the positions of which are found by considering the condition for constructive interference for a double slit, situated inside a single slit 'envelope'. It is possible for the narrow peak spacing to be such that one of the peaks falls in a minimum in the envelope.

We are trying to find which *peak* (m = ?) in the double-slit pattern coincides with a *minimum* in the single-slit pattern. Given this we can say that the angle to the nth minimum is the same as the angle to the mth maximum and so $\sin\theta_n = \sin\phi_m$

$$\sin\theta_n = \sin\phi_m$$
$$\frac{n\lambda}{D} = \frac{m\lambda}{d}$$
$$m = n\frac{d}{D}$$

which when using $n = 1$, $d = 360 \times 10^{-6}$ m and $d = 90 \times 10^{-6}$ m gives

$$m = 1 \times \frac{360 \times 10^{-6}\text{ m}}{90 \times 10^{-6}\text{ m}} = 4$$

The fourth double-slit maxima coincides with the first single-slit minima and so will be missing from the pattern produced.

32.7 Diffraction Gratings

A diffraction grating is a series of evenly-spaced slits, typically chosen to produce only a few narrow maxima. They are then useful for separating out different wavelengths of light. Again the position of the nth order maximum is given by

$$n\lambda = d\sin\theta \tag{32.13}$$

where d is the slit-to-slit distance, which is related to the number of slits per metre, N, by $d = 1/N$. The number N is usually many thousand lines per centimetre.

Diffraction gratings are useful for separating wavelengths of light, rather like a prism. They have some advantages, though. For example, a reflection-type grating can be constructed which will specularly reflect all wavelengths and act like a normal mirror (for the zeroth order), and also separate out the light into its component wavelengths over a range of angles (for the 1st-order maximum).

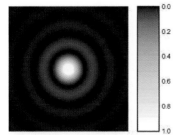

Figure 32.15 Intensity of diffraction pattern from a circular aperture. The scale at the right indicates the intensity, with white being the brightest. [Public domain picture courtesy of Wikipedia.]

32.8 Circular Apertures and Diffraction

The Airy Pattern

When the aperture is no longer a long slit, the pattern formed on a screen is slightly different. When a plane wave is diffracted through a circular aperture, the distribution of the wave intensity follows a characteristic shape called an **Airy pattern**. Figure 32.15 shows how the pattern would look viewed on a screen and Figure 32.16 shows the intensity profile from a crosssection through the centre.

Consider a situation where a circular aperture is uniformly illuminated by light from a distant point source; this means that instead of being able to form a corresponding sharp image point, an imaging system (such as the eye) really produces a smudge with rings. This limits the ability of a system to resolve objects that are too close together – the resulting smudges will overlap.

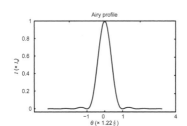

Figure 32.16 The Airy profile

The Rayleigh Criterion and Resolution

The sine of the angular distance (in radians) to the first minimum in the Airy pattern is in fact 1.22 times the wavelength divided by the diameter of the aperture, rather than

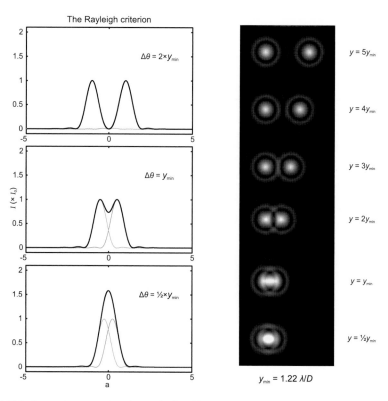

Figure 32.18 As the angular separation of two point-like objects viewed through some sort of circular aperture, be it a telescope or the human eye, gets smaller, the Airy patterns of the objects start to overlap. Eventually this results in an inability to definitively say that the two objects are separate from one another. (Left) Airy profile functions. (Right) Simulated Airy patterns through two small circular holes.

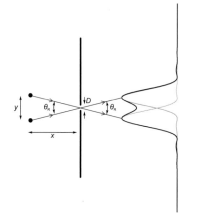

Figure 32.17 For small angles, the angle θ_R in radians is approximated well by y/x.

the 1 for a slit. This allows us to define a criterion by which we can specify when two objects are too close together to be resolved by the imaging system.

Imagine that we have two identical, distant, point sources of light that emit light at only one wavelength, λ. The light from these sources uniformly illuminates a circular aperture and we observe the (spread out) image of the two sources on a screen. The light intensity from the two sources will add together, and when the sources get closer, so will the images, until the central areas are so overlapped that it is no longer possible to tell that there are in fact two separate sources. (See Figures 32.17 and 32.18

Mathematically, we can express this as

$$\theta_R \sim \sin\theta_R = 1.22\frac{\lambda}{D} \tag{32.14}$$

where λ is the wavelength, D is the *diameter* of the aperture and θ_R is the minimum angular distance between the objects for them to be resolved.

This relationship tells us a great deal about the limits of any system designed to gather information of a wave nature. For example, for a telescope to gather information about objects in deep space (whether is the form of light or radio waves), the bigger D is made, the smaller the minimum angle of resolution. The diameter of the pupil of the eye sets limits on how good eyesight can be, although there are other considerations, such as the spacing of the light-sensing cells on the retina.

32.9 Visual Acuity

Visual acuity is the ability of the visual system to resolve details. Acuity is limited by diffraction, aberrations and the density of photoreceptors on the retina. Other factors, such as illumination and contrast, may affect image resolution.

One commonly used method of measuring this is the Snellen chart. The letters on the chart are constructed so that the details (width of stroke or gap) take up one-fifth of the height. In Snellen notation, a person's visual acuity is specified in terms of the ratio of two distances. The first is the viewing distance (usually 20 ft (feet) or 6 m). The second is the distance at which the smallest letters that could be distinguished would subtend an angle of a 12th of a degree. Normal acuity is 20/20 measured in feet, or 6/6 in metres, meaning that a person can read letters with a height of 8.8 mm at a distance of 20 feet. A person is legally blind if they cannot read the largest letter on the chart (an acuity of 20/200) while using the best corrective lenses available.

Example 32.2 *Resolution limit*

Problem: A telescope can just resolve features 5 km apart on the Moon's surface. Given that the moon is 3.85×10^8 m away from the Earth, what is the diameter of the telescope's aperture? Use a wavelength of 550 nm.

Solution: $s = r\theta$ therefore $\theta = \frac{s}{r} = \frac{5\times10^3}{3.85\times10^8} = 1.30 \times 10^{-5}$ radians. To be resolvable, the central maximum of one object's diffraction pattern must coincide with the first minimum of the other object's diffraction pattern. Therefore the angle to the first minimum of the diffraction pattern is 1.30×10^{-5} radians. Angle to the first minimum of the diffraction pattern of a circular aperture is (using the small angle approximation)

$$\theta \approx \sin\theta = \frac{1.22\lambda}{D}$$

Therefore

$$D = \frac{1.22\lambda}{\theta} = 0.052 \text{ m}$$

Example 32.3 *Resolution limit*

Problem: How close is normal visual acuity to the diffraction limit, assuming a pupil size of 5 mm?

Solution: To see the line on the chart that represents 20/20 vision, a person needs to be able to be able to resolve objects one-fifth of 8.8 mm apart at about 6 m distance. This requires an angular resolution of

$$\theta = \frac{s}{r} = \frac{8.8 \times 10^{-3} \times \frac{1}{5} \text{ m}}{6 \text{ m}} = 2.9 \times 10^{-4} \text{ radians}$$

The diffraction limit for an aperture of 5 mm with visible light ($\lambda \sim 500$ nm)

$$\theta_R = 1.22\frac{\lambda}{D} = 1.22\,\frac{500 \times 10^{-9} \text{ m}}{5 \times 10^{-3} \text{ m}} = 1.22 \times 10^{-4} \text{ radians}$$

The human eye is not so far from the limits of how good it can be in this respect.

32.10 Thin-Film Interference

Bands of colour in the light reflected from an oily puddle of water or a soap bubble are a familiar sight. When you look at a soap bubble in white light, you might notice that it is not really rainbow-like in its colouring, though. What you will see is mostly a mix of yellows, pinks and blues. This is because as the thickness and your viewing angle change, there is a particular wavelength range for which there is some destructive interference of the waves reflecting from the two water/air interfaces. If this is in the green part of the spectrum, the remaining colours will look much like a mix of red and blue – magenta. (See Section 31.9.)

In the chapter on Waves (Chapter 8) it was pointed out that a wave may or may not change phase upon reflection. For light waves:

- Waves reflecting off of a medium with a higher refractive index than the one they are travelling through have 180° phase change.

> **Iridescence**
>
> When some kind of interference phenomenon causes colours to change with viewing angle, this is called *iridescence*. This is the cause of some of the most beautiful displays of colour found in nature, such as the stunning blue wings of butterflies of the *Morpho* genus. Another example is the peafowl feather, where there are multiple kinds of interference involved, such as scattering off periodic nanostructures.

• Waves reflecting off of a medium with a lower refractive index than the one they are travelling through have no phase change.

For the waves reflecting off the front and back surfaces, the path difference will be twice the film thickness, t. If there is no phase change (or the same phase change at both surfaces), we would expect destructive interference to occur when $2t$ is 1/2, 3/2, 5/2 of a wavelength and so on. Similarly, if there is a phase change for one of the reflected waves (as for a soap bubble), then constructive interference will result instead.

32.11 Summary

Key Concepts

Huygens' principle An approach to wave propagation problems in which every point on a wave front is considered to be a source of forward-propagating secondary wave fronts. The combined effect of all the secondary wavelets gives the resultant advancing waves.

aperture An opening in something. The opening through which light passes to expose the film or hit the sensor in a camera. The diameter of the primary mirror or lens in a telescope.

diffraction The bending of wave fronts around obstacles or apertures. The effects of diffraction become significant as the wavelength approaches the obstacle or aperture size.

superposition (waves) When more than one wave propagates through a medium, the waves pass through one another unchanged, and the resulting disturbance is the sum of the individual displacements.

interference When multiple individual waves are superposed at the same position in a medium, this is termed interference. Constructive interference results when the displacements add to a disturbance larger than the individual waves, and destructive interference results when the sum of the disturbances is smaller than the individual waves.

Young's double-slit experiment A crucial experiment in the development of the wave theory of light, performed by Thomas Young, which proved conclusively that light exhibited wave properties. Monochromatic light passing through two narrow slits, closely spaced, shows an interference pattern on a screen.

single-slit diffraction When monochromatic light passes through a single, small aperture and falls on a screen some distance away, a series of dark minima and light maxima are observed with a bright central maximum.

resolution Due to diffraction, the ability of any optical system to produce distinct images of objects which are close together is limited. This limit is called the resolution or resolving power of the instrument.

Rayleigh criterion A criterion by which we can judge if two objects will be resolvable, which relates the diameter of the aperture to angle of resolution.

angular resolution The minimum angular separation of two objects, to then be distunguishable.

Equations

$$m\lambda = d\sin\theta \quad m \text{ integer (double slit, maxima)}$$
$$m\lambda = D\sin\theta \quad m = \pm1, \pm2, \pm3, ... (m \text{ not zero, single slit, minima})$$
$$\theta_m = 1.22\frac{\lambda}{D}$$

32.12   Problems

32.1 Light of wavelength 550 nm passes through a 10 μm wide slit on to a screen 1 m away from the slit.

(a) How far either side of the central maximum are the 1st, 2nd, and 3rd dark regions in the diffraction pattern?

(b) What is the maximum possible number of bright fringes that could be viewed either side of the central maximum in perfect conditions?

32.2 A high intensity source of microwaves used in a piece of medical diagnostic equipment is not adequately shielded. The microwaves produced by this equipment have a frequency of 150 GHz, and there is a gap in the shielding 1 cm wide. At what angles from the gap in shielding will the intensity of the microwave radiation be zero? (assume the microwaves are incident normally to the gap.)

32.3 A concert is to be held in a large hall and two speakers are placed 5 m apart at the front of and in the middle of the hall. During sound testing these speakers are producing a steady 1000 Hz tone (the speed of sound in air is $340 \, \mathrm{m \, s^{-1}}$). The hall is 50 m long and 30 m wide.

(a) The hall is 30 m wide and a person starts in the middle. As the person walks to their right they should notice the sound intensity vary. At what positions would you predict the person would be able to hear the 1000 Hz tone produced by the speakers *most clearly*?

(b) It is unlikely that the person will actually notice much difference in the sound intensity as they walk across the back of the hall. What is the single biggest reason for this difference between your prediction in (a) and the actual experience of the person?

32.4 A particular toy telescope produces a magnified virtual image of distant objects. This virtual image has a magnification of +3 and is located at the same distance from the telescope as the object. The diameter of the objective lens of the telescope is 3 cm. When looking through the telescope, the pupil of the eye dilates to 6 mm. What is the minimum separation between two objects that can be resolved at a distance of 150 m? (Assume light of wavelength $\lambda = 550$ nm.)

32.5 The primary mirror of the Hubble Space Telescope is 2.4 m in diameter. Suppose it is most sensitive to light of wavelength of 820 nm.

(a) What is the diffraction limited angular resolution of the Hubble Space Telescope?

(b) If the HST were used to look at the surface of the Moon, what is the minimum distance between two distinguishable points? (The Moon orbits at a mean distance of 384000 km from the center of the Earth which has a radius of 6380 km, the HST orbits at an altitude of 569 km.)

(c) If the HST were used to look at the surface of Jupiter, what is the minimum distance between two distinguishable points? (The orbital radius of the Earth is 150×10^6 km, and the orbital radius of Jupiter is 779×10^6 km.)

(d) If the HST were used to try and find planets around a star that was only 20 light years away (very close in astronomical terms), what is the minimum distance between two distinguishable points? ($c = 3.00 \times 10^8 \, \mathrm{m \, s^{-1}}$)

32.6 A normal person's pupil diameter varies from a minimum of around 2 mm to a maximum of around 8 mm.

(a) What is the minimum and maximum diffraction limited angular resolution of a normal person's eye for light of wavelength 600 nm?

The light that falls on the retina, around 2 cm behind the pupil, is detected by photo-receptive cells. These cells are most closely spaced in a region called the fovea on which the image of objects directly in front of the eye is produced by the cornea and lens. The density of the photo-receptive cells in the fovea is approximately 3×10^5 cells per square millimeter and the cells are approximately circular which means that they have a diameter of around 2.06 μm.

(b) Is the theoretical maximum resolution of the eye for 600 nm light likely to be limited by the number of photo-receptive cells or the Rayleigh criterion for a wide open pupil?

(c) Given your answer in (b) what is the minimum distance between two objects that can be resolved by the unaided eye at distances of 1 m, 10 m, and 1 km? (Note that these are theoretical limits and the actual limits are somewhat worse than this.)

32.7 An eccentric physicist inexplicably makes his home in the bilge of a vodka delivery ship crossing the Atlantic ocean north of the arctic circle. As the physicist forgot to note down the wavelength of a new light source before throwing the packaging overboard he performs a single slit diffraction experiment using a slit of width 110.0 μm. The diffraction pattern is displayed on a screen 3.000 m from the slit and the 1st minima in the diffraction pattern are 1.420 cm either side of the central maximum. ($n_{\mathrm{air}} = 1.008$.)

Disaster strikes at midnight in the form of an iceberg pushed into the path of the ship by a team of vengeful teetotaler polar bears. As the ship sinks and the lab becomes inundated the single slit diffraction experiment becomes submerged in the briny water. The very cold but excited physicist notices that the spacing between the 1st minima and the central maxima of the diffraction pattern changes to 1.066 cm.

With the air in his lungs rapidly running out and the chilling cries of triumphant polar bears filling his ears he hurriedly scrawls (to four significant figures) both the wavelength of the light source in air and the refractive index of salty water on the wall for the edification of future salvage divers. What did the dead physicist write on the wall?

VI

Radiation and Health

Introduction

The discovery of X-ray emission by Röntgen in 1895 and nuclear radiation by Becquerel in 1896 led directly to the development of 'modern' physics in the twentieth century. The theoretical and experimental developments which arose from these discoveries led to the development of atomic and nuclear physics. Atomic, nuclear and radiation physics allow for an understanding of a diverse group of phenomena ranging from the workings of smoke detectors to the evolution of stars.

In this topic we will investigate the importance of ionising radiation to the health sciences by looking both at the health risks associated with exposure to sources of hazardous radiation, and at how radiation is used in the diagnosis and treatment of medical conditions. We will develop an understanding of the physical mechanisms underlying the phenomena, and with the various quantities used to measure radiation and its effects on biological material.

To understand the nature and effects of radiation, we need to first look at the structure of matter at both the atomic and nuclear level. This is covered in the first two chapters, where we will look at the constituents of atoms and nuclei and forces that hold those sub-atomic particles together.

In the third chapter we will look at different types of ionising radiation: α, β, gamma, and X-ray radiation. We will see which types of radiation are generated by atomic-level processes and which have their origin in the nucleus. We will also gain an understanding of the importance of the rate at which radiation is released and the energy transferred by this radiation.

The next chapter will then look at how radiation interacts with other matter, and how this differs for the types of radiation (particle and photon). This will lead on to an investigation of the effects of radiation on biological systems. We will look at the effects of radiation on both individual cells and on overall health: the relationship between dose and effect, the medical symptoms of exposure and the use of radiation in treatment and diagnosis of disease.

The goal of medicine is the understanding of the human body – its construction and its function, in health, sickness and cases of injury. The body is a very complex system, and images can provide a concise way of transmitting important information about the state of the human body. Several modern imaging techniques will be covered in varying depth in the final chapters.

ATOMS AND ATOMIC PHYSICS

33

33.1   Introduction

Atomic physics is the physics of how whole atoms behave, and is largely concerned with how atomic electrons are arranged around the atomic nucleus and how they interact with other atoms and fields. The sub-structure of the nucleus is of little importance to the electron configuration, and so the nuclear structure will be covered in the next chapter.

Most of the major advances in understanding the structure of the atom were made in the early twentieth century. A key breakthrough in understanding atoms was the development of the Böhr model of the atom. While modern atomic physics has moved far beyond the over-simplified picture of electrons moving in planet-like orbits, we will use this model of the atom to gain some useful insight into the electronic structure of atoms, and use it as a basis for predicting X-ray energies in Chapter 35.

Key Objectives

- Understand the structure of the atom.

- To understand the basic electronic configuration of the atom and how energy is absorbed and emitted by electrons.

- To be able to use the Böhr model of the atom to calculate electron energies.

33.2   Parts of the Atom

An atom can be regarded as the smallest unit of matter which retains the chemical properties of an element, by which we mean that an atom of, say, carbon loses its 'carbon-ness' if it is broken into its constituent sub-atomic particles. Most of an atom is space. It is made of protons, neutrons (in the nucleus) and electrons (in the space around the nucleus), and has no net charge. If the nucleus was the size of the full stop at the end of this sentence, the outer edges of the atom would be over 50 m away.

The basic atom consists of a central, positively charged **nucleus** surrounded by a cloud of negatively-charged **electrons**. The nucleus (covered next chapter) is composed of two types of particles – **protons** and **neutrons**.

An atom is electrically neutral. The electron and proton carry electric charge of equal magnitude but opposite sign, so in an atom there are equal numbers of protons in the nucleus and orbiting electrons. Therefore, the number of protons determines the number of electrons bound to the nucleus in an electrically neutral atom, and it is the electron configuration that determines the chemical properties of an atom. Each **chemical element** has a specific, unique number of protons in the nucleus. For example, all carbon atoms have six protons, and oxygen always has eight.

Antimatter

Antimatter is not just something made up by science-fiction writers. Each massive fundamental particle has an antimatter equivalent, for example, the electron and positron are a particle/antiparticle pair. If a particle and its antiparticle meet, they both cease to exist, so most antimatter doesn't hang about for long in our universe.

33.3   Orbitals and Energy Levels

Electrons

An **electron** is what is known as a **fundamental** or **elementary** particle. This means that is not believed to have any underlying substructure; there are no smaller parts from which it is made. Electrons belong to a group of fundamental particles called **leptons**, a grouping it shares with the electron neutrino (which will be discussed later in the section on β decay). It is believed to be a point-like object with no spatial extent.

The electron has a charge of -1.602×10^{-19} C and has a mass of 9.109×10^{-31} kg. Its antiparticle equivalent is the **positron**, which has the same mass and same charge magnitude, but it is positively charged instead, hence the name.

Electrons are often bound to positively charged nuclei to form atoms and molecules, but they can also exist as free particles.

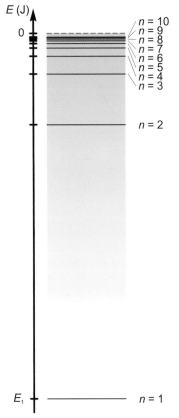

Figure 33.1   A typical energy-level diagram for a single electron bound to a hypothetical atomic nucleus.

Atomic dimensions		
Radius of the atom	$\sim 10^{-10}$ m	
Radius of the nucleus	$\sim 10^{-15}$ m	
	Mass	*Charge*
Neutron	1.675×10^{-27} kg	0
Proton	1.673×10^{-27} kg	$+1.602 \times 10^{-19}$ C
Electron	9.109×10^{-31} kg	-1.602×10^{-19} C

Table 33.1 Summary of atomic parameters.

Orbitals and Energy Levels

An **orbital** is a path described by an object under the influence of a central potential. For example, the motion of satellites about the Earth, or of the planets around the Sun are called orbitals because the gravitational forces which cause these objects to move in this way all point towards a central point. This label is often used for the electrons in an atomic system also; their motion is determined by an electrostatic force which always points towards the nucleus of the atom.

Consider the case of a single electron and a positively charged nucleus in isolation. When the negatively charged electron is near the nucleus, its potential energy is lower than it is when it is further away. When the electron is in 'orbit' about the nucleus, energy must be added to the system to remove it from its orbit about the atom. The electron is in a **bound state** – it is bound to the nucleus. Work must be done on the electron to remove it from the atom. This means that it has less energy while bound to an atom than it does when it is free. A free electron could be at rest in empty space, i.e., its kinetic and potential energy may be zero. This means that we should consider particles in bound states as having negative energies.

It is a general result in quantum-mechanical theory that when bound to an atomic nucleus an electron can only have certain specific total energies. This is called **energy quantisation**, and means that the kinetic and potential energy of the electron can only sum to certain specific values. These allowed energies dictate which electron orbitals are possible, and which are not. The electrons inside atoms may exist only in certain *states*. These allowed states give the range of **energy levels** of an atom.

The lowest allowed energy state is known as the **ground state**, and all the other energy states are called **excited states**. Figure 33.1 shows how we can illustrate the allowed energy levels for a simple system like a single-electron atom. The lines show the energies of the allowed states, which are numbered, with energy increasing up the vertical axis. Note that an energy diagram such as Figure 33.1 is not a graph of any function, it is a schematic representation of the energy levels of a bound particle. In particular the horizontal axis has no meaning at all. The allowed states get closer together in energy as n increases, and there is a zero energy level past which the electron is no longer bound, but becomes free.

Emission and Absorption Spectra

When an electron in an atom is in an excited state, and there is a lower-energy state which is allowed and vacant, the electron can spontaneously make the transition to the lower energy state by emitting the extra energy as a bundle of electromagnetic radiation – a **photon**. This process is called **spontaneous emission**. The energy of the photon that is emitted is the difference between the energies of the two levels.

In the Optics section, we emphasised the wave nature of electromagnetic radiation. However, in some experiments it is observed that such radiation is absorbed and emitted in 'bits', and the size (the magnitude of the energy) of those bits, the photons, is proportional to the frequency. If a photon is emitted with energy E, then it will have a particular frequency:

$$E = hf \tag{33.1}$$

The energy and the frequency are proportional to one another, and the proportionality constant, h, is known as **Planck's constant**, and has the value $h = 6.626 \times 10^{-34}$ J s.

The range of possible transitions that the electrons in a particular type of atom may make correspond to specific frequencies of electromagnetic radiation. This is called the **emission spectrum** of an element. An electron may make a transition to a higher energy state by absorbing precisely the right amount of energy to do so. By measuring the frequencies at which a collection of atoms absorb radiation, an **absorption spectrum** can be obtained. These processes are illustrated schematically in Figures 33.2 and 33.3.

Because the way such frequencies were observed was by separating out emitted or transmitted light with a prism and observing the coloured or dark lines, the terms emission and absorption *line* (or just spectral line) are often used to refer to a particular frequency emitted or absorbed by an element. Note that the discrete lines of atomic absorption and emission spectra are due to the quantisation of the energy levels of the electrons orbiting the atom and thus the spectrum emitted or absorbed depends on the number and arrangement of electrons around the atomic nucleus, i.e., on the element observed. A familiar example of an emission line is the sodium D line – when sodium atoms are excited, the most intense emissions occur at around 589 nm. This gives sodium street lights their characteristic orange tinge. Absorption and emission of photons by atoms is the fundamental process which underlies the interaction of matter with electromagnetic radiation. Phenomena as diverse as the operation of lasers, the colour of materials and fluorescence are explained by these processes.

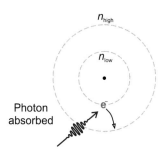

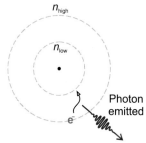

Figure 33.2 (Top) An atom can absorb the energy of a photon that strikes it by elevating one of its electrons to a higher energy level. (Bottom) Likewise an excited atom can lose energy by emitting a photon as an electron jumps to a lower energy level.

> **Stimulated emission**
>
> In addition to spontaneous emission, a electron can undergo **stimulated emission** when a photon of the same energy passes by. This is the process that allows lasers to operate, and *laser* is an acronym for *light amplification by stimulated emission of radiation*.

33.4 The Böhr Model of the Atom

Circular Orbits and Quantisation

The first model of the atom to give an explanation of emission and absorption spectra in terms of quantisation of energy levels was developed in the early twentieth century by the Danish physicist Niels Böhr (1885–1962). While it is not the best model we have of how electrons in atoms behave, it provides useful insight into how the wave nature of matter affects atomic structure.

Böhr's model was proposed (in 1913) for the simplest types of atoms and ions which have a single electron: hydrogen, singly ionised helium, etc. Böhr's proposal was that the electrons in the atom moved in circular orbits around the nucleus and could only have particular values of **angular momentum** (given the symbol L). Böhr hypothesised that the angular momentum of the electron in orbit around the nucleus was *quantised*. For such a circular orbit the angular momentum would be

$$L = mvr \tag{33.2}$$

where m is the mass of the electron, v is its velocity and r is the radius of the orbit. The allowed values suggested by Böhr were integer multiples of a fundamental quantity,

$$L_n = n\frac{h}{2\pi} = n\hbar \tag{33.3}$$

> **Key equations from mechanics and electricity**
>
> Any object travelling in a circular path with constant speed is kept in this path by a (centripetal) force directed towards the centre of the circle with magnitude
>
> $$F = \frac{mv^2}{r}$$
>
> Though not covered in this book, it can be shown that the electrostatic potential energy contained in a system of two charges separated by distance r is equal to the work done moving one charge toward the other from infinity, which is
>
> $$U = k\frac{q_1 q_2}{r}$$

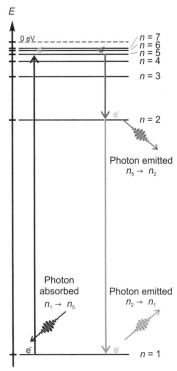

Figure 33.3 Atomic electrons make transitions up or down the energy level structure of an atom by emitting or absorbing photons of the appropriate energy.

where h is Planck's constant and n is a non-zero integer. $\hbar$ is commonly used as shorthand for $h/(2\pi)$, and is known as the reduced Planck constant.

The potential energy of an electron at a distance r from a nucleus with one proton is

$$U = -\frac{ke^2}{r} \tag{33.4}$$

(See the note in the margin for useful equations from mechanics and electricity.) The total energy, the sum of the kinetic and potential energies of this electron, is

$$E_{\text{total}} = \frac{1}{2}mv^2 - \frac{ke^2}{r} \tag{33.5}$$

The force on the electron is the Coulomb attraction between the electron and the atomic nucleus. This is the centripetal force that keeps the electron moving in a circle, so

$$F = \frac{ke^2}{r^2} = \frac{mv^2}{r}$$

or

$$mv^2 = \frac{ke^2}{r} \tag{33.6}$$

Putting this expression for mv^2 into Eq. (33.5) gives the total energy

$$E_{\text{total}} = -\frac{1}{2}\frac{ke^2}{r} \tag{33.7}$$

We can also rearrange Eq. (33.6) to show that

$$v = \sqrt{\frac{ke^2}{mr}} \tag{33.8}$$

If the angular momentum given by Eq. (33.2) is quantised, then only certain orbital radii, r, are permitted. Thus, quantised angular momentum implies that only certain electron orbitals, with the right radius, angular momentum and total energy are allowed. These orbits will be labelled with the index n. This label n is our first example of a quantum number. The nth allowed orbit has angular momentum $L_n = mvr_n = n\hbar$, and radius, r_n. Substituting in the value of the velocity from Eq. (33.8)

$$L_n = n\hbar = \sqrt{ke^2mr_n} \tag{33.9}$$

and so the allowed radii are

$$r_n = n^2\frac{\hbar^2}{ke^2m} \tag{33.10}$$

These specific allowed radii place a constraint the possible energy levels:

$$E_n = -\frac{ke^2}{2r_n} = -\frac{1}{n^2}\frac{k^2e^4m}{2\hbar^2} = -\frac{13.6\,\text{eV}}{n^2} \tag{33.11}$$

For cases other than hydrogen this formula needs to be slightly modified. For a helium ion, the two charges involved, the electron and nucleus, now have charges $+2e$ and $-e$, so the Coulomb attraction is stronger. For a nucleus with charge $Z \times e$ (Z protons), the energy levels are

$$E_n = -13.6\,\text{eV}\,\frac{Z^2}{n^2} \tag{33.12}$$

The electron volt, eV

This will be covered in detail in the next chapter, see Section 34.3. It is an energy unit, and is equivalent to 1.6×10^{-19} J.

Hydrogenic atoms

A hydrogenic atom is one with a single electron. Thus hydrogen is the only neutral hydrogenic atom. All other hydrogenic atoms have more protons than electrons and are thus positively charged. Some examples of hydrogenic atoms: C^{5+}, Li^{2+}, U^{91+}.

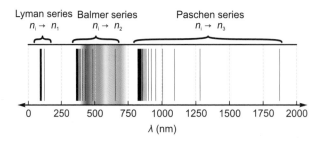

Figure 33.4 The emission lines from hydrogen. The black lines show observed emission and absorption lines, and the visible light part of the spectrum is shown for reference.

www.wiley.com/go/biological_physics

This result gives excellent agreement with the observed absorption and emission spectra of the hydrogen atom when the atomic number is set to one, i.e., $Z = 1$, in the above equation. These spectral lines are shown in Figure 33.4 and the atomic transitions which give rise to this spectrum are illustrated in Figure 33.5. Note that Eq. (33.12) is a good approximation for the transition energies of hydrogenic atoms (see margin note) with higher atomic numbers.

Eq. (33.12) gives the same results as the well-known Rydberg formula, determined empirically from spectroscopic data

$$\frac{1}{\lambda} = R_{\text{H}} \left(\frac{1}{n_1^2} - \frac{1}{n_2^2} \right) \tag{33.13}$$

where λ is the vacuum wavelength of the spectral line, n_1 and n_2 are the integers corresponding to the principal quantum numbers of energy levels and $R_{\text{H}} = 1.1 \times 10^7 \, \text{m}^{-1}$ is the Rydberg constant for hydrogen. (Confirming this agreement is left as an exercise for the reader. Recall that $E = hf$, $c = f\lambda$ and note that $hcR_{\text{H}} = 13.6 \, \text{eV}$.)

de Broglie and Waves

A few years after Böhr introduced this model, the French physicist Louis de Broglie (1892–1987) proposed a radical explanation for the quantisation of angular momentum. By the turn of the twentieth century, the wave-like properties of light were well known, but Einstein's explanation of a phenomenon called the photoelectric effect established that a particle-like model was also needed. de Broglie's idea was that if objects that were traditionally modelled as waves also had particle-like properties, then the converse might be true – particles might have wave properties. This idea, that everything requires a wave and a particle description when we are discussing objects on an atomic scale, is termed **wave-particle duality**.

With any particle (including photons) we can associate a wave, and the wavelength, λ, (called the **de Broglie wavelength**) is

$$\lambda = \frac{h}{p} \tag{33.14}$$

where p is the momentum and h is Planck's constant. The electron's small mass means that the de Broglie wavelength is usually significant, as the following examples illustrate.

This insight, that matter behaves in wave-like ways at very small distances, moves us away from the 'solar system' model of the atom. Electrons should be treated as waves rather than tiny little billiard balls orbiting a much larger ball (the nucleus). This wave-particle model may be used to explain the quantisation of angular momentum and the separation of atomic electrons into allowed orbitals.

Assume that an atomic electron moves in a circular path, radius r_n. The angular momentum of this electron is $L_n = p_n r_n$, so the momentum of the electron in the nth orbital is

$$p_n = \frac{L_n}{r_n} \tag{33.15}$$

The wavelength associated with a particle in the n-th energy level is

$$\lambda_n = \frac{h}{p_n} = 2\pi\hbar \frac{r_n}{L_n} = \frac{2\pi\hbar r_n}{n\hbar} = \frac{2\pi r_n}{n} \tag{33.16}$$

In other words, $2\pi r_n$, the circumference of the circular orbit, is an integer number of particle wavelengths, $n\lambda_n$. Thus the quantisation of angular momentum is equivalent to the claim that only orbits in which this particle-wave *constructively interferes* are allowed. Constructive interference of de Broglie waves is illustrated in Figure 33.6; orbits with greater or lesser radii than these result in destructive interference of de Broglie waves as shown in Figure 33.7.

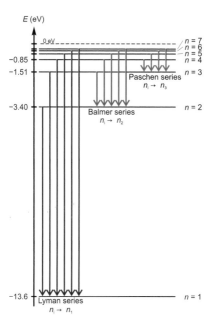

Figure 33.5 The electronic transitions which give rise to the hydrogen emission lines shown in Figure (33.4).

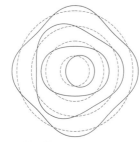

Figure 33.6 Electrons in an orbital whose radius is an integer multiple of the electron's de Broglie wavelength will constructively interfere. This means that this is an allowed electron orbital.

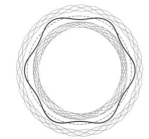

Figure 33.7 Orbits whose radii are not integer multiple of the electron de Broglie wavelength are not allowed. The electron wave would destructively interfere with itself on each successive orbit around the atom, thus cancelling itself out.

Example 33.1 *The Balmer series*

Problem: What are the photon energies and wavelengths of the first five transitions in the Balmer series for hydrogen?

Solution: Transitions in the Balmer series are those that start on an orbital where $n > 2$ and finish on the $n = 2$ orbital. The first five of these transitions are: $n = 3 \rightarrow n = 2$, $n = 4 \rightarrow n = 2$, $n = 5 \rightarrow n = 2$, $n = 6 \rightarrow n = 2$, $n = 7 \rightarrow n = 2$. The energy of a photon released in the transition is equal to the difference in the energy of each orbital (from Equation (33.11))

$$E_{\text{photon}} = -13.6 \,\text{eV} \left(\frac{1}{n_1^2} - \frac{1}{n_2^2} \right)$$

which for the $n = 3 \rightarrow n = 2$ transition gives

$$E_{3 \rightarrow 2} = -13.6 \,\text{eV} \left(\frac{1}{3^2} - \frac{1}{2^2} \right) = -13.6 \,\text{eV} \times -1.39 = 1.89 \,\text{eV}$$

Similarly, the other transitions give $E_{4 \rightarrow 2} = 2.55 \,\text{eV}$, $E_{5 \rightarrow 2} = 2.86 \,\text{eV}$, $E_{6 \rightarrow 2} = 3.02 \,\text{eV}$ and $E_{7 \rightarrow 2} = 3.12 \,\text{eV}$.

We can obtain each of these energies in joules if required by multiplying by $1.6 \times 10^{-19} \,\text{J eV}^{-1}$. This is required if we wish to use these energies to find the wavelength of each photon via $E_{\text{photon}} = hf = \frac{hc}{\lambda}$, which gives $\lambda = \frac{hc}{E_{\text{photon}}}$. (Alternatively we could use the Rydberg formula).

For the $n = 3 \rightarrow n = 2$ transition this gives $\lambda_{3 \rightarrow 2} = \frac{6.626 \times 10^{-34} \,\text{J s} \times 3.00 \times 10^8 \,\text{m s}^{-1}}{1.89 \,\text{eV} \times 1.6 \times 10^{-19} \,\text{J eV}^{-1}} = 657 \,\text{nm}$, which lies in the visible part of the spectrum and is bright red.

Similarly for the other transitions $\lambda_{4 \rightarrow 2} = 487 \,\text{nm}$ (blue-green), $\lambda_{5 \rightarrow 2} = 434 \,\text{nm}$ (deep blue/violet), $\lambda_{6 \rightarrow 2} = 411 \,\text{nm}$ (violet) and $\lambda_{7 \rightarrow 2} = 398 \,\text{nm}$ (ultraviolet, not visible).

Example 33.2 *de Broglie wavelength of an electron I*

Problem: What is the de Broglie wavelength of an electron travelling at 100 km h^{-1}?

Solution: First convert 100 km h^{-1} into SI units

$$100 \,\text{km h}^{-1} = \frac{100 \times 10^3 \,\text{m}}{3600 \,\text{s}} = \frac{100}{3.6} = 27.8 \,\text{m s}^{-1}$$

We then use this velocity and the mass of the electron ($9.11 \times 10^{-19} \,\text{kg}$) to calculate the momentum of the electron

$$p_e = m_e v = 9.11 \times 10^{-19} \,\text{kg} \times 27.8 \,\text{m s}^{-1}$$
$$= 2.53 \times 10^{-29} \,\text{kg m s}^{-1}$$

Finally we use the de Broglie relationship to calculate the wavelength of the electron

$$\lambda_e = \frac{h}{p_e} = \frac{6.63 \times 10^{-34}}{2.53 \times 10^{-29}} = 2.62 \times 10^{-5} \,\text{m} = 26.2 \,\mu\text{m}$$

Example 33.3 *de Broglie wavelength of an electron II*

Problem: What is the speed of an electron whose de Broglie wavelength is equal to r_1, the radius of the first energy level of the hydrogen atom?

Solution: We first calculate the value of r_1. To do this we substitute the appropriate physical quantities into Eq. (33.10)

$$r_1 = 1^2 \frac{\hbar^2}{ke^2 m} = \frac{\left(1.05 \times 10^{-34}\right)^2}{9.0 \times 10^9 \times \left(1.60 \times 10^{-19}\right)^2 \times 9.11 \times 10^{-31}} \,\text{m} = 5.25 \times 10^{-11} \,\text{m}$$

Note that this quantity is known as the Böhr radius and is often given the symbol a_0. The value we have calculated here is slightly different to the accepted value of $a_0 = 5.29 \times 10^{-11}$ m; this is due to the fact that we have used values for $\hbar$ and e (the charge of the electron) which are only accurate to three significant figures.   We now use this value as the required de Broglie wavelength, and rearrange the de Broglie relation to find the momentum of this electron and then the velocity of the electron

$$m_e v_e = \frac{h}{\lambda_e} \Rightarrow v_e = \frac{h}{m_e \lambda_e}$$

$$= \frac{6.63 \times 10^{-34}}{9.11 \times 10^{-31} \times 5.25 \times 10^{-11}}$$

$$= 1.39 \times 10^7 \text{ m s}^{-1}$$

This is about 5% of the speed of light.

33.5 Multielectron Atoms

We have touched on the idea that within an atom, an electron may exist only in certain allowed states. In order to understand the chemical properties of an element, it is necessary to describe these states and how the electrons are arranged in these allowed states.

The states which are allowed are ones which have allowed values of energy and angular momentum, and we use **quantum numbers** to specify the values that a state has. The most important quantum number is the **principal quantum number**, n, which we have already met in looking at the Böhr model of the atom. It specifies which basic **energy shell** a state belongs to. The quantum number n can take on only positive integer values, and for historical reasons these also have letter names. The $n = 1$ level is called the K shell, $n = 2$ is the L shell, and so on.

Each shell has $2n^2$ allowed sub-states in which an electron can exist. These states all have different **angular momentum** quantum numbers. There are two types of angular momentum: the **orbital angular momentum** that an electron has because it is moving around the nucleus, and the **intrinsic angular momentum** (also known as **spin**) which is a property that all electrons have.

The allowed values of the orbital angular momentum depend on n, and are described by the **azimuthal quantum number**, l. The allowed values of l are zero up to $n - 1$ in integer steps. Thus $n = 1$ shell can only have an azimuthal quantum number $l = 0$. The $n = 2$ shell can have azimuthal quantum numbers $l = 0$ and $l = 1$. The different l values also have letter symbols: $l = 0$, 1, 2, 3 and 4 are known as the s, p, d, f and g orbitals. There are different spatial orientations possible for most values of l and these are specified by the **magnetic quantum number**, m_l. The values of m_l can range from $-l$ to $+l$.

The last quantum number is the **spin quantum number**, m_s. It can have two values: $+\frac{1}{2}$ and $-\frac{1}{2}$ (which are often referred to as 'spin up' and 'spin down').

We can now give an example of how the quantised values of energy and angular momentum specify what states can exist. For the L shell ($n = 2$), there are $2 \times 2^2 = 8$ allowed states. These are listed in Table 33.2.

The **Pauli exclusion principle** provides the basis for understanding how these allowed states determine the electronic configuration and hence chemical nature of elements. It states that

Key concept:
No two identical fermions can occupy the same quantum state at the same time

n	l	m_l	m_s
2	0	0	$-1/2$
2	0	0	$+1/2$
2	1	-1	$-1/2$
2	1	-1	$+1/2$
2	1	0	$-1/2$
2	1	0	$+1/2$
2	1	$+1$	$-1/2$
2	1	$+1$	$+1/2$

Table 33.2 Allowed quantum numbers for the L shell.

Fermions are one of the two broad categories of into which particles can be placed, the other being **bosons**. Fermions are particles which obey the Pauli exclusion principle; bosons don't and many bosons can have the same quantum state. (One example of a boson is the photon.) The bound electrons in atoms are fermions, and therefore, inside a single atom, only one electron may be in each of the allowed states we have listed. In a later chapter we will find that the protons and neutrons which constitute the nucleus of the atom are also fermions.

When an atom is in its *ground state*, the occupied states are easily found. A common way of writing the electron configuration is to use **spectroscopic notation**. This lists the principal quantum number, a letter label (*s*, *p*, *d* etc.) determined by the azimuthal quantum number indicating which sub-shells are occupied, and a superscript which indicates how many electrons are in the sub-shell (unless that number is one, in which case it is often omitted). For example, sodium has 11 electrons, so its ground-state configuration can be written $1s^2 2s^2 2p^6 3s$.

The chemical properties of elements depend on the electron configuration, and in particular how close to being full the shells and sub-shells are. Elements with full electron shells tend to be chemically unreactive. Elements with a lone electron in an outer shell tend to lose that electron (e.g., sodium) to form positively charged ions. Elements that are one electron short of a full shell tend to grab an electron from other atoms (e.g., fluorine) to form negatively charged ions.

Example 33.4 *Quantum numbers*

Problem: An electron occupies the $n = 5$ orbital of a hydrogen atom and has an azimuthal quantum number of $l = 3$. What are the possible values of the magnetic (m_l) and spin (m_s) quantum numbers it might have?

Solution: If the principal quantum number is $n = 5$, then the azimuthal quantum number l can range from 0 to 4 ($l = 0, ..., n - 1$). It has an azimuthal quantum number of $l = 3$.

The magnetic quantum number can have values $m_l = -l, -l + 1, ..., l - 1, l$. For an electron with $l = 3$ then m_l could be $-3, -2, -1, 0, 1, 2,$ or 3. In each of these seven cases the spin quantum number can be either $+\frac{1}{2}$ or $-\frac{1}{2}$ giving a total of 14 possible states: $(n, l, m_l, m_s) = (5, 3, -3, +\frac{1}{2}), (5, 3, -3, -\frac{1}{2}), (5, 3, -2, +\frac{1}{2}), (5, 3, -2, -\frac{1}{2}), (5, 3, -1, +\frac{1}{2}), (5, 3, -1, -\frac{1}{2}), (5, 3, 0, +\frac{1}{2}), (5, 3, 0, -\frac{1}{2}), (5, 3, 1, +\frac{1}{2}), (5, 3, 1, -\frac{1}{2}), (5, 3, 2, +\frac{1}{2}), (5, 3, 2, -\frac{1}{2}), (5, 3, 3, +\frac{1}{2}), (5, 3, 3, -\frac{1}{2})$

33.6 Quantum Mechanics

Quantum mechanics is a highly successful mathematical framework for explaining the behaviour of systems of small particles, and is the theory underpinning modern chemistry, and atomic, nuclear, and condensed-matter physics. Quantum mechanics allows us to calculate the possible outcomes when we try to measure a particular quantity and to assign probabilities to those outcomes.

One of the key principles of quantum mechanics is the **Heisenberg uncertainty principle** – the idea that there are certain quantities that cannot be measured to infinite precision simultaneously. An example is the position and momentum of an object. The more precisely the position is known, the less well-determined is its the momentum.

Another key idea is that of wave-particle duality – that matter particles also require a wave description to account for their behaviour. One formulation of quantum mechanics (wave mechanics) associates with a particle a **wave function**. This is a complex mathematical function, the squared amplitude of which gives the probability of finding the particle at each location.

The Böhr model of the atom is based on the quantisation of the angular momentum of atomic electrons and this may be interpreted in terms of the de Broglie wavelengths of these electrons. However, the Böhr model specifies the exact radius of this orbit, and the momentum of the electrons. Quantum mechanics requires a more sophisticated model which retains many of the important features of the Böhr model, but only specifies the *probability* of finding an electron at each spatial location. Models such as these are not necessary for an understanding of the phenomena which we will investigate in the remaining chapters of this section and will not be investigated further.

The Heisenberg uncertainty principle

The Heisenberg uncertainty principle places a limit on the accuracy with which some pairs of observable quantities can be measured. For example, the product of the uncertainty in the position, Δx, and the uncertainty in the momentum, Δp, must obey

$$\Delta x \Delta p \geq \hbar / 2$$

33.7 Summary

Key Concepts

electron An electron is a fundamental particle. It has charge -1.6×10^{-19} C, mass 9.1×10^{-31} kg and spin $\frac{1}{2}$.

photon A quantum of electromagnetic radiation (and the carrier particle of electromagnetic force), exhibiting both wave and particle properties.

atom An atom can be regarded as the smallest unit of matter which retains the chemical properties of an element. It is made up of electrons surrounding a nucleus of neutrons and protons, and is electrically neutral.

nucleus The central part of atoms and ions, consisting of positively charged protons and uncharged neutrons.

electron volt (eV) The change in potential energy for a charge of magnitude e when it is moved through a change in electrical potential of 1 V. 1 eV $= 1.602 \times 10^{-19}$ J.

orbital The path followed by an object under the influence of a central potential. The allowed positions in space for an electron in a particular state.

energy level A state or set of quantum states with a specific energy that a particle may occupy.

photon energy The energy of a photon is proportional to its frequency, $E = hf$.

Planck's constant (h) A constant that has a crucial role in quantum mechanics $h = 6.63 \times 10^{-34}$ J s.

quantisation of energy Particles in bound states may only have particular allowed energies.

de Broglie wavelength All objects have wave-like properties and the wavelength associated with an object is inversely proportional to the momentum. $\lambda = \frac{h}{p}$.

Equations

$$E = hf$$
$$L_n = n\hbar$$
$$E_n = -13.6\,\text{eV}\frac{Z^2}{n^2}$$
$$\frac{1}{\lambda} = R_\text{H}\left(\frac{1}{n_1^2} - \frac{1}{n_2^2}\right)$$
$$\lambda = \frac{h}{p}$$
$$n\lambda_n = 2\pi r_n$$

33.8 Problems

33.1 A photon is emitted by an atom when one of the electrons orbiting the atom drops from an energy level of $E_i = -10.64$ eV to an energy level of $E_f = -12.70$ eV.

    (a) What is the energy of this photon (in eV)?
    (b) What is the energy of this photon (in J)?
    (c) What is the frequency of this photon?
    (d) What is the wavelength of this photon?
    (e) What is the momentum of this photon?

33.2 The electromagnetic radiation emitted from the Sun is most intense at around 502 nm.

    (a) What is the energy per photon (in J) for light of this wavelength?
    (b) What is the energy per photon in electron volts?
    (c) What is the momentum per photon?
    (d) How fast would an electron ($m_{electron} = 9.1 \times 10^{-31}$ kg) need to be traveling to have the same momentum as this photon?
    (e) What would the de Broglie wavelength of such an electron be?

33.3 What is the de Broglie wavelength of:

    (a) an electron ($m_{electron} = 9.1 \times 10^{-31}$ kg) travelling at $15\,\mathrm{km\,s}^{-1}$?
    (b) an electron with a kinetic energy of 1 eV?
    (c) a proton ($m_{proton} = 1.67 \times 10^{-27}$ kg) travelling at $15\,\mathrm{km\,s}^{-1}$?
    (d) a proton with a kinetic energy of 1 eV?
    (e) an elephant ($m_{elephant} = 10$ tonnes) travelling at $15\,\mathrm{km\,h}^{-1}$?
    (f) an elephant with a kinetic energy of 1 eV?

33.4 Which of the following atomic transitions in hydrogen (labelled (i) to (v)) will:

    (a) release a photon of the highest energy?

    (b) release a photon of the longest wavelength?

    (c) release a photon of wavelength 433 nm

    (d) release a photon of energy 0.661 eV

    (i) $n = 2 \to n = 1$
    (ii) $n = 5 \to n = 1$

    (iii) $n = 5 \to n = 2$

    (iv) $n = 4 \to n = 3$

    (v) $n = 10 \to n = 5$

33.5 What are the wavelengths of the $n = 2 \to n = 1$, $n = 3 \to n = 1$, and $n = 4 \to n = 1$ transitions for a singly charged Helium nucleus?

33.6 A muon is a elementary particle whose properties are similar to those of an electron (a negative charge and a spin of 1/2) with the exception of its mass. Because of this it is possible to replace one or more electrons in an atom with muons. A muon is 207 times more massive than an electron (and so has mass 1.88×10^{-28} kg). If the electron in a hydrogen atom was replaced with a muon then an exotic 'muonic hydrogen' atom is created.

    (a) By what factor would the Bohr radius of the 'muonic hydrogen' atom change?

    (b) By what factor will the energy of a particular electronic energy level change for 'muonic hydrogen'?

    (c) What will the wavelengths of the first three lines in the Balmer series ($n_i \to n_f = 2$) be for this exotic 'muonic hydrogen' atom?

    (d) For ordinary hydrogen, the Balmer series falls in the visible region of the electromagnetic spectrum. In what region of electromagnetic spectrum would you search for the absorption/emission lines of the Balmer series of 'muonic hydrogen'?

33.7 One of the most compelling demonstrations of wave-particle duality is the wave-like interference pattern displayed by electrons (which otherwise behave like a particle) when passing through a pair of double slits. If a beam of electrons is created by accelerating them from rest through a potential difference of just 500 V, and this beam is trained on a pair of slits 10 μm apart with a detector 1 m behind the slits, what is the separation between adjacent bright spots on the screen?

33.8 List the possible states of an electron in the $n = 3$ shell of a hydrogen atom (i.e, reproduce Table 33.2 for $n = 3$)

THE NUCLEUS AND NUCLEAR PHYSICS

34.1 Introduction

The bulk of the mass of the atom is concentrated in the nucleus, which is made up of protons and neutrons. In this chapter we will cover the basic structure of the nucleus, how nuclei differ from one another and the stability of the nucleus. There are around 300 known stable nuclear configurations, and here we will give an explanation of why some nuclei are stable and others unstable.

The chapter will end with a brief description of fission and fusion, both processes that alter the structure of the nucleus. This discussion of the processes by which nuclei change will continue into the next chapter on the production of ionising radiation.

Key Objectives

- Understand the structure of the nucleus.

- Understand the relationship between the mass of the nucleus and its binding energy and stability.

- Understand how the binding energy determines which nuclei will undergo fission, which will undergo fusion and how much energy is released in these processes.

34.2 Nuclei and Isotopes

Protons and Neutrons

The **nucleus** (plural *nuclei*) of the atom contains two types of particle – **protons** and **neutrons**. Unlike the electron, these are not fundamental particles, but themselves contain fundamental particles known as **quarks**. Both the neutron and the proton are made up of three quarks, and are classified as **hadrons** (particles made of quarks) and **baryons** (particles made of three quarks) by particle physicists. Protons and neutrons are also collectively referred to as **nucleons**.

Table 33.1 in the previous chapter lists the mass and charge of the proton and neutron. Note that the charge of the proton has the same magnitude as the charge of the electron, but the opposite sign. Thus an electrically neutral atom has one electron for every proton. The neutron has no charge and very nearly the same mass as the proton.

Atomic Number

In a neutral atom there are the same number of protons and electrons. The number of electrons, and their arrangement into energy levels is called the **electron configuration**. The chemical properties of the atom are determined by the electron configuration, thus the chemical properties depend entirely on the number of protons in the nucleus. The number of protons in the nucleus is given the name **atomic number** and the symbol Z.

Atomic Mass Number

While the chemical properties, and hence the **element**, are determined by the atomic number, the number of neutrons can vary without significantly changing the chemistry. To completely specify what kind of nucleus we have, we have to give the number of neutrons also. The total number of protons and neutrons is called the **atomic mass number**, A.

Symbols and Terminology

An element whose chemical symbol is X is written

$$^A_Z X$$

For example, uranium-235 is written

$$^{235}_{92} U$$

In this case the notation tells us that there are 92 protons (in the nucleus), 92 electrons (orbiting the nucleus in the shells), 235 nucleons (protons and neutrons) and $235 - 92 = 143$ neutrons (in the nucleus).

An added m beside the atomic mass number and/or a star superscript on the right ($^{Am}_Z X^*$) indicates that the given nucleus is **excited**, that is, not in its lowest energy state. The m stands for **metastable state** meaning that the nucleus may persist in this excited state for a relatively long time.

A list of some terms used in nuclear physics:

nucleon A nuclear constituent, either a proton or a neutron.

nuclide Another name for a nucleus or an atom with a specific nuclear makeup (i.e. number of neutrons is important).

isotopes Atoms with the same number of protons, but different numbers of neutrons.

isotones Atoms with the same number of neutrons, but different numbers of protons.

isobars Atoms with the same number of nucleons, but different numbers of protons (i.e., A doesn't change, but Z does).

isomers Atoms having the same number of protons and the same number of neutrons. They differ in their nuclear energy states. For example, $^{131m}_{54} X$ and $^{131}_{54} X$ are isomers, but $^{131m}_{54} X$ is in a metastable state.

34.3 Energy and Mass Units

Equivalence of Mass and Energy

In 1905, Einstein was able to show that energy and mass are related by the expression,

$$E = mc^2 \tag{34.1}$$

Using this relationship, we can specify masses in terms of their energy equivalent. This will make the task of calculating the energy changes involved in nuclear processes much easier.

The Electron Volt

Energies of the scale that are relevant for describing atomic and nuclear energy states and mass equivalents are extremely small when measured in joules (on the order of 10^{-10} to 10^{-20} J). Keeping track of these large exponents is unnecessarily cumbersome and we can simplify things by defining a new unit of energy.

It is worthwhile thinking about the SI unit of energy, the joule, to get some hints as to how we should go about defining our new unit. To begin with we will think about the size of the joule as a unit of energy. One joule of kinetic energy is the energy of a 2 kg mass traveling at 1 m s^{-1}, and one joule of gravitational potential energy is the potential energy lost by a 1 kg mass that falls near the Earth's surface through a distance of 10 cm. These examples show that the joule is a useful measure of energy for systems with velocities around 1 m s^{-1} and masses around 1 kg, or systems with masses around 1 kg acted on by gravitational forces. The joule is a perfectly sensible unit when working with everyday macroscopic systems; in these situations measuring energy in joules will often yield results on the order of 0.1–100 or so joules.

It is equally clear however that there is no good reason to use masses of this size or gravity to define the size our new atomic unit of energy. Gravity is not normally an important factor in atomic physics, so we should define an energy unit in terms of a force that is. Thus we use electrostatic potential energy instead of gravity and we will use the behaviour of the electron in this potential to define our new unit.

We define the **electron volt** (eV) this way:

> **Key concept:**
> An electron volt is the change in the electrostatic potential energy of an electron when it is moved through a potential difference of 1 V.

In joules, the size of the electron volt is

$$
\begin{aligned}
1\,\text{eV} &= e \times (1\,\text{V}) \\
&= (1.602 \times 10^{-19}\,\text{C}) \times (1\,\text{V}) \\
&= 1.602 \times 10^{-19}\,\text{J} \qquad (34.2)
\end{aligned}
$$

so $1\,\text{J} = \frac{1}{1.6 \times 10^{-19}}\,\text{eV} = 6.2 \times 10^{18}\,\text{eV}$. As another example, consider Planck's constant, h. In SI units this constant is

$$
h = 6.626 \times 10^{-34}\,\text{J s}
$$

In terms of the electron volt this is

$$
\begin{aligned}
h &= \frac{6.63 \times 10^{-34}\,\text{J s}}{1.6 \times 10^{-19}\,\text{J eV}^{-1}} \\
&= 4.14 \times 10^{-15}\,\text{eV s}
\end{aligned}
$$

The modified Planck's constant, $\hbar$ is

$$
\begin{aligned}
\hbar &= \frac{1.05 \times 10^{-34}\,\text{J s}}{1.6 \times 10^{-19}\,\text{J eV}^{-1}} \\
&= 6.58 \times 10^{-16}\,\text{eV s}
\end{aligned}
$$

These are both very small numbers, even when we use our new energy unit. However, it is useful to have these constants in these units when performing calculations in which the predominant energy unit is the electron volt.

The Atomic Mass Unit

The mass of an atom expressed in kilograms is on the order of $\sim 10^{-27}$ kg. This is also a very small number and is also cumbersome to work with. Chemists and physicists therefore employ a more appropriate unit called the **atomic mass unit** (amu or sometimes just u). The atomic mass unit is defined so that the atomic mass of the carbon-12 atom is exactly 12 amu. Avogadro's number, N_A, is the number of atoms in exactly 0.012 kg of carbon-12. Thus each atom of carbon-12 has a mass of

$$
\frac{0.01200\,\text{kg per mole}}{6.022 \times 10^{23}\,\text{per mole}} = 1.993 \times 10^{-26}\,\text{kg}
$$

This is *exactly* 12 amu, so

$$
1\,\text{amu} = \frac{1.993 \times 10^{-26}\,\text{kg}}{12} = 1.661 \times 10^{-27}\,\text{kg} \qquad (34.3)
$$

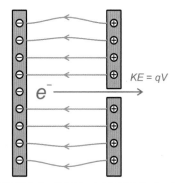

Figure 34.1 An electron gains 1 eV of kinetic energy when accelerated by a potential difference of 1 V.

> **A Brief Reminder**
> The change in electrostatic potential energy for a charge moving in an electric field is given by
> $$\Delta U = q\Delta V$$
> If the charge is free to move under the influence of the electric field and there are no other forces acting on the charge then this change in electrostatic energy will correspond to a change in kinetic energy as shown in Figure 34.1

The are many similar-sounding terms in use for masses. The **atomic mass**, A, is the mass of an atom, usually given in amu. For example, helium-4 has an atomic mass of 4.002602 amu. Another term used is the **relative atomic mass**, unfortunately this is also generally given the symbol A. This is usually a measure of the average atomic mass, weighted by isotopic abundance, and relative to 1/12th of carbon-12, so it is a dimensionless quantity. The **molar mass** or **gram atomic mass** is the mass of one mole (a sample containing Avogadro's number of atoms or molecules) of a substance.

Mass in 'MeV'

As mentioned earlier, mass and energy and related by Einstein's famous equation, Eq. (34.1). It is thus natural to ask what the energy equivalent of a 1 amu particle is. This will also give us a reference point when we discuss the energies involved in nuclear processes. Using Eq. (34.1), we are able to calculate the energy equivalent to the mass of a 1 amu particle as follows

$$E = mc^2$$
$$= \left(1.661 \times 10^{-27}\ \text{kg}\right) \times \left(2.998 \times 10^8\ \text{m s}^{-1}\right)^2$$
$$= 1.492 \times 10^{-10}\ \text{J}$$

This is the energy equivalent to a mass of 1 amu in SI units. It will be more useful to know this equivalent energy in electron volts. This is calculated as follows

$$\frac{1.492 \times 10^{-10}\text{J}}{1.602 \times 10^{-19}\text{J eV}^{-1}} = 931500000\ \text{eV} = 931.5\ \text{MeV}$$

$$1\ \text{amu} = u = 1.661 \times 10^{-27}\ \text{kg} = 931.5\ \text{MeV}/c^2 \qquad (34.4)$$

It is fairly common to hear physicists talk of a particle having 'a mass of x MeV'. What they mean by this is that the mass is *equivalent* to an energy of x MeV. Technically, $m = E/c^2$ so the mass is *equal* to x MeV/c^2, i.e., the correct mass unit is MeV/c^2.

34.4   Nuclear Forces

There are four known fundamental forces or fundamental interactions. These are gravity, the electromagnetic force and two forces that are only important on the scale of the nucleus: the strong and weak nuclear forces. It is remarkable that all of the interactions in nature ultimately reduce to just these four forces. We have already discussed the electrical force in some detail. We will now consider the two other forces which are important for an understanding of atomic and nuclear physics, the strong and weak forces.

The Strong Force and the Nucleus

The **strong nuclear force** is an attractive force that acts between protons and protons, neutrons and neutrons, and protons and neutrons. There is no difference between the strength of the p–p, p–n or n–n interactions. The strong nuclear force is a very short-range force which acts rather like glue or Velcro on the surface of nucleons, holding all the nuclear particles together.

The strong nuclear force is strong enough to overcome the repelling electrostatic force between protons that tries to push the nucleus apart, allowing the creation of some stable nuclei. In radioactive nuclides, the repulsive electrostatic forces are large enough to make the situation unstable, and the nucleus may break apart, or change the number of protons and neutrons to a more stable configuration.

The colour force

Protons and neutrons are composite particles. They are composed of three fundamental particles called **quarks**. Quarks are held together inside the nucleon by the **colour** force. This is a very unusual force in that it acts between three distinct charges or colours. It is stranger even than this in that it gets stronger the further apart the quarks get rather than weaker like the electric force. The strong nuclear force is actually a residual force much like the Van Der Waals force between adjacent atoms, and is the result of attractions between the quarks of adjacent nucleons. Thus strictly speaking the colour force is one of the four fundamental forces and the strong nuclear force is not. However, the colour force is well outside the scope of this book and we will treat the strong nuclear force as a fundamental force in the interests of simplicity and clarity.

Example 34.1 *Atomic energy*

Problem: A litre of gasoline (0.75 kg) contains around 32×10^6 J of chemical energy. This energy can be obtained by allowing chemical changes in the gasoline, e.g., burning it. How much energy would be obtained if the mass of this gasoline could be converted directly into energy?

Solution: By using $E = mc^2$ we find that there is $0.75\,\text{kg} \times \left(3 \times 10^8\,\text{m s}^{-1}\right)^2 = 6.75 \times 10^{16}$ J of energy contained in a litre of gasoline. This is around 2×10^9 times higher than the amount of chemical energy that can be obtained from the gasoline (even under ideal conditions)!

The Weak Nuclear Force

The weak force is responsible for the phenomenon of nuclear β decay. The detailed behaviour of this force is extremely complex and we will not investigate this behaviour any further in this book.

34.5 Nuclear Decay and Stability

Binding Energy

Key concept:
If it takes energy to pull the atom apart, then the constituent parts must lose energy when they form an atom. So the energy, and consequently mass, of the atom is less than that of the individual parts. This energy is the binding energy. *The greater the binding energy of the atom, the more energy we have to supply to break it apart, and the more stable the atom is.*

The energy of an electron in an atom is lower than the energy of a free electron. In order to remove an electron from an atom, work must be done on the electron. In other words, energy must be transferred to the electron from some external energy source. This will free the electron from the atom and at the same time increase the electron's total energy.

The same is true of any bound system. Work must be done to unbind such a system so that the total energy of all of the unbound constituents will be greater than the total energy of the bound system.

In the case of atomic systems we would expect that the total energy of the unbound protons, neutrons and electrons would be greater than the total energy of the bound atom. This increase in energy is manifest as an increase in the mass of the atomic constituents. Thus the total mass of the constituent parts of the atom, i.e., the mass of all of the protons, neutrons and electrons, is greater than the mass of the original atom.

For example the mass of the carbon-12 atom is 12 amu (by definition). The mass of its constituent parts may be found using the mass of the proton (1.00728 amu), the mass of the neutron (1.00866 amu), and the mass of the electron (0.00055 amu) as follows:

$$\text{Mass of constituent parts} = 6 \times (1.00728)\,\text{amu} + 6 \times (1.00866)\,\text{amu} + 6 \times (0.00055)\,\text{amu}$$
$$= 12.09894\,\text{amu}$$

The mass of the carbon-12 nucleus, and the sum of the masses of its component sub-particles are clearly not the same. The difference in mass between the atom and its constituent parts is called the **mass defect**, Δm. For carbon-12 the mass defect is

$$\Delta m = 12.09894 - 12.000 = 0.09894\,\text{amu}$$

This extra mass is due to the greater energy of the unbound constituent atomic parts. Using Eq. (34.1), we can find the energy equivalent of this mass defect (converting mass to SI units first)

$$0.09894 \text{ amu} = (0.09894) \times \left(1.6605 \times 10^{-28}\right) \text{ kg}$$
$$= 1.6429 \times 10^{-27} \text{ kg}$$
$$\Delta E = 1.6429 \times 10^{-28} \text{ kg} \times \left(2.998 \times 10^{8} \text{ m s}^{-1}\right)^{2}$$
$$= 1.477 \times 10^{-11} \text{ J}$$
$$= 92.2 \text{ MeV (about 7.7 MeV per nucleon)}$$

Binding energy and ionisation energy

Given the relatively small size of the electrostatic force binding the electrons to the nucleus compared to the forces holding the nucleus together, most of the energy deficit calculated here would be used to separate the components of the nucleus. For this reason binding energy is generally considered a nuclear phenomenon and the 'binding energy' of the electrons is normally called the ionisation energy of the atom. In general ionisation energies are of the order of a few tens of electron volts, whereas the binding energy per nucleon is of the order of 7–8 million eV.

This is the amount of work which must be done to separate the carbon-12 atom into its component parts. It is also the energy which would be released should six isolated protons, neutrons and electrons combine to form a carbon-12 atom. This difference in energy is called the **binding energy** of the nucleus.

The binding energy for a single nucleus is given by

$$BE = \Delta E = \Delta mc^2 = (m_s - m_b)\,c^2 = \left(Zm_p + (A-Z)m_n - m_b\right)c^2 \qquad (34.5)$$

where m_s is the mass of the separate nucleons, m_b is the mass of the bound nucleus, m_p is the mass of a proton and m_n is the mass of a neutron. The atomic mass (A) and number (Z) are defined as previously.

This equation is somewhat impractical to use, however. The information that is most easily obtained is not the *nuclear* mass, but is the *atomic* mass, which includes the electrons. Another form of Eq. (34.5) is

$$BE = \left(\left[Z\,m(^1H) + (A-Z)m_n\right] - m(^A_Z X)\right)c^2 \qquad (34.6)$$

In this case, the masses are the atomic masses of both hydrogen and the isotope in question. We have in effect added Z electrons in the hydrogen mass and then subtracted Z electrons in the atomic mass of the isotope, so this equation gives the same result as Eq. (34.5) (to within the difference in electron binding energies).

Example 34.2 *Binding energy*

Problem: Calculate the mass defect of an α particle (mass = 4.00153 amu). What is the binding energy per nucleon?

Solution: The mass of the constituent parts is 2× the mass of a proton + 2× the mass of a neutron

$$\text{Combined mass} = 2 \times (1.00728) \text{ amu} + 2 \times (1.00866) \text{ amu}$$
$$= 4.03188 \text{ amu}$$

The mass defect is 4.03188 amu − 4.00153 amu = 0.03035 amu.
1 amu is equivalent to 931.5 MeV, so the binding energy per nucleon is

$$\text{BE per nucleon} = \frac{0.03035 \text{amu}}{4} \times \frac{931.5 \text{ MeV}}{1 \text{ amu}} = \frac{28.27 \text{ MeV}}{4} = 7.06 \text{ MeV} \qquad (34.7)$$

This is in agreement with Figure 34.2.

Example 34.3 ^{14}C decay

Problem: ^{14}C is a radioactive isotope of carbon, notable for having uses in dating organic matter up to around 60 000 years. ^{14}C decays into ^{14}N via the emission of an electron and an antineutrino. ($m_{14C} = 14.00324$ amu, $m_{14N} = 14.00307$ amu, $m_{e^-} = 0.000545$ amu, $m_{1H} = 1.00794$ amu, and while an antineutrino has mass, this mass is so small that you can ignore it here.)

(a) What is the binding energy of the ^{14}C atom?

(b) What is the binding energy of the ^{14}N atom?

(c) How much energy is released during the decay?

Solution: (a) We can calculate the binding energy of the *atom* by using Eq (34.6)

$$BE = \left(\left[Z\, m(^1H) + (A - Z) m_n \right] - m(^A_Z X) \right) c^2$$

where the number of protons in ^{14}C is 6 and the number of neutrons must be 8.   This gives

$$BE = ([6 \times 1.00794\, \text{amu} + 8 \times 1.00866\, \text{amu}] - 14.00324\, \text{amu}) \times 1.6605 \times 10^{-27}\, \text{kg amu}^{-1} \times \left(2.998 \times 10^8\, \text{m s}^{-1} \right)^2$$

$$= 1.700 \times 10^{-11}\, \text{J}$$

(b) A similarly for ^{14}N, where there are 7 protons and 7 neutrons BE = 1.688×10^{-11} J. (c) The energy released in the decay is the difference between these two numbers. This is 1.2×10^{-13} J = 749 keV. The mass of the electron that is emitted is not needed, as we are working with atomic masses, and the atomic mass of nitrogen already includes 14 electrons.

The Liquid Drop Model

Due to the magnitude of the forces involved in holding the nucleus together, it is very difficult to determine the properties of the nucleus as a whole from the behaviour of the individual protons and neutrons from which it is made. It is possible however to understand some of the behaviour of the nucleus using much simpler approximate models.

One of the earliest and most successful of these simple models is the **liquid drop model**. In this model we treat the nucleus as though it was a droplet of charged liquid. To turn a drop of liquid into a gas (i.e., break it into its constituent parts) we must give the drop some energy; this extra energy is the latent heat of vaporisation. This energy is required to break inter-molecular bonds. The binding energy of the nucleus is the energy we would need to add if we wanted to break the nucleus into its constituent parts. The liquid drop model is based on the insight that the latent heat of vaporisation is a direct analogue of the binding energy of the nucleus.

We will not go into any further detail regarding this model other than to state that it may be used to predict the stability of nuclei and the energy released in nuclear reactions such as nuclear decay, and nuclear fission and fusion.

The Nuclear Stability Chart

For small nuclei, adding extra nucleons increases the binding energy. In this case the nucleons are all very close together and the strong force binding nearest neighbour nucleons dominates the repulsive electrostatic forces between the small number of positively charged protons which are all sitting in very close proximity to each other. Once the nucleus is large enough that it is about four nucleons in diameter (mass number around 60), adding more protons begins to reduce nuclear stability. Nuclei with atomic mass around 60, such as iron and nickel, are the most stable and thus have the highest binding energy per nucleon. After this point the electrostatic repulsion caused by the long-range electromagnetic force begins to overwhelm the binding from the short-range nuclear force. By the time the nucleus is around six nucleons in diameter, it is too large to be stable, and nuclei above $A = 209$ are all unstable.

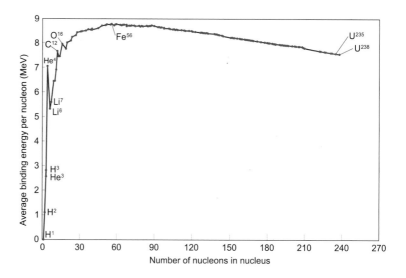

Figure 34.2 The binding energy per nucleon versus atomic mass. [Courtesy of NASA.]

This behaviour may be seen in the nuclear 'stability diagram', shown here as Figure 34.2. A stability diagram plots the binding energy per nucleon of a nucleus against the number of nucleons in that nucleus. The binding energy per nucleon is a measure of how much work must be exerted to remove a single nucleon from a nucleus and is calculated by dividing the total binding energy by the atomic mass number for that nucleus. The stability diagram shows the increase in binding energy as the number of nucleons increases to about 60, and then the slow decrease in binding energy per nucleon as the number of nucleons increases past this point. The stability diagram also indicates that the binding energy per nucleon is relatively constant at about 8 MeV per nucleon. This is due to the fact that the short-range strong nuclear force does not increase with atomic size, the slow decrease in binding energy per nucleon is due to the fact that the long-range electrostatic repulsion between protons does increase with increasing atomic number.

There exist certain 'magic numbers' of either protons or neutrons that are more stable than would be predicted from the general trend in Figure. 34.2. This is believed to be because the nucleus has a shell structure that is analogous to the arrangement of electrons in shells. In the same way that closed electron shells are more stable, so are closed shells in the nucleus. The magic numbers known at present are: 2, 8, 20, 28, 50, 82, and 126. Nuclei with proton and neutron numbers that are magic are particularly stable. For example, the helium-4 nucleus (two protons, two neutrons) has a binding energy per nucleon that is significantly higher than the other nuclei around it in Figure 34.2, as has oxygen-16 (8 protons and 8 neutrons). These are the second and third most abundant nuclei in the universe. The anomalously large binding energy per nucleon of the helium-4 nucleus is responsible for the fact that excess nucleons are generally emitted from unstable nuclei as α particles, i.e., helium-4 nuclei.

Fission

Fission is the process of breaking a nucleus into smaller parts. Spontaneous fission occurs more often in the heavy elements (mass numbers from around 200 and up) such as uranium. Alpha decay (covered in the next chapter) can be thought of as a type of spontaneous fission where one of the fragments is a helium-4 nucleus, but larger fragments occur in most fission reactions.

When a large nucleus fissions it breaks into two smaller nuclei. These daughter nuclei are closer to $A = 60$ on the stability chart than the parent nucleus and thus have greater total binding energy than the parent nucleus. This extra binding energy is released in fission, ultimately as heat, sound and light.

Fission can also be caused by the bombardment of the nucleus with neutrons. In this situation, if enough neutrons are freed by each fission, a chain reaction can be

Naturally occurring chain reaction

In the early 1970s in Oklo, Gabon (in West Africa), scientists found evidence of natural, sustained, fission reactions. Typical samples of uranium contain about 0.77% U-235, an isotope that releases enough neutrons per fission to induce more fissions and cause a chain reaction. The samples taken from local mines had only half the U-235 that was expected. However, the most common isotopes of uranium have different decay rates, and about 2 billion years ago, the proportion of U-235 would have been more like 3% — enough for sustained fission that would reduce the amount of this isotope to what we see now. Such a reaction would not be possible with the present levels of U-235.

started – this is how the simplest nuclear weapons and reactors work. Large amounts of radiation are also released in fission reactions and the daughter products of fission reactions are generally also radioactive.

Fusion

In some cases it is energetically favourable for two nuclei to fuse together in a process called **fusion**. This is the case for light nuclei such as the isotopes of hydrogen and helium. Fusion reactions occur in stars like the sun which combine nuclei with atomic masses all the way up to iron and nickel. Heavier elements are produced in the extreme conditions occurring in supernovae explosions.

As for fission, energy may be released in fusion reactions. In fusion, two nuclei which are lighter than $A = 60$ combine to form a heavier nucleus. The resulting nucleus thus has greater total binding energy than the reactant nuclei, and again the excess binding energy is released during the reaction ultimately as heat, sound and light.

The most powerful nuclear weapons utilise fusion reactions and are known as thermonuclear weapons. The fusion process can release vast amounts of energy, but to overcome the electrostatic repulsion of the nuclei and get them close enough to fuse, a fission explosion is used.

It has long been hoped that fusion could provide a comparatively clean and abundant energy source. At the time of writing, a sustained fusion reaction which generates more energy than it consumes has not been produced.

34.6 Summary

Key Concepts

nucleon A proton or neutron.

binding energy The energy required to separate the parts of a bound system, the energy required to separate a nucleus into its constituent nucleons.

mass defect The difference in mass between the atomic mass and the mass of its constituent parts.

atomic mass unit (u or amu) A non-SI unit of mass, defined such that the atomic mass of carbon-12 is exactly 12 amu.

strong nuclear force One of the four fundamental forces. The attractive force that binds nucleons together in the nucleus.

weak nuclear force One of the four fundamental forces. Responsible for β decay.

fission The induced or spontaneous splitting of a single nucleus into multiple parts.

fusion The fusing together of two smaller nuclei to make one larger nucleus.

Equations

$$1 \text{ eV} = 1.602 \times 10^{-19} \text{ J}$$

$$1 \text{ amu} = 1.661 \times 10^{-27} \text{ kg} = 931.5 \text{ MeV}/c^2$$

$$BE = \Delta E = \Delta m c^2 = (m_s - m_b) c^2 = \left(Z m_p + (A - Z) m_n - m_b \right) c^2$$

$$BE = \left(\left[Z\, m(^1\text{H}) + (A - Z) m_n \right] - m(^A_Z X) \right) c^2$$

34.7 Problems

34.1 If a particular atom has a mass of 6.644×10^{-26} kg and has a total of 20 electrons when neutral (uncharged) what element is it (use a periodic table to answer this question)?

34.2 (a) If all the mass of a carbon-12 atom was converted into energy, how much would this be in both joules and electron volts?

(b) How much energy (in J) would be released by the conversion of 1 kg of carbon to energy?

(c) A kilotonne of TNT releases 4.184×10^{12} J of energy. How much carbon would you need to convert to energy to create an explosion the size of the largest hydrogen bomb test at Bikini Atoll (equivalent to about 15,000 kilotonnes of TNT)?

34.3 An atom of uranium-235 (atomic mass: $^{235}_{92}U - m(^{235}_{92}U) =$ 235.04392 amu) decays to thorium-231 (atomic mass: $^{231}_{90}Th - m(^{231}_{90}Th) = 231.03630$ amu) via the emission of an α particle (nuclear mass: $^{4}_{2}\alpha - m(^{4}_{2}\alpha) = 4.00151$ amu). Use $c = 2.998 \times 10^8$ m s^{-1}. (Note: use the mass information on in Section 34.4 to solve this problem.)

(a) What is the binding energy of an $^{235}_{92}U$ nucleus (in J)?

(b) What is the binding energy of a $^{231}_{90}Th$ nucleus (in J)?

(c) What is the binding energy of an α particle (in J)?

(d) What is the maximum possible kinetic energy (in J) of the α particle emitted during this decay (Hint: what is the difference between the mass of $^{235}_{92}U$ and the total mass of $^{231}_{90}Th$ and an α particle)?

(e) What is the maximum velocity of the emitted α particle?

34.4 There is a rare, but naturally occurring, isotope of helium called helium-3. A helium-3 nucleus has two protons and just one neutron and has an atomic mass of 3.01603 amu.

(a) What is the binding energy per nucleon of helium-3?

(b) How does this compare with the binding energy per nucleon of an alpha particle (helium nucleus)?

34.5 What is the binding energy of each of the following nuclei?

(a) Silicon-28, $m(^{28}Si) = 27.976927$ amu

(b) Iron-56, $m(^{56}Fe) = 55.934939$ amu

(c) Selenium-80, $m(^{80}Se) = 79.916520$ amu

PRODUCTION OF IONISING RADIATION

35.1 Introduction

Ionising radiation is most often produced in one of two ways: acceleration of charged particles or radioactive decay.

An atomic nucleus may undergo spontaneous or induced changes. Spontaneous changes to the atomic nucleus occur when the nucleus is initially unstable and are called nuclear or radioactive decay processes. Radioactive decay processes are named for the type of radiation produced in the decay: α (alpha), β (beta) and γ (gamma). Other possibilities are proton or neutron expulsion, and spontaneous fission. In practice, an unstable nucleus may decay in several different ways, and may undergo several of these processes in rapid succession. Bombarding nuclei with other particles can cause a change in the nuclear structure also, and may cause the nucleus to fission or expel protons or neutrons.

The main production technique for X-rays is the acceleration of electrons using an electric field. These fast-moving electrons generate X-rays in multiple ways, which we will examine. Two kinds of X-ray-producing tubes will also be described.

Another process which produces ionising radiation that is highly relevant to the medical sciences is particle–antiparticle annihilation – the destructive interaction of matter and antimatter.

Key Objectives

- To understand the nature and origin of the three main types of radiation: α, β and gamma radiation, and the decay processes which produce them.

- To understand the origin of *characteristic* X-rays and the generation of a continuous spectrum by *bremsstrahlung*.

- To understand exponential decay processes.

- To understand the concepts of nuclear activity and half-life.

35.2 Nuclear Decay Processes

There are several processes by which an unstable nucleus transforms into a more stable nucleus. During these processes, termed **nuclear decay**, energetic particles or electromagnetic radiation are emitted. The main processes we will describe here are given the names alpha, beta and gamma decay.

Alpha Decay

Some unstable nuclei can become more stable by ejecting two neutrons and two protons in an asymmetric spontaneous fission process. The emitted particles are bound together as a highly stable α **particle**, denoted α, or $^4_2$He, as it is identical to the nucleus of a helium-4 atom. The charge on the α-particle is +2 times the magnitude of the charge on an electron, and this is sometimes written explicitly, e.g., $^4_2$He^{2+}.

Introduction to Biological Physics for the Health and Life Sciences   Franklin, Muir, Scott, Wilcocks and Yates
©2010 John Wiley & Sons, Ltd

Alpha decay is most commonly seen in the case of large nuclei, where the electrostatic repulsion between protons is large enough to destabilise the nucleus. Reducing the amount of positive charge in the nucleus would thus tend to increase stability and emitting an α particle is one way to do this. For example, radium-222 will emit an α particle to decay into radon

$$^{222}_{88}\text{Ra} \rightarrow {}^{218}_{86}\text{Rn} + {}^{4}_{2}\alpha \tag{35.1}$$

Note that the mass numbers of the products (the daughter nuclei) add up to 222 and the atomic numbers add up to 88. In α decay the charge and nucleon number are conserved. We may represent a general α decay process as follows

$$^{A}_{Z}X \rightarrow {}^{A-4}_{Z-2}Y + {}^{4}_{2}\alpha \tag{35.2}$$

Some examples of α emitters are $^{210}_{84}\text{Po}$ (famous for its use in the poisoning of Alexander Litvinenko), and the naturally occurring isotopes of uranium (atomic number 92, mass numbers 234, 235 and 238).

The daughter nucleus produced in α decay is often either in an excited nuclear state or is still unstable. In the first case the daughter nucleus will de-excite by gamma radiation (see below), and in the second case the daughter nucleus will decay further by α or β emission.

Beta Decay

There are three separate processes that come under the heading of β **decay**: β^- decay, β^+ decay, and electron capture. As with α decay, the daughter nucleus that is created is often still unstable and the β decay is followed rapidly by gamma emission or another decay process. (For some nuclei, two β emissions can occur at the same time.) Beta-decay processes do not change the number of nucleons in the nucleus – the product and the parent are *isobars*.

β^- *decay*

β^- ('β minus') decay occurs when a neutron decays and a proton is created, with the accompanying emission of a β^- particle and an **antineutrino**, symbol $\overline{\nu}_e$. The antineutrino is the antiparticle of the neutrino, and the bar over the ν indicates that it is an antimatter particle. The β^- particle is just an electron. The process responsible for β^- decay is the transformation of a neutron into a proton, a process mediated by the weak nuclear force.

$$n \rightarrow p^+ + e^- + \overline{\nu}_e \tag{35.3}$$

The neutrino is a particle from the same group of fundamental particles as the electron (the group called the leptons) and interacts very little with other matter (the neutrino interacts with other matter only via the weak force and gravity). It is believed to have mass, but the upper limits placed on this mass by β^- decay experiments is very low, on the order of 1 eV/c^2 compared with 0.511 MeV/c^2 for the electron.

The β minus decay process occurs inside the atomic nucleus: the electron ejected comes from inside the nucleus and is not one of the atomic electrons. Since neutrons transform into protons, the atomic mass number of the parent nucleus remains the same but the atomic number is increased by one.

The generic equation for β^- decay of $^{A}_{Z}X$ is

$$^{A}_{Z}X \rightarrow {}^{A}_{Z+1}Y + {}^{0}_{-1}\beta + \overline{\nu}_e \tag{35.4}$$

Note that this process can also occur outside the nucleus and is responsible for the instability of lone neutrons, which will decay by β emission. (This is quite rapid: half a sample of single, unbound neutrons will decay in 10.3 minutes.)

The symbols e^- and β^- are used interchangeably, and sometimes $^{0}_{-1}\beta^-$ is written to make it simpler to see that the charge and nucleon number are conserved. (Remember

---

Useful advice

Can't remember any examples for writing a β-decay equation? Pick potassium-40. $^{40}_{19}\text{K}$ is naturally occurring and undergoes all three kinds of β decay.

Neutrinos

The neutrino/antineutrino here is called the *electron* neutrino or antineutrino as there are other sorts, but we will not encounter them in this book. This is the reason for the e subscript.

---

that the upper-left-hand number is the number of protons and neutrons and the lower gives the quantity of positive charge.)

An example of β decay is the decay of caesium-137

$$^{137}_{55}\text{Cs} \rightarrow ^{137}_{56}\text{Ba} + e^- + \bar{\nu}_e \tag{35.5}$$

β^+ *decay*

β^+ ('beta plus') decay is the conversion of a proton into a neutron. Unlike β^- decay, energy is required for the process and so it happens only inside a nucleus where the energy is available, and comes from the creation of a daughter nucleus that is more tightly bound. Free protons are stable. (At least, proton decay has never been observed, and experiments designed to detect the radiation from such an event in vast underground tanks of liquid put the half-life at over 10^{35} years.) In β^+ decay, a proton in the nucleus decays, creating a neutron a β^+ particle and a neutrino. The β^+ particle is the antiparticle of the electron, and is also known as the **positron**, e^+

$$\text{energy} + p^+ \rightarrow n^0 + e^+ + \nu_e. \tag{35.6}$$

The generic equation for β^+ decay of $^A_Z X$ is

$$^A_Z X \rightarrow ^A_{Z-1} Y + ^0_{+1}\beta + \nu_e \tag{35.7}$$

For example

$$^{22}_{11}\text{Na} \rightarrow ^{22}_{10}\text{Ne} + e^+ + \nu_e \tag{35.8}$$

Again there are several notation variants in use for the particle: β^+, e^+ and $^0_{+1}\beta$.

Electron capture

Electron capture is essentially the reverse of β^- decay. Instead of a neutron transforming into a proton and a β^- particle (i.e., an electron), a nuclear proton captures an orbiting electron and transforms into a neutron. This is alternative to β^+ decay, as it is also results in a proton $\rightarrow$ neutron conversion, but by a different route.

The generic equation for electron capture by $^A_Z X$ is

$$^A_Z X + e^- \rightarrow ^A_{Z-1} Y + \nu_e \tag{35.9}$$

For example

$$^{22}_{11}\text{Na} + e^- \rightarrow ^{22}_{10}\text{Ne} + \nu_e \tag{35.10}$$

Gamma Decay

Nuclei, like atoms, can exist in excited states. Just as an atom with its electrons in a high-energy configuration can transition to a lower-energy configuration by emitting a photon of the appropriate energy (and thus frequency), a nucleus in an excited state can transition to a lower-energy state by emitting a photon. The photons emitted in this process are much more energetic than their counterparts in atomic transitions due to the much greater energies involved in holding the nucleus together. These high-energy photons are called **gamma radiation**, denoted γ. The frequency of the γ radiation emitted by a nucleus is determined by the possible excited states of the nucleus which are unique to the particular nucleus concerned and the γ-ray spectrum emitted by a nucleus may be used to identify that nucleus.

The generic equation for γ decay of $^A_Z X^*$ is

$$^A_Z X^* \rightarrow ^A_Z X + \gamma\,(+\gamma + \dots) \tag{35.11}$$

Nuclei in excited states are generally indicated by the symbol '*' as in $^A_Z X^*$. Another commonly used notation is $^{Am}_Z X$ where the 'm' indicates that this is a relatively long lived, or metastable, excited state. Note also that the number of γ photons emitted is

> **Technetium**
>
> Technetium-99m is by far the most commonly used isotope in nuclear medicine. Its parent nucleus ^{99}Mo has a half-life of about 66 hours, which means that a sample that has a useful lifetime of about a week can be manufactured and shipped to a hospital. The technetium is easily chemically separated from the molybdenum.

not clearly specified in this generic equation as there may be a chain of de-excitations occurring as a single excited nucleus decays to its lowest-energy configuration.

For example, molybdenum-99 is created artificially and decays into technetium-99m. This ^{99m}Tc is a gamma emitter (each gamma photon having 140 keV energy) with a half-life of 6 h

$$^{99m}\text{Tc} \rightarrow ^{99}\text{Tc} + \gamma \tag{35.12}$$

Example 35.1 *Decay products*

Problem: Yttrium-90 is used in radioisotope therapy to treat various neuroendocrine tumors. Yttrium-90 decays into zirconium-90 as shown in the following decay scheme

$$^{90}_{39}\text{Y} \rightarrow ^{90}_{40}\text{Zr} + X_A + X_B$$

What are X_A and X_B?

Solution: The decay does not involve a change in the total number of nucleons: there are 90 nucleons before and after the decay. This rules out α decay, as this would result in the number of nucleons being reduced by four.   The atomic number of the atom goes up by one during the decay which rules out γ decay in which the atomic number does not change.   The remaining possibilities are β^- or β^+ decay. The atomic number has increased during the decay, indicating that a neutron has decayed into a proton, and expelled an electron and an antineutrino

$$^{90}_{39}\text{Y} \rightarrow ^{90}_{40}\text{Zr} + ^{0}_{-1}e^- + \overline{\nu}_e$$

35.3 Activity and Half-Life

Activity

Nuclear decay is an essentially random process: we cannot accurately predict when a given unstable nucleus will decay. However, we can determine the *probability* that a nucleus will decay within a given time period. Thus we can predict, given a large number of nuclei, how many will have decayed after a specified time even though we cannot predict beforehand which of the nuclei will decay. If there is a 50% chance that a particular type of unstable nucleus will decay in 1 s, and we have a sample containing 1 million such nuclei, we can predict that close to 500 000 of these nuclei will have decayed after 1 s. Note that the remaining 500 000 undecayed nuclei will still each have a 50% chance of decaying in the next 1 s period, so that one second later we would expect that 250 000 of the remaining nuclei will have decayed, and so on. This results in the characteristic exponential decrease in both the number of undecayed atoms and the 'activity' of the sample over time (see Figure 35.1).

Key concept:
The theory of radioactive decay depends on one fact: *The number of atoms which decay in a given time is proportional to the number of atoms present at the beginning of that time.*

Mathematically, if N is the number of a nuclei of a particular type in the original sample and ΔN is the change in the number of nuclei present after a given time, Δt then

$$\frac{\Delta N}{\Delta t} \propto N$$
$$\frac{\Delta N}{\Delta t} = -\lambda N \tag{35.13}$$

The negative number appears because the number decreases with time. λ is called the **decay constant**. Eq. (35.13) implies the decay constant has units of inverse time, such

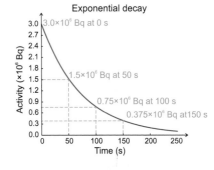

Figure 35.1 An exponential decay curve for a half-life of 50 seconds. Notice that after 50 s, the activity drops to a half the original. After 100 s, or two half-lives, it is a quarter of the original and so on.

as s^{-1}. The quantity $\left|\frac{\Delta N}{\Delta t}\right|$ is the time rate of change of the number of nuclei, and tells us the number of nuclei which decay in each time period, which is called the **activity**, A, and $A = N\lambda$.

A radioactive sample with a high activity may have a large number of moderately unstable nuclei (small decay constant), or a small number of highly unstable nuclei (large decay constant).

The SI unit of activity is the **becquerel**, symbol Bq. A becquerel is equivalent to one disintegration per second. Another unit still in use, particularly in the US, is the **curie**, symbol Ci.

$$1\ \text{Ci} = 3.7 \times 10^{10}\ \text{Bq} \tag{35.14}$$

Half-Life

Eq. (35.13) is a special type of equation. Any quantity for which the rate of change is proportional to the original quantity follows an exponential law, so

$$N(t) = N_0\, e^{-\lambda t} \tag{35.15}$$

where N_0 is the number of radioactive nuclei present at time $t = 0$ and λ is again the decay constant. (The number of nuclei remaining is written as $N(t)$ as a reminder that the number is not constant, but a function of time.) As $A = N\lambda$, we can multiply each side of Eq. (35.15) by λ and obtain an expression for the activity

$$A(t) = A_0\, e^{-\lambda t} \tag{35.16}$$

Eq. (35.15) for $N(t)$ tells us the *number of radioactive nuclei remaining* at time t. So if we started with N_0, the number of nuclei which have decayed after a time, t, is

$$N_{\text{decayed}}(t) = N_0 - N(t) = N_0(1 - e^{-\lambda t}) \tag{35.17}$$

The probability that a nucleus will decay in a time t is the number which decay divided the number which we started off with (N_{decayed}/N_0). This equation tells us that the probability of a nucleus decaying in time t is $1 - e^{-\lambda t}$.

The decay constant tells us something about the probability of nuclear decay. However, another perhaps more useful quantity is the **half-life**, $T_{1/2}$. This is the length of time it takes for half the number of a sample of identical unstable nuclei to decay, or equivalently, the time period over which the probability (P) that a particular nucleus will decay is exactly $\frac{1}{2}$

$$P = 1 - e^{-\lambda T_{1/2}} = \frac{1}{2}$$

so

$$e^{-\lambda T_{1/2}} = \frac{1}{2}$$

$$-\lambda T_{1/2} = \ln(\frac{1}{2}) = -0.693$$

with the result

$$T_{1/2} = \frac{0.693}{\lambda} \tag{35.18}$$

Another timescale sometimes used in exponentially-decaying systems is the **mean life**, τ, which is $1/\lambda$.

The half-life is handy for getting an idea of how long it will take for a given sample to reduce in number or activity. The number or activity will reduce to half the original value after one half-life, to one quarter after two, to one eighth after three, to one sixteenth after four half-lives and so on.

Example 35.2 *Decay rates and half-lives*

Problem: Yttrium-90, which featured in Example 35.35.1, has a half-life of 64 h. A sample of ^{90}Y has an activity of 2.5×10^7 Bq when first measured.

(a) **What is the decay constant for ^{90}Y? (in s^{-1})?**

(b) **How many atoms of ^{90}Y must be in the sample initially?**

(c) **What will the activity of the sample be after 256 h?**

(d) **What fraction of the atoms of ^{90}Y will be left in the sample after just 1 hour?**

Solution: (a) The decay constant is directly related to the half-life; both quantities are a measure of how likely an atom is to decay in a given unit of time

$$T_{1/2} = \frac{0.693}{\lambda}$$

$$\lambda = \frac{0.693}{T_{1/2}} = \frac{0.693}{64\,\text{h} \times 3600\,\text{s}\,\text{h}^{-1}} = 3.0 \times 10^{-6}\,\text{s}^{-1}$$

(b) Because we know the activity of the sample (how many atoms are decaying in a second) and the decay constant (what fraction of a given sample size will decay in a given second) we can find the number of atoms in our sample

$$A = \lambda N$$

$$N = \frac{A}{\lambda} = \frac{2.5 \times 10^7\,\text{s}^{-1}}{3.0 \times 10^{-6}\,\text{s}^{-1}} = 8.31 \times 10^{12}$$

which is actually quite a small amount (just 1.4×10^{-11} mol or 1.2 ng). (c) A period of 256 hours corresponds to exactly four half-lives. Over each half-life the activity of the sample will drop by a factor of two, so after four half-lives that activity will be $0.5^4 = 1/16 = 0.0625$ of the initial activity, or $0.0625 \times 2.5 \times 10^7$ Bq $= 1.56 \times 10^6$ Bq.

(d) The number of atoms left after 1 h (3600 s) is

$$N = N_0\,e - \lambda t$$

$$\frac{N}{N_0} = e - \lambda t$$

$$= e - 3.0 \times 10^{-6}\,\text{s}^{-1} \times 3600\,\text{s} = 0.989$$

so only 1.1% of all of the atoms will have decayed.

Radon hazard and local soil composition

The naturally occurring decay sequences from various uranium and thorium isotopes all produce isotopes of radon, which is an inert gas. The most important is the decay of uranium-238, which produces radon-222. This has a half-life of 3.8 days, which means that there is sufficient time for it to escape from uranium-containing soil and pose a health hazard, particularly in enclosed unventilated spaces like basements. The α-emitting decay products of radon can adhere to dust particles and be inhaled. The dose received this way varies greatly according to local soil composition and building design.

Most Likely Decay Mode and Examples of Decay Series

The most likely mode of decay for a given unstable nucleus can be predicted from its composition. A Segré chart shows which nuclei are stable. (An interactive version can be found at the website of the International Atomic Energy Agency (IAEA): `http://www-nds.iaea.org/relnsd/vchart/index.html`.)

There are three main regions that can be identified on this plot: nuclei below the stable region have too many neutrons so that β^- is the most likely decay process; nuclei above have too few neutrons so that the most likely decay process is β^+ decay or electron capture; and nuclei with masses over about 208 amu are too big and will most likely decay via α-particle emission.

In nature there exist three main chains of decays, known as **decay series**, by which a nucleus that is far from the stable region can undergo a sequence of decays that eventually result in a stable nucleus. Most naturally occurring radioactive nuclides are a member of such a series. For example, the following series starts from thorium-232 and ends on a stable isotope of lead; for this reason it is known as the thorium series

$$^{232}\text{Th} \xrightarrow{\alpha} {}^{228}\text{Ra} \xrightarrow{\beta^-} {}^{228}\text{Ac} \xrightarrow{\beta^-} {}^{228}\text{Th} \xrightarrow{\alpha} {}^{224}\text{Ra} \xrightarrow{\alpha} {}^{220}\text{Rn} \xrightarrow{\alpha}$$

$$^{216}\text{Po} \xrightarrow{\alpha} {}^{212}\text{Pb} \xrightarrow{\beta^-} {}^{212}\text{Bi} \xrightarrow{\beta^-} {}^{212}\text{Po} \xrightarrow{\alpha} {}^{208}\text{Pb}$$

$$\text{or} \qquad {}^{212}\text{Bi} \xrightarrow{\alpha} {}^{208}\text{Tl} \xrightarrow{\beta^-} {}^{208}\text{Pb}$$

Two other sequences from isotopes of uranium to stable isotopes of lead exist. Another series of decays known as the 'neptunium series' features relatively short-lived parent nuclei, so is no longer naturally occuring.

35.4   X-ray Production

X-rays are electromagnetic radiation with wavelengths that are below about 10 nm. The production of X-rays is an atomic process, not a nuclear one. The lower energy end of the X-ray range (the 'soft' end) overlaps with ultraviolet, and the high energy ('hard' end) overlaps with gamma rays. The distinction between parts of the electromagnetic spectrum in this wavelength range is not so much one of wavelength or energy, but one of generation. An X-ray photon that is in all ways indistinguishable from a gamma photon once it has been created is called an X-ray because it was not generated in a nuclear process.

There are two key processes which are used to generate X-rays: electronic transitions into inner shells, and the deceleration of fast-moving free electrons. Electronic transitions within an atom can produce only certain, discrete X-ray energies, so the X-rays created this way are known as **characteristic X-rays** or **characteristic radiation**; the energies produced in these transitions are characteristic of a particular atom. When fast-moving charged particles are decelerated and lose kinetic energy, this lost energy can be emitted in the form of electromagnetic radiation. This process is known as **bremsstrahlung**. If the charged particles are sufficiently fast, the highest-energy radiation will lie in the X-ray region of the electromagnetic spectrum. The combination of these two processes produces an X-ray spectra like that shown in Figure 35.2.

Characteristic Radiation

The energy levels of an atom are labelled by several **quantum numbers** which give all the essential information about the energy, angular momentum etc. of the electron in that particular orbital. The orbitals corresponding to **principal quantum number** $n = 1, 2, 3$ and 4 are traditionally labelled the K, L, M and N orbitals. We will use these letters to label the radiation produced by transitions to those particular levels, along with a Greek letter subscript. For example, transitions from the L shell ($n = 2$) to the $n = 1$ level are labelled K_α. K_β radiation is produced by transitions from the M shell to the K shell. Transitions to the $n = 2$ level are labelled L_α, L_β ... The K and L transitions are illustrated schematically in Figure 35.3.

In the case of the heavier elements, the transition energies involved when a electron falls into the lower K and L shells may fall into the X-ray area of the electromagnetic spectrum. In 1913, Henry Moseley discovered a relationship between the emitted wavelength and the atomic number of the element for these characteristic X-rays. The following empirical formula, known as **Moseley's Law** was found to give a very good fit to the data

$$\lambda \propto \frac{1}{(Z-\sigma)^2} \qquad (35.19)$$

where Z is the atomic number, 13.6 eV is the ionisation energy for hydrogen, and σ is an experimentally determined constant which must be determined for each spectroscopic series. For the K series, $\sigma = 1$, and for the L series, σ is found to be 7.4.

The allowed energies for the K-series transitions are given by

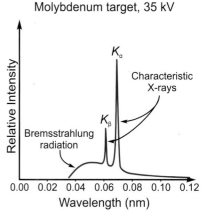

Molybdenum target, 35 kV

Figure 35.2 A typical X-ray spectrum, showing the continuum bremsstrahlung radiation and the characteristic X-rays.

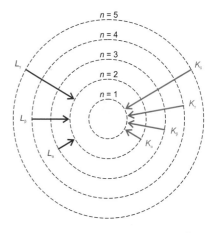

Figure 35.3 Spectroscopic labelling of electronic transitions.

X-ray nNomenclature

The X-ray labelling notation that has historically been used (Siegbahn notation) is more complicated than the simplified scheme we have used here, which is that most often found in introductory texts. The energy levels of electrons in multielectron atoms depend on more than just the principal quantum number, so in reality there are many more possible X-ray energies.

$$\Delta E = (Z - 1)^2 \left[1 - \frac{1}{n_2^2} \right] 13.6 \, \text{eV} \tag{35.20}$$

where n_2 is the principal quantum number of the shell which the electron leaves when it jumps to the K shell. For example, the energy of the K_α spectral line is given by

$$E_{K_\alpha} = \frac{3}{4} (Z - 1)^2 \, 13.6 \, \text{eV} \tag{35.21}$$

The formula for the characteristic X-rays resembles the formula we found for hydrogen and hydrogen-like ions in our section on the Böhr model of the atom (Section 33.4). In this case, however, the atomic number Z is replaced with $Z - 1$. This is due to the presence of another electron in the K shell (which can take a maximum of two electrons with opposite spins). This electron effectively shields the outer electrons from the full Coulomb attraction of the nucleus, making it appear to contain $Z - 1$ protons rather than Z protons.

A similar formula exists for the (L_α lines), with ($Z - 7.4$) appearing instead of ($Z - 1$) to take into account the effective amount of shielding from the inner electrons.

Bremsstrahlung

When a fast-moving electron travels through a material, interactions with the electric fields produced by local nuclei in the material can deflect the electron from its straight-line path. Such a deflection amounts to an acceleration of the electron, that is, a change in its speed and/or direction of travel. This means that a force has been exerted on the electron, work has been done on it, and its kinetic energy will change. Generally these deflections will slow the electron down and thus it will continually lose kinetic energy as it travels through any material. For energy to be conserved, the electron has to emit any energy lost in some form, and this energy is emitted as a photon, i.e., the decelerating electron emits electromagnetic radiation. This type of radiation is known as **bremsstrahlung** which is German for 'braking radiation'.

The amount of energy emitted by the decelerating electron will depend on how much kinetic energy the electron loses in a given deflection. However, there is a maximum amount of energy that the electron can lose. If the electron is brought to a complete stop on its very first interaction with the nuclei of a material, it will give up all of its kinetic energy at once as a single bremsstrahlung photon. Thus the maximum energy of X-rays produced by bremsstrahlung is determined by the decelerating electron's initial kinetic energy.

As we will see in the next section on the construction of X-ray tubes, electrons are accelerated by a large potential difference. Because the maximum kinetic energy of the electrons is a function of the electric field used to accelerate them, this maximum X-ray energy is fixed only by the applied voltage. If the applied voltage is V, then the maximum kinetic energy of an electron accelerated by this potential difference is

$$\text{KE} = qV \tag{35.22}$$

If all of this kinetic energy is given up at once then the emitted photon will have this amount of energy, i.e.

$$h f_{\text{max}} = qV \tag{35.23}$$

From this expression we are able to calculate the maximum frequency produced by bremsstrahlung for a given accelerating potential

$$f_{\text{max}} = \frac{qV}{h} \tag{35.24}$$

and the minimum wavelength

$$\lambda_{\text{min}} = \frac{hc}{qV} \tag{35.25}$$

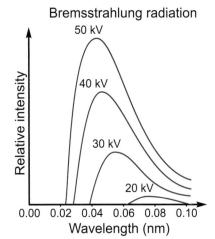

Bremsstrahlung radiation

50 kV

40 kV

30 kV

20 kV

Figure 35.4 Variation in X-ray spectra for different accelerating voltages.

Bremsstrahlung in the Lab

Some radioactive isotopes routinely used in the life sciences, such a phosphorus-32, emit high-energy beta particles. For many situations, shielding from radiation is achieved with layers of dense materials, like lead, but for such beta-emitters the production of secondary radiation through bremsstrahlung can be problematic. Lower density materials like plastic and water are typically used for shielding in such cases, as the proportion of the incident energy that is converted to bremsstrahlung radiation is proportional to atomic number.

The characteristic X-rays produced by an X-ray tube are determined by the type of metal in the target which the accelerated electrons strike. The accelerating voltage will determine which characteristic X-rays may be generated, but it will not change their energies.

Example 35.3 *X-ray production*

Problem: A thermionic X-ray tube accelerates electrons over a potential of 100 kV into a tungsten (W) target.

(a) **What is the minimum X-ray wavelength produced by this tube?**

(b) **What is the wavelength of the characteristic spectral line produced by electrons making the transition from the $n = 3$ to the $n = 2$ orbital in the tungsten target in this X-ray tube?**

Solution: (a) The minimum wavelength (maximum energy) photons produced by the X-ray tube will occur when all of the energy of an electron hitting the target is converted into a single photon. This energy does not depend on the target material, but only on the energy of the electrons hitting it

$$\lambda_{min} = \frac{hc}{qV} = \frac{6.626 \times 10^{-34} \text{ J s} \times 2.998 \times 10^8 \text{ m s}^{-1}}{1.6 \times 10^{-19} \text{ C} \times 100 \times 10^3 \text{ V}} = 0.0123 \text{ nm}$$

(b) The $n = 3$ to the $n = 2$ transition corresponds to the L_α transition. Using Moseley's law we can say that resulting photons emitted from the tungsten target ($Z = 74$) will have an energy of

$$\Delta E = 13.6 \text{ eV} (Z - 7.4)^2 \left[\frac{1}{n_1^2} - \frac{1}{n_2^2} \right] = 13.6 \text{ eV} \times 66.6^2 \times \left[\frac{1}{2^2} - \frac{1}{3^2} \right] = 8380 \text{ eV}$$

which corresponds to a wavelength of

$$\lambda_{L_\alpha} = \frac{6.626 \times 10^{-34} \text{ J s} \times 2.998 \times 10^8 \text{ m s}^{-1}}{8380 \text{ eV} \times 1.6 \times 10^{-19} \text{ J eV}^{-1}}$$
$$= 0.15 \text{ nm}$$

X-ray Tubes

Crookes Tube

The Crookes tube is an evacuated glass tube and electrode setup developed by Sir William Crookes. It paved the way for the proper scientific exploration of X-rays, and was used by many experimental physicists in the early investigation of the nature of matter.

A Crookes tube is made of glass, contains gas at low pressure and has two built-in electrodes. When an electrical potential difference is applied between the electrodes, the electrode at higher potential (the positive electrode) is called the anode and the electrode at lower potential (the negative electrode) is called the cathode. When the applied voltage difference is large enough, some of the gas that remains in the tube is ionised. The positive ions are accelerated towards the cathode, and when they strike the cathode they liberate electrons. All the electrons created by these collisions and the ionisation of the gas are accelerated toward the anode. If the anode is not blocking the path of these electrons, they will overshoot and fly straight past anode, colliding with the glass end of the tube or any target that is placed there. With sufficiently high potential difference between the electrodes, these electrons will produce X-rays when they are suddenly decelerated by the glass end of the tube or the target material placed there.

By trial and error it was found that the best target material for generating X-rays was tungsten (chemical symbol W).

While the Crookes tube is of great historical importance and clearly displays the processes involved in the generation of X-rays, it has largely been replaced in modern devices by the thermionic tube.

Thermionic Tube

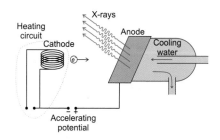

Figure 35.5 Schematic diagram of the modern thermionic X-ray tube.

In the **thermionic** (or Coolidge) tube, a heated filament is used to produce electrons. A filament is heated to the point were electrons in the metal gain enough energy to leave the metal and form a cloud, or space charge around the filament. An electric field is used to accelerate these electrons toward a target material (most commonly tungsten), which also acts as the anode. The accelerated electrons collide with the anode and produce X-rays either by bremsstrahlung or by causing the emission of characteristic X-rays. In the Crookes tube, increasing the accelerating potential increased the ionisation of the gas in the tube, which increased the number of positive ions colliding with the cathode and thus increased the amount of X-ray photons produced. Increasing the accelerating potential in the Crookes tube would thus increase both the number of X-ray photons produced and their maximum energy. In the thermionic tube, on the other hand, the current used to heat the filament and thus to generate the electrons which produce the X-rays can be adjusted independently of the accelerating potential. Thus the number of X-rays produced may be varied independently of the maximum energy of the X-rays. This allows for greater flexibility in the clinical use of X-rays than was possible with the Crookes tube.

The generation of X-rays is not an efficient process. For example, with a tungsten target and an accelerating voltage of 100 kV, about 0.8% of the energy used to accelerate the electrons is ultimately converted into X-rays. The rest of the energy is lost heating the target, so only materials that have large heat capacity and high melting point are suitable target materials.

35.5 Other Sources of Radiation

Pair Annihilation

When a particle and its antiparticle meet (e.g., an electron and a positron) they **annihilate** each other and the energy equivalent of their mass is released as energy in the form of γ photons. In these interactions, all conservation laws are obeyed. The particle and antiparticle are opposite in the sense that whatever property the particle has the antiparticle has the opposite. If the particle has positive charge, the antiparticle has equal but opposite charge, etc. However, energy and momentum must also be conserved. The γ photons produced in the annihilation will have total energy equal to the total rest mass energy of the particle–antiparticle pair. To conserve the total momentum of the particle–antiparticle pair, two photons of equal energy, travelling in roughly opposite directions, must be created. These photons have a characteristic energy that makes them useful in medical diagnostics (0.511 MeV for electron–positron annihilation).

Cosmic Rays

Cosmic rays might sound like something you would find in comic books, but they are very real. The term is used to describe high-energy particles that stream into the Earth's atmosphere from space. The origins and nature of these particles vary, from protons originating in the Sun, to higher-energy particles from galactic supernovae, neutron stars and black holes, and extremely high-energy particles coming from other galaxies.

The highest-energy particles to hit the Earth's atmosphere have energies in excess of several billion billion electron volts. This is equal to the kinetic energy of a well-hit tennis ball. To put this in perspective, the most powerful particle accelerator ever built on Earth (the Large Hadron Collider), will accelerate particles to energies of around seven trillion (seven thousand billion) electron volts, a million times less energetic than the most energetic cosmic rays. For many years, scientists could only speculate on the origin of most of these energetic particles, largely because they are usually charged, and are therefore deflected a lot by interactions with electromagnetic fields. Recent work on detection of the highest energy of these cosmic rays – the ones that are moving so fast that they are not so easily deflected – suggests they originate in black holes that exist in active galactic centres.

When such high-energy particles reach the Earth's atmosphere, they interact to produce showers of lower-energy particles. The number of particles in a single cascade can number in the billions, and it is this cascade which enables the detection of these events.

Cosmic rays hitting the upper atmosphere are responsible for the creation of some unstable nuclei, such as carbon-14. This process keeps the amount of carbon-14 in the environment roughly constant, and allows the use of carbon-14 dating to determine the time since death for some organic matter.

The expected radiation dose from cosmic rays in the absence of the Earth's shielding atmosphere and magnetic field is a big issue for future space missions to other planets. The kind of exposures astronauts on a mission to Mars might get are around 200 times the level of exposure on the Earth.

35.6 Summary

Key Concepts

ionising radiation Particles or electromagnetic waves which have sufficient energy to ionise atoms and molecules.

X-ray A type of ionising electromagnetic radiation with wavelengths from about 0.01 nm to around 10 nm. The low energy end of the X-ray spectrum (the soft X-rays) overlaps with the extreme ultraviolet.

gamma (γ) ray A form of ionising electromagnetic radiation with wavelengths less than about 0.01 nm. Gamma radiation is generated by processes within the nucleus and by antimatter annihilation.

bremsstrahlung (German for 'braking radiation'.) The continuum X-ray radiation produced by the braking of fast-moving electrons when they interact with matter.

characteristic X-rays The X-ray photons produced by electronic transitions to tightly bound inner shell orbitals. The transitions occur after the removal of an inner shell electron, usually by collision with an externally produced fast electron. The photon energies produced are characteristic of the target atom.

annihilation The process in which a particle meets its antiparticle and both particles cease to exist, their rest mass energy being converted to gamma radiation. The reverse process is *pair production*.

electron neutrino (ν_e) An elementary particle produced in some nuclear processes that travels at close to the speed of light and has zero charge. The mass is not known, but the upper limit of the possible mass range is very small.

antimatter Most fundamental particles (the exception being some massless particles like photons) have an antiparticle equivalent with the same mass and opposite charge (and other quantum numbers). When a particle and its antiparticle meet, annihilation occurs.

α-decay The emission of a helium-4 nucleus (α particle) from a larger, unstable nucleus, leaving a daughter nucleus that has two fewer protons and two fewer neutrons.

β-decay One of three different processes (β^-, β^+ and electron capture) that result in a change in nuclear composition, but not nucleon number. A neutron is converted into a proton, or viceversa, with the accompanying creation of a positron, or the creation/loss of an electron and the production of a neutrino or antineutrino.

γ-decay A nucleus in an excited state can emit energy as a photon of electromagnetic radiation, known as a gamma photon.

half-life ($T_{1/2}$) The time taken for half the unstable particles in a pure sample to decay. Also the time taken for the activity of a sample to halve.

activity The measure of the rate of decay of a radioactive sample. The SI unit of activity is the becquerel (Bq), with 1 Bq being equivalent to one decay per second.

Solar activity andcosmic rays

Over the 50-year period that scientists have been monitoring it, the pressure from the *solar wind*, the constant stream of particles (mostly protons and electrons) ejected from the Sun, has been decreasing. According to NASA, this has led to an increase in the cosmic rays reaching Earth, though fortunately not the number reaching the surface. This extra radiation dose could affect manned space missions to other planets and could cause more damage to satellites in high Earth orbit. Strangely enough, when the Sun ejects material in the direction of Earth, this is a 'bad thing' the charged particles can destroy satellites but when one of these solar flares appears and is not pointing in our direction, it can be a good thing, as the resulting magnetic fields deflect galactic cosmic rays – ones from outside the solar system – away from us.

decay constant (λ) For an exponential decay process, the rate at which the quantity decreases is proportional to the quantity, with the constant of proportionality being the decay constant.

exponential A quantity is said to change *exponentially* when the rate of change of that quantity is proportional to the original value of that quantity.

Equations

$$N(t) = N_0 e^{-\lambda t} \quad A(t) = A_0 e^{-\lambda t}$$

$$A = \lambda N \quad T_{1/2} = \frac{0.693}{\lambda}$$

$$h f_{\text{max}} = qV \quad \lambda_{\text{min}} = \frac{hc}{qV}$$

$$\Delta E \quad = (Z - \sigma)^2 \left[\frac{1}{n_1^2} - \frac{1}{n_2^2} \right] 13.6\,\text{eV}$$

$$^A_Z X \longrightarrow\ ^{A-4}_{Z-2} Y + ^4_2 \alpha$$

$$^A_Z X \longrightarrow\ ^A_{Z+1} Y + ^0_{-1} \beta + \overline{\nu}_e$$

$$^A_Z X \longrightarrow\ ^A_{Z-1} Y + ^0_{+1} \beta + \nu_e$$

$$^A_Z X + e^- \longrightarrow\ ^A_{Z-1} Y + \nu_e$$

$$^A_Z X^* \longrightarrow\ ^A_Z X + \gamma + \gamma + \gamma + \dots$$

35.7 Problems

You will need to use a periodic table to answer the three following questions. **35.1** What is the decay product $^A_Z X$ in the following nuclear decay process?

$^{238}_{92} U \rightarrow ^A_Z X + ^4_2 \alpha$

35.2 What is the decay product $^A_Z X$ in the following nuclear decay process?

$^{237}_{92} U \rightarrow ^A_Z X + e^- + \bar{\nu}_e$

35.3 What is the decay product $^A_Z X$ in the following nuclear decay process?

$^{11}_{6} C \rightarrow ^A_Z X + e^+ + \nu_e$

35.4 What is the decay product $^A_Z X$ in the following nuclear decay process (gold-196 by electron capture)?

$^{196}_{79} Au + e^- \rightarrow ^A_Z X + \nu_e$

35.5 Cadmium-107 has a half life of 6.52 hours. If you start off with sample which has an activity of 1.0×10^{10} Bq, what will the activity in Bq be after the following times (note: you can answer these questions without using Equation (35.16)?

(a) 6.52 hours;

(b) 19.6 hours;

(c) 3 days;

(d) 6 days.

35.6 Iodine-120 has a half life of 1.35 hours. If you start off with a sample which has an activity of 1.0 Ci, how long is it before the activity drops to the following values (note: you can answer these questions without using Equation (35.16)):

(a) 0.50 Ci;

(b) 0.125 Ci;

(c) 9.77×10^{-4} Ci;

(d) 9.31×10^{-10} Ci.

35.7 You find an old radioactive source in the back of the cupboard which is labeled 'Caesium-137 - $^{137}Cs \rightarrow ^{137}Ba + e^-$, $T_{\frac{1}{2}} =$ 30.2 years, $A = 1.4$ mCi *as of 01/01/1954'*. If the date is now 30/06/2009, what is the current activity of this sample (in mCi)?

35.8 Some X-rays are produced by accelerating a beam of electrons across a potential difference of 120 kV into a tungsten ($^{184}_{74}$W) target.

(a) What are the lowest wavelength (highest energy) X-rays produced when the beam of electrons hits the target?

(b) What is the wavelength of the K_α characteristic X-rays produced in the target?

(c) What is the wavelength of the K_β characteristic X-rays produced in the target?

35.9 Some X-rays are produced by accelerating a beam of electrons across a potential difference of 15 kV into a nickel ($^{58}_{28}$Ni) target.

(a) What are the lowest wavelength (highest-energy) X-rays produced when the beam of electrons hits the target?

(b) What is the wavelength of the K_α characteristic X-rays produced in the target?

(c) What is the wavelength of the K_β characteristic X-rays produced in the target?

35.10 You wish to design an X-ray tube that will produce a 1 W beam of 0.07217 nm X-rays.

(a) If your X-rays are produced by the K_α characteristic transition, what element is the target made out of? (Consult a periodic table.)

(b) You design your X-ray tube such that a beam of electrons is accelerated through some potential difference into a target and the maximum photon energies produced are twice that of the useful 0.07217 nm beam. Through what potential difference do you accelerate the electrons?

(c) If 0.5% of the energy deposited in the target is converted into X-rays, and only 3% of the X-rays are the useful 0.07217 nm X-rays, what is the power in the electron beam?

INTERACTIONS OF IONISING RADIATION

36

36.1 Introduction

The way in which ionising radiation interacts with matter depends on the type of radiation and its energy. Particulate radiation (α and β radiation) interacts with matter by colliding with the atoms or molecules and ionising them. Antiparticles collide with and are completely annihilated with their corresponding matter particles. Photons may be completely absorbed by an electron, involved in scattering collisions, or be the cause of particle–antiparticle pair production.

This chapter also includes some of the commonly used devices for detection of ionising radiation.

Key Objectives

- To understand the concept of an interaction or scattering cross section.

- To understand some ways in which high – energy photons can interact with matter: Compton scattering, the photoelectric effect and pair production.

- To understand some ways in which particulate radiation can interact with matter.

- To estimate the penetrating ability of different kinds of radiation.

- To examine the operation of some commonly used radiation detectors.

36.2 Attenuation and Cross Section

When a beam of radiation (of any type) enters matter it is normally *attenuated*, that is, the amount of energy in the beam decreases with distance into the material. The processes by which the radiation interacts with matter will be discussed in the next sections. Without knowing the specific kind of interaction, we are still able to make some general statements about the rate of attenuation.

Consider a block of some material with area ΔA and depth Δx; the volume of the block is then $\Delta V = \Delta x \times \Delta A$. Suppose also that the block contains ΔN atoms or molecules with which radiation can interact. If a beam of radiation strikes the top surface of this block the chances of an incoming particle striking an atom in the block is twice as great if there are two target atoms per unit area instead of one – the probability that a particle in the beam will be absorbed is proportional to the total number of interaction centres and inversely proportional to the total area over which they are spread. To simplify the discussion we will define a quantity called the particle *fluence*, Φ which is the number of incoming radiation particles per unit area. The number of incoming particles lost in a volume ΔV is then $-\Delta\Phi$. The number of particles per unit area in the incoming beam that are absorbed in the volume ΔV is then proportional to the total number of absorbing centres and inversely proportional to the total area over which these centres are spread

$$-\frac{\Delta\Phi}{\Phi} \propto \frac{\Delta N}{\Delta A}$$

We can convert the proportionality into an equality by introducing a constant of proportionality

$$\frac{\Delta\Phi}{\Phi} = -\sigma\frac{\Delta N}{\Delta A}\tag{36.1}$$

The constant of proportionality, σ, is called the **scattering crosssection** and has the units of area. It can be thought of as the effective area of the target particles – the area that an incoming particle needs to hit for an interaction to occur. The scattering cross-section is measured in units called *barn*, named by scientists working on the Manhattan Project based on colloquial expressions about hitting targets that involve barns, such as 'such a bad shot, he couldn't hit the broad side of a barn'. The barn is a unit of area: a very small one, but large in terms of the typical size of a nucleus. One barn equals 10^{-28} m^2.

If we note that the concentration of the absorbing target particles, n, is $\Delta N/\Delta V$, so that $\Delta N/\Delta A = n\Delta x$, it follows from Eq. (36.1)that

$$\frac{\Delta\Phi}{\Phi} = -\sigma\frac{\Delta N}{\Delta A}$$
$$= -\sigma n\Delta x$$

so that (with a little rearranging)

$$\frac{\Delta\Phi}{\Delta x} = -\sigma n\Phi\tag{36.2}$$

In other words, the change in particle fluence with distance ($\Delta\Phi/\Delta x$) is proportional to the particle fluence (Φ). As we have seen before in radioactive decay, when the rate of change of a quantity is proportional to that quantity, then that quantity is described by an exponential function.

> **Key concept:**
> The intensity of beam of particles or photons, all having the same energy, decreases *exponentially* with distance in an isotropic material.

For example, if the beam intensity is reduced by half after 1 cm, that does not mean that the beam will be completely blocked by 2 cm – the beam intensity will be reduced to one quarter. The real situation is often far more complex because of all the secondary radiation from the absorption and scattering of the original beam particles.

36.3 X-rays and Gamma Radiation

Most of the attenuation of high-energy electromagnetic radiation (gamma and X-rays) in matter is due to interactions between the incoming photons and **orbiting electrons in the target atoms**. There are several interaction mechanisms:

- Complete absorption of the energy of the photon.

- Partial absorption of the energy of the photon, which results in a new, lower-energy photon travelling in a different direction (Compton effect).

- The ionisation of the atom and the freeing of an orbital electron (Photoelectric effect).

- The creation of new particles.

Since the chance of a collision depends on the density of electrons, high-atomic-number materials, e.g., lead, tend to be more attenuating. This is the reason that lead is so often used as radiation shielding. There are other elements with similar atomic number, but these are either radioactive themselves or relatively expensive. The level of attenuation of X-rays and gamma radiation depends very much on photon energy – higher energy photons are typically more penetrating. X-ray and γ-ray photons are removed from a beam by the absorption and scattering processes listed above, and the likelihood of the different interactions depends on photon energy. We will consider these different attenuation mechanisms in the following sections.

The Photoelectric Effect

The **photoelectric effect** is the name given to the process in which a photon is completely absorbed by a bound electron, giving the electron enough energy to escape whatever binding potential is holding it, generating a free electron called a photoelectron. The bound electron is generally either bound to an atom or is held in a crystalline solid by the collective action of a number of atoms (as in the case of conduction electrons in a metal). Whether or not the photoelectric effect will occur depends on the binding energy of the electron. The binding energy is the amount of energy that the electron needs to completely escape from the binding potential.

The photoelectric effect can be observed when light in the visible or UV part of the spectrum irradiates certain metals. In this case, the energy of the incident photons may provide enough energy to allow the escape of electrons from a crystalline solid (the metal). Here the conduction electrons are not bound to a single nucleus, and the threshold energy for electrons to be emitted is called often called the *work function* of the material.

The energy of a photon is related to its frequency by $E = hf$. Thus the higher the frequency of the incident light the more energy each photon has, and this energy must be greater than the work function for the absorption of a photon to result in the electron gaining enough energy to escape. There is a cut-off frequency below which none of the photons absorbed will be able to provide bound electrons with enough energy to escape the binding potential. Thus below the cut-off frequency, no electrons will be emitted from the metal surface.

Increasing the intensity of the radiation without changing the frequency has the effect of increasing the number of photons striking the material, but does not increase the energy that each of these photons has. Hence increasing the intensity of the incident light will increase the number of photoelectrons emitted, but only if the frequency of the light is above the cut-off frequency.

If the frequency of the incident light is greater than the cut-off frequency then there will be some energy left over after a bound electron absorbs the photon and escapes the binding potential. This extra energy appears as the kinetic energy of the emitted photoelectron. Some of this kinetic energy may be absorbed by collisions with the crystal lattice as the photoelectron makes its way to the surface of the metal and escapes, however it is possible that the photoelectron is already at or close to the surface so that it does not lose any of this kinetic energy. Thus the maximum kinetic energy of photoelectrons emitted by the photoelectric effect is simply the photon energy minus the binding energy of the metal. The maximum kinetic energy of photoelectrons depends solely on the frequency of the incident light and not at all on the intensity of this light

$$\text{KE}_{\text{max}} = hf - B \qquad (36.3)$$

where f is the frequency, h is Planck's constant and B is the binding energy or work function.

The photoelectric effect is historically very significant as it was Einstein's explanation of the process that showed that electromagnetic radiation came in packets which were named photons. The theories of electromagnetic waves that had been postulated previously predicted results that were in part at odds with experiment. In particular, they suggested that any frequency of incident light should produce photo-electrons – it would just take longer to free electrons if the light intensity was low. This is because the classical model of electromagnetic radiation predicted that an electron would continuously absorb energy from the incident waves until they had sufficient energy to escape. The experimental evidence indicates that this is not what happens; no electrons are emitted when the incident light is below the cut-off frequency. Furthermore the emission of electrons is nearly instantaneous and the time taken for emission does not depend on the intensity of the incident light.

The photoelectric effect is more complicated in the case of high-energy radiation. X-rays and gamma rays have sufficient energy to liberate not just the loosely bound electrons in metals, but also more tightly bound atomic electrons, and generally are able to liberate electrons from inner atomic orbitals. When photo-electrons are ejected from inner atomic orbitals, the resulting vacancy is quickly filled by an outer-shell atomic electron which drops down into this vacancy. This process, an outer-shell electron dropping down into an inner-shell vacancy, results in the emission of the energy difference between these two shells as another (X-ray) photon.

Attenuation of X-rays and gamma radiation via the photoelectric effect occurs because the ejected photoelectron carries away some of the energy of the incident X-ray or gamma-ray photon as kinetic energy. The X-ray which is emitted as a result of the dropping of an outer-shell electron into the inner-shell vacancy will generally be less energetic than the original incident radiation, and can be emitted in any direction. Thus the energy of the incident beam has been reduced.

Pair Production

Sufficiently high-energy gamma-ray photons may spontaneously convert into a electron – positron pair. This is more likely in the vicinity of a nucleus. The energy of the photon is converted into matter and antimatter. Because energy is conserved, the mass–energy equivalence relationship $E = mc^2$ determines the total mass that can produced. There is a minimum photon energy necessary for pair production to proceed, which is equivalent to the rest mass of the two electrons.

The rest mass energy of the electron may be calculated as follows

$$E_{\text{electron}} = m_e c^2$$
$$= 9.108 \times 10^{-31} \times \left(2.998 \times 10^8\right)^2$$
$$= 8.187 \times 10^{-14} \text{J}$$

or in eV:

$$= 511.0 \text{ keV}$$

Thus we must have $E_{\text{photon}} > 1.022$ MeV to produce an electron and positron since they each have a rest mass of $511 \text{ keV}/c^2$. This means that only photons with frequencies greater than about 2.5×10^{20} Hz will be able to produce electron–positron pairs. This is a frequency in the gamma-radiation range.

All conservation laws must be satisfied in pair production. For example, charge is conserved by the simultaneous creation of a positive and negative charge of equal magnitude. Energy and thus mass is clearly conserved, but momentum must also be conserved. This means that the pair of particles must have equal and opposite momenta.

The reverse of the pair production process is particle–antiparticle annihilation, which was discussed in Section 35.5.

The energy of an incoming beam of gamma radiation will be attenuated if pair production occurs. The positron produced in this process will ultimately undergo annihilation with an electron within the medium and this will produce two photons. Each of these photons will have less energy than the original gamma photon responsible for the pair production, and the direction will be different. In this way the process of pair production will act to reduce the energy of a beam of electromagnetic radiation.

The Compton Effect

The Compton effect, or Compton scattering, is an interaction between electromagnetic radiation and matter in which photons display their particle-like nature. In determining what happens when a high-energy photon is scattered by a particle (such as an electron) without being absorbed, it is necessary to analyse the event rather like a collision between two billiard balls, and to look at conservation of energy and momentum to fully explain the way the photon wavelength and energy change. Such a collision demonstrating the Compton effect is shown Figure 36.1

When an incoming photon is incident on a nearly free electron which is more or less at rest, energy is transferred from the photon to the electron. The electron gains

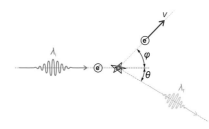

Figure 36.1 The Compton Effect. An incoming photon is scattered from a nearly free electron, giving energy and momentum to the electron and moving off at some other angle with a reduced frequency, i.e., in increased wavelength.

kinetic energy and moves off in some direction, and a lower-energy, longer-wavelength photon. The energy and momentum of the electron and photon after the collision are such that both energy and momentum are conserved in the process.

A full mathematical treatment of Compton scattering is outside the scope of this book, but there are a few points that are worth mentioning. Firstly, even though photons have no rest mass, as previously mentioned, they do have momentum, and this is related to their wavelength by $p = \frac{h}{\lambda}$, where p is the momentum, h is Planck's constant and λ is the wavelength. Another observed result is that the change in wavelength is related to the angle at which the photon is scattered. The maximum energy transfer from photon to electron occurs when the scattered photon returns in the opposite direction to the incident photon.

When Compton scattering occurs, the incoming photon loses energy to the scattered electron, so a beam of gamma or X-ray photons will also be attenuated by Compton scattering in the medium.

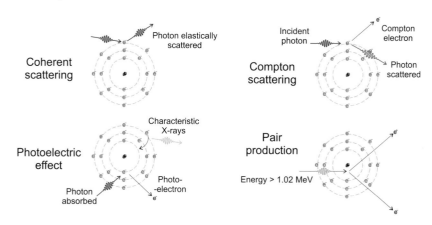

Figure 36.2 A summary of the ways photons interact with matter.

36.4 Particles

We will not discuss the mathematical details involved the particle–particle collisions which are responsible for the attenuation of beams of particulate radiation. These collisions are often between charged particles moving at relativistic speeds and the detailed analysis is beyond the scope of this book. Here we will content ourselves with a summary of the kinds of interactions that occur.

Neutrons

Neutrons are uncharged, and so interact predominantly with the nuclei in attenuating material in the following ways:

elastic collision Kinetic energy conserving collision with another particle.

non-elastic collision The neutron interacts with a nucleus and is re-emitted with a different (normally reduced) kinetic energy.

capture The neutron is captured and becomes part of a nucleus.

spallation/fission The neutron is captured, but the increase in energy causes fragmentation of the nucleus. Spallation is a term for the production of fragments when an object is subjected to an impact or stress. Fission is the splitting of an object into two parts.

The probability of an interaction between a neutron and a particular nucleus is energy dependent. Except for very low energies (less than 100 keV) where capture is important, elastic collisions dominate. Inelastic collisions are more likely above a few MeV, and spallation occurs above about 20 MeV.

Ions

This category includes protons, α particles and heavy nuclei.

capture of electrons The energetic incoming ion captures one or more electrons from the absorbing material and becomes a neutral atom. In the process the ion loses kinetic energy and ionises the surrounding material. This occurs for low-energy radiation.

collisions with electrons The energetic incoming ion collides with atoms of the absorbing material but does not capture electrons and become neutral. In the process the surrounding material is ionised and the atoms of the absorbing material may be lifted into excited atomic states. Again the energy to ionise or excite the surrounding material is provided by the incoming ion, which therefore loses kinetic energy in the process.

nuclear collisions The incoming ion collides directly with the nucleus of an atom of the absorbing material. This occurs only when the incoming ions are at very high energies. In these collisions processes such as spallation and fission may be induced. Particle accelerators are sometimes used to produce a beam of protons which are then directed into a target material to produce neutrons by spallation.

Electrons/Positrons

Some of the interactions of energetic electrons have already been covered in the X-ray production section (see Chapter 35). The current discussion is more generalised, however, and also applies to positrons.

annihilation Positrons may collide with an electron and be annihilated.

collisions with atomic electrons Energetic incoming electrons may collide with and eject electrons from various atomic shells, in the atoms of the absorbing material. The atomic shells involved will depend on the energy of the incoming electron.

bremsstrahlung The deceleration of energetic incoming electrons and the subsequent emission of the lost kinetic energy as electromagnetic radiation (up to X-ray energy) has been described earlier (see Section 35.4).

Cerenkov radiation Cerenkov radiation is electromagnetic radiation emitted when electrons travel through a material at speeds greater than the speed of light in that material (though still lower than the speed of light in a vacuum, c). The effect may produce damaging ultraviolet radiation. Occurs for electrons with kinetic energy above about 500 keV when travelling in water.

36.5   Detection of Ionising Radiation

The Geiger–Müller Tube

The **Geiger–Müller tube** (often just called a Geiger tube) is a type of gas detector which is most useful for the detection of α and beta radiation. They can be used for gamma detection, but the detection efficiency is usually quite low, as the gas inside is at quite low pressure. The ionising particle passes through a thin window at the end of the tube (usually made of the mineral mica) and ionises the gas. The charged particles that are created are then accelerated by an electric field, and collide with further gas particles, creating an avalanche of charged particles. This current is detected and recorded as an audible click or a spike in the output signal.

Geiger–Müller tubes can detect only the presence of radiation and not its energy. They are also only useful up to count rates of about 1000 counts per second. The ionised gas takes some time to *recombine* and return to a state where another particle event can be detected (this is called the dead time of the detector). As the count rate increases, so does the probability that simultaneous events will be missed.

The Photomultiplier

Photomultipliers, as the name suggests, amplify the signal generated by the detection of a photon. When a photon hits a *photocathode* inside a vacuum tube, it triggers the release of a photoelectron. This electron is accelerated towards a nearby electrode (called a dynode) that is at a higher potential than the cathode, and when it strikes the dynode, more electrons are emitted. This process is repeated several times to produce a burst of up to $\sim 10^6$ electrons from a single photon. Voltages of several thousand volts are typically required. A simplified diagram of how a photomultiplier works is shown in Figure 36.3.

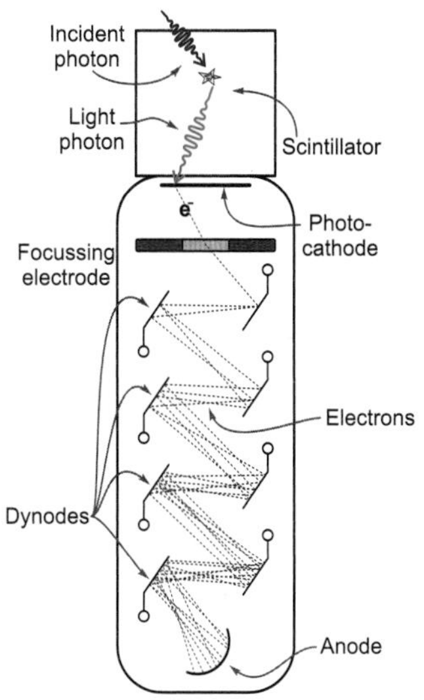

Photomultipliers can be used as part of *scintillation counters*. A scintillating material, one that generates photons when struck by ionising radiation, is used, and then these photons are converted to electrical pulses with a photomultiplier. Such detectors have the advantage of producing information about the energy of the incident particles, as the number of photons generated in the scintillating material is proportional to the energy. Sodium iodide crystals are commonly used as scintillators in such detectors, since they provide good efficiency due to the high atomic number and thus electron density of their component atoms. Germanium crystals have very good energy resolution for gamma detection, but require low temperatures (achieved with liquid nitrogen) for their operation.

Figure 36.3 A schematic of a photomultiplier with scintillating material.

Photographic Emulsions

The earliest type of detection equipment was the **photographic film**. It was the discovery of the darkening of photographic films by nearby uranium samples that originally led to the discovery of ionising radiation. X-rays affect the film in much the same way as visible light does; silver halide salts are converted to metallic silver by incident photons.

Photographic film is routinely used for the detection of X-rays in a medical diagnostics. To improve the sensitivity of the film and reduce the exposure needed, the film is often coated on both sides with a fluorescent material that increases the effect of X-rays on the film, and sometimes a layer of lead is used under the photographic emulsion to backscatter the X-rays through the film a second time.

36.6 Summary

Key Concepts

photoelectric effect The process in which electrons are emitted from a material when electromagnetic radiation is incident on the surface.

Compton effect A process in which a photon is scattered off an electron such that it undergoes a change in direction and a corresponding reduction in frequency.

photon momentum While photons have 'zero rest-mass', they do carry momentum $p = \frac{h}{\lambda}$.

pair production The production of a particle–antiparticle pair, usually from energetic gamma photons.

Geiger–Müller tube A gas-filled tube for detecting and counting radiation. It is most efficient for α and beta radiation.

photomultiplier A light-detection device. Incident light produces electron emission by the photoelectric effect, and these electrons are amplified by a series of electrodes at increasing potentials to produce a detectable current.

scintillating material A material, such as sodium iodide, that emits a flash of light when it absorbs ionising radiation. These are often used in conjunction with photomultiplier for radiation detection.

Equations

$$\frac{\Delta \Phi}{\Phi} = -\sigma \frac{\Delta N}{\Delta A}$$

$$\frac{\Delta \Phi}{\Delta x} = -\sigma n \Phi$$

$$p = \frac{h}{\lambda}$$

$$KE_{max} = hf - B$$

BIOLOGICAL EFFECTS OF IONISING RADIATION

37.1 Introduction

In this chapter we will investigate the interaction of ionising radiation with biological tissue. The specific effects produced by this interaction depend on the type of radiation, and how the exposure occurs. The concept of radiation dosage will be discussed and we will look at the risks and medical symptoms associated with different types of radiation exposure in humans. The typical NZ citizen's yearly radiation equivalent dose is given in Figure 37.1.

Key Objectives

• To understand absorbed and equivalent dose.

• To be able to predict the possible or likely effects of a particular full-body radiation dose.

• To understand how damage occurs at a cellular level and how this translates into particular medical symptoms.

37.2 Mechanisms of Cell Damage

Ionising radiation causes damage to molecules, occasionally by a direct hit on a molecule, but more often indirectly by the creation of 'free radicals'. Free radicals are uncharged atoms or fragments of molecules possessing an unpaired electron. They are formed by the symmetrical breaking of a covalent bond.

Free radicals may cause damage to cellular proteins by breaking molecular bonds and rendering protein molecules non-functional or even harmful. If there are many undamaged copies of the protein, this may not adversely affect cellular function, but too much damage may result in cell death.

Direct hits and free-radical production by radiation may damage cellular DNA. If this occurs there are several possible outcomes:

• DNA damage which the cell can detect and fix.

• DNA damage that cannot be fixed, causing the cell to undergo *apoptosis*, a form of programmed cell death.

• Non-lethal damage that is passed on as a *mutation* in subsequent cell divisions.

In addition, neighbouring cells which are not directly damaged by radiation may experience damage by communication with the damaged cells (the 'bystander effect').

DNA damage is described as either **somatic** or **genetic**. Somatic damage is non-inheritable. The serious adverse effect of somatic damage is to increase the risk that a cancer will develop. Genetic damage is inheritable and results from the mutation of DNA in the reproductive cells, and these changes may thus be passed on to future generations.

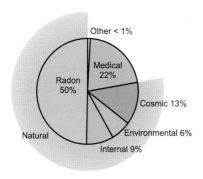

Figure 37.1 A typical NZ citizen's yearly radiation equivalent dose. The typical dose is 2300 µSv y^{-1} [Copyright (c) 2010 National Radiation Laboratory, New Zealand]

Introduction to Biological Physics for the Health and Life Sciences   Franklin, Muir, Scott, Wilcocks and Yates
©2010 John Wiley & Sons, Ltd

37.3 Dose and Dose Equivalent

Absorbed Dose

The **absorbed dose**, D, is a physical quantity which quantifies the amount of energy absorbed by some material. The absorbed dose is not specific to the absorption of radiation by biological material: an absorbed dose could, in principle, be calculated for the absorption of sunlight by a metal plate, or the absorption of sound waves by a wall in a house. However, the absorbed dose will be particularly useful in the quantification of the effect of radiation on biological material. The SI unit of absorbed dose is the gray, symbol Gy. One gray is equivalent to one joule of energy being delivered to one kilogram of matter. An older (much smaller) unit of absorbed dose is the rad. The relationship between these units is

$$1\,\text{Gy} = 1\,\text{J}\,\text{kg}^{-1} = 100\,\text{rad}$$

The absorbed dose is given by

$$D = \frac{\Delta E}{m} \tag{37.1}$$

where ΔE is the energy lost from the radiation beam, and m is the mass of material into which the energy is absorbed. As stated earlier, the absorbed dose is a general concept. It therefore applies to all kinds of radiation and all types of absorber.

Absorbed dose is an important concept, since the effect of radiation on biological tissue depends directly on the amount of energy absorbed by that tissue. The damaging effects of radiation occur when molecular bonds in important biological molecules are broken or when free radicals are formed. Both of these processes require energy, and thus the amount of damage produced by radiation is proportional to the amount of energy from the radiation that is absorbed. However, as we have seen and as shown in Figure 37.2, different types of ionising radiation lose energy in matter in different ways; this is also an important factor in analysing the biological effects of the radiation. In order to quantify this variation of effect for the same energy absorbed, we introduce the concept of *dose equivalent*.

Dose Equivalent

The **dose equivalent** can be thought of as an expression of the dose in terms of its biological effect. To estimate this, we define the **relative biological effectiveness** or RBE of a particular kind of radiation (also known as the Quality Factor, Q). The RBE varies from one kind of radiation to another and quantifies the damage produced by each kind of radiation in biological tissue.

> Approximate RBE values (summarized from ICRP (1991)) X and Gamma rays, RBE = 1 Electrons, RBE = 1 Neutrons (energy dependent), RBE = 5-2-0 Protons, RBE = 5 Alpha Particles, RBE = 20

The distance travelled by a particular kind of radiation is dependent on the rate at which that type of radiation deposits energy in matter. Electrically charged particles interact strongly via the electric force and thus deposit their energy relatively quickly. The larger particles will also lose energy rapidly because they are more likely to simply collide with the molecules of the material through which they are travelling. The α particle has twice the charge of the β particle and is also much larger. Thus energetic α particles will deposit their kinetic energy more rapidly and over a smaller distance than β particles with the same initial energy. Similarly, β particles will deposit their kinetic energy more rapidly and over a smaller distance than γ-ray photons with the same initial energy.

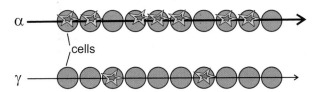

α cells

γ

Figure 37.2 Even if the amount of energy ultimately deposited by different types of radiation is the same, more concentrated damage is caused by larger particles like α particles.

The fact that the range of the α particle is less than that of β, γ or X-rays, results in α radiation depositing its energy in a smaller area. The greater concentration of ionisation and cell damage results in greater (adverse) biological effect, and thus a larger value of the RBE. Similarly, β radiation has a larger value of the RBE than γ radiation. The RBE of γ radiation is set at 1. This means that radiation with a RBE of 7 (for example) is seven times as damaging as the same absorbed dose of γ radiation.

The dose equivalent is measured in sievert (Sv). The dose equivalent is obtained by multiplying the absorbed dose by the RBE

$$Sv \equiv Gy \times RBE \tag{37.2}$$

The non-SI unit still in use is the **rem** (which stands for Röentgen equivalent man):

$$\text{dose equivalent in rem} = \text{absorbed dose in rad} \times RBE \tag{37.3}$$

$$1\,Sv = 100\,rem \tag{37.4}$$

Effective Dose

As different tissue types are affected differently by radiation, the risks associated with lower doses of radiation can be better quantified by taking into account a weighting factor that depends on tissue type. When this weighting is introduced, it is then called the **effective dose**. In this book, we will avoid going into this much detail, and we will concern ourselves only with the risk factors associated with a whole-body dose.

Example 37.1 *Dose and RBE*

Problem:

(a) A 5 g tumour is irradiated with high energy X-rays and absorbs a total of 0.2 J of energy. What is the absorbed dose in gray and rad, and the dose equivalent in sievert and rem?

(b) An alternate treatment for the same tumour is to administer a chemical solution containing a radioactive isotope which is preferentially absorbed by the tumour. If the isotope involved is an alpha emitter with an RBE of 20 and the tumour absorbs 0.05 J of energy, what is the absorbed dose in gray and rad, and the dose equivalent in sievert and rem?

Solution: (a) The dose in gray is just the number of joules absorbed per kilogram

$$D_{Gy} = \frac{\Delta E}{m} = \frac{0.2\,J}{0.005\,kg} = 40\,Gy$$

The dose in rad is a multiple of this

$$D_{rad} = 40\,Gy \times 100\,rad\,Gy^{-1} = 4000\,rad$$

The dose equivalent in sievert and rem can be found from the dose in gray and rad by taking into account the relative biological effectiveness of the type of radiation. The RBE of X-rays is 1 and so the dose equivalent in each unit is the same as its dose counterpart

$$D_{Sv} = D_{Gy} \times RBE = 40\,Gy \times 1 = 40\,Sv$$

$$D_{rem} = D_{Sv} \times 100\,rem\,Sv^{-1} = 40\,Sv \times 100\,rem\,Sv^{-1} = 4000\,rem$$

(b) Again, the dose in Gy is the number of joules absorbed per kilogram

$$D_{Gy} = \frac{\Delta E}{m} = \frac{0.05\,J}{0.005\,kg} = 10\,Gy$$

The dose in rad is a multiple of this

$$D_{rad} = D_{Gy} \times 100\,rad\,Gy^{-1} = 10\,Gy \times 100\,rad\,Gy^{-1} = 1000\,rad$$

The absorbed dose in this case is less than in the previous part of the example.

$$D_{Sv} = D_{Gy} \times RBE = 10\,Gy \times 20 = 200\,Sv$$
$$D_{rem} = D_{Sv} \times 100\,rem\,Sv^{-1} = 200\,Sv \times 100\,rem\,Sv^{-1} = 20000\,rem$$

Because α radiation has such a large effect in biological tissue it has a much larger effect per unit energy absorbed than X-rays.

Example 37.2 *Activity and dose*

Problem: A careless 75 kg radiopharmacologist accidentally ingests a small amount of a 1.7 MeV beta emitter with a half-life of 24 days. The activity of the sample ingested is 13 mCi. What is the equivalent dose in rem received by the pharmacologist in the first minute of his exposure to the radioactive sample, if 20% of all beta particles emitted are absorbed in his body?

Solution: The half-life of the sample is much longer that the period of 1 min over which we are asked to calculate the dose, so we can make the simplification that the activity will be a constant 13 mCi over this 1 min period. This corresponds to an activity of $13 \times 10^{-3}\,Ci \times 3.7 \times 10^{10}\,Bq\,Ci^{-1} = 4.81 \times 10^8\,Bq$. As only 20% of these particles will be absorbed, there will be 0.96×10^8 β particles absorbed per second.

Each one of the β decays releases $1.7\,MeV = 1.7 \times 10^6\,eV \times 1.6 \times 10^{-19}\,J\,eV^{-1} = 2.72 \times 10^{-13}\,J$ of energy. We can assume that, for those β particles absorbed by the pharmacologist, all of this energy is deposited in the pharmacologist. The total amount of energy deposited in 1 s is the number of β particles absorbed per second multiplied by the energy per β particle

$$P = 0.96 \times 10^8\,\beta\,s^{-1} \times 2.72 \times 10^{-13}\,J\beta^{-1}$$
$$= 2.61 \times 10^{-5}\,J\,s^{-1}$$

So in 1 min a total of $2.61 \times 10^{-5}\,J\,s^{-1} \times 60\,s = 1.56 \times 10^{-3}\,J$ is absorbed by the pharmacologist. This corresponds to a dose (in Gy) of

$$D_{Gy} = \frac{1.56 \times 10^{-3}\,J}{75\,kg} = 2.09 \times 10^{-5}\,Gy$$

which, taking into account the RBE for β emission above 0.03 MeV of RBE=1.7 corresponds to an equivalent dose of

$$D_{Sv} = D_{Gy} \times RBE = 2.09 \times 10^{-5}\,Gy \times 1.7 = 3.55 \times 10^{-5}\,Sv$$
$$D_{rem} = D_{Sv} \times 100\,rem\,Sv^{-1} = 3.55 \times 10^{-3}\,rem$$

The absorbed dose is 3.55 mrem which is around 1/68th of the dose received from background radiation over the course of one year. While this might not sound like much, this is the dose absorbed in just 1 min. The total dose from the ingestion of this radioactive material will be much higher.

37.4 Types of Effect

Biological effects from ionising radiation have already been categorised as somatic or genetic in terms of the effect of this radiation on DNA. The biological effects of radiation can also be divided into two other categories: **deterministic** and **stochastic**. These categories refer to large-scale physiological effects of radiation exposure rather than the effects of this exposure on DNA.

Deterministic effects are produced by radiation doses that are high enough to denature proteins or to cause cell death. These effects are therefore definite, noticeable and fairly immediate. When the radiation dose is smaller, there may not be any obvious damage to cells or organs, but the *risk* of a disease like cancer is increased. It is because of the probabilistic nature of the consequences of these smaller doses that we refer to the effects of such doses as stochastic (meaning a process that is governed by the laws of probability).

There are a number of distinct characteristics of deterministic and stochastic effects of radiation exposure. These characteristics provide a useful contrast between the two categories of effect. The defining characteristics of the deterministic effects of radiation exposure are:

- These are early effects: they appear very quickly after the radiation dose is received.

- They are the result of 'lethal damage' to tissue. The cells of the tissue are killed by the radiation exposure.

- The killing of cells by radiation is extensive enough that it reduces or destroys at least some organ function.

- The lethal damage required to produce cell death occurs only for radiation exposure which results in doses above some minimum value, i.e., there is a 'threshold' dose below which deterministic effects do not occur.

- The *severity* of these effects increases with increasing dose.

Some examples of deterministic effects are the formation of cataracts, infertility and erythema (skin reddening). These are all the result of cell death in the organs concerned.

Stochastic effects, because they can only be discussed in terms of increased risk of disease or of inheritable mutation, are much harder to definitely characterise. However, they are distinct from deterministic effects in several ways:

- These are 'late' effects, they do not appear immediately, but occur some time after the radiation exposure has occurred (sometimes a generation later).

- They are effects caused by cellular or DNA damage, but damage which is not immediately lethal.

- The fact that a cell, while damaged, is not killed by radiation exposure means that the cellular repair mechanisms come into operation. This cellular repair process is sometimes imperfect and can lead to cellular mutation, or abnormal changes in cell function. The most likely result of such a change in cell function is the induction of cancer at some later stage in the life of the individual.

- Since these effects are dependent on mechanisms such as cellular repair and the inheritance of mutations, it is only possible to estimate the *probability* of harm given a particular dose level.

- The severity of the effect is not dependent on the dose, again because of the complexity of the causal mechanisms leading from the original radiation exposure to the ultimate appearance of disease or disfunction.

- The *probability* of harm increases with increasing dose.

37.5 Medical Effects and Risk

Large radiation doses may cause **radiation sickness**. Radiation sickness is a catch-all phrase referring to a group of deterministic effects which have been observed to appear soon after very large radiation doses. The first symptoms of radiation sickness to occur are typically, nausea, vomiting, diarrhoea and fatigue. Other symptoms that may occur are skin burns (redness, blistering), weakness, fainting, dehydration, anemia, dry cough, inflammation of exposed areas (along with redness, tenderness, swelling, bleeding), hair loss, ulceration of the oral mucosa, ulceration of the esophagus, stomach or intestines, vomiting blood, bloody stools, bleeding from the nose, mouth, gums, and rectum, bruising, open sores on the skin and headache.

The cells that are most susceptible to death from radiation are those in the intestinal lining, white blood cells and the cells that make red and white blood cells. Many of the symptoms listed, such as dehydration, vomiting, and diarrhoea, are the result of damage to the intestinal tract.

The time taken for a person to die from a lethal radiation dose is around two to four weeks. Patients who receive a high full-body dose of radiation and are still alive after six weeks are likely to recover.

The treatments available are largely aimed at ameliorating the symptoms: antinausea drugs, painkillers, antibiotics to help fight infections and blood transfusions for anemia.

Exposure to smaller amounts of radiation may cause no noticeable effects at the time. Long-term epidemiological studies on populations exposed to varying doses of radiation, such as atom-bomb survivors in Hiroshima and Nagasaki, suggest that the *probability* of developing cancer from radiation exposure increases linearly with the accumulated dose, and there is no minimum threshold of exposure below which there is no risk. This is known as the *linear no-threshold (LNT)* hypothesis. Based on this hypothesis and the available data, about 1% of the population could be expected to develop cancer due to exposure to normal background levels of radiation.

Determining the exact risk of cancer from the available information is difficult. It is not ethical or even possible to experimentally determine how much risk is associated with low doses of radiation. Interpretation of the data that is available is further complicated by the fact that radiation-induced cancers are indistinguishable from cancers resulting from other known risk factors like diet and smoking.

37.6 Ultraviolet Radiation

Ultraviolet has been included here as a separate section, away from the higher-energy forms of electromagnetic radiation. Exposure to ultraviolet radiation may well result in damage to biological tissue, but it is not dangerous in the same way as γ or X-ray radiation.

Ultraviolet (UV) photons are those in the wavelength range between X-rays and visible light, around 10 nm to 400 nm. The most important source of ultraviolet radiation in everyday life is the Sun, which emits radiation from wavelengths of about 200 nm up. This range, the 'near UV' is subdivided into three categories:

- UVA – 400 nm to 320 nm.

- UVB – 320 nm to 280 nm.

- UVC – 280 nm to 200 nm.

Ozone in the upper atmosphere is a very good absorber of ultraviolet light, so practically no radiation from the Sun below about 300 nm reaches sea level in most places. In any case, the air is largely opaque at wavelengths below 200 nm due to absorption by oxygen. Thus, most of the UV we are exposed to is UVA. Note that some UVB exposure is needed for the production of vitamin D by the human body.

Ultraviolet radiation is mostly not energetic enough to interact with any but the valence electrons in the atoms of the matter it passes through. This can still disrupt biological molecules however. All three types of ultraviolet light reaching the Earth's surface will damage collagen in the skin, causing premature aging. The least harmful is UVA. UVB on the other hand, causes the most skin reddening (erythema) and can cause DNA damage by disrupting covalent bonds which can lead to skin cancer, and also causes cataracts.

Short-wavelength UV radiation can be used for antibacterial sterilisation, as it causes DNA damage that can inhibit the ability of bacteria to replicate. Even if the bacterium is not killed, it is the ability to replicate that makes its dangerous, so this technique is useful when used in combination with other germ-killing methods.

Sunscreen ingredients protect the skin in one of two ways – by blocking (reflecting) the UV light (titanium dioxide is a common ingredient in sunscreens of this kind), or by absorbing and re-radiating the energy at much longer wavelengths (which is what another common ingredient, avobenzone, does).

37.7 Summary

Key Concepts

deterministic effects The definite, observable effects that result from doses of radiation above a threshold. The severity of the effect increases with dose.

stochastic effects The effects that may or may not occur due to small radiation doses. The probability of an effect is proportional to the long-term accumulated dose.

dose A measure of radiation exposure.

absorbed dose (D) The amount of energy absorbed from ionising radiation per kilogram of tissue. Measured in grays or rad.

equivalent dose The absorbed dose weighted to account for the radiation type by multiplying by the RBE. Measured in sieverts or rem.

relative biological effectiveness (RBE) A weighting factor which depends on the type of ionising radiation. The RBE is higher for radiation types that deliver their energy to a smaller number of cells.

rad Non-SI unit of absorbed dose. One hundred rad equals one gray.

gray (Gy) SI unit of absorbed dose. One gray equals one joule per kilogram.

rem Non-SI unit of equivalent dose. One hundred rem equals one sievert.

sievert (Sv) SI unit of equivalent dose, symbol Sv. The equivalent dose in sieverts is found by multiplying the absorbed dose by the RBE.

background radiation The radiation exposure experienced from everyday unavoidable sources, such as cosmic rays and naturally occurring radioisotopes. Around 2 mSv per year.

linear no-threshold model (LNT) Hypothesis that the probability of developing cancer from exposure to radiation increases linearly with dose, and there is no threshold below which the risk is zero.

UV Ultraviolet radiation. The part of the electromagnetic spectrum between the blue end of the visible part and X-rays.

Equations

$$D = \frac{\Delta E}{m}$$

$$1 \, \text{Gy} = 1 \, \text{J kg}^{-1} = 100 \, \text{rad}$$

$$\text{Sv} \equiv \text{Gy} \times \text{RBE}$$

$$1 \, \text{Sv} = 100 \, \text{rem}$$

37.8 Problems

37.1 A 65 kg person undergoing a series of X-rays receives a dose of 12 rem.

(a) What dose does he receive in sieverts and rad?

(b) How much energy was deposited in the person's body?

37.2 A 14 g ovarian tumor is treated using a sodium phosphate solution in which the phosphorus atoms are the radioactive ^{32}P isotope with a half life of 14.3 days and which decays via beta emission with an energy of 1.71 MeV. Half of the sodium phosphate solution is absorbed by the tumor and deposits 9.00 J of energy into it. The other half of the solution is dispersed throughout the patients tissues, also depositing 9 J of energy into the 50.0 kg of body tissues.

(a) What is the dose (in Gy and rem) that the tumor receives?

(b) What is the dose (in Gy and rem) that the rest of the patient receives?

37.3 A person is exposed to ionizing radiation which deposits 10 J of energy in their tissue.

(a) What dose (in Gy) would an 80 kg adult and a 15 kg child receive under these circumstances?

(b) What dose (in rem) would the adult and child each receive if the radiation were low energy (< 0.03 MeV) β radiation?

(c) What dose (in Sv) would the adult and child each receive if the radiation were low energy α radiation? (RBE)=10.)

37.4 A person with lymphoma receives a dose of 35 Gy in the form of γ radiation during a course of radiotherapy. Most of this dose is absorbed in 18 g of cancerous lymphatic tissue.

(a) How much energy is absorbed by the cancerous tissue?

(b) If this treatment consists of five 15 minute sessions per week over the course of 5 weeks and just 1% of the γ photons in the γ ray beam are absorbed, what is the power of the γ ray beam?

(c) If the γ ray beam consists of just 0.5% of the γ photons emitted by the γ source, each of which has an energy of 0.03 MeV, what is the activity (in Ci) of the γ ray source?

37.5 A 60 kg person accidentally ingests a small source of alpha particles (RBE=15). The activity of the source is 0.04 Ci, the half life of the source is 110 years, and each alpha particle emitted has an energy of 0.586 MeV. It takes 12 hours for the alpha source to pass through the persons digestive system and exit the body.

(a) How many alpha particles are absorbed by the person (you may assume that 100% of the alpha particles emitted by the source are absorbed by the person)?

(b) How much energy is deposited in the person by the source (in J)?

(c) What is the absorbed dose (in rad)?

(d) What is the absorbed dose (in rem)?

37.6 A radioactive contaminant gives an unfortunate 0.5 kg lab rat a dose of 1500 rem in just 1 minute. Assuming that the half life of the radioactive isotope in the contaminant is much longer than 1 minute, what would the activity (in Bq) of the contaminant be if ...

(a) the contaminant is a 5 MeV alpha emitter (RBE = 15)?

(b) the contaminant is a 1.1 MeV beta emitter?

(c) the contaminant is a 0.01 MeV gamma emitter?

MEDICAL IMAGING

38

38.1   Introduction

Many of the current technologies in use in the field of medical diagnostic imaging have their origin in pure research, or in military and defense research. The development of sonar for military purposes led to medical ultrasound. The development of the nuclear reactors that are used to produce radioisotopes for nuclear medicine was originally part of the push towards building the first atomic weapons at the end of World War II. Until recently, it has largely been a case of technologies developed in other fields being adapted for use in medicine. This is changing. There is an increasing emphasis on developing, or improving on, the imaging and diagnostic techniques that are available specifically for medical purposes.

The changing relationship between technological research and medical applications of this research is the result of a number of factors, and we will here mention only a few of the more striking ones. Firstly, the increase in understanding of human biological systems that has resulted from imaging technologies, like magnetic resonance imaging (MRI) and computerised tomography (CT) scanning, has lead us to ask increasingly sophisticated questions. Secondly, rapid advances in computer processing power and communications allow us to collect and store far more data, visualise structures in three dimensions, overlay data from different sources and transmit this data to others. Finally, the clinical and research staff now entering the health sciences are much more adept at understanding and using the new technology, having grown up in a computer-rich environment.

There is no doubt that the non-invasive imaging techniques now routinely available in hospitals have become indispensable, and that the technology will continue to advance. Any student wishing to work in the health sciences would do well to develop some understanding of the basic physics that lies behind the technology, to aid them in understanding the possible applications, limitations and risks involved.

There are a number of imaging techniques commonly in use in the health sciences today. Here we will cover the most important: X-ray photography, computerised X-ray scanning tomography (CT), emission computerised tomography (PET and SPECT), and ultrasound. Magnetic resonance imaging (MRI) will be covered in a separate chapter.

The imaging techniques mentioned above are covered here because they are the most likely to be encountered by those working in a clinical setting in the health sciences. They are all useful for gathering anatomical or physiological data. Increasingly, researchers in the biological sciences are turning to imaging on a molecular or cellular scale, using a host of new methods: laser-scanning confocal microscopy, fluorescent labelling, multiphoton microscopy, electron energy loss spectroscopic imaging ...this list really does go on.

Key Objectives

- To understand the basic principles of the major medical imaging technologies.

38.2 X-ray Imaging

X-ray images are widely used in medicine and dentistry, and have been used to produce images of the internal structure of objects that are opaque to visible light since soon after the discovery of X-rays was announced by Röentgen on the 28th of December, 1895.

To produce an **X-ray radiograph** of the human body, the area of interest is exposed to X-rays while a photographic film is placed beneath the body. This produces an image where the areas of greatest exposure correspond to the areas of the body that are the most transparent to X-rays. Instead of recording reflected light, as in a photograph, it records degree of transmission of X-rays as a shadowgraph.

The attenuation of X-ray photons depends on the photon energy (which determines which type of interaction occurs – Compton scattering and the photoelectric effect are the dominant processes at X-ray and γ-ray energies as shown in Figure 38.1), and the atomic number of the key elements present in the materials being imaged.

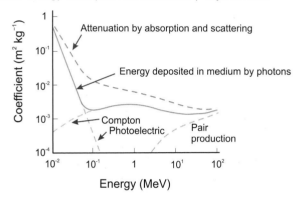

Mass energy absorption coefficient for γ–rays in water

Figure 38.1 The amount of electromagnetic energy absorbed is dependent on photon energy, as different interaction processes dominate at different photon energies. Gamma and X-ray absorption coefficients are available from the National Institute of Standards and Technology (NIST) public databases. [Reprinted with permission from Biological Radiation Effects, Jurgen Kiefer. Copyright (1990) Springer.]

For low-energy (below about 35 keV) X-ray photons, the predominant interaction with atoms in the target is through the **photoelectric effect**. The probability of a photoelectric interaction falls rapidly with increasing photon energy (varying roughly in proportion to the inverse of the energy cubed), and the attenuation is stronger for higher effective atomic number (it varies with Z_{eff}^3). Soft tissue and body fluid are largely made up of water and have an effective Z of about 7.5. Fat has a slightly higher concentration of low-atomic number elements like hydrogen, so, overall, has a slightly lower Z_{eff} of around 6. Bone has more high-Z trace elements, so has an effective Z up around 13. This means that for 15 keV X-ray photons, the attenuation is about four times as much for bone as it is for tissue by the photoelectric effect. Lower X-ray energies are used to improve the contrast when imaging areas for which the inherent contrast between the different tissue types is low, like the breast.

Higher-energy X-rays (more than 30–40 keV) interact with matter primarily through the **Compton effect**. The probability of Compton scattering has little dependence on atomic number, but does depend on electron density. X-ray images from high-energy photons are best suited for showing differences in physical density. They can be useful for reducing the shadowing effect of bone to view the tissue behind.

Contrast agents can be used to improve the contrast of an X-ray image. Introducing high-Z elements such as barium and iodine will make those organs which absorb them more opaque to X-rays. Iodine is used for imaging the circulatory system, such as in angiography. Barium is used as a contrast agent in imaging the gastrointestinal tract.

38.3 CT Scan

CT stands simply for *computed tomography*. (Sometimes the word 'axial' is added and the term CAT scan is used.) The word tomography is derived from the Greek for 'slice' (tomos), and is used to describe any imaging technique that produces cross-sectional images. While there are several techniques that are a kind of computed tomography, unless otherwise specified, CT is usually taken to refer specifically to X-ray transmission computed tomography obtained using an apparatus like that shown in Figure 38.2.

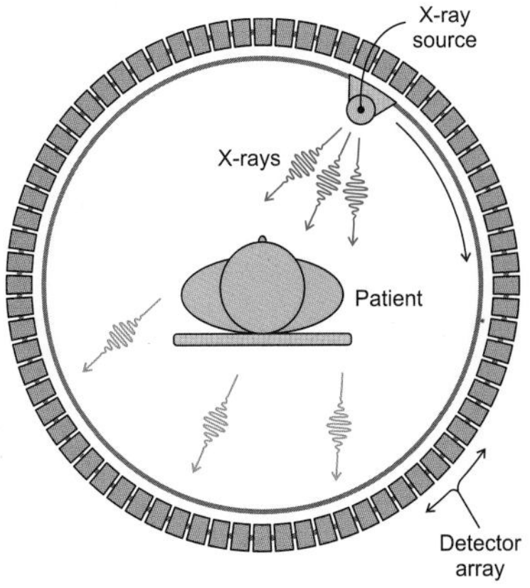

A limitation of X-ray radiography is the projection of a three-dimensional structure onto a two-dimensional film. There is also a limit to the contrast that film can provide and differences in intensity of a few percent cannot be detected. In CT scanning, the orientation of the path of the X-rays through the body is varied, and a computer is used to reconstruct a picture of the cross section from the transmission data. The reconstruction techniques have their origins in many branches of physics – radio astronomy, electron microscopy and holography.

The first generation of CT scanners, introduced in the early 1970s, used a thin, pencil beam of X-rays and a single detector (sodium iodide) in line on the far side of the patient. A series of transmission measurements (over 100) were taken in translational scan, and then the angular orientation was varied by a degree and the translation scan was repeated, eventually moving through 180°. Around 30 000 separate transmission measurements were used to reconstruct a 'slice' image. This took minutes and was prone to motion blurring.

Modern CT scanners use instead a fan X-ray-beam geometry and a circular array of 600–700 fixed detectors to reduce scan times for a single slice to about 1 s. The more recent generations of scanners also use a helical path of the X-ray source about the patient, rather than a slice-by-slice approach.

The patient dose for a CT scan is higher than for a traditional X-ray. A scan involves an effective radiation dose of about 2 – 10 mSv.

Figure 38.2 Fourth-generation CT scanners employ a circular array of detectors and a rotating source.

38.4 PET scan

PET stands for **positron emission tomography**. When a nucleus undergoes positive β decay, a positron is emitted. The positron is the antiparticle of the electron, and with ordinary matter being very full of electrons, the positron quickly collides with an electron and both are annihilated. This produces two photons, each with 511 keV of energy (equivalent to the rest mass of an electron/positron), and for momentum conservation these two photons are emitted in nearly exactly opposite directions as shown in Figure 38.3. By using a circular array of detectors and looking for the simultaneous detection of photons with the correct energy, the location of the decay event can be narrowed down to the line joining the two detectors that were triggered.

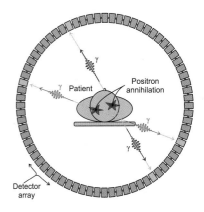

One of the great benefits of PET scanning is that it can be used to get information on metabolic processes instead of producing only structural information. Positron-emitting radionuclides can be included in molecules involved in a particular process, and areas of the body where those molecules accumulate will be shown.

For example, fluorine-18 ($T_{1/2}$ about 110 min) can be included in the glucose-analogue molecule fluorodeoxyglucose (FDG). The molecule is taken up by cells and undergoes a specific metabolic process – phosphorylation by hexokinase enzymes. The resulting molecule is not metabolised further, and is trapped inside the cell. Because cancer cells have elevated hexokinase, tumours will collect more of the tagged molecules and show up strongly on the PET scan. The oncological use of FDG is the most common medical application of PET.

Figure 38.3 A PET scanner works by detecting simultaneously-emitted pairs of photons at about 511 keV.

Another commonly used isotope is nitrogen-13 (half-life around 10 min), which can be incorporated into ammonia molecules. Its main use is to study the supply of blood and nutrients to tissues (perfusion studies). The use of PET to study blood flow is not limited to medical studies. For example, it is being used by researchers in clinical psychology to measure blood flow to the brain to determine levels of activity in studies of memory and post-traumatic stress disorder (PTSD).

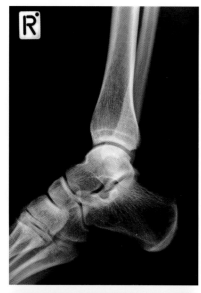

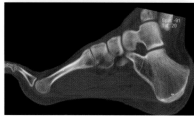

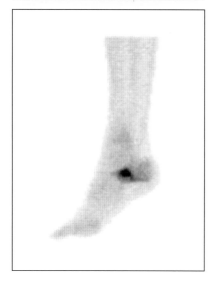

It is becoming more common to use PET in conjunction with other scanning techniques, such as CT, to get structural and functional information.

The biggest limitation on the use of PET is the need for short-lived radioisotopes that have to be produced in a cyclotron. Most hospitals and indeed many countries do not have the facilities to produce isotopes for medical use and have to import these on a regular basis.

The dose from a single PET scan is similar to that from a CT scan.

38.5   Gamma Camera and SPECT

The **gamma camera** (also called a scintillation camera or Anger camera) is used for detection of radiation in **nuclear medicine** studies.

Nuclear medicine involves inhaling, injecting or ingesting radionuclides that emit γ radiation. Depending on what body system is being investigated, several different radionuclides are used. One example is gallium-67. The body treats Ga^{3+} like Fe^{3+} (an iron ion), and it concentrates in areas of inflammation and rapid cell division, making it useful for cancer diagnosis. Radio-isotopes of iodine are used for thyroid studies. Technetium-99m is the most widely used γ-emitter; there are about 30 radiopharmaceuticals into which it is included for imaging studies for many areas of the body – the brain, myocardium, liver, lungs, kidneys, skeleton and gallbladder, to name some.

Radionuclides are useful not only for obtaining **structural** information; by detecting the accumulation of a particular pharmaceutical, they can also provide **functional** information about the *rate* of metabolic processes (see Figure 38.4).

The function of the gamma camera is to detect the γ radiation from the radionuclides that have been introduced into the patient's body and to pinpoint the point of origin of the radiation in the body. The detection of the γ radiation is achieved with a scintillating crystal such as sodium iodide, and photomultiplier tubes. To determine the direction from which the radiation originated, lead collimators are used. These ensure that only radiation coming from a narrow angular range will get through each gap in the lead, thus determining the direction to the source of the radiation.

SPECT stands for **single-photon emission computed tomography**, which is similar in many respects to X-ray transmission computed tomography. An image is reconstructed by computer from information gathered at multiple times and locations showing the distribution of the radionuclide that was administered. However, in SPECT the detector is rotated about a stationary radiation source – the patient.

38.6   Diagnostic Procedures: Dose

Table 38.1 shows the typical doses and the associated cancer risk for a number of diagnostic procedures. The risk is higher for pediatric patients (multiply by about two) and lower for geriatric patients (divide by about five). This data was provided by the Health Protection Agency, UK (http://www.hpa.org.uk).

38.7   Ultrasound Sonography

Sonography using ultrasound is a medical diagnostic imaging technique that does not use ionising radiation. Instead it utilises high-frequency acoustic vibrations, above the limit of human hearing (which is about 20 kHz). When a wave is travelling through a medium and it reaches a boundary with another medium, some of the wave is transmitted through the boundary and some is reflected. In the case of sound waves, the amount of reflection depends on the difference in a property of the media called the *acoustic impedance* – the greater the impedance mismatch between two media, the more reflection of ultrasound there will be. Acoustic impedance depends on the speed at which sound travels through a medium, and this in turn depends on the density of the material.

The reflection coefficient, R, is the proportion of the energy of an incoming sound wave that is reflected at a boundary between two media. The reflection coefficient

Figure 38.4 (Top) A plain X-ray image of a fractured foot. The fracture is difficult to see in this image. (Middle) A CT scan of the same foot. The image is a slice through the middle of the foot. The fracture is also difficult to see in this image. (Bottom) The same foot imaged using a nuclear medicine bone scan. It shows increased Technetium uptake in the fracture. [Images courtesy of Professor Terry Doyle, University of Otago School of Medicine.]

Diagnostic procedure	Typical effective doses (mSv)	Equivalent period of natural background radiation	Lifetime additional risk of fatal cancer per examination
X-ray examinations:			
Limbs and joints (except hip)	< 0.01	< 1.5 days	1 in a few million
Teeth (single bitewing)	< 0.01	< 1.5 days	1 in a few million
Teeth (panoramic)	0.01	1.5 days	1 in 2 million
Chest (single PA film)	0.02	3 days	1 in a million
Skull	0.07	11 days	1 in 300 000
Cervical spine (neck)	0.08	2 weeks	1 in 200 000
Hip	0.3	7 weeks	1 in 67 000
Thoracic spine	0.7	4 months	1 in 30 000
Pelvis	0.7	4 months	1 in 30 000
Abdomen	0.7	4 months	1 in 30 000
Lumbar spine	1.3	7 months	1 in 15 000
Barium swallow	1.5	8 months	1 in 13 000
IVU (kidneys and bladder)	2.5	14 months	1 in 8000
Barium meal	3	16 months	1 in 6700
Barium follow	3	16 months	1 in 6700
Barium enema	7	3.2 years	1 in 3000
CT head	2	1 year	1 in 10 000
CT chest	8	3.6 years	1 in 2500
CT abdomen/pelvis	10	4.5 years	1 in 2000
Nuclear medicine studies:			
Lung ventilation (Kr-81m)	0.1	2.4 weeks	1 in 200 000
Lung perfusion (Tc-99m)	1	6 months	1 in 20 000
Kidney scan (Tc-99m)	1	6 months	1 in 20 000
Thyroid scan (Tc-99m)	1	6 months	1 in 20 000
Bone scan (Tc-99m)	4	2 years	1 in 5000
Dynamic cardiac (Tc-99m)	6	2.7 years	1 in 3300
Myocardial perfusion (Tl-201)	18	8 years	1 in 1100

Table 38.1 Dose data provided by the Health Protection Agency in the UK. Based on average dose of 2.2 mSv per year. Risk assessment for 16–69 year-olds.

depends on the acoustic impedances of the materials on each side of the boundary, if we label the acoustic impedance of the first medium as Z_1 and the acoustic impedance of the second medium as Z_2, then the proportion reflected at the boundary is

$$\text{proportion reflected} = \frac{(Z_1 - Z_2)^2}{(Z_1 + Z_2)^2} \tag{38.1}$$

so depends strongly on the differences in Z. (The reflection coefficient varies with the frequency of the incoming sound wave, as do the acoustic impedances of the two media).

In order to generate an image in a medical setting, a device called a piezoelectric transducer is used to produce acoustic waves with frequency in the low MHz range. The waves that are produced are reflected back from the boundaries between tissue types. The ultrasound signal is not emitted continuously, but in pulses. By detecting the time delay for the echo of the pulses, and the signal strength, a picture can be built up of the location of the boundaries.

Because the impedances of air and skin are so different, if there is an air gap between the skin and the transducer then most of the intensity is lost by the waves being reflected before penetrating the body. It is possible to resolve this problem using a gel layer between the probe and the skin. This procedure, ensuring a small difference in impedance to ensure minimal reflections, is called *impedance matching*.

Ultrasonography is generally considered to be safe, and in many countries it is used routinely to monitor pregnancy. However, the ultrasound signal consists of mechanical pressure waves, which can have a heating effect on the tissue.

38.8 Summary

Key Concepts

tomography From the Greek for 'slice' (tomos). Used to describe any imaging technique that produces cross-sectional images.

PET Positron emission tomography. A radio-imaging technique which uses positron-emitting radionuclides. The emitted positrons meet nearby electrons and are annihilated, producing two γ photons which have a characteristic energy (0.511 MeV), which are emitted in opposite directions. Coincidence counting, i.e. detecting simultaneous photons, enables reconstruction of the location where the positron was annihilated.

SPECT Single photon emission computed tomography. Computer reconstruction of multiple images from a gamma camera to produce a 3-D image.

CT or CAT scan Computed tomography or computerized axial tomography. The reconstruction of a series of X-rays to form images of slices through the body.

X-ray radiograph A shadowgraph formed by exposure of a photographic film to transmitted X-rays.

MAGNETISM AND MRI

39.1 Introduction

The great Danish physicist Niels Böhr is quoted as saying about magnetic resonance methods, 'You know, what these people do is really very clever. They put little spies into the molecules and send radio signals to them, and they have to radio back what they are seeing'. This is the idea at the heart of magnetic resonance imaging: transmitting radio waves of the right kind into the body, in the right circumstances, forces the molecules in the human body to radio back information on their location and their local environment. Decoding these transmissions gives us the means to re-create a map of the inside of the body. Additionally, because the process uses radio waves, MRI avoids the damaging effects on living tissue caused by ionising radiation.

Magnetic resonance imaging or MRI has become one of the most important medical imaging techniques currently employed by medical professionals. Furthermore, MRI in the form of functional MRI has become a central technology in a range of research areas in the health sciences, particularly in neuroscience. MRI is based on the phenomenon of nuclear magnetic resonance (NMR), which is currently increasing in importance in many areas of science, not in the health sciences – areas of interest range from the characterisation of materials to quantum computing.

This technique has become so important to modern medical science that it warrants its own chapter in this text. Here we will cover the most important aspects of the physics of magnetism, NMR and MRI. The level of technical detail is beyond that of most of this book – even if the main learning outcomes for many students will focus on telling what type of image they are looking at and which best distinguishes tissues types, it seems best to present a fuller picture of how the echo sequence parameters can be altered to achieve certain goals.

Key Objectives

- To understand the physical processes involved in nuclear magnetic resonance, particularly the phenomenon of Larmor precession.

- To understand the connection between a static magnetic field and the Larmor frequency of a precessing nucleus.

- To understand the sequence of processes leading to the production of an MRI signal.

- To understand spin–lattice and spin–spin processes leading to the decay of an MRI signal.

- To understand the T1, T2 and T2* times in an MRI signal.

- To understand free induction decay and the spin echo pulse sequence.

- To understand the production of T1-weighted and T2-weighted images.

39.2 Magnetism

The word **magnetism** is used to describe a set of phenomena associated with the movement of charged particles. Magnetism is caused by the same fundamental force as that which causes the electric phenomena described in the Electricity and DC Circuits topic – the electromagnetic force – so the two are inextricably linked.

We have previously avoided a discussion of magnetism in favour of topics with more obvious relevance to biological systems. However, magnetism is central to the physics of nuclear magnetic resonance (NMR) and magnetic resonance imaging (MRI). We will therefore briefly cover the most important aspects of this subject. There are some key points we wish to introduce regarding magnetism:

1. Magnetic fields are generated by moving electrical charges, such as in an electrical current.

2. Moving electrical charges experience a force in a magnetic field.

3. A time-varying magnetic field will induce a current in a loop of wire if that magnetic field passes through the loop.

4. Sub-atomic particles such as electrons and protons have intrinsic angular momentum. They thus behave like tiny current loops and produce a magnetic field even when they are stationary.

We will now cover these points in more detail.

The Magnetic Force and Field

In the chapters about electricity, we saw that electric forces result from a fundamental property of matter called charge, which is non-zero for some particles. Charged particles exert a force on one another, and the strength and direction of this force is described by Coulomb's law. An alternate model we use is the field description, where we picture how a collection of charges (the source charges) create an electric field, and then look at how other charges are influenced by this electric field.

Magnetism is a little more complicated than this because there is no simple magnetic equivalent of charge. Instead, the phenomenon we call magnetism has its origins in the **movement of electrical charges**. This can be demonstrated by examining a current-carrying wire. It is well known that some metals, iron in particular, are affected by magnetic fields. If you take small shards of iron (iron filings) and put them near a magnet, then they align in particular directions. Putting iron filings near a wire through which charge is flowing will show that they experience a force causing them to arrange themselves into patterns when the current is on, but no such force when it is off.

However, **permanent magnets** – objects that possess the ability to attract iron when stationary and not connected to an electrical circuit – are a common, everyday item. While there is no externally driven movement of charge as there is in a current-carrying wire, there are still moving charges. The electrons in the atoms are moving about the nucleus. Additionally, the electrons and nuclei have a property called 'spin' or 'intrinsic angular momentum', and behave rather as though they are permanently rotating. We will come back to sub-atomic particles and spin in much greater depth in a later section (Section 39.4). Both the orbital motion of these particles and their intrinsic angular momentum produce magnetic fields.

A basic property of a simple permanent magnet is that is has two **poles**, which we refer to as north and south. Two magnets exert a force on one another. Opposite poles attract and like poles repel. We can visualise the effect of a magnet on the region around it, its **magnetic field**, by picturing the influence it would have on a tiny bar-shaped magnet at each point in space, in much the same way as we imagined a test charge for visualising an electric field. If our imaginary tiny bar magnets are free to rotate, they will move just like a compass needle. We can draw **magnetic field lines** showing which direction the north pole of a tiny bar magnet would point, as shown in Figure 39.2. Note that the north pole of our test magnet/compass needle points to the south pole of the magnet, so the arrows on magnetic field lines point from north pole to south.

Magnetic moment

The **magnetic moment** of a system which produces a magnetic field is a measure of the strength (and direction) of the system's magnetism. It differs from the magnetic field strength in the same way as electric charge (or strictly speaking, the electric dipole moment) differs from electric field strength.

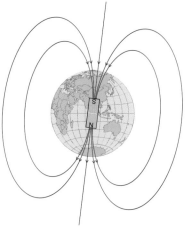

Figure 39.1 The Earth has a magnetic field that is shaped as though the Earth contained a giant bar magnet with its north pole somewhere under Antarctica. This is caused by currents in the Earth's iron core. This field is very important for protecting us from charged particles streaming in from space.

Magnetism and animal navigation

It is known that many animals are able to navigate long distances using the Earth's magnetic field, though the exact mechanisms by which they achieve this vary, and are still largely a source of speculation. Various theories have been proposed, such as strings of magnetic particles selectively opening ion channels through cell membranes, and electric potentials being created across cell membranes by the oscillations of the sensory organ with respect to the field direction.

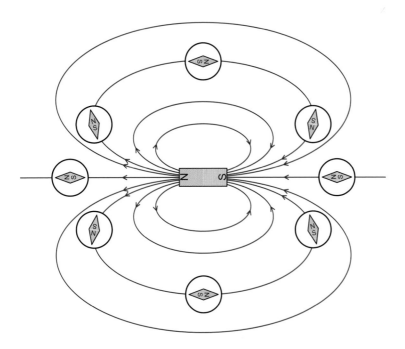

Figure 39.2 A permanent bar magnet is shown with the north and south poles labeled. The magnetic field lines are shown starting at the north pole and ending at the south pole. Small compasses are placed around the bar magnet and line up with the magnetic field lines so that the north pole of the compass points towards the south pole of the magnet and vice versa.

The most common permanent magnets (such as standard bar magnets and compass needles) are made mostly of iron. This is because iron is what we call **ferromagnetic**. A ferromagnetic material acts like a whole lot of tiny little bar magnets that tend to line up the same way as their neighbours. These form into little regions with the same alignment, called **magnetic domains**. The magnetisation of the material depends on the degree of alignment of the domains. In a permanent magnet these domains are largely pointing in the same direction and stay that way; in unmagnetised iron, the alignment of the domains is random.

The **magnetic field strength** (also known as the magnetic flux density) is usually denoted by the symbol **B**, and it is a vector quantity. The term B-field is often used as well, to reduce the confusion between this and a slightly different quantity, called the magnetising field (referred to as **H**). The SI unit of magnetic field strength is the **tesla**, symbol T. Another commonly used unit is the gauss and 1 T is equivalent to 10 000 gauss. The Earth's magnetic field strength varies from place to place on the surface, but the strength is on the order of 0.5 gauss or 50 microtesla. A 1 T magnitude field is strong indeed, and field strengths of 1–2 T (and, increasingly, higher fields) are used in MRI.

> **Quick reference:**
> 1 tesla = 10 000 gauss
> Earth's field varies, but is around 50 µT

Magnetic Field Examples

The magnetic field at a point in space can be found by adding all the contributions from all the currents in the region, just as the electric field can be found from summation of contributions from all the nearby charge. We will not give any mathematical proofs or explanations here, but will merely give some examples of the fields created in certain situations. There are two key examples to be familiar with: the straight, current-carrying wire; and the bar magnet/current-carrying coil. We will also describe two more-advanced examples that are relevant for magnetic resonance imaging machines and for controlling fields in the lab: Helmholtz and anti-Helmholtz coils. These are illustrated below.

Current-Carrying Wire

The magnetic field around a current-carrying wire is circular and the direction can be determined from the right-hand rule: point the thumb of your right hand in the di-

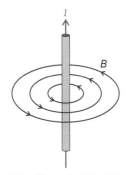

Figure 39.3 The magnetic field around a current-carrying wire. The magnetic field lines form closed circles around the wire.

rection of the current (and remember, that means the direction of the conventional current, from positive to negative potentials) and your fingers will curl in the direction of the magnetic field lines. An example of the magnetic field around a wire is shown in Figure 39.3.

Current Loop/Bar Magnet

The field lines around a bar magnet have already been discussed, but we'll add another important fact here: a current-carrying coil of wire like that shown in Figure 39.4 produces a magnetic field that looks the same from a distance. If you take your right hand and curl your fingers in the direction of the current, and your thumb will point in the direction of the field lines through the coil.

Helmholtz and Anti-Helmholtz Coils

The last examples we'll give here are some particular coil pairs. By taking two identical, flat, circular coils of wire and placing them a particular distance apart – for circular coils this is the radius – we can produce a field that points along the central axis and is nearly uniform by ensuring that the currents in the two coils are equal and in the same direction. Such coils are known as Helmholtz coils, and they can be used for applications such as cancelling out the Earth's magnetic field. If the direction of current in one loop is reversed, these are known as anti-Helmholtz coils, and have a uniform field *gradient*. The exact field properties of such current configurations are beyond the scope of this textbook, but are mentioned here to give readers some idea of how more complex magnetic fields can be created. In particular, anti-Helmholtz coils are used in MRI machines to provide a gradient to the static field. Both Helmholtz and anti-Helmholtz coils are shown in Figure 39.5.

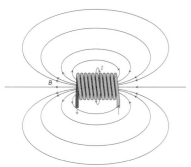

Figure 39.4 The magnetic field in the region around a current carrying coil. The magnetic field lines look much like those produced by a bar magnet.

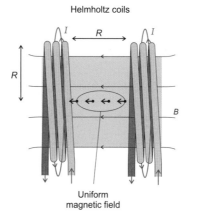 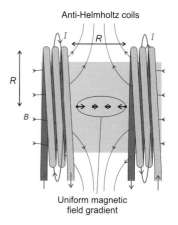

Figure 39.5 (Left) A pair of Helmholtz coils. The magnitude and direction of the current in each identical coil is the same. The distance between the coils is the same as the coil radius. (Right) Anti-Helmholtz coils. The arrangement of the coils is the same as for a pair of Helmholtz coils except that the currents through the coils are in the opposite directions. The magnetic field lines in the vicinity of each set of coils and the magnetic field vectors at several points are shown.

Force on Charges in a Magnetic Field

We have discussed how magnetic fields are created by moving charge. They in turn also affect moving charge. The magnitude of the force F on a charge moving in a static magnetic field B is given by

$$F = qvB\sin\theta \qquad (39.1)$$

where B is the magnetic field strength, q is the charge, v is the velocity of the charge, and θ is the angle between the magnetic field vector and the velocity vector. The direction of the force is perpendicular to both these vectors (see Figure 39.6). A useful tool for remembering which direction the force is in is the **'right-hand slap' rule**: point the

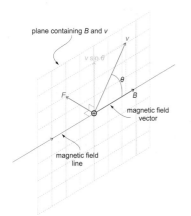

Figure 39.6 The force on a charge moving in a magnetic field is perpendicular to the plane that contains the velocity and magnetic field vectors. The direction that the force acts can be found by using the 'right-hand slap rule'.

thumb of your right hand out at right angles to your fingers and point it in the direction of positive-charge movement, then point your fingers in the direction of the field lines. The direction you would slap your palm is the direction of the force.

Some examples of the motion of a charged particle in a magnetic field are shown in Figure 39.7. When the charge is moving in the direction of the field, there is no force, and the velocity of the particle is unaffected. In the case where the charge is moving across the field, θ is 90°, the force is the maximum and the charge will move in a circle (assuming a uniform field).

A combination of these two motions happens for other angles. The component of the velocity in the direction of the field is unchanged; the component perpendicular to the field has its direction, but not magnitude, modified. The result is the charge spiralling around the magnetic field direction.

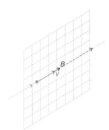 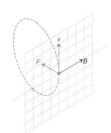 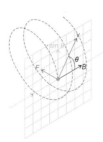

Figure 39.7 (Left) When a charge is moving along (or against) a magnetic field line it experiences no magnetic force ($\sin 0° = \sin 180° = 0$) and will travel in a straight line (provided no other forces are acting upon it.) (Middle) A charge moving perpendicular to magnetic field lines will travel in a circle as the magnetic force on will always be at right angles to the direction of travel. The diagram shows a postive charge. (Right) A charge moving at an angle to the magnetic field will have force components that are parallel and perpendicular to its velocity. The charge will travel in a spiral in such an instance. Again, the diagram shows the force on a positive charge.

Induced Currents: Faraday's Law and Lenz's Law

There is one final key law of magnetism that has relevance to MRI: **Faraday's law**. A charge moving in a static field experiences a force, but it is also the case that a *changing field* will affect charges. Most often we are interested in how this will affect charges in simple circuits, like loops of wire.

Key concept:
Faraday's Law:
The induced electromotive force in any closed circuit is equal to the time rate of change of the magnetic flux through the circuit.

The magnetic flux, Φ, is proportional to the number of field lines through a surface. In the simple case of a uniform field and planar area

$$\Phi = BA\cos\theta \qquad (39.2)$$

where B is the field strength, A is the area and θ is the angle between the normal to the surface and the field direction (see Figure 39.8). The flux can be altered (and an emf and hence current induced) if the field changes strength or direction, or if the area of the circuit is changed. This is how MRI machines receive information from the object being scanned: a current is induced in receiver loops by a changing magnetic field, and the frequency and decay time of this is recorded and decoded.

The direction of the induced current is given by another rule:

Key concept:
Lenz's Law:
A changing magnetic field will induce a current. The direction of the induced current will be that which produces a magnetic field which opposes the changes to the magnetic field, i.e, in the same direction if it is decreasing, and opposed to otherwise.

The Aurora Australis and Aurora Borealis (the Northern Lights) are caused by the magnetic force. Charged particles from space follow paths that spiral around the field lines from the Earth's field. This funnels them towards the poles, and so large quantities of fast, charged particles will hit the upper atmosphere at high latitudes, sometimes creating a visible glow in the sky.

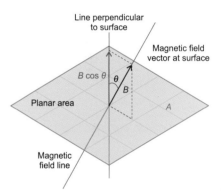

Figure 39.8 The flux through an area A is proportional to the component of the magnetic field vector that is perpendicular to the surface ($B\cos\theta$.

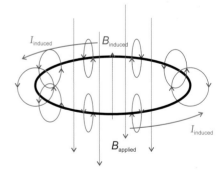

Figure 39.9 The current induced in a loop of wire by a changing magnetic field will act to oppose the change. In this case, the applied field is increased, producing an induced current that would create a upward field through the plane of the loop.

In other words, a magnetic field, B_1 induces a current, I, in a loop of wire like that shown in Figure 39.9. The current in the wire will itself produce a magnetic field, B_2. The direction of B_2 is determined by the direction of the current in the loop. Lenz's law says that B_2 will always be in a direction opposite to the direction of changes in B_1. The current, I will flow in a direction that ensures this.

Types of Magnetic Materials

We can classify the magnetic properties of materials into three distinct groups: paramagnetic, diamagnetic, and ferromagnetic (see Figure 39.10). There are implications for MRI scanning with all three material types.

Paramagnetism

Paramagnetic materials contain unpaired electrons, and so the molecules have a permanent non-zero magnetic moment. In metals, this can arise from the magnetic moments associated with the conduction electrons. Usually, the electron spins are random in orientation. In the presence of an external magnetic field, the material becomes magnetised and produces a magnetic field in the same direction as the external magnetic field. This effect does not persist once the external field is removed.

Ferromagnetism

Ferromagnetic materials can retain a net magnetic moment in the absence of an external field. As in the case of paramagnetism, they have unpaired electrons, but while normally the lowest energy state is for neighbours to anti-align, in ferromagnetic materials it is energetically favourable, over short distances, to line up with magnetic moments parallel. Iron and nickel are examples of ferromagnetic materials.

Diamagnetism

Diamagnetic materials are pretty much opposite to the above: materials with nicely-paired electrons that do not exhibit the more normal responses to external fields. However, the external field influences the electron orbits in such a way that they become magnetised in the opposite direction to the external field. In most materials, the effect is quite weak, but in superconductors the effects can be quite interesting.

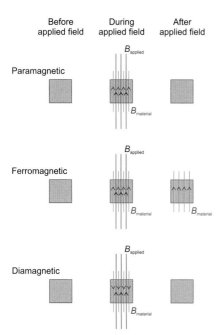

Figure 39.10 Paramagnetism, ferromagnetism, and diamagnetism.

39.3 A Brief Outline of MRI

The physics of NMR and MRI is very complex. For this reason we begin the discussion of MRI with a brief outline of the actual process of producing an MRI image without the details of the physics involved.

The essential practice of MRI is as follows. A patient, or a biological sample to be investigated, is placed in a large magnetic field. This magnetic field is of constant magnitude and is pointed in a single direction, i.e., it is uniform and homogeneous. For the purpose of our discussion, we will call the direction that this magnetic field is pointed in the z-direction. This field is often called the longitudinal field and is about 0.5–1.5 T in magnitude in most machines currently used, although machines with a longitudinal field of 3 T are beginning to appear.

With the patient lying in the longitudinal magnetic field, an oscillating electromagnetic field is turned on. This field is oscillating in the radio-frequency range of the electromagnetic spectrum and is tuned to transfer energy to the protons which are the nuclei of hydrogen atoms in the patient. The energy of the RF field is absorbed by the hydrogen nuclei and this energy is then re-radiated by these nuclei as another RF electromagnetic field. This second RF field is detected by antennas in the MRI machine and the signal produced is analysed by powerful computers to produce images like those shown in Figure 39.11. By carefully engineering the characteristics of the RF field sent into the patient, we are able to ensure that the RF field returned by the protons in the patient contains information about the position and composition of the tissues in which the protons reside. In particular, the frequency (and phase) of the emitted signal can indicate where the signal came from, and the decay time gives information about the tissue composition.

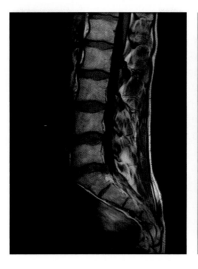 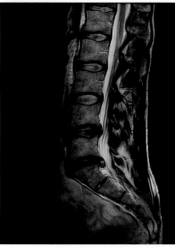

Figure 39.11 T1 (left) and T2 (right) weighted MRI images of the same spine. These images show a lumbosacral intervertebral disc protrusion (slipped disc). [Images courtesy of Professor Terry Doyle, University of Otago School of Medicine.]

39.4 Nuclear Magnetic Resonance

Angular Momentum, Rotation and Precession

Magnetic resonance imaging (MRI) employs **nuclear magnetic resonance** – a technique which exploits the magnetic properties of the nucleus to map the interior of a sample. The word 'nuclear' is now usually left out of the term nuclear magnetic resonance (NMR) when referring to imaging due to confusion with nuclear decay and explosions (which are entirely different processes), as this caused unnecessary fear and anxiety. To understand NMR, we need to re-acquaint ourselves with some technical terms and introduce some new ones – we will need to understand a little about angular momentum. An excellent tool for familiarising ourselves with the necessary quantities is the spinning top, and we will use the top analogy extensively.

Many of the physical quantities that describe translational motion have rotational analogues. For example, the analogue of the displacement is the angular displacement, the velocity has the angular velocity as its companion, and we can also assign a value to the angular acceleration as we do to linear acceleration. Newton's third law, the familiar $F = ma$, tells us the effect of a force on an object: a is the rate of change of the velocity, making ma the rate of change of mv, the momentum. The equivalent quantities when discussing rotational motion are the torque and the angular momentum. The torque is the time rate of change of angular momentum; applying a torque to a body will change its rotational motion.

Consider the case of a top that is not spinning, but merely standing on its point. There are two forces at play here: the downward force through the centre of mass and the upwards force through the point. If these forces are not acting exactly along the same line, then there will be a torque on the top that will rotate it. It will acquire angular momentum by falling over. A stationary top is thus extremely unstable since these forces do not generally line up exactly and the top quickly falls over.

But a *spinning* top like that shown in Figure 39.12 doesn't fall over. Why not? The spinning top already possesses angular momentum: the torque resulting from a mismatch between the gravitational and support forces still changes the top's angular momentum, but the effect is now a little different. The result is that the axis about which the top is spinning is no longer up-down (we'll call this the z-axis), but at some angle to the *z*-axis, and this direction *will change with time*. The axis of rotation will describe a circle around the *z*-axis. This movement is called **precession**. The angle between the *z*-axis and the rotation axis is called the **precession angle**. The top is quite resistant to efforts to push it over – applying such a force will merely change the precession angle. The term for a change in precession angle is **nutation**.

There are two directions of importance here: the *z*-axis and the axis of rotation of the top. The *z*-axis is defined by the direction of the gravitational field the top is in, here provided by the Earth. The axis of rotation of the top defines the direction of the top's **angular momentum vector**. The spinning top analogy is a very useful one for describing the effects of a magnetic field on a collection of nuclei because the *average* direction of the spin (intrinsic angular momentum) of the particles behaves similarly, precessing about the direction of magnetic field.

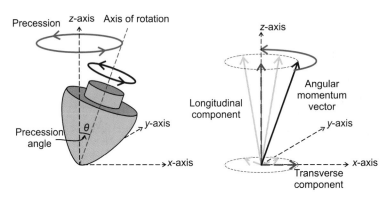

Figure 39.12 The axis of rotation of a spinning top precesses about the vertical. The angular momentum vector of the top has a vertical longitudinal component which is constant and a horizontal transverse component which rotates about the vertical.

Classical Picture Versus Quantum Mechanics

At this point, we will take a slight digression in order to clarify why the word 'average' was necessary in the previous section. This has to do with the 'classical' and 'quantum-mechanical' views of the world. In the classical world view, properties that a body might possess, such as energy, have allowed values which form a continuous range. In the quantum view, some properties may only take on certain discrete values. A example of quantisation which we have already encountered is that of the allowed energies of the electrons bound in atoms.

While quantum mechanics is a very successful framework for predicting how the world behaves on a small scale, where classical mechanics fails, the classical view is adequate much of the time because *the predicted average behaviour of a large collection of quantum systems* gives the same result. We mention this here because we will at times be talking about specifically quantum-mechanical particle properties like spin, and at others talking about the bulk magnetisation resulting from the accumulated effects of all the spins. Our spinning top analogy is very good for describing the bulk magnetisation. It is actually a poor model for the behaviour of a single particle, but we will still borrow it for this from time to time.

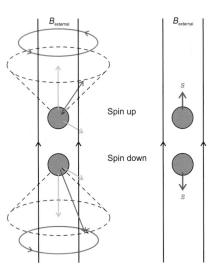

Figure 39.13 The component of the magnetic moment of a proton is measured with respect to an external field. The measured component of the magnetic moment is always a multiple of some fundamental value.

For an object where the classical picture is adequate, like a spinning top, the angular momentum vector can have any magnitude, and can be described completely – you could write equations that show the direction of the vector at all times, and predict the result of a measurement of the angular momentum component in a chosen direction at some future time. This classical picture does not work for particles like electrons, whose behaviour is fundamentally quantum-mechanical. Most importantly for our discussion of MRI, the spin of an electron has a set of fixed possible values, i.e., the spin of the electron is quantised. This has some unusual consequences. These consequences are quite counterintuitive, but rest assured that they are firmly based on experimental evidence.

The central experiment which was the historical source of our understanding of quantum-mechanical spin was the Stern–Gerlach experiment (see margin box). If you design an experiment to measure the value of the component of the spin of some sub-atomic particle, in a given direction, your measurements of spin will *only ever* yield an integer number of values, rather than a continuous range of values. For an electron, for example, the component of spin in the chosen direction will be $+\frac{\hbar}{2}$ J s or $-\frac{\hbar}{2}$ J s, no matter what that direction is. Furthermore, if you change your mind and redo the experiment, measuring the component of angular momentum again but in a different direction, you will find that the value you measure is again $+\frac{\hbar}{2}$ J s or $-\frac{\hbar}{2}$ J s.

Interaction of Nuclei With Static Magnetic Fields

In atoms, there are three important sources of angular momentum. The circular motion of an electron in its orbital about the atomic nucleus contributes to the angular momentum of this electron and to the total angular momentum of the atom. This is called (not surprisingly) the **orbital angular momentum** of an electron. As discussed in the previous section, a particle like an electron also has **spin**. This spin is a quantum-mechanical form of angular momentum and thus contributes both to the total angular momentum of the electron and the total angular momentum of the atom. Finally, the nucleus is composed of nucleons and each nucleon has both orbital and spin angular momentum. The angular momenta of the nucleons combines to give a total angular momentum for the **nucleus** as a whole and this contributes to the total angular momentum of the atom.

For our discussion of MRI we will only need to concern ourselves with the behaviour of the hydrogen nucleus – a single proton. Protons are spin-1/2 particles: there are only two possible values for the component of the spin or intrinsic angular momentum vector, just as in the case of the electron. The component of the spin of the proton in a particular direction is either aligned with the chosen measurement direction or aligned against it. We often call these spin values 'spin up' and 'spin down' for convenience.

In the section on magnetism, it was noted that current moving in a loop acts like a bar magnet. We have now determined that protons (and electrons etc.) have an intrinsic angular momentum (spin) and since the proton is charged this implies that the proton has some kind of intrinsic circulating current. This in turn implies that protons will behave a bit like little bar magnets – that they will have a magnetic moment. This magnetic moment vector points in the same direction as the axis of spin of the proton.

In the presence of a static external magnetic field, the two different spin states have slightly different energies, and more protons will be in the lower energy state (with z-component of the magnetic moment in the field direction). Overall, then, if we have many protons and add up all the little magnetic moments from their spins, the net **magnetisation vector** of the collection will be in the direction of the magnetic field (see Figure 39.14); the sample is *longitudinally magnetised* in the field direction (which we will also refer to as the positive z-direction).

Interaction of Nuclei With a Resonant Electromagnetic Field

In an external magnetic field, there is an energy difference between the spin up and spin down states of a collection of protons. If we are able to supply the correct amount of energy somehow, a number of the protons in the lower energy state will flip from one spin state to the other. Fortunately we have a handy mechanism which we can use to supply protons with the correct energy. We can irradiate the sample with *photons* having the correct energy, or, in other words, we need to irradiate the protons with photons having the correct *frequency*. To flip protons, this correct frequency is

$$f = \gamma B \tag{39.3}$$

where γ is a number called the **gyromagnetic ratio of the proton** and is equal to $42.5\,\mathrm{MHz\,T^{-1}}$. The frequency defined by Eq. (39.3) is called the **Larmor frequency**, and the fact that this frequency depends on the size of the external magnetic field is central to the physics

The Stern–Gerlach experiment

Theory predicts a force on a magnetic moment in an inhomogeneous magnetic field; a gradient in the z-direction will produce a force which depends on the component of the magnetic moment in the z-direction. This provides a way to test quantum theory against classical mechanics. Classical theory suggests that an atom with angular momentum can have a continuous range of z-component values, while quantum theory predicts only discrete allowed values. In 1921 Stern and Gerlach performed an experiment of this kind with a beam of silver atoms, which split into two distinct beams in the magnetic field, supporting the quantum world view.

No external field

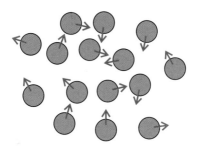

External magnetic field

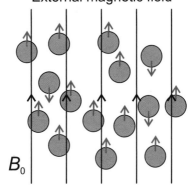

B_0

Figure 39.14 In the presence of a static magnetic field, more protons align with the field than against it.

Gyromagnetic ratio

Books on this topic use inconsistent notation for the gyromagnetic ratio. Some write that $\gamma/2\pi = 42.5$ MHz/T instead, if they are using angular frequency rather than frequency.

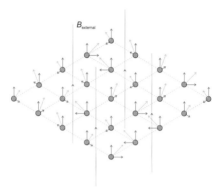

Figure 39.15 Initially the protons mostly have magnetic moment aligned with the field. The transverse components of the magnetic moment are all out of phase however.

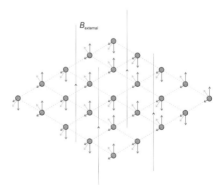

Figure 39.16 After the application of a specific RF field the longitudinal components of the protons magnetic field cancel each other out (half spin-up, half spin-down) but the transverse components are all in phase with each other. This is called a 90° pulse.

of MRI. In a field on the order of 2 T, this frequency is in the radio part of the electromagnetic spectrum, in the region used for FM radio broadcasting.

If we send resonant radio-frequency photons into a sample containing hydrogen nuclei then the sample will temporarily change its longitudinal magnetisation (see Figure 39.15). This change in magnetisation occurs because protons in the low-energy state with their magnetic moment aligned with the external magnetic field can now absorb a photon and flip so that their magnetic moment points in the opposite direction. The total magnetisation of the sample is the sum of all of the magnetic moments of all of the protons, so if enough of the protons flip into the opposite direction then the longitudinal magnetisation of the sample will be cancelled out. If even more of these lower energy protons flip, the longitudinal magnetisation of the sample can even be reversed. It will eventually revert to its original aligned state if the radio-frequency (RF) radiation is stopped.

The RF field has another important effect, though. We have previously talked only about the longitudinal magnetisation. What about the other directions, i.e., the transverse plane perpendicular to the magnetic field? In the presence of a purely static magnetic field, there is no magnetisation in the transverse plane, because the transverse components of the magnetic moments of all the protons are not synchronised. This means that when the transverse component of the moment of one proton is pointing one way, the next one could be pointing anywhere in the transverse plane: when all the transverse components of all the protons are added up they cancel each other out. Even though the protons are all precessing about the z-direction, there is no synchronisation between neighbours they are not in phase at all – so overall the average value of the transverse magnetisation is zero.

This changes with the addition of a resonant (transverse) electromagnetic field, which has the extra effect of creating *phase coherence* the phase of the spins of the protons are synchronised by the RF field (see Figure 39.16). The principle is the same for all kinds of systems: if you apply a force at the resonant frequency, the energy transfer is the most efficient, and the energy source and the recipient rapidly end up in a fixed phase relationship. If you want to push a child on a swing, you need to always push at the same part of the swing's cycle to get them to go higher. If you were able to push 100 children on swings at the same time, they would all end up swinging in sync no matter how they started out. Since the spins (and thus the magnetic moment vectors) of all the protons are now synchronised, they will all add up rather than cancelling out. Thus there will now be a component of the magnetisation which is rotating in the transverse plane. Furthermore it will be rotating at the Larmor frequency. These rotating magnetic moments will produce a changing magnetic field, which will induce a current in a wire loop placed nearby.

To summarise, a resonant, radio-frequency electromagnetic field will reduce (or even reverse) the magnetisation in the longitudinal (z) direction at the same time as introducing a rotating magnetisation in the transverse plane.

If the RF field is applied for the just the right amount of time, then the longitudinal magnetisation can be completely cancelled out – there will be as many protons with aligned spins as there are protons with anti-aligned spins. The magnetic-moment vectors of all of the protons will also be synchronised. They will all be pointing in the same direction at the same time and rotating at the same frequency (the Larmor frequency). In effect, the magnetisation of the sample has gone from pointing in the same direction as the external longitudinal magnetic field to rotating in the plane transverse to the longitudinal field. The magnetisation of the sample has been rotated 90° into the transverse plane. For this reason the pulse of RF field that achieves this is called a **90° pulse**.

Relaxation Processes and Times

After a 90° RF pulse has been sent into a sample, the magnetisation of the sample is rotated 90° into the transverse plane and rotates at the Larmor frequency. Eventually, though, if the sample is left unperturbed, the magnetisation will *relax* back to its original longitudinal direction (see Figure 39.17). There are two basic relaxation processes involved, and they each occur on a different time scale:

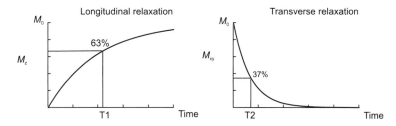

Figure 39.17 The longitudinal and transverse components of the bulk magnetisation return to their original values following a resonant RF pulse.

- Loss of coherence in the transverse plane as the spins of the protons lose their synchronisation.

- The total magnetisation of the sample rotates back into the longitudinal direction as the spins of the protons flip until the original number of protons in each spin state is achieved.

The characteristic length of time taken until the longitudinal magnetisation recovers, that is, the time it takes the longitudinal magnetisation to go from zero back to within 37% of its old value following a 90° pulse, is called the **T1 time**. This time is governed by how long it takes for protons in the higher-energy state to give up the energy they gained from the RF field. The protons can lose this energy only by transferring it to something else; in this case the only place that they can transfer the energy to is the surrounding lattice of atoms and molecules. The T1 time is therefore dependent on the environment in which the protons find themselves (the density of the surrounding medium, the particular form of the molecules in which the proton is embedded etc.), as this will determine the ways in which they can give up this energy. The T1 time is thus also known as the **spin–lattice** or *longitudinal* relaxation time.

The transverse magnetisation is a result of the synchronisation of the spin of a proton with its neighbouring protons. As they lose this synchronisation (or *phase coherence*), the transverse magnetisation is reduced: the individual proton spin vectors don't quite point in the same direction so their sum is slightly smaller. We have already seen that a proton can transfer energy to the surrounding lattice. A proton can also transfer energy to other protons. In this process one proton will lose energy and another will gain energy. This transfer of energy results in slight changes in frequency between neighbouring protons and thus will kick them out of synchronisation. As an analogy, imagine a row of ballerinas *en pointe*, twirling about in perfect unison. Now suppose that one reaches out to her neighbour and they press hands together as they spin past. One of them will get pushed ahead by this, increasing her rotational speed and frequency. The other ballerina will get pushed back the other way and her rotational speed and frequency will decrease. They won't be exactly in time any more. This is an analogy, but it is a reasonably good one. In actual fact the magnetic field of one proton interacts with the magnetic field of a nearby proton, changing the Larmor frequency of each of them (see Eq. (39.3)). These *spin–spin interactions* cause a decay in the total rotating transverse magnetisation of the sample with a characteristic time called the **T2 time**, or the *spin–spin* relaxation time. This is the time it takes the transverse magnetisation to decay to 37% of its original value.

The T2 time is, like the T1 time, dependent on the background material in which the protons are embedded. In particular, the T2 time depends on the density of protons in the background material. The more protons there are nearby, the more spin–spin interactions there will be and the faster the synchronisation of the transverse magnetic moments of the protons will disappear.

There is another process which causes loss of phase coherence. The spin–spin interaction is ultimately caused by the interaction of the magnetic field of a proton with the magnetic field of a nearby proton. A similar process will result from a local inhomogeneity in the external longitudinal magnetic field. This will cause the magnetic field to vary slightly from place to place, and hence the precession rate of different protons will differ very slightly. This combines with the spin–spin interaction of neighbouring pro-

Tissue type	T1 at 0.5 T(ms)	T1 at 1.0 T (ms)	T1 at 1.5 T (ms)	T2 (ms)
Fat	210	240	260	80
Liver	350	420	500	40
Muscle	550	730	870	45
Kidney	440	590	700	58
Heart	560	750	890	57
White brain matter	500	680	780	90
Gray brain matter	650	810	900	100
Cerebrospinal fluid	1800	2160	2400	160

Table 39.1 Approximate T1 and T2 times. [Reprinted with permission from *The physics of Diagnostic Imaging 2nd ed., Dowsett et.al.* Reproduced by permission of *Edward Arnold (Publishers) Ltd.* Copyright 1998.]

tons to cause the transverse magnetisation to decay with a characteristic time called T2* ('tee two star'). The T2* decay time is shorter than the T2 decay time as it involves the fluctuations in the large external field, which, although small, are large compared to the magnetic fields produced by individual protons. To go back to our ballerina analogy, imagine that at some places on the stage the ballerinas can spin faster since the floor is more highly polished there. Pretty soon they won't be spinning in time with their neighbours. This is similar to the loss of coherence due to variations in the background magnetic field, as opposed to variations due to the interaction of the protons. It is possible to mitigate the effects of these local variations in the background magnetic field using clever RF pulse sequences, and we will cover this in the next section.

The fact that different types of biological tissues have different T1 and T2 times is the central idea of MRI imaging. The T1 time is due to spin–lattice interactions which will vary depending on the nature of the underlying lattice in which the proton is sitting. The T2 time depends on spin–spin interactions, which are dependent on the density of protons in the material. Thus these two decay times will vary from material to material. Table 39.1 lists some characteristic T1 and T2 times for different biological materials and tissues; the values are quoted at three different values of the external longitudinal magnetic field.

Liquids and Solids

Solids tend to have short T2 times whereas liquids tend to have relatively long T2 times. The reason for this is the presence of inhomogeneities in the local magnetic field. These inhomogeneities will cause variations in the Larmor frequencies of neighbouring protons, causing them to rapidly fall out of synchronisation. In solids these inhomogeneities are 'frozen' in place by the background lattice of the solid in which the proton is embedded. Any inhomogeneities in the local magnetic field will not change with time as the background lattice does not change with time. In a liquid, on the other hand, any local inhomogeneities in the background magnetic field will wash out as a proton moves about in the fluid. In general the inhomogeneities experienced by a proton in a fluid will average out. If the field increases at one point, it will tend to decrease at another point. Thus though the Larmor frequency of the proton increases at one point it will decrease at another point so that on average it will have the same Larmor frequency as all the other protons in the fluid, since they are also moving through the same local inhomogeneities.

In ice, water protons are never more than a short distance from another proton. The field strength due to this proton will be on the order of 1×10^{-4} T which gives a dephasing time on the order of $\frac{1}{\gamma B} = 2.4 \times 10^{-4}$ s. This makes the T2 time in a solid like ice very short, and the T2 times for most solids are on the order of $\sim 10^{-4}$ s. In a liquid, however, the random tumbling motion of the other nearby protons tends to subject a proton to a field that averages out to zero, and the T2 time is much larger (about 2 s for water).

39.5 Magnetic Resonance Imaging

We now have the tools we require to understand diagnostic magnetic resonance imaging (MRI).

The human body is about 75% water. Each water molecule contains two hydrogen nuclei, i.e., single protons, with spin $\frac{1}{2}$. Large biologically important molecules also contain a large amount of hydrogen bound to other elements such as carbon in very specific configurations. When placed in an external longitudinal magnetic field and then stimulated with an RF field, these protons will then radiate an RF field of their own. This RF field can be detected as a radio-frequency voltage signal in a detector antenna loop. This signal will depend on the tissue parameters (proton density, T1 and T2 times) and machine parameters that can be altered to change how the signal depends on T1 and T2.

In the early, pioneering days of MRI it was hoped that all body tissues, normal or diseased, would have specific T1 or T2 relaxation times that would act as a signature of those tissues and which could be measured precisely. As discussed above solids tend to have shorter T2 decay times than liquids. We would thus hope that bone for example would have a much shorter T2 time than, say, blood and would thus be distinguishable from blood and other tissue simply by looking at the T2 decay times of the MRI signal. Unfortunately, things did not prove to be quite that simple. There is considerable variation in the measured T2 (and T1) times from a given tissue type. Again taking bone as an example, bone is porous to a greater or lesser degree (depending on the particular bone tissue considered) and thus will contain greater or lesser quantities of blood and other biological fluids. Thus the T2 signal from bone will not be characterised by a single clear signature T2 time. For this reason, accurate measurement of T1 and T2 is not generally attempted; an MRI image is a map of MRI signal strength at each location across the sample. The MRI machine is set up so that the signal strength depends strongly on the T1 or T2 times, or the density of protons contributing to the signal. This is the reason that most MRI images are black and white – they are just recording the intensity of the RF signal received from the sample as shown in Figure 39.18.

The character of the signals measured by an MRI machine depends on the types of RF pulse sequences that are sent into the sample and the nature of the longitudinal static field in which the sample is held. We will begin our discussion of MRI signals and their interpretation by describing the **free induction decay (FID)** pulse sequence, and the signal that can be detected from this. FID is the simplest case, and we should perhaps not even call it a pulse sequence. We will then examine more complex RF pulse sequences to see how they can be used to produce signals that depend most heavily on T1, T2 or hydrogen (proton) density. We will then look at how the point of origin of each component of the RF field can be encoded into the measured MRI signal with the clever use of extra magnetic fields which add various gradients to the large longitudinal static field.

Free Induction Decay

In the previous section we pointed out that an MRI image is generally a black and white image in which shades of gray represent the intensity of the MRI signal. We also mentioned that the intensity of the MRI signal can be linked to either the T1 or the T2 decay times. In this section we will briefly indicate how this is done. We will require some technical machinery to do this, and we will introduce it without much explanation, as the derivation of these equations is unnecessary for our purpose here.

In a free induction decay sequence, an RF pulse (not necessarily a 90° pulse, but we will use this as our example) is applied to a sample in a static, z-directed magnetic field. As previously discussed, this type of pulse rotates the magnetisation vector away from the z-axis and into the xy-plane. The bulk magnetisation vector of the sample, $\boldsymbol{M}$, has had its longitudinal component reduced from its initial value to zero by the pulse and this $\boldsymbol{M}$ vector then rotates in the xy-plane at the Larmor frequency. Once the stimulating RF field has been switched off, the magnetisation of the sample will return to its initial value.

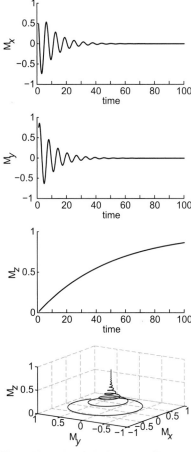

Figure 39.18 Free induction decay: The components of the magnetisation vector. The transverse component of the magnetisation decays at a different rate to the recovery of the longitudinal magnetisation. The oscillation frequency shown here is much smaller than in reality for clarity – the frequencies are on the order of 40–120 MHz, and the decay times on the order of 100–2000 ms.

The strength of the MRI signal emitted by the sample, which we will call S, will decrease exponentially with two time constants, T1 and T2*. We can express this using the following equation

$$S = S_0 e^{-t/\text{T1}} e^{-t/\text{T2}^*} \qquad (39.4)$$

The T2* time is a lot shorter than the T1 time so it will dominate Eq. (39.4). That is, term $e^{-t/\text{T2}^*}$ in this equation will become very small much faster than the term $e^{-t/\text{T1}}$, and this in turn will mean that the reduction in signal strength will be largely due to the short T2* time. Getting information about the T1 and T2 times from the free induction decay signal is therefore tricky. However, using more complex sequences of RF pulses, information about the T1 or T2 times can be extracted. We will now introduce one such pulse sequence, for the purpose of demonstrating how images can be weighted to emphasise different features.

Spin Echo Pulse Sequence

The much shorter T2* time is a serious limitation on the ability of MRI scans to extract T1 and T2 time information. The **spin echo pulse** sequence uses an additional RF pulse to get around this problem.

To understand how the spin echo pulse system works it is important to remember the source of the T2* decay time. The T2* time is due to variations in the local background magnetic field which this results in variations in the Larmor frequencies of neighbouring protons. This in turn results in a rapid loss of synchronisation between these protons. However, these background field inhomogeneities are relatively stable so that a proton with a slightly lower Larmor frequency will *always* have a slightly lower Larmor frequency. Over time those protons with higher Larmor frequencies will move a little ahead of the protons with the average Larmor frequency, and the protons with lower Larmor frequencies will start to lag. This results in a spread of magnetisation vectors across the sample as the faster protons pull ahead and the slower protons fall behind. If we could flip this spread around so that the faster protons were at the back and the slower protons were in the front then the spread would bunch up again.

As an analogy, suppose a group of runners, capable of running at slightly different speeds, all set off from the start line at the same time. A little while later they will be different distances away. However, if the runners all turn around at the same time and run back towards the start line, maintaining their original speed, they will get back there at the same time. This is the idea behind the **spin echo** pulse sequence. The process by which the proton magnetic moments are bunched up again is called **re-phasing** as it amounts to correcting the small phase differences which have developed between the proton magnetic moments. A typical spin echo sequence would go through the following steps (also shown in Figure 39.19):

1. A 90° pulse rotates the magnetisation of the sample into the transverse plane.

2. The transverse magnetisation decays with characteristic time T2*, as the individual proton magnetic moments spread out in the transverse plane.

3. At time T a second RF pulse flips the sample magnetisation vector (and thus all of the individual proton magnetic moments) through 180°, this puts the spins that were ahead behind and vice versa. This is (unsurprisingly) known as a 180° pulse.

4. After a second time period of T, the spins have become **re-phased** – the proton magnetic moments are bunched up again and the transverse magnetisation of the sample is restored. The MRI signal strength peaks when the dephasing effect of the magnetic field inhomogeneity has been exactly reversed (i.e., at time $2T$). The MRI signal strength is still dependent on T1 and T2, and so the signal will still be decaying but at a reduced rate. The time at which the peak in the MRI signal caused by the spin echo pulse sequence occurs is called the **time to echo, TE**.

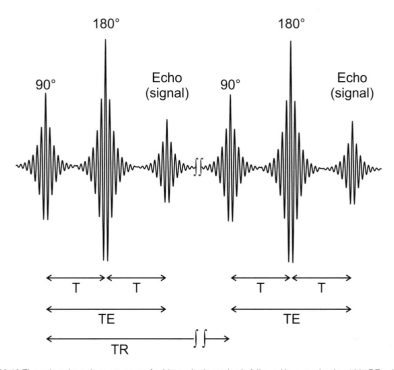

Figure 39.19 The spin echo pulse sequence. An 90° excitation pulse is followed by a re-phasing 180° RF pulse which reverses the effects of the magnetic field inhomogeneities on the precession frequencies. The effects of spin–spin interactions which cause the transverse magnetisation to decay with characteristic time T2 are not reversible.

5. The spin echo sequence is then repeated. The time between repetitions of the spin echo sequence is called the **time to repetition, TR**.

If the TE and TR times are adjusted, the echo signal strength will be altered. If the repetition time (TR) is short compared to the T1 time, the longitudinal magnetisation will not have fully recovered when the second (or third) spin echo sequence begins. The next 90° pulse will tip a smaller magnetisation vector into the xy-plane as the magnetisation vector has not fully recovered, and the resulting measured MRI signal will be reduced. The smaller the value of T1, the more the longitudinal magnetisation will have recovered by the time the next spin echo sequence arrives and the stronger the resulting MRI signal will be. In other words, when a spin echo pulse sequence is used to produce an MRI signal, a short T1 time will correspond to a strong signal and a long T1 time will correspond to a weaker MRI signal. We have thus encoded the T1 time as MRI signal intensity.

In MRI images produced in this way, short T1 means more signal, and a brighter spot on the image. An image where the TR and TE times are chosen to emphasise differences in T1 by choosing a short TR is known as **T1 weighting**. Of course the TE time must be shorter than the TR time, so both times must be short to produce a T1 weighted image.

If the TR time is very long, then the longitudinal magnetisation of the sample will be nearly fully recovered no matter what the value of the T1 time is. In this case, the strength of the MRI signal will not be so strongly dependent on the T1 time. In this regime, however, the MRI signal strength can be made to depend on the length of the T2 time. Increasing the time to echo (i.e., the time until the 180° pulse) means that signals with short T2 times will have dephased more. The 180° pulse only compensates for the T2* dephasing and does not change the T2 dephasing since this is irreversible. Thus samples with shorter T2 times will produce a smaller MRI signal because they dephase more before the next sequence begins (with a 90° pulse). Thus the bright areas on an MRI image made in this way will be the places with larger T2 values. Such an image is known as **T2 weighted**.

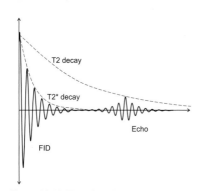

Figure 39.20 The spin echo sequence produces an echo signal from the re-phasing of the spins. The strength of the signal can provide information on the T2 time that had been obscured by dephasing from a slightly non-uniform field.

Key concept:
Short TR and TE values emphasise differences in T1, and bright image areas have shorter T1. Long TR and TE values emphasise differences in T2, and darker areas on an image have shorter T2 values.

If TR is long, and TE is short, the signal will be quite strong regardless of the T1 and T2 times, and will emphasise differences in proton-density instead. This is known as **proton density weighting**.

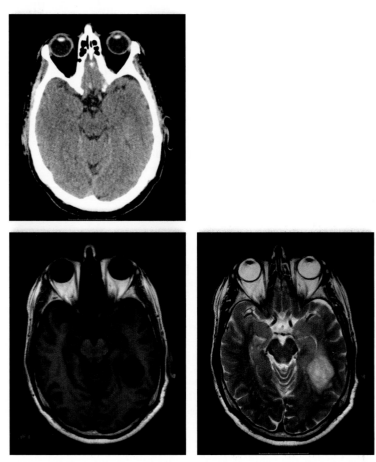

Figure 39.21 (Top) A CT scan of a patient's head. The patient had an unexplained accident but the CT scan showed no abnormality. (Bottom left) T1-weighted MRI image of the same patient located in the same place as the CT image. This image shows a low contrast image of a tumour in the left temporal lobe. (Bottom right) A T2-weighted image in which the large water signal in the tumour is clear. [Images courtesy of Professor Terry Doyle, University of Otago School of Medicine.]

One of the easiest ways to tell if an MRI image is T1 or T2 weighted is to look for areas in the sample where liquid is expected. Liquids have much longer T2 times than solids, so water and water-rich substances like cerebrospinal fluid or vitreous humour will appear bright on a T2-weighted image.

T1 and T2 in MRI

T1

The T1 time is determined by how long it takes to transfer the energy gained from an RF excitation pulse to the surrounding material lattice in which the protons are embedded. Molecules undergo various types of motion (vibration, rotation, and translation) and the frequency of these motions relative to the Larmor frequency will determine the efficiency of the energy transfer from protons to lattice. The frequency of these motions therefore determines the T1 time. Smaller molecules (like water) tend to have higher natural frequencies, while larger proteins tend to have lower motional frequencies due to their greater size and mass. Typically, both very large and very small molecules have frequencies far from the Larmor frequency, and so have long T1 times. In contrast, medium-sized molecules like cholesterol have frequencies close to those used in MR scanning, so T1 times in the region of such molecules are short. Thus, on a T1-weighted image, water is dark and fat tissue is bright. The myelin in the white matter of the brain is bright for this reason, also.

Water normally has a long T1 time, but this can be altered when a water molecule sits near large biological molecules, such as the surface of proteins. The water molecules are weakly bonded to the proteins, and this reduces their motion, decreasing the T1 time.

T2

The T2 is not a measure of the rate of energy loss to the surrounding material, but measures the dephasing or decoherence rate that leads to the loss of synchronisation between the spins of the protons in the sample. Inhomogeneities in the local magnetic field in the substance influence the T2 time, generally shortening this time considerably. In liquids, the rapid motions of molecules tend to average out these inhomogeneities as we discussed earlier, and this will mean that the T2* is much longer. T2 times are also longer, as individual protons are not as close together as they are in solids and hence do not dephase each other as much. In a T2-weighted image, long T2 times appear as bright patches, thus water shows up strongly in T2-weighted images.

Differences in protein binding of water molecules allows some distinction between white and grey brain matter: white matter will appear darker on a T2-weighted image.

> **Key concept:**
> On T1-weighted images, water is dark and fat tissue is white. T1-weighted images are provide useful structural/anatomical information.
> On T2-weighted images, water is bright. T2-weighted images are used where water sensitivity is important, such as when looking for oedema (accumulation of fluid).

Solids

Bone tissue, or any stones (gall stones, kidney stones etc.) that may be present, have short relaxation times and do not contribute significantly to the signal in either T1 or T2 weighted images. They will appear dark on an MR image. MRI is not very useful for examining the bone in detail, and where information on fractures etc. is needed, other techniques, such as CT scanning, are often used instead of, or in conjunction with, MRI images.

Spatial Information

So far we have shown how the signal strength can transmit information about the T1 and T2 times and proton density of the sample being examined. In order for this information to be used to create an image, it is also necessary to know where each component of the signal received originated from inside the sample. To achieve this, spatial information is encoded in the frequency and phase of the MRI signal. In the following sections we will show how this is done.

Slice Selection

When imaging the human body, images of a slice through the body a few millimetres in width are produced. This is done by ensuring that the magnetic field, and hence the resonant (Larmor) frequency of the slice is different from the rest of the body. This means that the 90° RF pulse is only absorbed in that particular slice. To produce this effect, a gradient field is added to the strong (1 T or more) static field that is always on during the MRI imaging procedure, and the magnetic field at position z will be

$$B(z) = B_0 + \frac{\Delta B}{\Delta z} z \qquad (39.5)$$

where B_0 is the magnetic field at $z = 0$, and $\frac{\Delta B}{\Delta z}$ is the field gradient.

This extra gradient field is directed along the z-axis like the static field, but it varies linearly in strength along this axis. This ensures that the precession frequency of the spins will also vary longitudinally and a cross-sectional slice can be excited by an appropriately tuned RF field, while the rest of the body is not.

Frequency Encoding

The frequency of the signal *emitted* by protons in a sample is dependent on the local value of the Larmor frequency, and this is dependent on the local value of the longitudinal magnetic field. By making the magnetic field strength vary along a particular axis, the detected signal now has components at a range of different frequencies which correspond to particular locations along that axis. It is common to use this kind of frequency encoding on one axis, say the x-axis, and to use phase encoding for the perpendicular (y) direction.

The gradient field used for this frequency encoding does not need to be on during the whole imaging sequence; in fact, this would be unhelpful, as the T1 times are field dependent. The extra field is only switched on during the signal detection part of the sequence, and so is often referred to as the **readout gradient**.

Phase Encoding

Phase encoding is a method which encodes spatial information in the relative phase of protons within a slice that has been selected by some other method. Initially, all of the protons in a particular slice have the same Larmor frequency and are all synchronised with each other. A gradient field in a transverse direction is then turned on for a short period. Suppose that we chose the y-direction as the direction in which we wish to use phase encoding. The new gradient field will then have a gradient in the y-direction. While this new gradient field is on, the Larmor frequency of the protons in the slice will vary in the chosen direction. The new gradient field is then turned off so that now the protons in the slice are all precessing at the same Larmor frequency again. Now, however, the transverse magnetic moment of the protons at a particular point in the y-direction will be slightly ahead of the magnetic moment of the protons to one side and slightly behind that of the protons on the other side. The *phase* of the magnetic moment of the protons now encodes their position in the y-direction. The phase encoding will be present in the MRI signal and can be used to locate points in the chosen slice.

Magnetic Resonance Spectroscopy

Magnetic resonance spectroscopy is the analysis of the frequency spectrum of an MRI signal. This spectrum can supply information about the chemical environment in which protons in the sample are embedded. In diagnostic applications, this can provide valuable information about metabolic processes in the sample tissue. In standard MRI images, hydrogen nuclei consisting of a single proton are the source of the MRI signal. However, any nuclei with a non-zero magnetic moment (i.e., non-zero spin) will produce an MRI signal, these other nuclei are commonly used in magnetic resonance spectroscopy. The gyromagnetic ratios of some nuclei used are shown in Table (39.2).

Element	γ, MHz/T
^{1}H	42.58
^{19}F	40.1
^{31}P	17.2
^{23}Na	11.3
^{13}C	10.7

Table 39.2 Elements used in magnetic resonance spectroscopy. [Reprinted with permission from *The physics of Diagnostic Imaging 2nd ed., Dowsett et.al.* Reproduced by permission of *Edward Arnold (Publishers) Ltd.* Copyright 1998.]

Chemical Shift

The exact magnetic field which a nucleus experiences depends on the amount of shielding from the molecular electron cloud. This magnetic field will cause a shift in the Larmor frequency for the nuclei. The Larmor frequency, and thus the frequency of the MRI signal, is chemical-species dependent, and so the frequency spectrum of an MRI signal will show peaks that correspond to particular molecules. The area under the peak indicates concentration. As T1 and T2 times are unimportant, the FID signal alone gives all the required frequency information.

Hydrogen MR spectroscopy can be used to observe metabolites like lactate and glucose. For example, increased lactate concentrations indicate anerobic conditions. The strong signal from the water molecules can tend to obscure the peaks from chemicals of interest. There are a number of techniques which can be used to suppress the water signal, but these techniques are beyond the scope of this text.

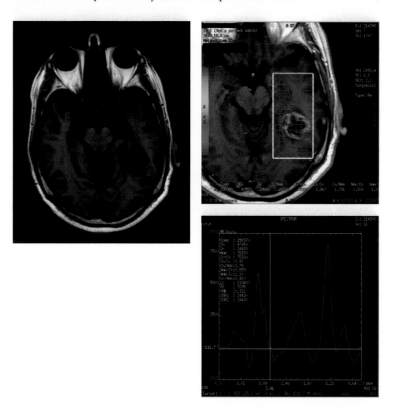

Figure 39.22 (Left) The same T1-weighted image as in Figure 39.21. (Top right) An MRI spectroscopy image of the area around the tumour. (Bottom right) The MRI signal spectrum showing choline and creatinine peaks characteristic of a tumour. [Images courtesy of Professor Terry Doyle, University of Otago School of Medicine.]

Contrast Agents

Contrast agents are chemical substances that affect the decay rate of the MRI signal. They are useful for manipulating features in the images as hown in Figure 39.23.

Paramagnetic materials contain unpaired electrons. These electrons can interact with nearby precessing nuclei, increasing the longitudinal relaxation rates and shortening T1 times. Thus, the presence of a paramagnetic substance will change the response of nearby tissue and can be used to provide contrast on an MRI image. The most commonly used paramagnetic contrast agents contain the metal gadolinium.

Superparamagnetic iron oxides (SPIO) consist of very small crystals of iron oxide. Each crystal is a single magnetic domain so that each crystal behaves like a particle with a very strong magnetic moment. The direction of this magnetic moment would normally flip about randomly due to random temperature fluctuations in the medium. In an external magnetic field they act like paramagnetic materials, since the tiny crystals tend to line up with the field. In low doses, such agents will decrease the T1 time of blood. At higher doses, the iron oxides have a noticeable effect on the T2 relaxation times. SPIO contrast agents are of particular use for imaging the liver and spleen. Because the iron oxides accumulate in certain liver cells which are not present in hepatic tumours, improved contrast results from the signal loss in the healthy tissues.

Contrast agents can be classified as positive or negative, depending on whether they make the signal stronger or weaker. Decreasing the T1 time will make an area of the image brighter, and decreasing T2 relaxation times will make it darker.

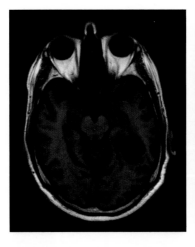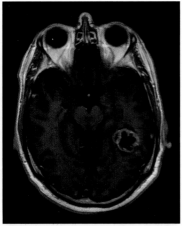

Figure 39.23 (left) The same T1-weighted image as in Figure 39.21. (right) A T1-weighted image after after gadolinium administration. This image shows increased signal around the periphery of the tumour. [Images courtesy of Professor Terry Doyle, University of Otago School of Medicine.]

Functional MRI

Functional MRI, or fMRI, refers to any MRI technique which allows the imaging of function rather than just structure. An example of fMRI is the use of MRI images of blood to determine levels of functioning in the brain.

Deoxygenated and oxygenated blood have different magnetic properties. Haemoglobin is diamagnetic when oxygenated, but becomes paramagnetic when deoxygenated, and the difference can be distinguished on an MRI scan. Changes in blood flow and oxygenation are linked to areas of activity in the brain, since neurons lack an internal energy and oxygen store and so must have this supplied by the blood. Areas of the brain with actively firing neurons will correlate with lowered blood oxygenation levels.

Instrumentation

There are four main parts to an imaging system: the primary magnet, the gradient magnets, the RF coils and the computer system.

It is not feasible to generate the required magnetic fields of several tesla with a conventional permanent magnet. The field is instead generated with a large electromagnet, which is the most expensive component in the system. The magnet is composed of several kilometres of superconducting wire; once current begins to flow in the coil, it will remain in place more-or-less indefinitely, as long as the wire is maintained at a sufficiently low temperature.

Several types of room-temperature coils are used to create the gradient fields. The magnetic field gradient in the z-direction is created using a pair of anti-Helmholtz coils, which have oppositely directed currents. Figure-eight-shaped coils generate field gradients in the x- and y-directions.

There are several different kinds of RF field coils: some are transmit-only and produce the pulse sequences which rotate the magnetisation, some are for receiving only and produce a voltage when current is induced in them by the RF fields coming from the sample and some do both. Different coil types can be used for imaging different parts of the body.

The analysis of the signals received is done with sophisticated electronics which perform the calculations necessary to convert the frequency and phase information in the MRI signal into 2-D position information and generate on-screen images showing signal strength versus position, with increased brightness representing increased signal strength.

Safety

If standard safety procedures are followed, MRI is a very safe imaging technique. However, there is some risk associated with the use of MRI equipment.

The magnetic fields generated by the superconducting coils are *very* large. Any ferromagnetic materials that are in the same room as the MRI apparatus are likely to be lifted into the air and pulled into the magnet's bore. The force with which such an object hits the magnet casing may cause damage that affects the field homogeneity, cause the leakage of helium, or even destroy the RF coils. If there is a person in the scanner, the results can be tragic. In 2001, in New York State, a six-year-old boy was fatally injured when an oxygen tank that had been left in the room was pulled into the magnet, striking him in the head.

The US Food and Drug Administration (FDA) has set guidelines for the safe operation of MRI imaging facilities. These are largely to do with setting limits to the amount of heating permissible and the largest field gradients that a person may be exposed to. Energy is being sent into the patient in the form of radio waves. This energy ultimately has to go somewhere, and not all of it returns to the system in the form of an RF signal. Inevitably there is some degree of heating as energy is transferred to the surrounding tissue by spin–lattice interactions. The set of US FDA guidelines issued in 2003 places an upper limit of 8 T on the field strength for adults, children and infants aged 1 month or more, and 4 T for younger infants.

There have been reported cases of moderate to severe burns resulting from the RF fields. Malfunctioning RF coils can cause nasty burns, and so if the patient complains at all about any burning sensation, the scan should be stopped. Patients with tattoos or wearing drug-release patches containing aluminium have also reported mild burns.

39.6 Summary

Key Concepts

magnetism A group of phenomena associated with the movement of charged particles, caused by the electromagnetic force.

Faraday's law The induced electromotive force in any closed circuit is equal to the time rate of change of the magnetic flux through the circuit.

Lenz's law A changing magnetic field will induce a current in loop. The direction is such that the field generated by the induced currents will oppose the change in the external field.

MRI Magnetic resonance imaging. An imaging technique utilising radio-frequency EM radiation and magnetic fields to produce images of the body for medical diagnostic purposes.

magnetisation The magnetic moment per unit volume, which characterises the degree of magnetic polarisation of the sample.

longitudinal magnetisation The component of the magnetisation vector in the z-direction, the direction of the large, static magnetic field, directed horizontally through the bore of the main magnet.

transverse magnetisation The component of the magnetisation in the xy-plane.

T1 time The time taken for the longitudinal magnetisation to recover to 63% of its original value. It depends on the time taken for the nuclei to interact with the local molecules to lose energy, so it also called the *spin–lattice* relaxation time.

T2 time The time taken for the transverse magnetisation to decay to 37% of its maximum value. It depends on how quickly the nuclei become out of phase and is also called the *spin–spin* relaxation time. Dephasing from spin–spin interactions is not reversible.

T2* time Due to local inhomogeneity in the longitudinal magnetic field, the spins dephase at an increased rate, and the characteristic timescale for this is called the T2* time. This dephasing is reversible.

TE time Time to echo. The time taken for a rephasing pulse to re-synchronise the nuclei, generating an echo signal.

TR time Time to repetition. The time between the start of successive pulse sequences.

frequency encoding Using additional gradient magnetic fields, the local Larmor frequency and hence emitted signal frequency will have unique value which indicates its point of origin.

phase encoding By varying the magnet field along an axis for a short period, the nuclei will lag or lead those in the neighbouring columns, and this phase difference can be used to locate the point of origin of the emitted signal.

90° pulse An RF pulse which rotates the magnetisation vector through 90°, flipping it from the longitudinal direction to rotating in the transverse plane at the Larmor frequency.

free induction decay (FID) The signal emitted after a single excitation pulse as the sample returns to its original magnetisation state.

spin echo pulse sequence A pulse sequence which reverses the dephasing caused by local field inhomogeneity with a 180° pulse.

T1 weighting Using a short TR time, the longitudinal magnetisation will only have recovered enough for the next pulse sequence to produce a strong signal if T1 was short. Strong signal therefore correlates with short T1. T1-weighted images are useful for showing anatomic detail.

T2 weighting With long TR and TE, the signal will be the weakest where the spin–spin dephasing is most rapid, so the image will be darkest for short T2 times. Water has a long T2 time, so T2 weighted images are bright in water-rich areas.

Equations

$$F = qvB\sin\theta$$
$$f = \gamma B$$
$$B = B_0 \frac{\Delta B}{\Delta z} z$$

39.7 Problems

39.1 A β^+ particle is moving at $19\,500$ m s^{-1} parallel to the ground and due east through a region of space in which there is a uniform 0.05 T magnetic field. If the magnetic field lines point upwards in the vertical direction, what is the direction and magnitude of the magnetic force on the β^+ particle?

39.2 In a region of space there is a uniform magnetic field of magnitude 0.25 mT pointing vertically straight down.

(a) If an electron is moving in the horizontal plane at a speed of 550 km s^{-1}, what will the radius of the resultant circular path be?

(b) Will the electron be moving clockwise or counter-clockwise when viewed from above?

(c) If a positron (same mass as an electron but with a charge of $+q_e$) is moving in the horizontal plane at speed of 550 km s^{-1}, what will the radius of the resultant circular path be?

(d) Will the positron be moving clockwise or counter-clockwise when viewed from above?

(e) How fast would an α particle ($^4_2\alpha^{2+}$) need to be traveling to have a path of the same radius as the electron in part (a)?

39.3 An electron travelling parallel to the ground and due north enters a region of space in which there is uniform magnetic field of $0.5\,\mu$T pointing straight up. The electron is travelling at a speed of 1000 km s^{-1}.

(a) What is the magnitude of the magnetic force on the electron (in N)?

(b) In which direction is the magnetic force on the electron as it enters the region in which there is a magnetic field?

(c) As the electron curves in the magnetic field the direction of the magnetic force on the electron changes. How many seconds (after entering the region in which there is a magnetic field) before the magnetic force on the electron is pointing due south?

39.4 An electron enters a region in which there is a uniform electric field of $2.25\,\mu$T in the z–direction. The electron moves through the region in a 'corkscrew' pattern as shown in Figure 39.24. The radius of the corkscrew path is 0.15 m while the 'pitch' is 0.05 m. What is the velocity of the electron?

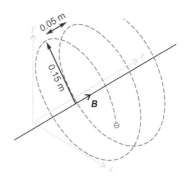

Figure 39.24 An electron moving in a region of uniform magnetic field.

39.5 Earths magnetic field at the surface near the poles is around 60×10^{-6} T. What is the Larmor frequency of protons at this location?

39.6 A patient is placed in an MRI machine and as the MRI is taken the magnetic field level with the patients eyes is 1.5 T. The magnetic field is largest at the persons head and the field gradient in the machine is 0.05 T m^{-1}. The patients heart is located 30 cm below their eyes, the patients liver is located 60 cm below their eyes, and the patients bladder is located 70 cm below their eyes. Assuming the magnetic field gradient remains constant across the whole body, what is the Larmor frequency at the patients eyes, their liver, and their bladder (in MHz)?

39.7 The MRI apparatus in Problem 39.6 can distinguish between signals whose frequencies differ by just 0.005 MHz. What is the resolution of this machine along the persons body?

39.8 The Larmor frequency of protons at the top of a patient's head is 54.30 MHz. The Larmor frequency of protons at the top of the spine 18.0 cm below is 55.15 MHz.

(a) What is the magnetic field strength in the MRI apparatus at the top of the head (in T)?

(b) What is the magnetic field gradient in the MRI apparatus (in T m^{-1})?

(c) The base of the patient's spine is 1.05 m below the top of their head. What will the Larmor frequency at the base of patients spine be assuming that the magnetic field gradient remains constant across the whole body (in MHz)?

www.wiley.com/go/biological_physics

VII

Appendices

PHYSICAL CONSTANTS

A.1   High Precision Mass Values

$m_{\text{electron}} = 9.10938215(45) \times 10^{-31}$ kg
$m_{\text{proton}} = 1.67262171(29) \times 10^{-27}$ kg $= 938.272029(80)$ MeV/$c^2 = 1.00727646688(13)$ u
$m_{\text{neutron}} = 1.67492729(28) \times 10^{-27}$ kg
$u = 1.660538782(83) \times 10^{-27}$ kg $= 931.494028(23)$ MeV/c^2

A.2   Useful Constants

$M_{\text{Earth}} = 5.974 \times 10^{24}$ kg
$R_{\text{Earth}} = 6.378 \times 10^6$ m
$g = 9.807$ m s^{-2}
$G = 6.673 \times 10^{-11}$ m^3 kg^{-1} s^{-2}
$k_e = 8.988 \times 10^9$ N m^2 C^{-2}
$m_{\text{proton}} 1.673 \times 10^{-27}$ kg
$m_{\text{neutron}} = 1.675 \times 10^{-27}$ kg
$m_{\text{electron}} = 9.109 \times 10^{-31}$ kg
$e = 1.602 \times 10^{-19}$ C
$c = 2.998 \times 10^8$ m s^{-2}
$N_A = 6.022 \times 10^{23}$ mol^{-1}
$\sigma = 5.670 \times 10^{-8}$ W m^{-2} K^{-4}
$R = 8.314$ J K^{-1} mol^{-1}
$k = 1.381 \times 10^{-23}$ J K^{-1}
$a_0 = 5.292 \times 10^{-11}$ m
$h = 6.626 \times 10^{-34}$ J s
$\varepsilon_0 = 8.854 \times 10^{-12}$ F m^{-1}

BASIC MATHS AND SCIENCE SKILLS

Physicists, when they are in a facetious mood, will occasionally joke that 'all other sciences are *merely* applied physics'. While this is arguably quite true it would also be inordinately difficult to get a comprehendible, explanatory and predictive model of, lets say, the human brain and thought processes using only 'fundamental' physical properties and laws. Thank goodness then for neuroscience which does indeed have ever-improving models of not just bulk physical features of the brain but how they interact to give us our minds as we know them. As we saw in Part IV : Electricity and DC Circuits, some basic electrostatics can illustrate how components of a neuron work and as such electrostatics is a valuable tool for neuroscientists.

Mathematicians, when *they* are in a facetious mood will occasionally joke that 'physics is *merely* applied mathematics'. And to be fair, there is some truth to this. Mathematical skills are an essential part of developing, understanding and utilising physical models. It is possible, and indeed advisable to understand most of the physical concepts presented in this book separate from the mathematical implementation and manipulation associated with them. Mathematical skills do, however, allow us to make use of these concepts and apply them to specific problems in the real world.

The purpose of this section of the textbook is to provide a short mathematical review and to indicate the mathematical concepts and skills which are necessary to a complete understanding of the material in this book. It is not intended as an exhaustive and complete mathematics primer. Any students who find themselves struggling with the mathematical components of problem solving are advised to seek out an appropriate textbook or other resource with which they can improve their mathematics skills.

B.1   Measurement and Units

Physics is not mathematics, although a considerable amount of mathematics can be required when solving physics problems. Physics, like all sciences, is based on the verification of theories by experiments. This means that physicists must, at some point, engage in measurement, as theories of physics are about measurable phenomena. Physics is about quantities and not about numbers. A *distance* of 5 m is different from a *time* of 5 s, even though the number 5 is used in both cases. This is why physicists insist on the use of units.

Units

A unit is a carefully defined amount of some quantity. For example, the **second** is defined as the amount of time it takes for the cesium-133 atom to oscillate a total of 9 192 631 770 times between two carefully defined electronic states. The **kilogram** is defined as the mass of a particular platinum–iridium cylinder held in a vault in Paris. The metre is defined as the distance which light travels through a vacuum in a time interval of 1/299 792 458th of a second.

Almost all measurable quantities in physics may be expressed in terms of a small number of fundamental types of quantity. These fundamental quantities are: length,

Quantity	Unit	Symbol		
Velocity	metres per second			m s^{-1}
Acceleration	metres per second per second			m s^{-2}
Force	newtons	N		kg m s^{-2}
Momentum	newton seconds		N s	kg m s^{-1}
Energy	joules	J	N m	$\text{kg m}^2\,\text{s}^{-2}$
Power	watts	W	J s^{-1}	$\text{kg m}^2\,\text{s}^{-3}$
Pressure	pascals	Pa	N m^{-2}	$\text{kg m}^{-1}\,\text{s}^{-2}$
Charge	coulombs	C	A s	
Electric potential	volts	V	J C^{-1}	$\text{kg m}^2\,\text{A}^{-1}\,\text{s}^{-3}$
Electrical resistance	ohms	Ω	V A^{-1}	$\text{kg m}^2\,\text{A}^{-2}\,\text{s}^{-3}$

Table B.2 Commonly used SI units.

Number		Prefix	Symbol
0.000000001	10^{-9}	nano	n
0.000001	10^{-6}	micro	μ
0.001	10^{-3}	milli	m
1000	10^{3}	kilo	k
1 000 000	10^{6}	mega	M
1 000 000 000	10^{9}	giga	G

Table B.3 Commonly used SI system prefixes. A distance of 0.0000512 m would be written as 51.2 μm while a period of 295 000 000 seconds could be written as 295 Ms or 0.295 Gs.

Quantity	Unit	Symbol
Amount	mole	mol
Time	second	s
Length	metre	m
Mass	kilogram	kg
Temperature	kelvin	K
Current	ampere	A
Luminosity	calendula	cd

Table B.1 The seven base SI units. All other SI units can be expresses as some combination of these units

time, mass, electrical current, temperature, amount of substance and luminosity. Most properties of the physical world not included in this list may be constructed as a combination of these fundamental properties. This fact allows us to define systems of units.

A **system of units** is a convention which defines the standard amounts or units of a set of fundamental quantities. These are called *base units*. The system of units used almost exclusively in science is the SI system (SI is an abbreviation of the French name of the system, the Système International d'Unité). The SI system of units is a metric system, i.e., a system based on the number 10. In this system the unit of length is the metre (m), the unit of time is the second (s) and the unit of mass is the kilogram (kg). The other base units in the SI system can be found in Table B.1.

There are many other units in use for quantities other than those described by the base SI units (such as force and energy). Table B.2 has a short list of units for such quantities which feature in this text book. Some of these quantities have their own named unit such as the unit of force, the newton. Such units are often named after famous scientist. All of these other units can, however, be expressed as a combination of the six base SI units.

This system of units also specifies a set of prefixes which are prepended to the unit to indicate quantity in powers of 10. These prefixes make it easier to refer to very large or very small numbers. A sub-set of these prefixes are given in Table B.3; there are many more, but these are the ones which will be useful in this text.

Unit Conversion

In this textbook we use the SI system of units as much as possible. However, this is not the only system of units currently in use. Furthermore, it is common to see non-SI units used in everyday life. For example, time is often measured in hours or minutes rather than seconds, and the speed of cars is commonly given in kilometres per hour rather than metres per second. This raises the question, how do you convert a quantity from one system of unit into another?

We will begin this discussion by working through a pair of examples.

Example B.1 *Unit conversion I*

Problem: How many kilometres per hour is 100 miles per hour?

Solution: There are 1.609 kilometres in 1 mile, so 100 miles per hour is just

$$\frac{1.609 \text{ kilometres}}{1.000 \text{ mile}} \times 100 \text{ miles per hour} = 160.9 \text{ kilometres per hour}$$

Example B.2 *Unit conversion II*

Problem: How many miles per hour is 100 kilometres per hour?

Solution: There are 1/1.609 miles in 1 kilometre, so 100 kilometres per hour is just

$$\frac{1.000 \text{ mile}}{1.609 \text{ kilometres}} \times 100 \text{ kilometers per hour} = 62.15 \text{ miles per hour}$$

In these examples we are using a **conversion factor** to convert from one unit system to another without saying that this is what we are doing.

A conversion factor is a ratio of the same amount of some quantity in two different units, with the *amount* of the relevant physical quantity on the numerator and denominator always the same. For example, acceptable conversion factors include, $\frac{60 \text{ seconds}}{1 \text{ minute}}$, $\frac{1000 \text{ metres}}{1 \text{ kilometre}}$ and $\frac{1 \text{ day}}{24 \text{ hours}}$.

Each of these conversion factors includes the same *amount* of some quantity in the numerator and denominator (60 s is the same amount of time as 1 min), and each is expressed using a different *number* depending on the units used.

If the numerator and denominator do not represent the same amount of some quantity, or represent different quantities altogether, then it is not a valid conversion factor. Some examples would be $\frac{50 \text{ second}}{1 \text{ minute}}$ and $\frac{11 \text{ kilometres}}{1 \text{ hour}}$. In the first case 1 minute is *not* the same amount of time as 50 seconds, and in the second case 11 km represents a distance, while 1 h represents a time.

To convert a quantity from one unit to another, multiply by a conversion factor which has the original unit in the denominator and the desired unit in the numerator.

The conversion factor from minutes to seconds is $\frac{60 \text{ seconds}}{1 \text{ minute}}$. With this conversion factor we can find the number of seconds in 30 minutes

$$\frac{60 \text{ seconds}}{1 \text{ minute}} \times 30 \text{ minutes} = 1800 \text{ seconds}$$

Example B.3 *Unit conversion I*

Problem: Use conversion factors to calculate the number of seconds in a century. (Ignore the effects of leap years.)

Solution: We will convert the units of time from years to seconds, and to do this we will use conversion factors for: years to days, days to hours, hours to minutes and minutes to seconds.

$$\frac{60 \text{ seconds}}{1 \text{ minute}} \times \frac{60 \text{ minutes}}{1 \text{ hour}} \times \frac{24 \text{ hours}}{1 \text{ day}} \times \frac{365 \text{ days}}{1 \text{ year}} \times 100 \text{ years} = 315\,360\,000 \text{ seconds}$$

Example B.4 *Unit Conversion II*

Problem: Convert 100 km h⁻¹ into m s⁻¹.

Solution: In this example we will need to use a conversion factor for h⁻¹, we will need to convert h⁻¹ into min⁻¹, and then min⁻¹ into s⁻¹. We also need to convert km into m

$$\frac{1\text{ minute}}{60\text{ seconds}} \times \frac{1\text{ hour}}{60\text{ minutes}} \times \frac{1000\text{ metres}}{1\text{ kilometre}} \times 100\text{ km h}^{-1} = 27.8\text{ m s}^{-1}$$

Accuracy, Uncertainty and Significant Figures

The numerical values of physical quantities cannot be given with infinite precision. Depending on the nature of the calculation, the numerical value of a physical quantity will be given to some number of *significant figures*. The number of significant figures is, with some conditions, the number of digits given. Thus the acceleration due to gravity may be given as 9.8 m s⁻² or 9.81 m s⁻²; in the first case there are two digits so there are two significant figures (2 s.f.), in the second case there are three significant figures. Numbers with more significant figures are more precise than numbers with fewer significant figures.

Caution is required when one of the digits is zero. For example, if we were to use $g = 10$ m s⁻² it would appear that, since this number has two digits, it therefore represents the acceleration due to gravity to two significant figures. This would imply that $g = 10$ m s⁻² is just as precise as $g = 9.8$ m s⁻². This is clearly not the case. In the expression $g = 10$ m s⁻² the zero serves as a place-holder and conveys no more information that this. Thus even though there are two digits, only one of them is significant. In other words $g = 9.8$ m s⁻² is accurate to two significant figures and $g = 10$ m s⁻² is only accurate to one significant figure. Similarly 0.38 m is accurate to two significant figures not three, and 0.0038 m is also only accurate to two significant figures.

If a physical quantity is given to two significant figures there there is an implicit uncertainty in the final digit. Thus 9.8 m s⁻² may be read as 9.8 ± 0.05 m s⁻². This implicit uncertainty has implications for the use of inexact numerical values in calculations. For example, the solution to a calculation involving quantities given to two significant figures should not be given to four or five significant figures. How many significant figures should be quoted in solutions to calculations? What if the numbers used in the calculation have different numbers of significant figures? These are complex questions and sophisticated approaches to these problems are used in research contexts. However, for our purposes, three simple rules of thumb will suffice:

1. Use all available digits in the calculation. Round the solution and not the intermediate steps.

2. When multiplying or dividing two numbers the solution should be rounded to the same number of significant figures as the number in the calculation with the least significant figures. For example

$$\frac{1.193}{0.23} = 5.18695622 = 5.2 \text{ (2 s.f.)}$$

3. When adding or subtracting numbers the solution should be rounded to the same number of decimal places as the number in the calculation with the least number of decimal places. For example

$$10.1 + 12.367 + 0.459 = 21.926 = 21.9 \text{ (3 s.f.)}$$

Here the answer should be reported as 21.9, since 10.1 has the least number of decimal places.

Note, again, that these rules of thumb are appropriate in this, and other, similar textbooks. Much more sophisticated techniques are available to deal with uncertainty in other contexts, such as research.

Number	Rounded	
299 792 458 m s^{-1}	299 792 458 m s^{-1}	(9 s.f.)
	299 792 460 m s^{-1}	(8 s.f.)
	299 792 500 m s^{-1}	(7 s.f.)
	299 792 000 m s^{-1}	(6 s.f.)
	299 790 000 m s^{-1}	(5 s.f.)
	299 800 000 m s^{-1}	(4 s.f.)
	300 000 000 m s^{-1}	(3 s.f.)
	300 000 000 m s^{-1}	(2 s.f.)
	300 000 000 m s^{-1}	(1 s.f.)
0.023045 m	0.023045 m	(5 s.f.)
	0.02305 m	(4 s.f.)
	0.0230 m	(3 s.f.)
	0.023 m	(2 s.f.)
	0.02 m	(1 s.f.)

Table B.4 Two numbers (the speed of light and an arbitrary length) rounded to varying significant figures. Notice that the speed of light is the same *number* when rounded to 1, 2 and 3 significant figures. Also, when rounding the speed of light to 6 s.f., the result is *not* the same as rounding the speed 299 792 500 m s^{-1} to 6 s.f. (this would be 299 793 000 m s^{-1}).

B.2 Basic Algebra

An equation can be interpreted in two different ways. An equation tells us what the relationships between different quantities are and how to use these relationships to calculate numerical quantities.

To illustrate this point, consider the following application of Newton's second law,

$$a = \frac{F}{m}$$

$$a = \frac{0.059 \, \text{N}}{1.24 \, \text{kg}} = 0.048 \, \text{m} \cdot \text{s}^{-2}$$

The first equation, read as a sentence, states that, 'The rate of change of an object's velocity is proportional to the force applied to it and inversely proportional to the mass of the object'. This definition contains the essential meaning of Newton's second law and from it I can make predictions such as, 'If I double the net force applied to an object, I will double the acceleration of that object'.

The second usage above gives an example of the way an equation can be instructions. If I wished to know what acceleration a 1.24 kg object would experience if a 0.059 N net force were applied to it, this equation tells me what I should do. I should divide 0.059 N by 1.24 kg. I will then find that the acceleration of this object will be 0.048 m s^{-2}.

In this example we have rearranged the equation which expresses Newton's second law. We will now briefly review the steps involved in these manipulations.

Working With Equations

An equation has two halves separated by an equals sign, '='. The essential characteristic of an equation is that the right-hand half is the same as the left-hand half, both in magnitude and in units. We may manipulate an equation; move bits around, remove bits and add bits, so long as this one essential fact does not change. There are a small number of basic rules for manipulating equations which guarantee that the right- and left-hand sides are always equal. These rules are:

1. Add the same number or variable to each side.

2. Subtract the same number or variable from each side.

3. Multiply or divide each side by the same number or variable.

As an example of the application of these rules, consider the equation

$$y = 2x + 1$$

We will use the rules for manipulating equations to make x the subject of the equation

$$2x + 1 = y$$
$$2x + 1 - 1 = y - 1 \text{ subtract 1 from both sides so that}$$
$$2x = y - 1$$
$$\frac{2x}{2} = \frac{y-1}{2} \text{ divide both sides by 2 so that}$$
$$x = \frac{y-1}{2}$$

In the example given at the beginning of this chapter, the equation representing Newton's second law was manipulated as follows

$$F = ma$$
$$\frac{F}{m} = \frac{ma}{m} = a \text{ divide both sides by } m$$
$$\text{so } \frac{F}{m} = a$$

Note that, since the sides of the equation are equal in both magnitude and units, the units of acceleration are m s^{-2} or N kg^{-1} – these units are thus equivalent.

Problem Areas

Quadratic Equations

The quadratic equation is obtained by finding the square of a sum

$$(a + b)^2 = (a + b) \times (a + b)$$
$$= a(a + b) + b(a + b)$$
$$= a^2 + ab + b^2 + ab$$
$$= a^2 + 2ab + b^2$$

Adding Fractions

Adding fractions can achieved by manipulating each fraction such that the denominators are the same. This is achieved by multiplying each fraction by a ration which equals 1

$$\frac{1}{a} + \frac{1}{b} = \left(\frac{1}{a} \times \frac{b}{b}\right) + \left(\frac{1}{b} \times \frac{a}{a}\right)$$
$$= \frac{b}{ab} + \frac{a}{ab}$$
$$= \frac{a+b}{ab}$$

Division by a Fraction

Dividing by a fractional number is the same as multiplying by the inverse of that fraction

$$\frac{N}{\frac{a}{b}} = N \times \frac{b}{a}$$

If both numbers are fractions this rule still holds

$$\frac{\frac{a}{b}}{\frac{c}{d}} = \frac{a}{b} \times \frac{d}{c}$$

B.3   Exponentials and Logarithms

Exponentials and logarithms occur frequently in all areas of science. They often cause some difficulties and we will here briefly review their definition and basic properties.

Definition of the Exponential

To begin our description of the exponential we will begin by considering multiplication. Multiplication may be thought of as a shorthand notation for repeated multiplication, after all

$$n \times a = \underbrace{a + a + a + a + a + a + \ldots}_{n \text{ times}} \tag{B.1}$$

In a similar fashion, the exponential is little more than a self consistent notation for repeated multiplication

$$a^n = \underbrace{a \times a \times a \times a \times a \times a \times \ldots}_{n \text{ times}} \tag{B.2}$$

The great value of this notation is that it makes multiplication of exponentiated quantities much more straightforward. For example, the sum of a^n and a^m is

$$a^n \times a^m \begin{aligned} &= \underbrace{a \times a \times a \times \ldots}_{n \text{ times}} \\ &\times \underbrace{a \times a \times a \ldots}_{m \text{ times}} \\ &= \underbrace{a \times a \times a \ldots}_{n + m \text{ times}} \\ &= a^{n+m} \end{aligned}$$

Thus multiplication has been converted into the addition of exponentials. Note that this is only possible when the exponentials concerned have the same base, 'a' in this case.

A few special cases must be given definitions so that the exponential notation is complete and consistent. To begin with, negative exponentials are defined so that division corresponds to the subtraction of exponents, i.e.

$$\frac{a^n}{a^m} = a^{n-m} \tag{B.3}$$

and thus

$$a^{-1} = \frac{1}{a}$$
$$a^{-2} = \frac{1}{a^2}$$
$$a^{-3} = \frac{1}{a^3}$$

etc.

With this definition we are able to define a^0 since,

$$a^{n-n} = a^0$$
$$= \frac{a^n}{a^n}$$
$$= 1$$
$$a^0 = 1$$

Thus $a^0 = 1$, for all values of the case of exponentiation, a.

We can also see that the notation can easily be expanded to include fractional exponents if we first consider nested exponentials

$$\left(a^n\right)^m = \underbrace{a^n \times a^n \times a^n \times \ldots}_{m \text{ times}}$$
$$= \underbrace{a \times a \times a \times \ldots}_{n \times m \text{ times}}$$
$$= a^{n \times m}$$

It is then clear that

$$\left(a^{\frac{1}{n}}\right)^n = a^{\frac{n}{n}} = a$$

An alternate way of writing fractional exponents is to use the $\sqrt{}$ symbol

$$a^{\frac{1}{n}} = \sqrt[n]{a}$$

A variation to this general rule is the convention that in the specific case of $a^{\frac{1}{2}}$

$$a^{\frac{1}{2}} = \sqrt[2]{a} = \sqrt{a}$$

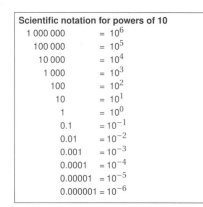

Scientific notation for powers of 10	
1 000 000	$= 10^6$
100 000	$= 10^5$
10 000	$= 10^4$
1 000	$= 10^3$
100	$= 10^2$
10	$= 10^1$
1	$= 10^0$
0.1	$= 10^{-1}$
0.01	$= 10^{-2}$
0.001	$= 10^{-3}$
0.0001	$= 10^{-4}$
0.00001	$= 10^{-5}$
0.000001	$= 10^{-6}$

Scientific Notation

Often in physics we are required to perform calculations with very large or very small quantities. Scientific notation makes use of exponentials to simplify these calculations. Central to this technique are the properties of exponentials outlined in the previous section. The list in the margin shows the first few powers of 10.

Note that for positive powers of 10, the number of zeros is the same as the power, i.e., one million is 10^6 or a 1 followed by 6 zeros. For negative powers of 10 there are one less zeros after the decimal point than the power, so 10^{-6} is a decimal point followed by 5 zeros and then a 1.

Non-integer values are dealt with by rewriting the number as a value between 1 and 10 and multiplying by the appropriate power of 10. The number 52 345 100 becomes 5.23451×10^7 while 0.0000897 becomes 8.97×10^{-5}.

Logarithms

Logarithms are essentially the opposite of exponentials. The definition of the logarithm is as follows

$$\text{If } y = a^x, \text{ then } \log_a y = x \tag{B.4}$$

Historically the logarithm was useful because instead of multiplying large numbers you may instead add smaller logarithms. To see how this works, suppose that we must find the product of two large numbers A and B. Instead of performing the laborious process of multiplying these two numbers by hand we make use of the fact that $A = 10^a$ for some exponent a, and $B = 10^b$ for some exponent b. If we knew what a and b were then the multiplication of A and B would involve adding a and b and then finding 10^{a+b}. The logarithm is exactly the function required here since $\log_{10} A = a$. It was to simplify the multiplication of large numbers that tables of logarithms were constructed.

Most commonly large numbers are expressed as powers of 10. For this reason logarithm tables were generally table of logarithms to base 10. Euler's number $e = 2.718\ldots$ is very important in mathematics and in all areas of science. This number is a common base for exponentiation and is thus also a common base of logarithms. The logarithm to base e is known as the natural logarithm and is represented as 'ln x' as opposed to 'log x', which is generally taken to mean the logarithm to base 10.

There are a number of useful identities involving logarithms, we will simply list the most useful here and leave their derivation to the interested reader.

$$\log(ab) = \log a + \log b \tag{B.5}$$

$$\log\left(\frac{a}{b}\right) = \log a - \log b \tag{B.6}$$

$$\log(a^n) = n \log a \tag{B.7}$$

$$\log\left(\sqrt[n]{a}\right) = \frac{\log a}{n} \tag{B.8}$$

B.4 Geometry

Geometry is the study of shapes in space. This section of the mathematics review will do no more than provide a list of formulae which may be useful in the body of the text. The following formulae are for the shapes shown in Figure B.1.

Triangles:

$$\text{area} = \frac{1}{2} \times \text{b} \times \text{h}$$

Circles:

$$\text{circumference} = 2\pi r$$

$$\text{area} = \pi r^2$$

Spheres:

$$\text{surface area} = 4\pi r^2$$

$$\text{volume} = \frac{4}{3}\pi r^3$$

Cylinders:

$$\text{surface area} = 2\pi\left(r^2 + lr\right)$$

$$\text{volume} = \pi r^2 l$$

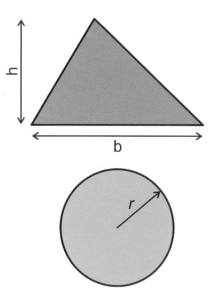

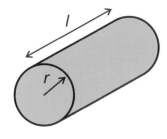

Figure B.1 A selection of basic shapes and their dimensions.

B.5 Trigonometric Functions

Trigonometry is the study of a group of functions, the sine, cosine and tangent functions. These functions are the result of a particular property of right-angled triangles. If a right-angled triangle is constructed as shown in Figure B.2, then there are three ratios which depend only the angle between two of the sides (the angle θ in Figure B.2). The length of the sides may increase or decrease, but if the angle θ does not change, and the triangle remains a right-angled triangle, then these ratios will not change. The ratios define the sine, cosine and tangent functions with are shown in Figures B.3 and B.4.

Figure B.2 A simple right-angled triangle. The longest side of the triangle (the hypotenuse) is labehled H. The side of the triangle opposite the corner labelled with the angle θ is O, and the side adjacent to this angle is labelled A. The angle between the sides O and A must always be 90° in a right-angled triangle.

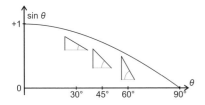

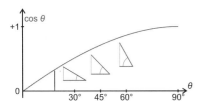

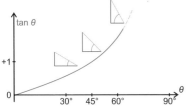

Figure B.3 The sine, cosine and tangent functions for angles between 0 and 90°. Each function is the ratio between two sides of a right-angled triangle. Small representations of such triangles are shown on each plot. These functions for a larger range of angles can be found in Figure B.4.

Basic Definitions

If a right-angled triangle is given as in Figure B.2, with the lengths of the sides as indicated, and the indicated internal angle is labelled θ, then the ratio of O to H is called the sine function of θ

$$\sin\theta = \frac{O}{H} \tag{B.9}$$

Similarly, the ratio of A to H is called the cosine function of θ

$$\cos\theta = \frac{A}{H} \tag{B.10}$$

Finally, the ratio of O to A is called the tangent function of θ (this is the ratio of the sine and cosine functions of θ),

$$\tan\theta = \frac{O}{A} = \frac{\sin\theta}{\cos\theta} \tag{B.11}$$

Some Important Identities

These identities may be demonstrated in a number of ways, however the proof is not necessary here. These identities are provided as a useful reference only.

$$\sin(-\theta) = -\sin(\theta) \tag{B.12}$$

$$\cos(-\theta) = \cos(\theta) \tag{B.13}$$

$$\tan(-\theta) = -\tan(\theta) \tag{B.14}$$

$$\sin(\theta) = \cos\left(\frac{\pi}{2} - \theta\right) \tag{B.15}$$

$$\cos(\theta) = \sin\left(\frac{\pi}{2} - \theta\right) \tag{B.16}$$

	$0°$	$30°$	$45°$	$60°$	$90°$
$\sin\theta$	0	$\frac{1}{2}$	$\frac{1}{\sqrt{2}}$	$\frac{\sqrt{3}}{2}$	1
$\cos\theta$	1	$\frac{\sqrt{3}}{2}$	$\frac{1}{\sqrt{2}}$	$\frac{1}{2}$	0
$\tan\theta$	0	$\frac{1}{\sqrt{3}}$	1	$\sqrt{3}$	?

Table B.5 Cosine, sine, and tangent values for common angles.

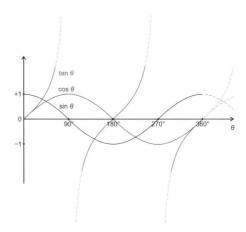

Figure B.4 The sine, cosine and tangent functions for angles between 0 and 360°. These functions repeat for angles greater than 360°.

Common Angles

Table B.5 gives the sine, cosine and tangent of frequently occurring angles. The '?' in the bottom right-hand corner indicates that the value of the tangent is undefined at an angle of $\frac{\pi}{2}$ radians. A proof of these values is relatively straightforward ,but, again, this proof not necessary here. These values are provided as a useful quick reference.

Graphs of Trigonometric Functions

The graphs of the sine, cosine and tangent functions are given in Figures B.3 and B.4. Figure B.4 shows the values of each function over an range of angles from 0 to 360°. Each function repeats itself at angles higher than this. Most of the angles that we will be interested in are between 0 and 90°, Figure B.3 shows each function in this restricted range of angles.

B.6   Vectors

Some physical quantities must be described as having both a magnitude (size) and a direction. If you were walking uptown one day and you were asked how to get to the nearest bank by a person on the street, you could tell them that it was 100 m from your current position. The person would of course not be satisfied with these instructions as there would be many places that were 100 m from your position. A more sensible response would be to say that the nearest bank was 100 m due east of your current position. This set of instructions has information about the direction in which the bank may be found, as well as how far away it is.

Clearly, direction is just as important as distance when describing changes or differences in position. Similarly physical quantities like velocity, acceleration and momentum are all characterised by a direction as well as a magnitude. Such quantities are described mathematically using vectors. A **vector** is a mathematical object that has a direction as well as a magnitude. In contrast, a quantity like temperature may be fully characterised by a single number. These are called scalar quantities, as **scalar** is essentially another name for a number.

Figure B.5 A simple representation of a vector. The length of the arrow indicates the magnitude of the vector while the direction in which the arrow points indicates the direction of the vector.

There are a number of ways of representing vectors. One of the most common ways of representing a vector is to draw them as arrows like that shown in Figure B.5. The length of the arrow represents the magnitude of the vector and the direction of the arrow represents the direction of the vector. This representation provides us with a useful visualisation of vectors, we will use this representation to describe the addition and subtraction of vectors and the multiplication of vectors by a scalar.

Addition and Subtraction of Vectors

Two vectors may be added to each other to produce a third vector, or one may be subtracted from the other to produce a third. To understand the addition of vectors consider the following example, a car drives 1 km due east and then 2 km due north. The total distance and the nett direction in which the car travelled is given by considering the triangle in Figure B.6. The distance travelled is the length of the hypotenuse of this triangle and the direction is the direction of that side of the triangle. To find the length of the vector sum we can in this case use Pythagoras' theorem

$$R = \sqrt{(x^2 + y^2)} = \sqrt{(1^2 + 2^2)} = \sqrt{(5)} \approx 2.2 \text{ km}$$

The direction may be found using one of the trigonometric relations discussed above. The definition of either the sine or cosine functions would work, in this case we will use the cosine

$$\cos\theta = \frac{\text{adjacent}}{\text{hypotenuse}} = \frac{1}{2.2} = 0.46$$

We then take the inverse cosine, $\cos^{-1}(0.46)$, to find that the sum vector is directed at an angle of 63° north of east.

From this we can see that the way to add two vectors, $A + B$, is to place the base of B at the tip of A and then draw a straight line from the base of A to the tip of B. The vector sum is then formed by placing an arrow head on the end of this new line, which is at the tip of B as in Figure B.7. The length of the vector sum may be found using Pythagoras' theorem, if the vectors A and B are perpendicular to each other, if this is the case then simple trigonometry may be used to find The direction of the sum vector relative to A or B.

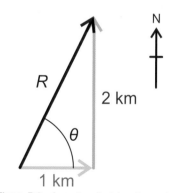

Figure B.6   A car travells 1 km due east and then 2 km due north. Adding a two vectros representing each part of the journey will enable us to find how far away and in what direction the car ends up.

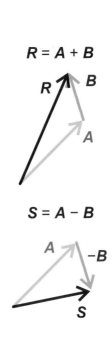

Subtraction of the vector B from the vector A may be acheived once we have defined the negative of a vector. The negative of a vector B is identical to B except that it points in the opposite direction (see Figure B.7). Now the subtraction of B from A is simply the addition of A and $-B$, as illustrated in Figure B.7.

Multiplication of a Vector by a Scalar

Having considered the addition and subtraction of vectors it is natural to now consider the multiplication of vectors. The multiplication of a vector with another vector is a defined operation. (In actual fact there are two possible ways that two vectors may be multiplied. However, this operation is not needed for the physics discussed in this textbook and so will not be reviewed here.)

The multiplication of a vector by a scalar is necessary, however. Newton's second law is an example of this operation. The force (a vector) is equal to the mass (a scalar) multiplied by the acceleration (a vector). Fortunately this operation is relatively simple to describe and use. The multiplication of a vector by a scalar simply changes the length of the vector. Using Newton's second law as an example, the vector F (force) is in the same direction as the vector a (acceleration), but it is m (mass) times longer.

The division of a vector by a scalar is simply the multiplication of that vector by the inverse of the scalar, i.e.

$$\frac{a}{m} = a \times m^{-1}$$

Figure B.7 Two vectors A and B are added together (top). Vector B is subtracted from vector A (bottom).

SELECTED REFERENCES AND FURTHER READING

A.W. T. Barenbrug. Psychrometry and Psychrometric Charts. Chamber of Mines of South Africa, 3rd edition, 1974.

Radiation sickness. http://news.bbc.co.uk/2/hi/health/medical_notes/461921.stm.

Why more people die in the winter. http://news.bbc.co.uk/1/hi/health/5372296.stm.

George B. Benedek and Felix M. H. Villars. *Physics With Illustrative Examples From Medicine and Biology*. Springer-Verlag, 2nd edition, 2000.

Gerald Carrington. *Basic Thermodynamics*. Oxford University Press, UK, 1994.

Adrian Cho. Universe's highest-energy particles traced back to other galaxies. *Science*, 318(5852):896–897, 2007.

Alan H. Cromer. *Physics For The Life Sciences*. McGraw-Hill, Inc., 2nd edition, 1977.

David Crystal. *How Language Works*. Penguin Books Ltd, UK, 2006.

Paul Davidovits. *Physics in Biology and Medicine*. Harcourt Academic Press, USA, 2nd edition, 2001.

David J. Dowsett, Patrick A. Kenny, and R. Eugene Johnston. *The Physics of Diagnostic Imaging*. Hodder Arnold, UK, 2nd edition, 2006.

Catherine Harold (Ed.). *Professional Guide to Diseases*. Lippincott Williams and Wilkins, USA, 9th edition, 2009.

A. L. Hodgkin and A. F. Huxley. A quantitative description of membrane current and its application to conduction and excitation in nerve. *J. Physiol.*, 117:500–544, 1952.

Howden-Chapman et al. *Effect of insulating existing houses on health inequality: cluster randomised study in the community*. http://www.bmj.com/cgi/rapidpdf/bmj.39070.573032.80v1, 2007.

John V. Forrester, Andrew D. Dick, Paul G. McMenamin, and William R. Lee. *The Eye: Basic Sciences in Practice*. W B Saunders, UK, 2nd edition, 2002.

Galileo Galilei. *Mathematical Discourse concering two new sciences relating to mechanicks and local motion, in four dialogues. . . .By Galileo Galilei, . . .With and appendix concerning the center of gravity of solid bodies. Done into English from the Italina, by Tho. Weston, . . .and now publishŠd by John Weston, ...* Samuel Baker, UK, 2nd edition, 1734.

Bonnie Guiton. Cold stress (hypothermia) and heat stress. http://www.hoptechno.com/book41.htm.

R. A. Helliwell. *Whistlers and Related Ionospheric Phenomena*. Stanford University Press, USA, 1965.

William R. Hendee and E. Russell Ritenour. *Medical Imaging Physics*. Wiley-Liss, Inc., USA, 4th edition, 1978.

Jürgen Kiefer. *Biological Radiation Effects*. Springer-Verlag, 1990.

Randall D. Knight. *Physics for Scientists and Engineers: A Strategic Approach*. Pearson Addison Wesley, USA, 2004.

Christopher W. Lawrence. *Cellular Radiobiology*. Edward Arnold Limited, UK, 1971.

Gary Linney. *MRI from A to Z: A Definitive Guide for Medical Professionals*. Cambridge University Press, UK, 2005.

Unexpected properties of hair. http://www.hair-science.com/_int/_en/topic/topic_sousrub.aspx?tc=root-hairscience^so-sturdy-so-fragile^properties-of-hair&cur=properties-of-hair.

R. Duncan Luce. *Sound and Hearing: A Conceptual Introduction*. Lawrence Erlbaum Associates, Inc., USA, 1993.

Ida Mann and Antoinette Pirie. *The Science of Seeing*. Wyman & Sons Ltd, UK, 1946.

Gary G.Matthews. *Cellular Physiology of Nerve and Muscle [electronic resource]*. John Wiley & Sons, Inc., Chichester, UK, 2009.

A. E. E.McKenzie. *Magnetism and Electricity*. Cambridge University Press, UK, 1961.

Radiation sickness: Medlineplus medical encyclopedia. http://www.nlm.nih.gov/medlineplus/ency/article/000026.htm.

T. C. A. Molteno and W. L. Kennedy. Navigation by induction-based magnetoreception in elasmobranch fishes. *Journal of Biophysics*, 2009.

Repair shops for broken DNA. http://science.nasa.gov/headlines/y2007/07nov_repairshops.htm?list3163.

Solar wind loses power, hits 50-year low. http://science.nasa.gov/headlines/y2008/23sep_solarwind.htm.

Hospital fined by health dept. in death of boy during M.R.I. New York Times, Saturday September 29, 2001.

X-rays: Moseley's law. http://www4.nau.edu/microanalysis/microprobe/Xray-MoseleysLaw.html.

Stephen E. Palmer. *Vision Science: Photons to Phenomenology*. Massachusetts Institute of Technology, USA, 1999.

K. C. Parsons. *Human Thermal Environments*. Taylor & Francis, London, 1993.

Daniel V. Schroder. *An Introduction to Thermal Physics*. Addison Wesley Longman, 2000.

C.Y. Shaw, D.Won, and J. Reardon. Managing volatile organic compounds and indoor air quality in office buildings – an engineering approach (rr 205). http://www.nrc-cnrc.gc.ca/obj/irc/doc/pubs/rr/rr205/rr205.pdf, 2005.

G. K. Strother. *Physics With Applications In Life Sciences*. Houghton Mifflin, USA, 1977.

H. N. V. Temperley and D. H. Trevena. *Liquids and Their Properties: A Molecular and Macroscopic Treatise With Applications*. Ellis Horwood, UK, 1978.

The silver institute – world silver survey 2006 summary. http://www.silverinstitute.org/publications/wss06summary.pdf.

J. A. Tuszynski and J.M. Dixon. *Biomedical Applications of Introductory Physics*. JohnWiley & Sons, Inc., USA, 2002.

Paul Peter Urone. *College Physics*. Brooks/Cole, USA, 2nd edition, 2001.

Wet-bulb globe temperature: A global climatology. http://www.dtic.mil/cgi-bin/GetTRDoc? AD=ADA229028&Location=U2&doc=GetTRDoc.pdf.

Michael M.Walker, Todd E. Dennis, and Joseph L. Kirschvink. The magnetic sense and its use in long-distance navigation by animals. *Current Opinion in Neurobiology*, 12, 2002.

Gordon Lynn Walls. *The Vertebrate Eye and Its Adaptive Radiation*. Hafner Publishing Company, USA, 1942.

Myopia. http://en.wikipedia.org/wiki/Myopia.

World gold council statistics. http://www.gold.org/value/stats/statistics/gold_demand/index.html.

Index